XXII International Conference on Spectral Line Shapes

(ICSLS 2014)

Journal of Physics: Conference Series Volume 548

Tullahoma, Tennessee, USA
1 – 6 June 2014

ISBN: 978-1-63439-662-2
ISSN: 1742-6588

Printed from e-media with permission by:

Curran Associates, Inc.
57 Morehouse Lane
Red Hook, NY 12571

Some format issues inherent in the e-media version may also appear in this print version.

Printed by Curran Associates, Inc. (2015)

Institute of Physics
Dirac House, Temple Back
Bristol BS1 6BE UK

Phone: 44 1 17 929 7481
Fax: 44 1 17 920 0979

techtracking@iop.org

Additional copies of this publication are available from:

Curran Associates, Inc.
57 Morehouse Lane
Red Hook, NY 12571 USA
Phone: 845-758-0400
Fax: 845-758-2634
Email: curran@proceedings.com
Web: www.proceedings.com

TABLE OF CONTENTS

ASTROPHYSICS

INVITED AND ORAL PRESENTATIONS

POSTER PRESENTATIONS AND LETTERS

BIOMEDICAL PHYSICS

INVITED AND ORAL PRESENTATIONS

MAGNETIC FUSION PHYSICS

INVITED AND ORAL PRESENTATIONS

NEUTRALS/ATOMIC-MOLECULAR-OPTICAL PHYSICS

INVITED AND ORAL PRESENTATIONS

POSTER PRESENTATIONS AND LETTERS

HIGH AND LOW TEMPERATURE PLASMA PHYSICS

INVITED AND ORAL PRESENTATIONS

POSTER PRESENTATIONS AND LETTERS

TECHNICAL/APPLIED PHYSICS

INVITED AND ORAL PRESENTATIONS

POSTER PRESENTATIONS AND LETTERS

Journal of Physics: Conference Series **548** (2014) 011001 doi:10.1088/1742-6596/548/1/011001

XXII International Conference on Spectral Line Shapes, ICSLS 2014

Figure 1. The 22nd ICSLS.

SUMMARY

The 22nd International Conference on Spectral Line Shapes (ICSLS) was convened at The University of Tennessee Space Institute (UTSI) at Tullahoma, Tennessee, USA, during June 1 to 6, 2014. A variety of topics of interest to the line shape community were addressed during invited and contributed oral and poster presentations. General categories of the ICSLS 2014 scientific contents included Astrophysics, Biomedical Physics, High and Low Temperature Plasma Physics, Magnetic Fusion Physics, Neutrals Atomic-Molecular-Optical (AMO) Physics, and Applied Physics. Research interests at UTSI and at the Center for Laser Applications (CLA) focus on Applied Physics and Plasma Physics areas such as laser-induced breakdown spectroscopy, spectroscopy with ultra-short light pulses, combustion diagnostics, to name a few. Consequently, the presentations during the conference addressed a variety of these topics.

Attendance at the conference included researchers from North America, Africa, Asia and Europe, with an international representation showing 250 authors and co-authors with over 25 different citizenships, and 100 participants at the Conference. Figure 1 shows a photo of Conference attendees. The schedule included 82 contributions, 41 oral and 41 poster presentations. The 29 invited, 12 contributed oral and 41 contributed poster presentations were selected following communication with the international organizing committee members. A smart phone "app" was also utilized, thanks to Elsevier, to communicate electronic versions of the posters during the conference. Special thanks go to the members of the international and local committees for their work in organizing the 22nd ICSLS. In addition, thank you notes also go to the peer reviewers for the proceedings. Following the success of the IOP: Journal of Physics Conference Series selected for the 21st ICSLS publication, the proceedings papers report ongoing research activities. Papers submitted amount to 68 in number, or 83% of the 82 papers contributed to the 22nd ICSLS conference will be published in the IOP: Journal of Physics proceedings.

The Executive Director of the University of Tennessee Space Institute welcomed all participants of the Conference on the first day of the technical sessions on Monday June 2, 2014. This welcome address is also included in the conference series publication, especially important for Physics and Engineering research at UTSI is the concurrent 50-year celebration of the Institute in 2014. Informal welcome occurred on Sunday June 1, 2014, and various social activities included a tour to the Jack Daniel's distillery in Lynchburg, Tennessee, followed by the conference dinner.

The international scientific committee met to look into various aspects of the ICSLS and future role of this conference for the spectral line shape community. The next meeting locations have been discussed, including the scheduling of the next 23rd conference in Torun, Poland, in June of 2016. Further meeting locations include hosting the conference in

Egypt in 2018, possibly in Luxor, Egypt. Communication regarding the 24th ICSLS in 2020 included mentioning of scheduling the Conference to occur in Dublin, Ireland.

Clearly, there is a wealth of interest in continuing the long standing tradition of communicating spectral signatures and line shapes at the biannual ICSLS meetings.

The 22nd International Conference on Spectral Line Shapes was supported by the Institute of Physics, the University of Tennessee Space Institute, the Center for Laser Applications, the Quantel Laser company, and by Elsevier. On behalf of the organizing committee, I greatly appreciate the support.

Figure 2. Photo of 22nd ICSLS participants

Christian G. Parigger

The University of Tennessee / The University of Tennessee Space Institute

Figure 3. 22nd ICSLS logo and sponsors.

Journal of Physics: Conference Series **548** (2014) 011002 doi:10.1088/1742-6596/548/1/011002

Welcome to the *22nd International Conference on Spectral Line Shapes*

Dear conference participants:

On behalf of Dr. Jimmy Cheek, Chancellor of the University of Tennessee Knoxville (UTK), I am pleased to welcome you to the University of Tennessee Space Institute (UTSI). The Space Institute is located in Tullahoma, Tennessee, approximately 160 miles southwest of the main UTK campus. The beautiful and inspiring UTSI campus occupies a peninsula extending into Woods Reservoir adjacent to the US Air Force Arnold Engineering Development Complex (AEDC). UTSI was established in 1964 with a primary mission of supporting AEDC by providing graduate education and research opportunities to AEDC employees. Over 500 AEDC employees have earned graduate degrees, and total graduate degrees earned exceed 2,000 with more than 250 PhDs. During the height of the U.S. space program, 12 astronauts attended the Institute, with 9 graduating with master's degrees awarded by UTK. Currently, courses of study and research offered at UTSI can lead to M.S. and Ph.D. degrees in mechanical, aerospace, and biomedical engineering, industrial and systems engineering, engineering science, and physics. Research programs at UTSI support national defense and the aerospace, automotive, and health industries.

The conference is truly international with scientists from over 20 different countries represented in the program; and UTSI, as the home of the Center for Laser Applications (CLA), is a highly appropriate site to host the ***22nd International Conference on Spectral Line Shapes.*** The CLA was chartered in 1984 and is recognized as an "established" Tennessee Higher Education Commission Center of Excellence. Current focus areas supported by the CLA include Bio/Nanophotonics, Materials Science, Laser/Materials Interaction, Spectroscopy, and Non-Equilibrium Fluid Physics.

The Space Institute is uniquely situated in Southern Middle Tennessee with easy access to Nashville, America's Music City, Chattanooga, a famous Tennessee River city. In addition, Southern Middle Tennessee is the home of world famous Tennessee walking horses and sipping whiskey with major distilleries within easy driving distance of UTSI. Also, please be aware that the international music festival, "Bonnaroo" is held the week after this conference in nearby Manchester, Tennessee.

We wish success for the Conference and are truly happy to have you visit our campus and region.

Sincerely yours,

Robert N. Moore, PhD
Executive Director and Professor
The University of Tennessee Space Institute

doi:10.1088/1742-6596/548/1/011003

Minutes of the International Program Committee, 22nd ICSLS

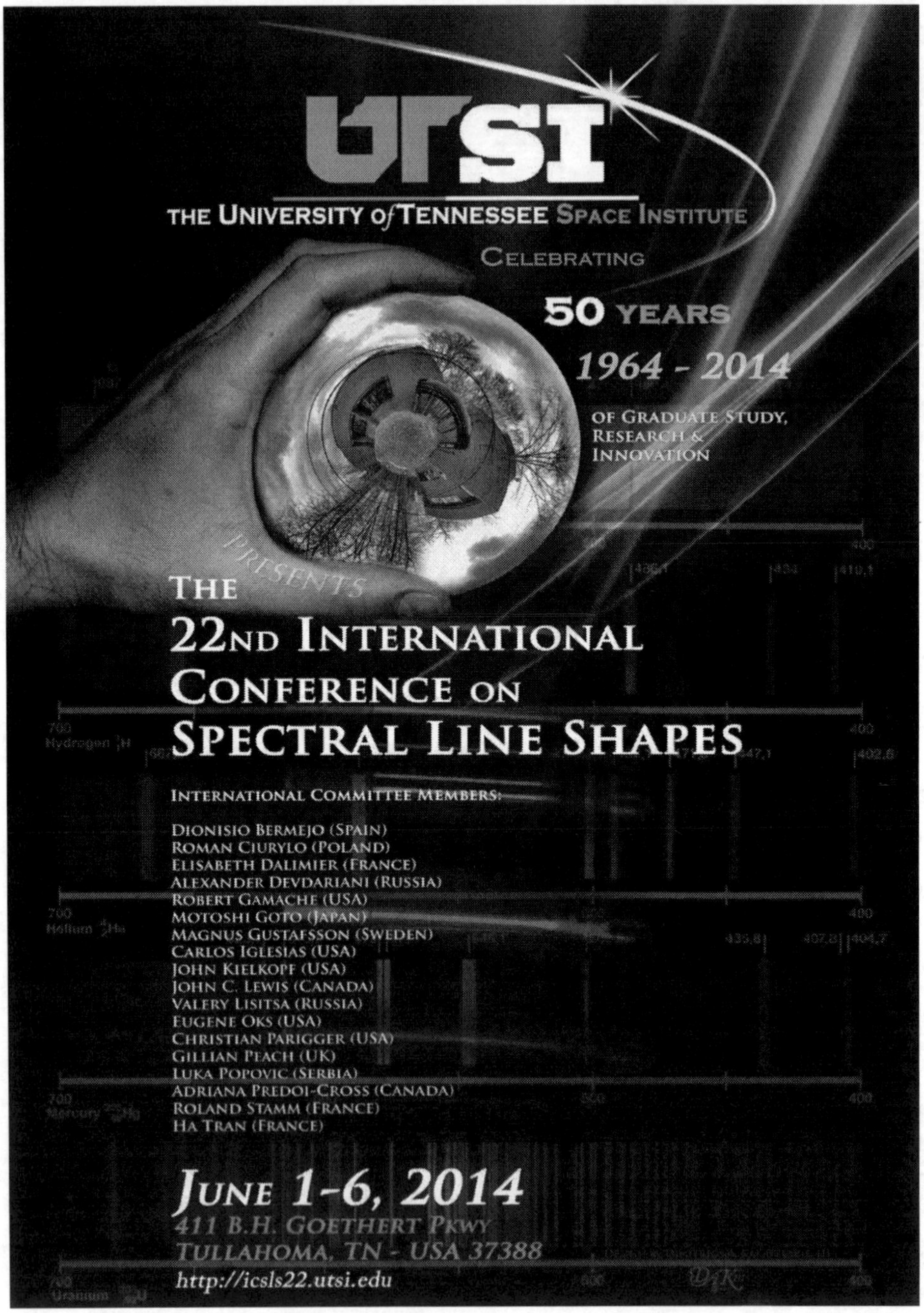

Figure 1. Poster for the 22nd ICSLS and current international committee members.

Journal of Physics: Conference Series **548** (2014) 011003 doi:10.1088/1742-6596/548/1/011003

The international committee members met during a working lunch from 12pm to 1pm on Tuesday June 2, 2014. Figure 1 shows the current members.

Members present at 22nd ICSLS: Dionisio Bermejo, Roman Ciurylo, Elisabeth Dalimier, Motoshi Goto, Magnus Gustafsson; Carlos Iglesias, John Kielkopf, John C. Lewis, Eugene Oks , Christian Parigger, Gillian Peach, Roland Stamm; or 12 of 18 members of the committee.

Agenda and Summary Report

- International Committee Membership. Three previous members completed their service on the International Committee, namely, Jean-Michel Hartman (France); Milan S. Dimitrijević (Serbia); Marco Antonio Gigosos (Spain). Thank you for your services over the years. Two new members were added between the 2012 (21st ICSLS) and the 2014 (22nd ICSLS) conferences, namely, Luka Popović (Serbia), and Ha Tran (France). Their membership was formally voted on and was unanimously in favor. The membership stands now at 18, some currently on the ICSLS committee indicated that they may serve only a little longer, but no further action was taken.
- 23rd Conference in Torun, Poland, in 2016. Roman Ciurylo presented plans for the 23rd conference in Torun, Poland, in late June of 2016. Following some discussion, unanimous support was expressed in favorably voting for the next conference to be held in Torun.
- 24th Conference in 2018 and 25th Conference in 2020. Following some communication about extending global presence of our line shape community interests, unanimous support is noted in a vote for a conference on the African continent in possibly Luxor, Egypt, in 2018. The specific conference month may move from the recent summer month meeting dates. Details about the 2018 conference will be communicated during the 23rd ICSLS conference in Torun. Also, interests for hosting the 25th Conference in Dublin, Ireland, were communicated but with further action deferred to the committee meeting in Torun in 2016.
- Summary of 22nd Conference at Tullahoma, Tennessee. Thanks to the outstanding support by our sponsors, especially UTSI and CLA, and of course IOP, Quantel, and Elsevier, significant reduction of fees for almost all students could be granted. Moreover, reduction and in part travel expenses could be provided to participants that expressed need of support. The schedule included 82 contributions, 41 oral and 41 poster presentations. There were 29 invited, 12 contributed oral and 41 contributed poster presentations. Over 250 authors and coauthors with over 25 different citizenships contributed to the 22nd ICSLS.

The meeting adjourned with a very positive outlook on future activities in spectral line shape research. For completion, the list of oral and poster presentations is collated on the next few pages, with invited speakers marked as ***I***, contributed speakers marked as ***C***, and contributed posters are listed using roman numerals I to XLI.

Peer review statement

All papers published in this volume of *Journal of Physics: Conference Series* have been peer reviewed through processes administered by the proceedings Editors. Reviews were conducted by expert referees to the professional and scientific standards expected of a proceedings journal published by IOP Publishing.

Journal of Physics: Conference Series **548** (2014) 012001 doi:10.1088/1742-6596/548/1/012001

Modelling the spectrum of Mg in cool stars

P S Barklem and Y Osorio

Theoretical Astrophysics, Department of Physics and Astronomy, Uppsala University, Box, 516, SE 751 20, Uppsala, Sweden

paul.barklem@physics.uu.se

Abstract. The astrophysical importance of Mg, together with its unique range of spectral features in late-type stars, plus its relative simplicity from an atomic physics point of view, makes it a prime target and test bed for detailed *ab initio* non-local thermodynamic equilibrium (NLTE) modelling in stellar atmospheres. In this paper, we present example first results for calculations of NLTE Mg line based on a new model atom with significant improvements in the collision data for neutral Mg. We perform calculations for excitation of the lower-lying levels due to electron impacts using the R-matrix method. Recent data for excitation and charge transfer due to hydrogen atom impacts involving low-lying levels are now employed. Further, we have made efforts to use physically-motivated methods for calculating radiative and collisional data involving Rydberg states. The results are compared with observed spectra and the impact of the new calculations briefly explored.

1. Introduction

Neutral magnesium creates a broad range of spectral features in late-type stars, including some of the strongest lines in their spectra. Consequently, Mg is detectable even in low-quality spectra and in old, metal-poor stars, making Mg an excellent tracer of alpha-element abundances. At $T > 5000$ K, Mg I is a minority species and as a result is expected to be sensitive to departures from local thermodynamic equilibrium (LTE) in stellar atmospheres due to photoionization by non-local radiation. While in the sun non-local thermodynamic equilibrium (NLTE) effects on lines in the optical are relatively small [1], significant departures from LTE are seen in calculations for metal-poor dwarfs and giants [2]. However, past studies such as these have often been forced to use atomic collision data for inelastic and line broadening processes of questionable quality, frequently approximate formulae, and this is a significant source of uncertainty in the NLTE modelling. Ref. [3] showed that such atomic data was the largest source of uncertainty in their calculations of the solar 12 micrometer lines (see also Ref. [4]). [1] demonstrated the particular sensitivity of the 8806 Å line to the rates for inelastic collisions with neutral hydrogen.

The astrophysical importance of Mg, together with the unique range of spectral features in late-type stars, plus its relative simplicity from an atomic physics point of view, makes it a prime target and test bed for detailed *ab initio* NLTE modelling in stellar atmospheres. Here we present some initial results, noting that a far more detailed study is underway [5].

Journal of Physics: Conference Series **548** (2014) 012001 doi:10.1088/1742-6596/548/1/012001

2. NLTE modelling using new collision data

We have performed NLTE Mg line formation calculations using the MULTI code [3] based on a new model atom with significant improvements in the collision data for neutral Mg.

Calculations for excitation of the lower-lying levels due to electron impacts are performed using the R-matrix method [5-12] based on configuration interaction calculations of the electronic structure employing polarization pseudostates [13-15]. In Figure 1 the resulting rate coefficients at a representative temperature are compared with the data from [16], which is the data employed in most recent studies. It is seen that the general agreement is reasonable, though there is a large scatter with some transitions differing by around 3 orders of magnitude. In particular, data for optically forbidden transitions are always larger in our calculations, often significantly so. Such processes correspond to spin-changing transitions, which are often important in NLTE modelling (see below). For transitions involving high-lying states, more approximate methods must be employed. The two most commonly used are the semi-empirical Bethe-Born formula of van Regemorter [17], and the impact parameter method of Seaton [18]. Figure 2 compares the rate coefficients from the two methods, which are seen to have significant differences for many transitions, often van Regemorter giving larger rates. As we will see in the next section, the choice has significant impact on the solar 12 micrometer lines.

Recent data for excitation and charge transfer involving low-lying levels due to hydrogen atom impacts from full quantum scattering calculations by some of us [19-21] are now employed. Prior to these calculations, the classical "Drawin formula" was often used, which has been shown to be completely unsatisfactory [22]. In particular, it cannot be used for charge transfer processes, or spin-changing processes, which are exactly those processes that have been found to be most important in NLTE modelling, e.g. [23, 24].

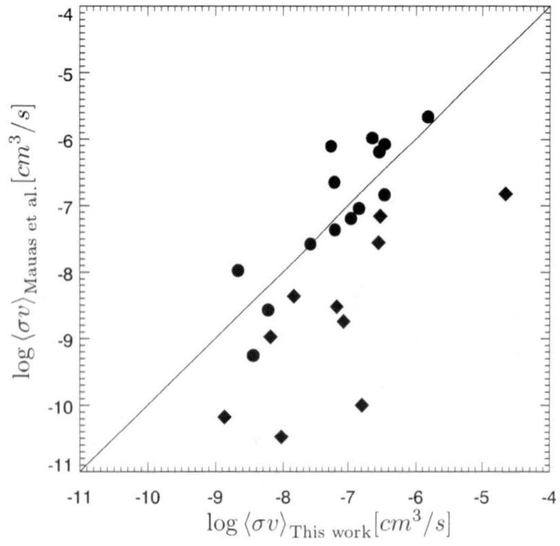

Figure 1. Comparison between the electron collisional rates at T=5000 K calculated by us and those used by [16]. Circles correspond to the optically allowed transitions, and diamonds are the optically forbidden transitions.

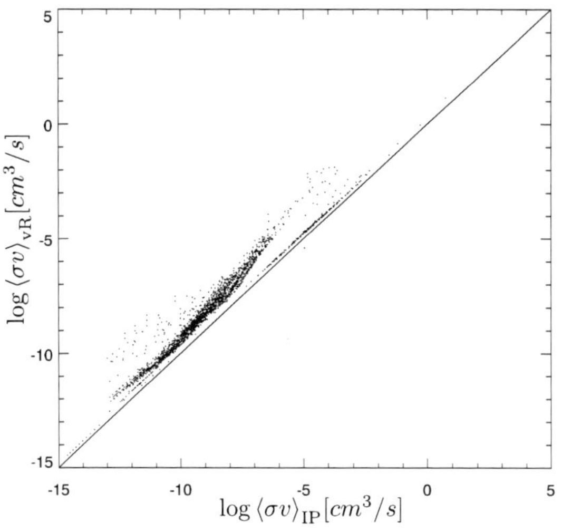

Figure 2. Comparison of electron collision rate coefficients at T=5000 K using the impact parameter (IP) approximation and the van Regemorter formula (vR).

3. Comparison with observed spectra

A few example comparisons of predicted and observed spectra are now shown to demonstrate the effects.

Figures 3 and 4 compare spectra for the 12 micrometer lines in the solar spectrum. Here we see that the predicted spectrum is rather sensitive to the choice of method for estimating the electron collision rates among high-lying states, as these transitions take place among Rydberg states. It is well known that the Born approximation overestimates cross sections at low energies and that the impact parameter method gives better results; see [18, 26]. Here we see the astrophysical comparison is also significantly improved by favouring the impact parameter method.

Figure 5 compares spectra of the quite strong 8806 Å absorption line with observations for 6 standard benchmark stars with relatively well-known stellar parameters. One sees directly that the NLTE models provide much better overall fits to the lines, especially in the cores of the lines which form high in the stellar atmospheres where collisions are relatively inefficient, and thus NLTE effects start to become very important.

Journal of Physics: Conference Series **548** (2014) 012001 doi:10.1088/1742-6596/548/1/012001

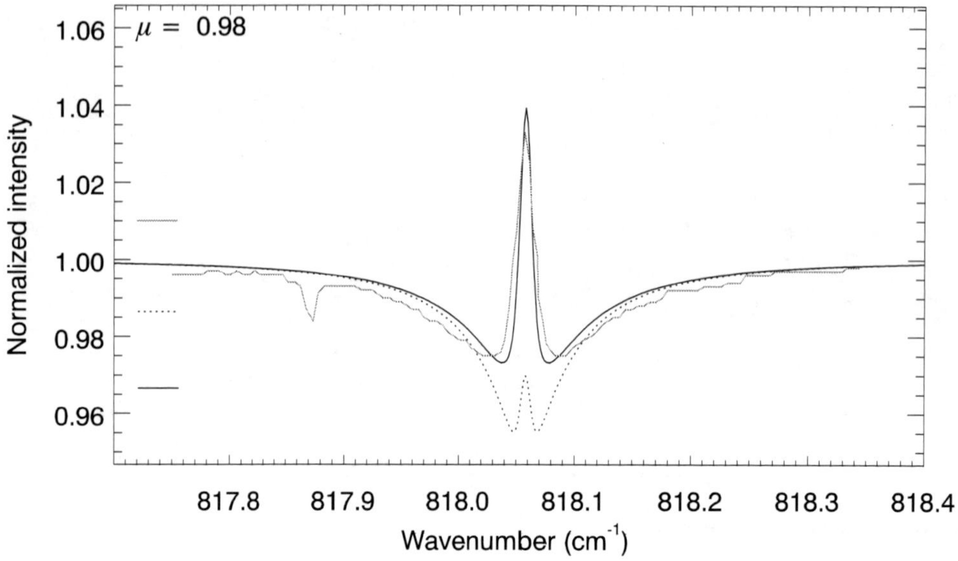

Figure 3. Infrared 12.2 micrometer line profile at solar disc centre calculated using all the available atomic data and using impact parameter method (solid black line) or the van Regemorter formula (dashed lines) formula where no R-matrix calculations were available. Observed spectrum (solid red line) is from [25].

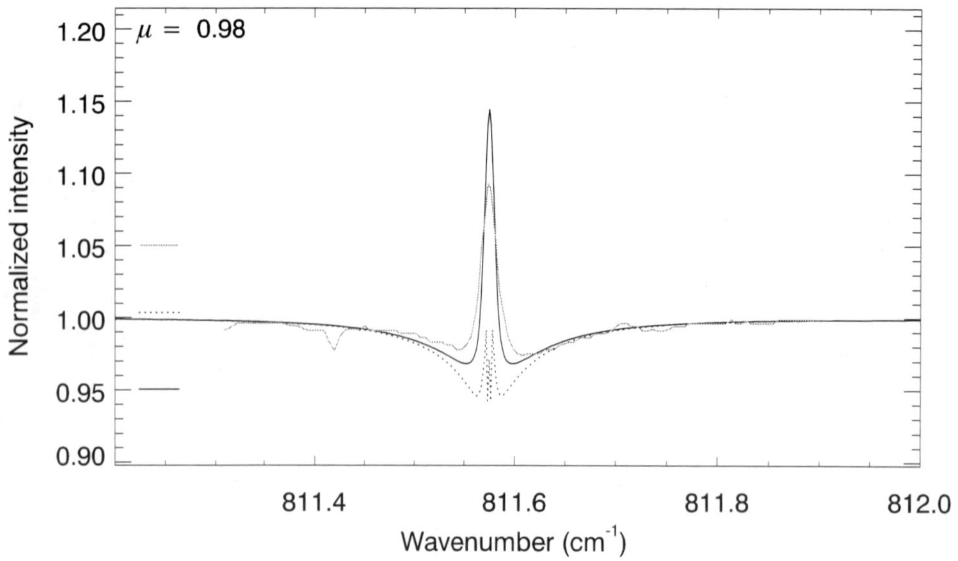

Figure 4. As for figure 3, but for the 12.3 micrometer line.

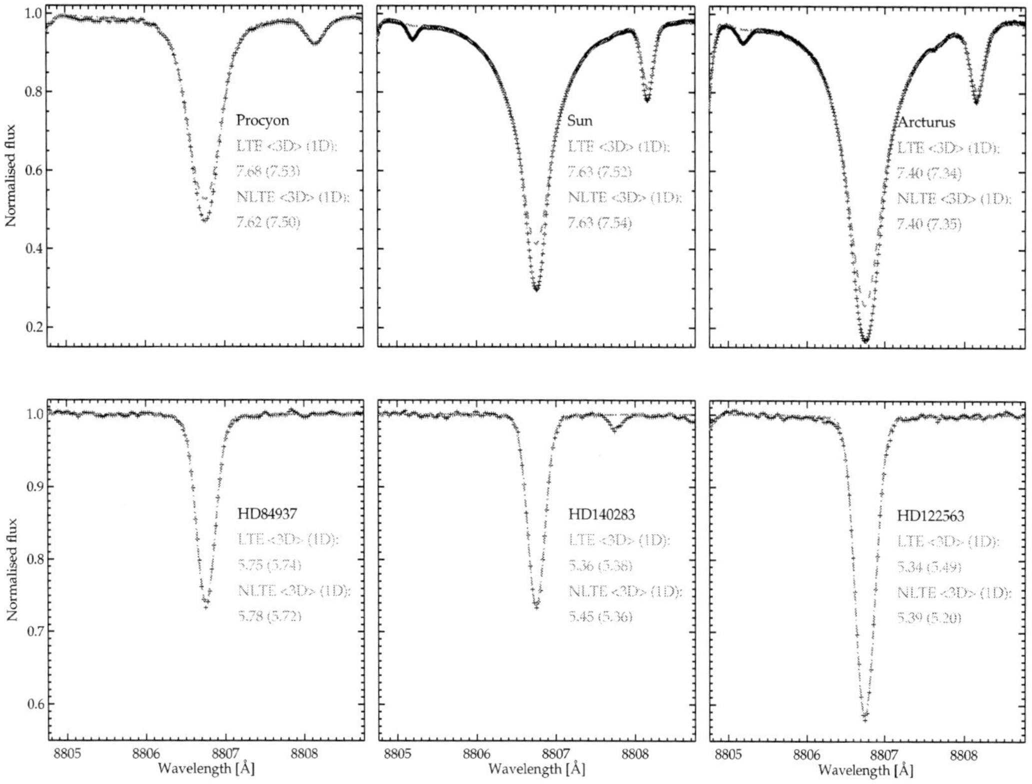

Figure 5. Observed spectra of the NLTE sensitive Mg I 8806Å line (pluses). The best fitting <3D> (3D averaged model atmosphere) LTE (red dashed line) and NLTE profile (blue dotted line) are overplotted and the corresponding Mg abundances stated in each panel. The abundances from classical 1D models are given in parentheses (spectra not plotted). Plot by Karin Lind, from [5].

4. Discussion

We have shown some preliminary example results of NLTE modelling of Mg I in stellar atmospheres. In particular, we have seen the diagnostic value of the 12 micrometer lines in differentiating between two possible descriptions for electron collision rates among high-lying states. We have also seen the general improvement of fits to the near infrared 8806 Å line from our NLTE modelling compared to LTE modelling. This work is ongoing, but we expect final and more extensive results, along with significantly more details of the model atom, to be published in the near future [5].

Acknowledgements

We acknowledge the important contributions of our collaborators to parts of this work, Andrey Belyaev, Nicole Feautrier, Marie Guitou, and Annie Spielfiedel who calculated hydrogen collision processes, and Karin Lind who calculated spectra in standard stars. We thank Karin Lind for permission to use figure 5. This work has been supported by Göran Gustafssons Stiftelse, the Swedish Research Council, the Royal Swedish Academy of Sciences, the Knut and Alice Wallenberg Foundation and the Wenner-Gren foundation.

References

[1] Zhao G, Butler K, Gehren T 1998 *Astron. Astrophys.* **333** 219
[2] Zhao G, Gehren T 2000 *Astron. Astrophys.* **362** 1077
[3] Carlsson M, Rutten R J, Shchukina N G 1992 *Astron. Astrophys.* **253** 567
[4] Sundqvist J O, Ryde N, Harper G M, Kruger A, Richter M J 2008 *Astron. Astrophys.* **486** 985
[5] Osorio Y, Barklem P S, Lind K et al. *in preparation*
[6] Burke P G, Hibbert A, Robb W D 1971 *J. Phys. B: Atom. Molec. Phys.* **4** 153
[7] Burke P G, Robb W D 1976 *Adv. Atom. Molec. Phys.* **11** 143
[8] Berrington K A, Burke P G, Butler K et al. 1987 *J. Phys. B: Atom. Molec. Phys.* **20** 6379
[9] Berrington K A, Burke, P G, Chang J J, et al 1974 *Comp. Phys. Comm.* **8** 149
[10] Berrington K A, Burke P G, Le Dourneuf M et al. 1978 *Comp. Phys. Comm.* **14** 367
[11] Berrington K A, Eissner W B, Norrington P H 1995 *Comp. Phys. Comm.* **92** 290
[12] Badnell N R 1999 *J. Phys. B: Atom. Molec. Opt. Phys.* **32** 5583
[13] Hibbert A 1975 *Comp. Phys. Comm.* **9** 141
[14] Hibbert A, Le Dourneuf M, Lan V K 1977 *J. Phys. B: Atom. Molec. Phys.* **10** 1015
[15] Plummer M, Noble C J, LeDourneuf M 2004 *J. Phys. B: Atom. Molec. Phys.* **37** 297
[16] Mauas P J, Avrett E H, Loeser R 1988 *The Astrophysical Journal* **330** 1008
[17] van Regemorter H 1962 *Ap. J.* **136** 906
[18] Seaton, M J 1962 *Proc. Phys. Soc.* **79** 1105
[19] Guitou M, Belyaev A K, Barklem P S, Spielfiedel A, Feautrier N 2011 *J. Phys. B: Atom. Molec. Phys.* **44** 5202
[20] Belyaev A K, Barklem P S, Spielfiedel A, Guitou M, Feautrier N, Rodionov D S, Vlasov D V 2012 *Phys. Rev. A* **85** 32704
[21] Barklem P S, Belyaev A K, Spielfiedel A, Guitou M, Feautrier N 2012. *Astron. Astrophys.* **541** 80
[22] Barklem P S, Belyaev A K, Guitou M, Feautrier N, Gadéa F X, Spielfiedel A 2011 *Astron. Astrophys.* **530** 94
[23] Barklem P S, Belyaev A K, Asplund M 2003 *Astron. Astrophys.* **409** L1
[24] Fabbian D, Asplund M, Barklem P S, Carlsson M, Kiselman D 2009 *Astron. Astrophys.* **500** 122
[25] Brault J, Noyes R 1983 *Ap. J.* **269** L61
[26] Bely O, van Regemorter H 1970 *Annu. Rev. Astro. Astrophys.* **8** 329

Supermassive black holes and spectral emission lines

D Ilić[1] and L Č Popović[1,2]

[1] Department of Astronomy, Faculty of Mathematics, University of Belgrade, Studentski trg 16, 11000 Belgrade, Serbia

[2] Astronomical Observatory of Belgrade, Volgina 7, 11000 Belgrade, Serbia

E-mail: `dilic@matf.bg.ac.rs`, `lpopovic@aob.rs`

Abstract. It is widely accepted that active galactic nuclei (AGN) are hosting a supermassive black hole in their center. The supermassive black hole is actively fueled by surrounding gas through an accretion disk, which produces a broad band continuum (from X-ray to radio emission). The hard photons from the accretion disk create the photoionized plasma around the central black hole, which emits a number of broad emission lines. Therefore, one of the signatures of the strong activity in galaxies is the emission of the broad spectral lines (line widths of several 1000 km/s), which are seen only in a fraction of AGN, so called Type 1 AGN. These broad emission lines often show very complex line profiles, usually strongly variable in time. Here we will describe the basic properties of the broad emission lines and how can we use them to derive the properties of the central supermassive black hole, i.e., the mass and spin, or see signatures of supermassive binary black holes.

1. Introduction

The X-ray, UV, optical, and IR broad emission lines (BELs) are seen in the spectra of some active galactic nuclei (AGN), the so-called Type 1 AGN. Their fluxes and profiles can give us the information about the geometry and physics of the region where they are forming, i.e., the broad-line region (BLR). The BLR is a photoionized gas heated by the radiation coming from the compact accretion disk which fuels the supermassive black hole in the center of an AGN, and produces the powerful radiation that makes AGN different from normal galaxies (see [1] for a review on AGN and their spectral properties).

One of the characteristics of AGN is a number of broad emission lines in a wide spectral range: from the X-ray to the infrared spectral range. In a number of AGN, the very broad spectral line Fe Kα in the X-ray spectrum has been observed (first time detected in MGC-6-30-15 by [2]). It was shown that this line originates in the innermost part of the accretion disk (see, e.g., [3]) and can give us the crucial facts about the supermassive black hole and the plasma conditions in its vicinity. The broad UV and optical lines are present in spectra of Type 1 AGN, and their width indicates a gravitationally bounded motion.

Nowadays it is of great importance to study the properties of the supermassive black holes, as it is believed that they evolve together with host galaxies, the so-called coevolution of galaxies and black holes, through mergers and accretion (for a review see [4]). It is crucial to determine the supermassive black hole masses to trace the coevolution across cosmic time. One of the powerful methods to obtain the mass is by reverberation mappings of the BLR in Type 1 AGN [5] and applying the virial theorem on the BLR properties assuming Keplerian motion (see, e.g., [6, 7, 8]). However, this method is still not precise enough, as the kinematics and structure of the BLR have not yet been mapped directly (see, e.g., [9]), and is, thus, important to study in detail the broad emission lines in order to constrain the physic and kinematics of the BLR [10]. Another possible method that can be used for the estimates of the supermassive black hole mass, is based on the spectropolarimetirc monitoring of broad emission lines [11], and was successfully applied in case of the Type 1 AGN Mrk 6 (see [11]). Finally, since massive galaxies grow through mergers, we should see the formation of supermassive binary black holes in the center of galaxies that have undergone major mergers [12]. We do not, however, observe them on small scales (see [19] for a recent review).

In this paper we give a short overview of the usage of the spectral lines (Fe Kα and UV/optical) for estimates of the black hole masses and spins. Mostly we will present the work from the Group for astrophysical spectroscopy at the Astronomical observatory in Belgrade [13, 14]. We will describe the basic properties of the broad emission lines and their application in determining the parameters of supermassive black holes, i.e the mass and spin. Moreover, we will describe how can we see signatures of supermassive binary black holes in the broad emission lines.

2. Broad emission lines and supermassive black holes

Broad emission lines are an important tool to study the properties of the BLR in AGN (see, e.g., [9, 15, 10]). However, one should note the observed fact that the BLR is a complex region showing multiple components (e.g., [9, 16]), thus the broad line properties show contributions from different regions with different kinematics [17]. For example, the X-ray Fe Kα line arise from the gas within a few gravitational radii of the center of supermassive black hole in the accretion disk [3], while the UV/optical broad lines originate further away in the accretion disk (e.g., broad double-peaked Hα and Hβ lines in 3C390.3, [18]) or complex geometry of the accretion disk with the additional component [16] or outflowing material [17]. Moreover, complex broad line profiles (e.g., asymmetric or showing two displaced peaks) could indicate a supermassive black hole binary and complex BLR [19]. Finally, the broad emission lines of AGNs often exhibit variability, which is probably caused by variation in the ionizing continuum strength and by dynamic evolution of the BLR gas. Studying the variability of the broad lines (fluxes, widths, profiles, asymmetrics) we can find the size, geometry and physical properties of the BLR [10].

2.1. Fe Kα line - spin of the supermassive black holes

The fluorescent/recombination iron Kα line is produced when the hard X-ray radiation cause the ejection of the K-shell electrons of an iron atom/ion following the photoelectric absorption of an X-ray [20]. For the neutral iron, the Fe Kα line energy is 6.4 keV and is usually a narrow line, but if it originates from a relativistic accretion disk of AGN it becomes wider and changes its profile due to kinematical effects: Doppler boosting and gravitational redshift, and that kind of line broadening is often observed in spectra of Seyfert galaxies [3]. Black holes have three measurable parameters: charge, mass and angular momentum (or spin), where in the case of supermassive black holes of AGN, only the latter two are of important as they are responsible for several observational effects detected in the observed Fe Kα line profile [21, 22, 23]. Figure 1 demonstrates the difference in the line emitting regions and the corresponding line profile in the case of the non-rotating (Schwarzschild metric) and rotating black hole (Kerr metric). It is clearly seen in Figure 1 (bottom panels), that for almost maximally rotating black hole, the red peak of the Fe Kα line is more intensive and the red wing is much more extended (see [23] for review). However, the angular momentum of the supermassive black hole has significant influence on the broad line profile and thus by comparing the observed and modeled Fe Kα line profiles some essential information about the spin of the supermassive black hole can be

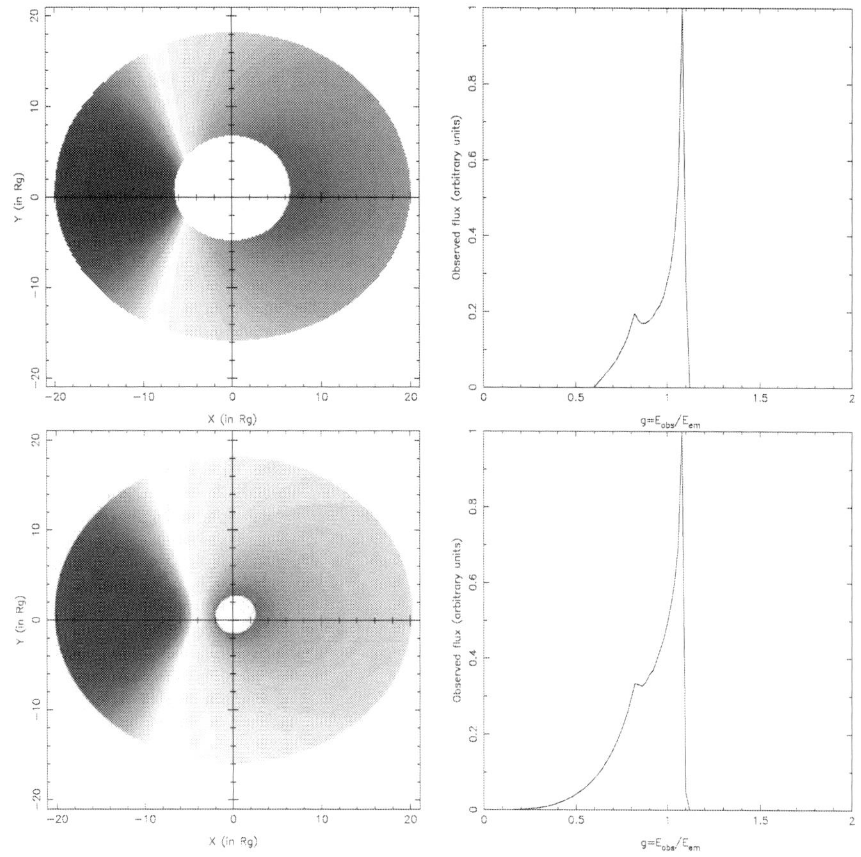

Figure 1. Illustrations of accretion disk (left panels) and the corresponding Fe Kα line profiles (right panels) in the case of Schwarzschild metric (upper panels) and Kerr metric with angular momentum $a = 0.998$ (bottom panels). The parameters of the disk in both cases are: inclination $i = 35°$, inner radius $R_{in} = R_{ms}$, and outer radius $R_{out} = 20$ R$_g$ (figure from [22]).

retrieved [21, 22, 23]. There are several more works in which the attempts to estimate the spin of the supermassive black hole using the observed Fe Kα were made (see, e.g., [24] for review).

2.2. UV/optical broad emission lines - mass of the supermassive black holes

The UV/optical broad emission lines are variable in most of AGN. A long-term monitoring of these lines, has revealed that changes in the broad lines fluxes are lagged to flux changes in the continuum (see, e.g., [25, 26, 27]). This is illustrated in Figure 2, which shows the cross correlation function between the light curves of the continuum and broad Hβ emission line fluxes for two AGN: 3C 390.3 and NGC 5548 (see [28] for details on the analysis). The objects have been monitored for more than a decade in [29, 30]. The time lag gives directly the distance of the BLR $R_{\rm BLR}$, which together with the gas velocity V, obtained from the line widths, gives the mass of the supermassive black hole $M_{\rm BH}$, with the assumption that the BLR gas is virialized (see [31] for review):

$$M_{\rm BH} = f\frac{R_{\rm BLR}V^2}{G},$$

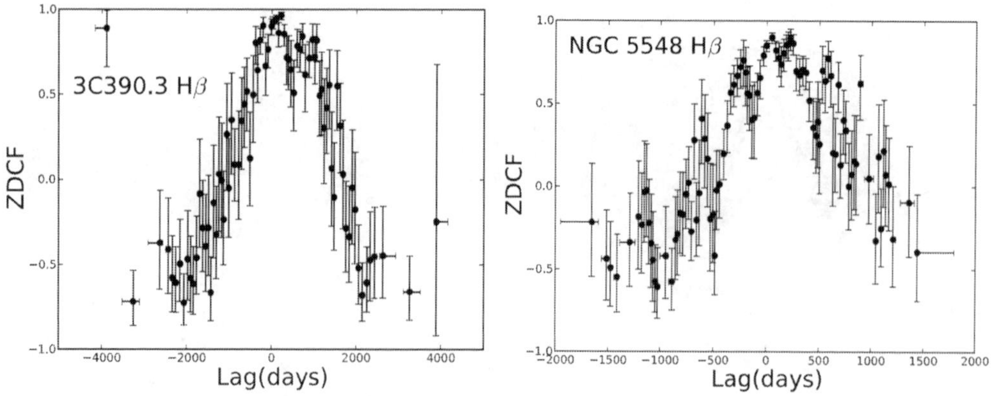

Figure 2. The Z-transformed discrete correlation function applied to the light curves of the continuum and Hβ emission line for 3C 390.3 and NGC 5548. The horizontal and vertical error bars correspond to 1σ uncertainties for a normal distribution (for details see [28]).

where f is the dimensionless parameter, so-called virial coefficient, and G is the gravitational constant. These are the basis of the reverberation mapping method (see, e.g., [5, 6, 7, 8]). It was also shown that the BLR distance $R_{\rm BLR}$ scales with the nuclear continuum luminosity L (see, e.g., [32, 33, 34, 35, 36]). This RL relationship gives a mass scaling relation that allows us to obtain an estimate of the mass of the supermassive black hole from an individual (i.e., single-epoch) spectrum of the AGN using the parameters of continuum and broad emission line (many different broad emission lines can be used, but the main one are Hβ, Mg II and C IV). This has been widely applied on different and large samples of data (e.g., [32, 33, 37, 38, 8, 39, 40, 41, 42, 43, 44, 45, 46, 47], and many others), even on very distant quasars $z > 6.5$ (e.g., [48]). Nevertheless, this method has many difficulties and is still not accurate (see, e.g., [49] for a review). First of all, the factor f depends on the BLR geometry that is still not known, and mostly the statistical values for f are used (e.g., [50, 51, 52]). Secondly, there are many systematic errors in the time lag determinations (see, e.g., [53, 49]), such as low data quality or undersampling of data, etc. Therefore, intensive and well-coordinate optical monitoring campaigns of Type 1 AGN are needed to investigate the structure (especially the geometry) of the BLR (such as presented in [54, 29, 55, 30, 56, 57]).

2.3. Broad emission lines and supermassive binary black holes

There are only several examples where supermassive binary black holes have been detected on spatially resolved imaging or/and spectroscopy (see [19] for a recent review). One possibility of a binary BLR detection (thus the supermassive binary black holes system detection) is by using the broad optical lines, even if the lines are single-peaked with some peculiar line profile. Recently, [58] analyzed the broad Hα line profile and flux variability during a period of more than 20 years of NGC 4151, one of the best studied Seyfert galaxy. They presented evidence for the first spectroscopically resolved subparsec orbit (with a 15.9 year period) of supermassive binary black holes. Moreover, [59] found 88 candidates for subparsec supermassive black hole binaries among Sloan Digital Sky Survey quasars, by searching for the broad Hβ emission lines that are displaced from the quasar rest-frame, since this would be expected if one of the two black holes accretes at a much higher rate than the other and carries with it the only BLR of the system.

In general, in case of the supermassive black hole binary system, we can have either one

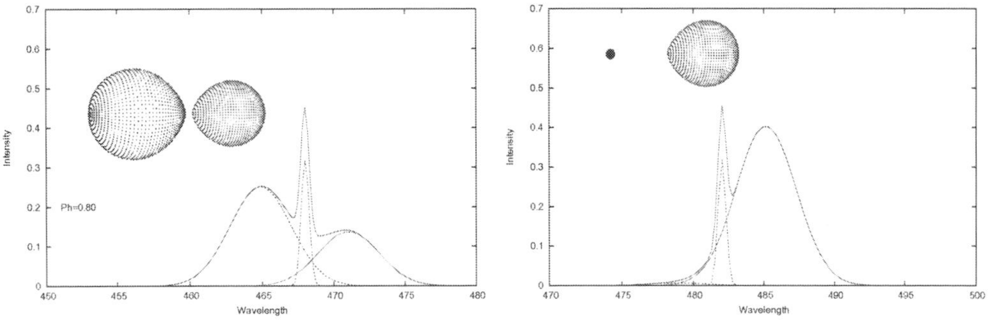

Figure 3. The Hβ line profile emitted from the supermassive binary black holes system for the cases: 1) both of black holes have the BLR (left) and only one component has the BLR (right). The scheme of supermassive binary black holes system is present on the plot.The Hβ line profile was simulated for a system with following parameters: mass ratio of $q = 0.5$, distance between black holes $R = 0.3$ pc, orbiting period of $P = 600$ years. A narrow component (narrow central Gaussian) is arbitrarily added. Figure from [19].

emitting region around two black holes, or two emitting regions clearly separated around each black hole. Thus the models of the binary BLRs around two supermassive black holes [60] give large diversity of line profiles, as shown in Figure 3. The left panel presents the case where both supermassive black holes have the BLR (as a Roche lobe around each of them), where the orbit is edge-on to the observer. The right panel presents the case when only one component has the BLR. The line profile is composite from emission of both regions in the case of binary BLR (Figure 3, left panel), but in some cases the line profile can show only one peak, e.g., if only one BLR is present (Figure 3, right panel). Even in the case of binary BLR, the line profile depends strongly on the orbital phase and may change from a very asymmetric (double peaked and highly shifted) to the symmetric profile (see Figure 4. in [19]). Thus, the variability in the line profile is expected to be detected.

3. Conclusions

Here we briefly presented the possibility to use broad emission lines in AGN to find the parameters of the central supermassive black hole. The main conclusion is that broad emission lines of Type 1 AGN and their properties (flux, profile, variability) can be used to obtain the properties of the BLR gas near the supermassive black hole. Using the parameters for the BLR (dimension and velocity field) we are able to determine the mass of the central supermassive black hole. On the other hand, the shape of the X-ray Fe Kα line can be used to find the spin of the central supermassive black hole. At the end, the complex and variable broad lines can indicate a presence of the supermassive binary black holes at sub-parsec scales. Therefore, the investigation of the spectral line shapes in general can help us to understand the structure of the most mysterious objects, as in this case of supermassive black holes in the center of AGN.

Acknowledgments

This work was supported by the Ministry of Education, Science and Technological Development of Republic of Serbia through the project Astrophysical Spectroscopy of Extragalactic Objects (176001).

References

[1] Osterbrock D E and Ferland G J 2006 *Astrophysics of Gaseous Nebulae and Active Galactic Nuclei (2nd ed.)* (Sausalito, California: University Science Books)
[2] Tanaka Y, Nandra K, Fabian A C *et al.* 1995 *Nature* **375** 659
[3] Nandra K, O'Neill P M, George I M and Reeves N 2007 *Monthly Noticice of Royal Astronomical Society* **382** 194
[4] Ferrarese L and Ford H 2005 *Space Science Reviews* **116** 523
[5] Blandford R D and McKee C F 1982 *Astrophysical Journal* **255** 419
[6] Gaskell C M 1988 *Astrophysical Journal* **325** 114
[7] Peterson M B 1993 *PASP* **105** 247
[8] Peterson B M, Ferrarese L, Gilbert K M *et al.* 2004 *ApJ* **613** 682
[9] Sulentic J W, Marziani P and Dultzin-Hacyan D 2000 *ARA&A* **38** 521
[10] Ilić D, Popović L Č, Shapovalova A I *et al.* 2012 *Journal of Physics Conference Series* **397** 012050
[11] Afanasiev V L, Popović L Č, Shapovalova A I, Borisov N V and Ilić D 2014 *Monthly Noticice of Royal Astronomical Society* **440** 519
[12] Begelman M C, Blandford R D and Rees M J 1980 *Nature* **287** 307
[13] Dimitrijević M S, Popović L Č, Jevremović D *et al.* 2009 *Publications of the Astronomical Observatory of Belgrade* **86** 271
[14] Popović L Č and Ilić D 2012 *Publications of the Astronomical Society "Rudjer Boskovic"* **11** 285
[15] Gaskell C M 2009 *New Astronomy Review* **53** 140
[16] Popović L Č, Mediavilla E G, Bon E and Ilić D 2004 *A&A* **423** 909
[17] Denney K D, Peterson B M, Pogge R W *et al.* 2009 *Astrophysical Journal Letters* **704** L80
[18] Popović L Č, Shapovalova A I, Ilić D *et al.* 2011 *A&A* **528** 130
[19] Popović L Č 2012 *New Astronomy Review* **56** 74
[20] Fabian A C, Iwasawa K, Reynolds C S and Young A J 2000 *PASP* **112** 1145
[21] Jovanović P and Popović L Č 2008 *Fortschr. Phys.* **56** 456
[22] Popović L Č and Jovanović P 2009 *American Institute of Physics Conference Series* **1121** 71
[23] Jovanović P 2012 *New Astronomy Review* **56** 37
[24] Reynolds C 2013 *Space Science Reviews* **81**
[25] Wanders I and Peterson B M 1996 *Astrophysical Journal* **466** 174
[26] Kollatschny W and Dietrich M 1997 *A&A* **323** 5
[27] Peterson M B 2008 *New Astronomy Review* **52** 240
[28] Kovačević A, Popović L Č, Shapovalova A I *et al.* 2014 *Advances in Space Research* **54** 1414
[29] Shapovalova A I, Doroshenko V T, Bochkarev N G *et al.* 2004 *A&A* **422** 925
[30] Shapovalova A I, Popović L Č, Burenkov A N *et al.* 2010 *A&A* **517** 42
[31] Peterson M B 2013 *Space Science Reviews* **60**
[32] Wandel A, Peterson B M and Malkan M A 1999 *Astrophysical Journal* **526** 579
[33] Kaspi S, Smith P S, Netzer H *et al.* 2000 *Astrophysical Journal* **533** 631
[34] Kaspi S, Maoz D, Netzer H *et al.* 2005 *Astrophysical Journal* **629** 61
[35] Bentz M C, Peterson B M, Pogge R W, Vestergaard M and Onken C A 2006 *Astrophysical Journal* **644** 133
[36] Bentz M C, Peterson B M, Netzer H, Pogge R W and Vestergaard M 2009 *Astrophysical Journal* **697** 160
[37] McLure R J and Jarvis M J 2002 *Monthly Noticice of Royal Astronomical Society* **337** 109
[38] McLure R J and Dunlop J S 2004 *Monthly Noticice of Royal Astronomical Society* **352** 1390
[39] Woo J-H, Urry C M, Lira P, van der Marel R P and Maza J 2004 *Astrophysical Journal* **617** 903
[40] Kollmeier J A, Onken C A, Kochanek C S *et al.* 2006 *Astrophysical Journal* **648** 128
[41] Vestergaard M and Peterson B M 2006 *Astrophysical Journal* **641** 689
[42] Salviander S, Shields G A, Gebhardt K and Bonning E W 2007 *Astrophysical Journal* **662** 131
[43] Shen Y, Greene J E, Strauss M A, Richards G T and Schneider D P 2008 *Astrophysical Journal* **680** 169
[44] Vestergaard M, Fan X, Tremonti C A, Osmer P S, and Richards G T 2008 *Astrophysical J. Lett.* **674** L1
[45] Vestergaard M. and Osmer P S 2009 *Astrophysical Journal* **699** 800
[46] Denney K D, Peterson B M, Pogge R W *et al.* 2010 *Astrophysical Journal* **721** 715
[47] Shen Y, Richards G T, Strauss M A *et al.* 2011 *Astrophysical Journal Supplement* **194** 45
[48] De Rosa G, Venemans B P, Decarli R *et al.* 2014 *Astrophysical Journal* **790** 145
[49] Vestergaard M, Denney K, Fan X *et al.* 2011 in *Narrow-Line Seyfert 1 Galaxies and their Place in the Universe, Proceedings of Science* **POS(NLS1)** 38
[50] Onken C A, Ferrarese L, Merritt D *et al.* 2004 *Astrophysical Journal* **615** 645
[51] La Mura G, Di Mille F, Ciroi S, Popović L Č and Rafanelli P 2009 *Astrophysical Journal* **693** 1437
[52] Graham A W, Onken C A, Athanassoula E and Combes F 2011 *Monthly Noticice of Royal Astronomical Society* **412** 2211

[53] Denney K D, Assef R J, Bentz M C *et al.* 2011 *Narrow-Line Seyfert 1 Galaxies and their place in the Universe, Proceedings of Science* **POS(NLS1)** 34
[54] Shapovalova A I, Burenkov A N, Carrasco L *et al.* 2001 *A&A* **376** 775
[55] Shapovalova A I, Popović L Č, Collin S *et al.* 2008 *A&A* **486** 99S
[56] Shapovalova A I, Popovic L Č, Burenkov A N *et al.* 2012 *Astrophysical Journal Supplement* **202** 10
[57] Shapovalova A I, Popović L Č, Burenkov A N *et al.* 2013 *A&A* **559** 10
[58] Bon E, Jovanović P, Marziani P *et al.* 2012 *Astrophysical Journal* **759** 118
[59] Eracleous M, Boroson T A, Halpern J P and Liu J 2012 *Astrophysical Journal Supplement* **201** 23
[60] Popović L Č, Mediavilla E G and Pavlović R 2000 *Serbian Astronomical Journal* **162** 1

The emission spectrum due to molecule formation through radiative association

Magnus Gustafsson[1] **and Gunnar Nyman**[2]

[1]Applied Physics, Division of Materials Science, Department of Engineering Science and Mathematics, Luleå University of Technology, SE-97187 Luleå, Sweden
[2]Department of Chemistry and Molecular Biology, University of Gothenburg, SE-41296 Gothenburg, Sweden.

E-mail: [1]magnus.gustafsson@ltu.se

Abstract. Quantum mechanical and classical methods for theoretical analysis of the emission spectrum due to radiative association are presented. Quantum mechanical perturbation theory is employed to obtain the spectra when the diatomic molecule HF forms by transitions within the electronic ground state and when it forms by transitions between two electronic states. We contrast these spectra with each other. The former peaks in the infrared, while the latter peaks in the ultraviolet. The classical spectrum, which concerns transitions within the electronic ground state, is also calculated and found to favorably compare with that from quantum mechanical perturbation theory. The emission stemming from resonance mediated radiative association is also discussed.

1. Introduction

Radiative association is a mechanism which contributes to the production of molecules in interstellar clouds. It is especially important in low-density and dust-poor regions where reactions due to ternary collisions are extremely rare [1]. On Earth the rates for radiative association reactions are typically orders of magnitude lower than those of other chemical processes, such as exchange reactions, and direct laboratory measurements have so far only been possible to carry out for certain ionic systems [2]. Thus astronomers who wish to model the chemical evolution in interstellar space often have to rely on theoretical calculations of radiative association rates. Not only molecular formation due to radiative association but also the corresponding emission spectrum has attracted attention. For example, radiative association of N and O atoms is a significant source of emission in the terrestrial nightglow and in the atmosphere of Venus [3]. In this work we analyze the emission spectrum due to formation of HF at different relative kinetic energies of the colliding atoms H and F. Some of the theory and the results are presented in more detail in Ref. [4].

2. Theory

In general it is desirable to Boltzmann average the cross sections for radiative association in order to obtain thermal rate constants. The focus here is on the spectra, however, which are conveniently analyzed through the energy dependent cross sections. In sections 2.1, 2.2, and 2.3 we will briefly outline three different approximations to obtain the cross sections. In section 2.4 some theoretical aspects of radiative association of HF are presented.

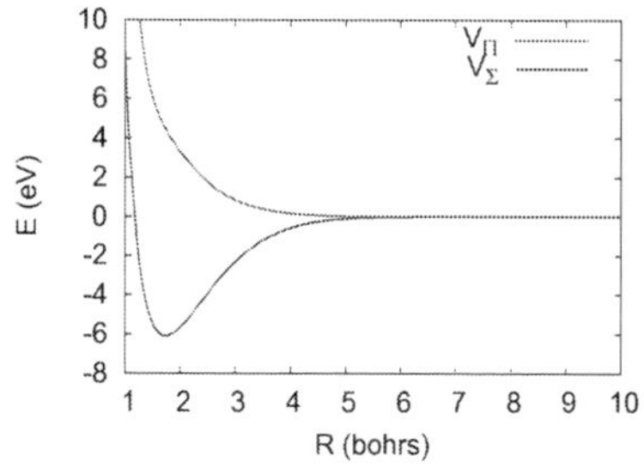

Figure 1. Potential energy curves for HF.

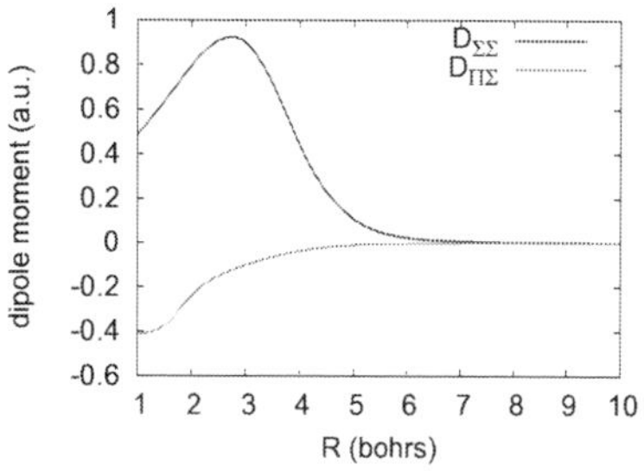

Figure 2. Electric dipole moment curves for HF.

2.1. Quantum mechanical perturbation theory

Perturbation theory for dipole transitions, or Fermi's golden rule, hinges on the evaluation of transition dipole matrix elements

$$M_{\Lambda EJ,\Lambda' v' J'} = \int_0^\infty F^\Lambda_{EJ}(R)\ D_{\Lambda\Lambda'}(R)\ \Phi^{\Lambda'}_{v'J'}(R)\ dR \tag{1}$$

where E is the asymptotic relative kinetic energy for the colliding particles, $F^\Lambda_{EJ}(R)$ is the continuum wave function, and $\Phi^{\Lambda'}_{v'J'}(R)$ is the bound state wave function. $D_{\Lambda\Lambda'}(R)$ is the electric transition dipole between electronic states Λ and Λ' if $\Lambda \neq \Lambda'$ and the permanent electric dipole moment if $\Lambda = \Lambda'$. The cross section for (spontaneous emission) radiative association of a molecule in rovibrational state v', J' is obtained through [5]

$$\sigma_{\Lambda\to\Lambda' v' J'}(E) = \frac{2}{3}\frac{h^2}{(4\pi\epsilon_0)c^3}\frac{1}{2\mu E}\ P_\Lambda\ \omega^3_{E\Lambda' v' J'} \sum_J S_{\Lambda J\to\Lambda' J'}\ M^2_{\Lambda EJ,\Lambda' v' J'} \tag{2}$$

where $\omega_{E\Lambda' v' J'}$ is the angular frequency of emitted photon and $S_{\Lambda J\to\Lambda' J'}$ are Hönl–London factors [6]. P_Λ is the statistical weight for approach in the electronic state Λ, i.e. the degeneracy of state Λ divided by the total degeneracy. Specific values for P_Λ are given in section 2.4. The total radiative association cross section for formation of a molecule by transition from electronic state Λ to Λ' is obtained by summation over all final vibrational and rotational quantum numbers, i.e.

$$\sigma_{\Lambda\to\Lambda'}(E) = \sum_{v' J'} \sigma_{\Lambda\to\Lambda' v' J'}(E)\ . \tag{3}$$

2.2. Classical theory (no electronic transition)

In the purely classical treatment of radiative association within one electronic state the radiation is given by the Larmor power [7]

$$\hat{I}(b, E, \omega) = P_\Lambda \frac{2\omega^4}{3c^3\pi(4\pi\epsilon_0)} \left| \int_{-\infty}^{\infty} dt\ e^{i\omega t}\ \mathbf{D}_{\Lambda\Lambda}(b, E, t) \right|^2 \tag{4}$$

with the time dependent dipole moment $\mathbf{D}_{\Lambda\Lambda}(b, E, t)$ evaluated along a classical trajectory with initial energy E and impact parameter b. The spectral density can be obtained as

$$\frac{d\sigma_{\Lambda\to\Lambda}(E)}{d\omega} = \int_0^\infty 2\pi b \frac{\hat{I}(b, E, \omega)}{\hbar\omega}\ db \tag{5}$$

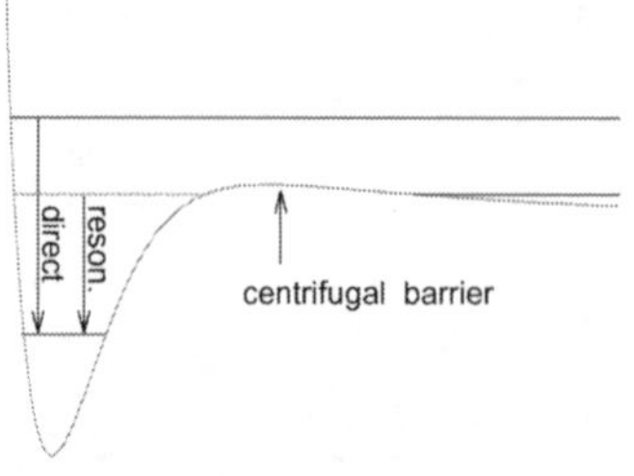

Figure 3. Schematic illustration of direct and resonance mediated radiative association. In this case the process takes place within one electronic state, as is the case for reaction (10). The resonance mediated process thus gives rise to shape resonances in the cross section. In general several electronic states may be involved, which can give rise to both shape and Feshbach resonances.

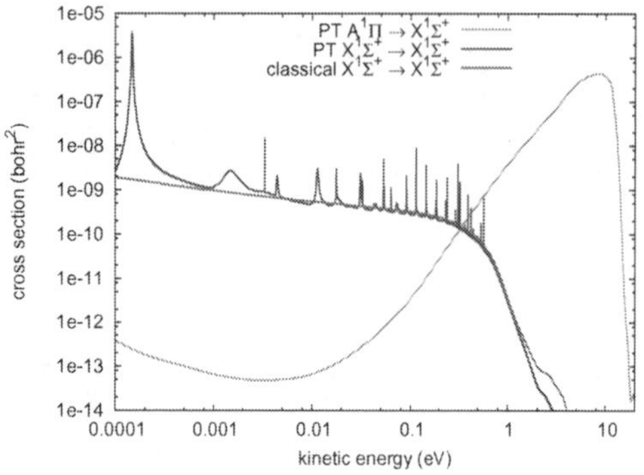

Figure 4. Cross sections for forming HF through radiative association, computed using quantum mechanical perturbation theory (PT), Eq. (3). The classical cross section, Eq. (6), for the $X^1\Sigma^+ \rightarrow X^1\Sigma^+$ transition is also shown. The classical result reproduces the baseline of that from quantum mechanics but completely misses the resonance structure which is due to shape resonances, which in turn stem from quantum mechanical tunneling.

and finally the radiative association cross section is computed by integration over photon energies [8]

$$\sigma_{\Lambda\rightarrow\Lambda}(E) = \int_{E/\hbar}^{\omega_2} \frac{d\sigma_{\Lambda\rightarrow\Lambda}(E)}{d\omega}\, d\omega \tag{6}$$

with the limits, $E/\hbar$ and ω_2, defined by the range of bound states.

2.3. Breit–Wigner theory

Radiative association cross sections are often rich in sharp resonance features. This makes it numerically hard to thermally average them to obtain rate constants. Breit–Wigner theory [9, 10] provides a convenient way to get around this problem. The cross section for radiative association, mediated through quasibound states with rotational and vibrational quantum numbers J and v, respectively, is

$$\sigma_{\Lambda\rightarrow\Lambda'}(E) = \sum_{vJv'J'} \sigma_{\Lambda vJ\rightarrow\Lambda'v'J'}(E) \tag{7}$$

where the state resolved cross is defined by

$$\sigma_{\Lambda vJ\rightarrow\Lambda'v'J'}(E) = \frac{\pi\hbar^2}{2\mu E}\, P_\Lambda\, (2J+1) \frac{\Gamma^{\rm rad}_{\Lambda vJ\rightarrow\Lambda'v'J'}\Gamma^{\rm tun}_{\Lambda vJ}}{(E - E_{\Lambda vJ})^2 + (\Gamma^{\rm tun}_{\Lambda vJ} + \sum_{v''J''}\Gamma^{\rm rad}_{\Lambda vJ\rightarrow\Lambda'v''J''})^2/4}\,. \tag{8}$$

In Eq. (8) $E_{\Lambda vJ}$ is the energy of the quasi-bound level, $\Gamma^{\rm tun}_{\Lambda vJ}$ is its tunneling width, and $\Gamma^{\rm rad}_{\Lambda vJ\rightarrow\Lambda'v'J'}$ is its radiative width for allowed $\Lambda vJ \rightarrow \Lambda'v'J'$ spontaneous emission processes. The $v''J''$ summation in the denominator of Eq. (8) could be extended to a summation over more than one final electronic state (Λ') but here we will only consider radiative decay to one electronic state. The resonance parameters $E_{\Lambda vJ}$, $\Gamma^{\rm tun}_{\Lambda vJ}$, and $\Gamma^{\rm rad}_{\Lambda vJ\rightarrow\Lambda'v'J'}$ can be calculated fast without solving the full scattering problem, e.g. with LEVEL [11].

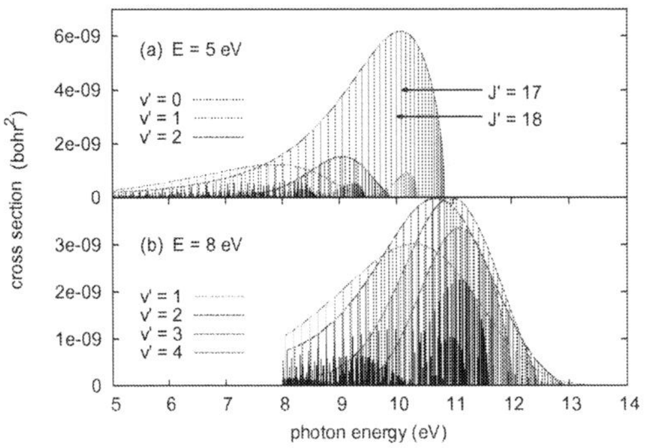

Figure 5. Emission spectra for $A^1\Pi \rightarrow X^1\Sigma^+$ radiative association of HF at collision energies of 5 eV and 8 eV. Reprinted with permission from [4]. Copyright (2014), American Institute of Physics.

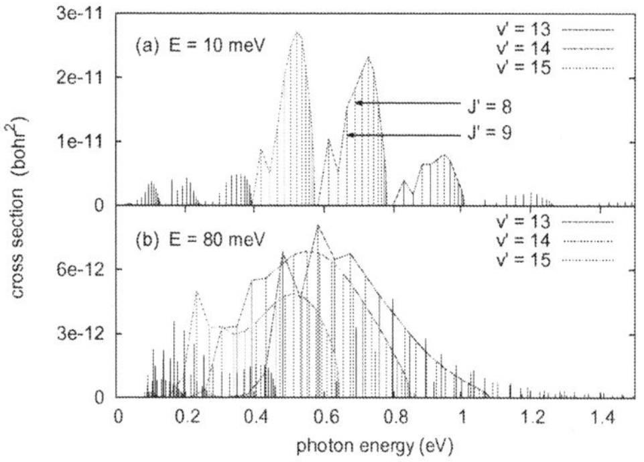

Figure 6. Emission spectra for $X^1\Sigma^+ \rightarrow X^1\Sigma^+$ radiative association at collision energies of 10 meV and 80 meV. Reprinted with permission from [4]. Copyright (2014), American Institute of Physics.

2.4. Radiative association of HF

In this study we consider the radiative association reactions

$$\mathrm{H}(^2S) + \mathrm{F}(^2P) \rightarrow \mathrm{HF}(A^1\Pi) \rightarrow \mathrm{HF}(X^1\Sigma^+) + \hbar\omega \tag{9}$$

$$\mathrm{H}(^2S) + \mathrm{F}(^2P) \rightarrow \mathrm{HF}(X^1\Sigma^+) \rightarrow \mathrm{HF}(X^1\Sigma^+) + \hbar\omega \tag{10}$$

for formation of HF in its ground electronic state. The factor P_Λ which appears in the expressions for the cross sections above is 2/12 and 1/12 for reactions (9) and (10), respectively. In the discussion below reactions (9) and (10) will sometimes be referred to as $A^1\Pi \rightarrow X^1\Sigma^+$ and $X^1\Sigma^+ \rightarrow X^1\Sigma^+$, respectively. The potential energies and electric dipole moments used in the calculations are shown in Figs. 1 and 2. The $X^1\Sigma^+$ state potential energy and dipole moment functions are based on the data of Ref. [12], and the $A^1\Pi$ potential energy and $A^1\Pi - X^1\Sigma^+$ transition dipole moment are based on the data of Ref. [13].

Radiative association can occur through two main classes of mechanisms commonly called direct and resonance mediated. The difference between them is illustrated in Fig. 3 and by inspecting the potential curves in Fig. 1 we conclude that reaction (9) only can occur through direct radiative association. This is a consequence of the fact that the $A^1\Pi$ potential is purely repulsive, except for a tiny van der Waals well, and therefore it does not support any quasibound states. Reaction (10) on the other hand occurs through both mechanisms and will thus have resonance structure.

3. Results

Fig. 4 shows the HF radiative association cross section versus kinetic energy from 0.1 meV to 20 eV. The cross section for reaction (10) dominates up to an energy of about 0.3 eV and at higher energies reaction (9) dominates. In our analysis of the spectra below, we focus on each reaction in the energy range where it dominates. HF radiative association emission spectra versus photon energy, $\hbar\omega_{E\Lambda'v'J'}$, computed quantum mechanically with Eq. (2), are shown in Figs. 5 and 6 for $A^1\Pi \rightarrow X^1\Sigma^+$ and $X^1\Sigma^+ \rightarrow X^1\Sigma^+$ transitions, respectively. The envelopes corresponding to vibrational quantum states, v', with dominating cross sections are indicated. The collision energies are chosen to be off-resonance in the $X^1\Sigma^+ \rightarrow X^1\Sigma^+$ case. There is a striking difference in photon energies emitted from the two reactions $A^1\Pi \rightarrow X^1\Sigma^+$, which emits mostly ultraviolet

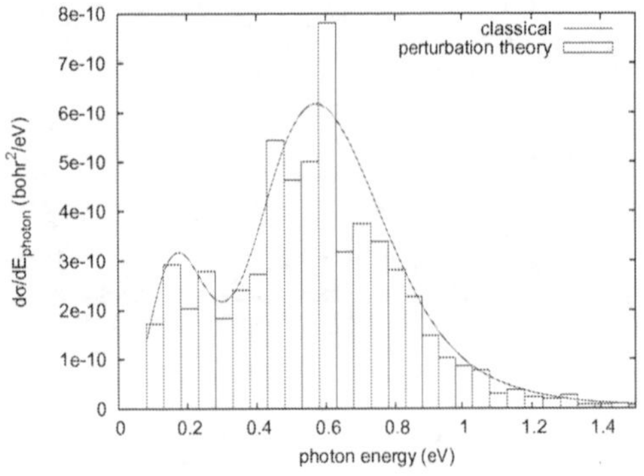

Figure 7. The spectral density for $X^1\Sigma^+ \to X^1\Sigma^+$ radiative association of HF at collision energy E=80 meV. The classical result is computed with Eq. (5). The perturbation theory result is computed from the data presented in Fig. 6 (b) by adding the cross sections in photon energy ($E_{\text{photon}} = \hbar\omega$) intervals of 0.05 eV and then dividing by that spacing. Reprinted with permission from [4]. Copyright (2014), American Institute of Physics.

Table 1. Major contributions to the cross section to the peak cross section for the v=18, J=13 resonance peak which appears at a collision energy of 3.304 meV.

v'	J'	σ (bohr^2)	E_{photon} (eV)
13	12	1.1	0.79
13	14	1.2	0.72
14	12	1.2	0.58
14	14	2.3	0.51

radiation in the range 8–12 eV, and $X^1\Sigma^+ \to X^1\Sigma^+$ which emits mostly infrared radiation in the range from 0.2 to 1.0 eV.

It is also interesting to see how well the classical spectral density given by Eq. 5 compares with its quantum mechanical counterpart. In Fig. 7 the comparison is shown and we see that for this system the quantum mechanical result shows more structure than that from a classical calculation. The general shape, however, is rather similar and, as was noted above in the discussion of Fig. 4, an integration over photon energy gives a classical cross section that agrees with the baseline of the quantum mechanical cross section. The former is only about 12% above the latter.

It cannot be denied that the values of the cross sections presented above are rather small, making it difficult to use measurement of photon emission as a probe of radiative association. At the peaks of the resonances, however, the cross sections can be large. As an example of this we analyze the narrow $X^1\Sigma^+ \to X^1\Sigma^+$ resonance peak at $E = 3.304$ meV in Fig. 4. The corresponding quasibound state has vibrational and rotational quantum numbers v=18, J=13, respectively. Using the Breit–Wigner formula, Eq. (7), to compute the peak value gives a cross section of 9.0 bohr^2. The peak is thus by no means resolved in Fig. 4. The contributions to this (9 bohr^2) cross section from formation of HF in different rovibrational states can be obtained through Eq. (8). The main contributions are presented in table 1 and we see that those four transitions account for about 2/3 of the cross section.

4. Conclusions

The emission spectrum due to radiative association of HF is highly dependent on the relative kinetic energy of the colliding H and F atoms. At collision energies below about 0.3 eV, where the molecule formation is due to $X^1\Sigma^+ \to X^1\Sigma^+$ transitions, the spectrum is most intense in the infrared with wavelengths of about 1–6 μm. At higher energies the molecule formation occurs primarily through $A^1\Pi \rightarrow X^1\Sigma^+$ transitions and the emission is dominated by ultraviolet radiation with wavelengths of roughly 100–150 nm. The resonance mediated emission can be rather strong compared to the off-resonance case. For the resonance case that we studied in detail there are four transitions, which all have cross sections that are larger than 1 bohr^2, that account for the majority of the peak cross section. Detecting the emitted photons for this kind of process is still expected to be difficult as the total photon count for a narrow resonance like the one we analyzed may be small. The small width of the resonance in turn stems from the fact that the corresponding quasibound state has a long lifetime.

Acknowledgments

This work was supported by the Swedish Science Council.

References

[1] Babb J F and Kirby K P 1998 *The Molecular Astrophysics of Stars and Galaxies* ed Hartquist T W and Williams D A (Oxford: Clarendon Press) p 11
[2] Gerlich D and Horning S 1992 *Chem. Rev.* **92** 1509
[3] Dalgarno A, Babb J and Sun Y 1992 *Planet. Space Sci.* **40** 243
[4] Gustafsson M, Monge-Palacios M and Nyman G 2014 *J. Chem. Phys.* **140** 184301
[5] Babb J F and Dalgarno A 1995 *Phys. Rev. A* **51**(4) 3021
[6] Hansson A and Watson J K 2005 *J. Molec. Spectrosc.* **233** 169
[7] Levine H B and Birnbaum G 1967 *Phys. Rev.* **154** 86
[8] Gustafsson M 2013 *J. Chem. Phys.* **138** 074308
[9] Breit G and Wigner E 1936 *Phys. Rev.* **49** 519
[10] Bain R A and Bardsley J N 1972 *J. Phys. B* **5** 277
[11] Le Roy R J 2007 *University of Waterloo Chemical Physics Research Report CP-663*
[12] Talbi D and Bacchus-Montabonel M C 2007 *Chem. Phys. Lett.* **443** 40
[13] Brown A and Balint-Kurti G G 2000 *J. Chem. Phys.* **113** 1870

Possible New Positronium Bound State

H rater[1] and Y Won [2]

[1]University of Tennessee Space Institute, 411 B.H. Goethert Parkway, Tullahoma, TN 37388, USA

[2]Physics Division, Oak Ridge National Laboratory, Oak Ridge, TN 37830, USA

E-mail: hcrater@utsi.edu, wongc@ornl.gov

Abstract. The Two-Body Dirac equations of constraint dynamics applied to QED yields an exact Sommerfeld-like solution for the spectrum of ${}^1J_J$ singlet positronium states which agrees with standard perturbative results through order α^4. At short distance the radial part of the wave function $u = r\psi$ has two solutions with probabilities near the origin behaving like $\psi^2 d^3r = u^2 dr d\Omega = r^{(1\pm\sqrt{(2J+1)^2-4\alpha^2})} dr d\Omega$. For $J \neq 0$ only the first sign is allowable but both signs for $J = 0$ are well behaved. The first sign in that case corresponds to ordinary positronium (with a binding energy of about 6.8 eV). The second sign corresponds to a new positronium state with a binding energy of about 300 keV and a root-mean-square radius on the order of a Compton wavelength. The ordinary $1S$ positronium state decays into this new $1S$ state by two photon emission with c.m. energy of about 300 keV. The peculiar $1S$ state then annihilates promptly into two photons with c.m. energy of about 700 keV. Thus the existence of this new positronium state would be a distinctive 4 photon decay signature of ordinary singlet positronium.

In this paper we report on the prediction of a new positronium bound state with a very large binding energy of about 300 keV. It results from a metastable two-photon decay of the usual positronium bound state which has a binding energy of about 6.8 eV. It has a size of about the Compton wave length of an electron. Once it is formed it annihilates promptly into 2 photons with a c.m. energy of about 700 keV. Thus the existence of this new positronium bound state would be a distinctive 4 photon decay signature of the usual singlet bound state.

The prediction arises from an exact solution of the Two Body Dirac equations (TBDE) of constraint dynamics applied to QED solutions for singlet $S-$ states [1], [2]. In the references just quoted one can examine the details of the predictions. Here we present the highlights. One finds that the TBDE can be recast in a Schrödinger-like form reminiscent of that for the hydrogen atom. The radial form of that equation displays an invariant form of the Coulomb potential and its relativistic square,

$$\begin{aligned}
\left\{-\frac{d^2}{dr^2}+2\varepsilon_w\mathcal{A}(r)-\mathcal{A}^2(r)\right\}u &= b^2u, \\
\mathcal{A}(r) &= -\frac{\alpha}{r},\ \alpha\text{ -fine structure constant}, \\
w &= c.m.\text{ total energy}, \\
b^2 &= \varepsilon_w^2-m_w^2;\text{ two-body relativistic eigenvalue}, \\
\varepsilon_w &= (w^2-2m^2)/2w,\text{ invariant energy of relative motion}, \\
m_w &= m^2/w,\text{invariant reduced mass of relative motion }.(1)
\end{aligned}$$

This equation has the short distance ($r << \alpha/2\varepsilon_w$) behavior

$$\left\{-\frac{d^2}{dr^2}-\frac{\alpha^2}{r^2}\right\}u=0, \tag{2}$$

with solutions

$$\begin{aligned}
u_+ &\sim r^{\lambda_++1},\text{ usual} \\
u_- &\sim r^{\lambda_-+1},\text{ peculiar} \\
\lambda_\pm &= (-1\pm\sqrt{1-4\alpha^2})/2.
\end{aligned} \tag{3}$$

With these behaviors, the probability for the relative location of the electron and positron is

$$\psi_\pm^2d^3r=\frac{u_\pm^2}{r^2}r^2drd\Omega=u_\pm^2drd\Omega=r^{(1\pm\sqrt{1-4a^2})}drd\Omega. \tag{4}$$

This is clearly finite for both signs. Thus both behaviors are physically acceptable. If we have a nonzero angular momentum with $J\neq 0$ such that $J(J+1)-\alpha^2>0$ then the second (peculiar) solution would not be physically acceptable. The peculiar solution would also not be acceptable if the electron and positron were not point particles ($\mathcal{A}\neq-\alpha/r$). Thus, finding these peculiar states would strongly support that the electron be a point particle.

Treating the $-\alpha^2/r^2$ terms as a negative angular barrier, the solutions can be obtained analytically for both the usual and peculiar bound states $^1S_0$. The corresponding sets of eigenvalues for the total invariant c.m. energies $w_{\pm n}$ in terms of the principle quantum number n are

$$w_{\pm n}=m\sqrt{2+2/\sqrt{1+\alpha^2/(n\pm\sqrt{1/4-\alpha^2}-1/2)^2}}, \tag{5}$$

For the usual states, the bound state eigenvalues w_{+n} agree with standard QED perturbative results through order α^4,

$$w_{+n}=2m-m\alpha^2/4n^2-m\alpha^4/2n^3(1-11/32/n)+O(\alpha^6),\ n=1,2,3,... \tag{6}$$

For the set of peculiar states, the ground state ($n = 1$) has mass

$$w_{-1} = m\sqrt{2 + 2/\sqrt{1 + \alpha^2/(1/2 - \sqrt{1/4 - \alpha^2})^2}} \sim \sqrt{2}m\sqrt{1+\alpha}. \tag{7}$$

This represents a very tightly bound state, one with a binding energy on the order of 300 keV for an e^+e^- state and a root mean square radius on the order of a Compton wave length instead of an angstrom.

We now give a brief overview of the constraint Two Body Dirac Equations [3], [4], and how they lead to the above form given in Eq. (1) and the tentative new bound state. In the covariant constraint formalism for two spin-one-half particles interacting by through four-vector potentials there is a Dirac-like equation for each particle,

$$\begin{aligned} \mathcal{D}_1\psi &\equiv (\gamma_1 \cdot (p_1 - \tilde{A}_1) + m_1)\psi = 0 \\ \mathcal{D}_2\psi &\equiv (\gamma_2 \cdot (p_2 - \tilde{A}_2) + m_2)\psi = 0. \end{aligned} \tag{8}$$

The tildas symbolize the restrictions on the potentials that come from the mathematical requirement that the constraints be compatible[5],

$$[\mathcal{D}_1, \mathcal{D}_2]\psi = 0. \tag{9}$$

and the connection to quantum field theory by way of what has been called the quantum mechanical transform[6] of the Bethe-Salpeter equation [7]. This leads to the four vector potentials $\tilde{A}_1$, $\tilde{A}_2$ that depend on an invariant function $\mathcal{A}(x_\perp)$ and its gradient as well as energy, momentum, and spin dependence through the gamma matrices,

$$\tilde{A}_i = \tilde{A}_i(\mathcal{A}(x_\perp), \partial\mathcal{A}(x_\perp), w, p_1, p_2, \gamma_1, \gamma_2). \tag{10}$$

The invariant $\mathcal{A}$ is a function of the separation four-vector,

$$\begin{aligned} x_\perp^\mu &= (x_1 - x_2)_\nu(\eta^{\mu\nu} - P^\mu P^\nu/P^2), \\ P &= p_1 + p_2, \\ P \cdot x_\perp &= 0. \end{aligned} \tag{11}$$

perpendicular to the total four-momentum. For QED in lowest order,

$$\begin{aligned} \mathcal{A}(x_\perp) &= -\frac{\alpha}{r}, \\ r &\equiv \sqrt{x_\perp^2},\ x_\perp = (0, \mathbf{r}) \text{ in c.m. } (P = (w, \mathbf{0})). \end{aligned} \tag{12}$$

The relative momentum is space-like

$$P \cdot p\psi = 0, \tag{13}$$

so that combined with Eq. (12) we see that the compatibility condition leads to a covariant three dimensional formalism. For equal masses $p = (p_1 - p_2)/2$.

The wave function that satisfies the simultaneous equations (8) has 16 components given in terms of 4 four component spinors

$$\psi = (\psi_1, \psi_2, \psi_3, \psi_4)\text{- sixteen component spinor.} \tag{14}$$

In the c.m. frame $\hat{P} = (1, \mathbf{0})$ and $\hat{r} = (0, \hat{\mathbf{r}})$. We defined [1] the 4 component wave functions $\psi_\pm, \eta_\pm$

$$\begin{aligned} \psi_\pm &= \exp(-\mathcal{F}(r) - \mathcal{K}(r)\sigma_1\cdot\hat{\mathbf{r}}\sigma_2\cdot\hat{\mathbf{r}})(\psi_1 \pm \psi_4) \\ \eta_\pm &= \exp(-\mathcal{F}(r) - \mathcal{K}(r)\sigma_1\cdot\hat{\mathbf{r}}\sigma_2\cdot\hat{\mathbf{r}})\,(\psi_2 \pm \psi_3) \end{aligned} \tag{15}$$

and performed a Pauli reduction to 4 - uncoupled four-component Schrödinger-like equations. For ψ_+ we found

$$\begin{aligned} &\mathcal{B}^2\psi_+ \equiv \{\mathbf{p}^2 + \Phi(\mathbf{r}, m_1, m_2, w, \sigma_1, \sigma_2)\}\psi_+ \\ =\ &\{\mathbf{p}^2 + 2\varepsilon_w\mathcal{A} - \mathcal{A}^2 + \Phi_D(\mathcal{A}, \nabla\mathcal{A}, \nabla^2\mathcal{A}) + \quad\cdot(\sigma_1{+}\sigma_2)\Phi_{SO}(...) \\ &+\sigma_1\cdot\hat{\mathbf{r}}\sigma_2\cdot\hat{\mathbf{r}} \quad\cdot(\sigma_1{+}\sigma_2)\Phi_{SOT}(...) + \sigma_1\cdot\sigma_2\Phi_{SS}(..) + (3\sigma_1\cdot\hat{\mathbf{r}}\sigma_2\cdot\hat{\mathbf{r}} - \sigma_1\cdot\sigma_2)\Phi_T(..) \\ &+ \quad\cdot(\sigma_1{-}\sigma_2)\boldsymbol{\Phi}_{SOD}(..) + i \quad\cdot\sigma_1{\times}\sigma_2\Phi_{SOX}(..)\}\psi_+ \\ =\ &b^2\psi_+ \end{aligned} \tag{16}$$

The Darwin and various spin-dependent terms are dependent on the invariant $\mathcal{A}$ and its derivatives. The attractive spin-spin and repulsive Darwin quasipotentials are quite strong (individually they overwhelm $-\mathcal{A}^2$) but for equal mass spin singlet states we have their exact cancellation $\Phi_D + \sigma_1\cdot\sigma_2\Phi_{SS} = 0$, leading to the bound state radial equation given in Eq. (1)

One finds that the lowest lying usual and peculiar states are not orthogonal with respect to one another. To see this we write their respective radial forms

$$\begin{aligned} u_+(r) &= c_+r^{\lambda_++1}\exp(-\kappa_+\varepsilon_{w_+}\alpha r), \\ \kappa_+ &= \frac{2}{1+\sqrt{1-4\alpha^2}} = \frac{1}{\lambda_++1}, \\ u_-(r) &= c_-r^{\lambda_-+1}\exp(-\kappa_-\varepsilon_{w_-}\alpha r), \\ \kappa_- &= \frac{2}{1-\sqrt{1-4\alpha^2}} = \frac{1}{\lambda_-+1}, \end{aligned} \tag{17}$$

Clearly since they are both zero node solutions we have

$$\langle u_-|u_+\rangle = \int_0^\infty dru_+(r)u_-(r) \sim (\alpha^{3/2} \sim 1/1000) \neq 0. \tag{18}$$

How do we reconcile this with the expected orthogonality of the eigenfunctions of a self-adjoint operator corresponding to different eigenvalues? One can show that the second derivative is not self-adjoint in this context! But we have that for the quasipotential of the type $-\alpha^2/r^2$ at short distances, both the set of usual states and the peculiar states are physically admissible states. There does not appear to be reasons to exclude one set as being unphysical, given the attractive interaction as it is near the origin.

Only for interactions with sufficient attraction at the origin (so that $-1/4 \le \lambda(\lambda+1) < 0$) can the peculiar states be pulled into existence and appear as eigenstates in the physically acceptable sheet, with regular non-singular radial wave functions at the origin. It is desirable to find ways to admit both types of physical states into a larger Hilbert space to accommodate both sets of states with the mass operator $\mathcal{B}^2$ to be self-adjoint and the states to be part of a complete set. It is reasonable to assign a quantum number which we call "peculiarity" for a states emerging into the physical sheet in this way as physically acceptable states. The introduction of the peculiarity quantum number enlarges the Hilbert space, allows the mass operator $\mathcal{B}^2$ to be self-adjoint, and the set of physically allowed states become a complete set.

We introduce a new peculiarity observable $\hat{\zeta}$ with the quantum number peculiarity ζ such that

$$\begin{aligned} \hat{\zeta}\chi_+ &= \zeta\chi_+ \text{ with eigenvalue } \zeta = +1, \\ \hat{\zeta}\chi_- &= \zeta\chi_- \text{ with eigenvalue } \zeta = -1, \end{aligned} \tag{19}$$

with the corresponding spinor wave function χ_ζ assigned to the states so that a usual state is represented by the peculiarity spinor χ_+,

$$\chi_+ = \begin{pmatrix} 1 \\ 0 \end{pmatrix}, \tag{20}$$

and a peculiar state is represented by the peculiarity spinor χ_-,

$$\chi_- = \begin{pmatrix} 0 \\ 1 \end{pmatrix}. \tag{21}$$

With this introduction, a general wave function can be expanded in terms of the complete set of basis functions $\{u_{+n}, u_{-n}\}$ as

$$\Psi = \sum_{\zeta n} a_{\zeta n} u_{\zeta n} \chi_\zeta, \tag{22}$$

where n represent all the spin and spatial quantum numbers of the state and ζ the peculiarity quantum number. The variational principle applied to

$$\langle \mathcal{B}^2 \rangle = \frac{\langle \Psi | \mathcal{B}^2 | \Psi \rangle}{\langle \Psi | \Psi \rangle}, \tag{23}$$

would lead to

$$\begin{aligned} \mathcal{B}^2 u_{+n}\chi_+ &= -\kappa_{+n}^2 u_{+n}\chi_+, \\ \mathcal{B}^2 u_{-n}\chi_- &= -\kappa_{-n}^2 u_{-n}\chi_-. \end{aligned} \tag{24}$$

It is clear that in this context the usual and peculiar wave functions are orthogonal, $\mathcal{B}^2$ is self-adjoint, and the basis states are complete. We see that the introduction of the peculiarity quantum number resolves the problem of over-completeness property of the basis states and the non-self-adjoint property of the mass operator.

The usual 1^1S_0 ground state of positronium (designated by $1S_u$) turns into the peculiar 1^1S_0 ground state (designated by $1S_p$) by emitting two photons similar to the metastable mechanism by which the $2S$ state of hydrogen decays into the $1S$ state. The Golden Rule in this case takes the form

$$d^3w = 2\pi|T_{fi}|^2 d^3k_1 d^3k_2 d^3p_{1S_p}\delta(E_{1S_u} - E_{1S_p} - \hbar\omega_1 - \hbar\omega_2)\delta(\mathbf{0} - \mathbf{p}_{1S_p} - \mathbf{k}_1 - \mathbf{k}_2), \quad (25)$$

for the emission of photons characterized by $(\mathbf{k}_1, \alpha_1)$ and $(\mathbf{k}_2, \alpha_2)$. In [2] is described in detail the metastable decay mechanism for the usual ground state into the the peculiar state. Our assumption above is that the peculiar and usual states by virtue of the introduction of the operator $\hat{\zeta}$ are orthogonal. As such it would be impossible for the $1S_u$ state to decay into the $1S_p$ state. In that case the usual state would undergo just two photon annihilation. We allow for the possibility that the peculiarity quantum number is not conserved for the full hamiltonian so that there is a nonzero admixture between the two sectors. Calling the admixture amplitude $M_{\zeta\zeta'}$ we find that the branching ratio between the metastable and annihilation channels is $0.075\left|M_{\zeta\zeta'}\right|^2$. We further find that once the peculiar state is formed, it will annihilate into two photons with a c.m. energy of 700 keV with a lifetime on the order of 10^{-21} sec.

To summarize, the TBDE of constraint dynamics applied to QED give a covariant bound state formalism reproducing by nonperturbative treatment the correct spectra through order α^4. Applied to the $^1S_0$ states, new peculiar solutions are uncovered for point e^-e^+ $(\psi_- \rightarrow r^{(-1-\sqrt{1-4\alpha^2})/2})$ in addition to the usual solutions $(\psi_+ \rightarrow r^{(1-\sqrt{1-4\alpha^2})/2})$. The peculiar bound states, to preserve self adjointness, are distinguished from the usual ones by a new quantum number called peculiarity. The four photon decay signature (with one set of two photons bunching in energy around 300 keV and one set around 700 keV) would strongly indicate the point-like nature of the electron.

e erences

[1] Crater H C and Wong C Y 2012 Phys. Rev. D **85** 116005

[2] Crater H C and Wong C Y 2014 On the Question of the Point-Particle Nature of the Electron arXiv:1406.7268

[3] Van Alstine P and Crater H C 1982 J. Math. Phys. **23** 1997; Crater H W and Van Alstine P 1983 Ann. Phys. (NY) **148** 57

[4] Crater H C, Becker R L, Wong C Y, and Van Alstine P 1992 Phys. Rev. D **46** 5117

[5] Dirac P A M 1950 Canad. J. Math. **2** 129; Proc. Roy. Soc. Sect. A 1958 **246** 326; *Lectures on Quantum Mechanics* (Yeshiva University, Hew York, 1964)

[6] Sazdjian H 1987 J. Math. Phys. **28** 2618: *Extended Objects and Bound Systems*, Proceedings of the Karuizawa International Symposium 1992 Eds. Hara O, Ishida S, and Nake S (World Scientific, Singapore, 1992) 117

[7] Salpeter E E and Bethe H A 1951 Phys. Rev. **84** 1232

Journal of Physics: Conference Series **548** (2014) 012005 doi:10.1088/1742-6596/548/1/012005

Vibration frequencies for searches of free triatomic molecules in interstellar space

R Hefferlin and J Barrow

Southern Adventist University, Collegedale, Tennessee 37315, USA

E-mail: hefferln@southern.edu

Abstract. More and more cold interstellar molecules are being discovered, the majority of them being organic. Perhaps it is time to consider the numerous small molecules that also await observation. We report progress in tabulating symmetric-stretch vibration frequencies for neutral main-group ground-state triatomic molecules, formed from period-2 atoms, which are not yet studied.

1. Introduction

This investigation presents data that can be used in searches for free triatomic molecules that might exist in interstellar space. It is assumed that any combination of three neutral main-group atoms can form a cold molecule that could exist in circumstellar regions. Considerations such as bond types; bond orders; spin multiplicities; and chemical or radioactive reactivity are ignored in this study. The work pertains only to molecules all of whose atoms are from period 2; much more remains to be done.

2. Theory

The ability to visualize the mathematical space in which triatomic molecules exist is essential for understanding what follows in this paper. First, imagine an 8×8×8-atom cubical subspace populated by main-group period-2 atoms with rare-gas molecules on three faces. Then generalize to many such cubes, ranging from $(R_1,R_2,R_3) = (2,2,2)$ to $(7,7,7)$, where R_i is the period number of atom i (figure 1). The diagram is symmetric with respect to reflection in a plane containing the closest and the farthest edges of the figure; hence, for example, the cube with periods $(R_1,R_2,R_3) = (4,2,2)$ contains the same molecules as does the cube (2,2,4). Aside from such subspaces where $R_1 \neq R_3$, there are 49 cubes with $R_1 = R_3$, each containing $\{[(8\times9)/2]\times8\}$non-redundant molecules.

3. Data acquisition

3.1. Simple molecular mechanics

Comparison of the symmetric-stretch vibration frequency of a linear, symmetric, triatomic molecule ABA and the vibration frequency ω_e of a diatomic molecule BA (which may be given in handbooks [1,2]) shows that ν_1 for these particular molecules can be obtained by

$$\nu_1 = \omega_e \,\mathrm{sqrt}(\mu/m) \quad , \tag{1}$$

Where μ is the diatomic reduced mass and m is the mass of either of the outer atoms.

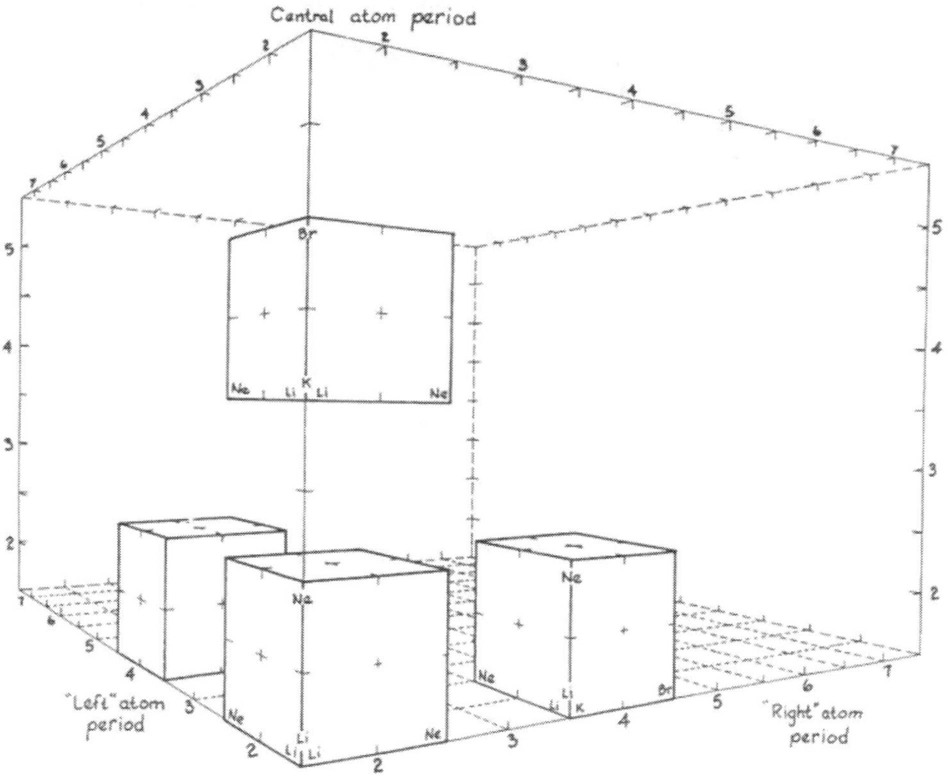

Figure 1. The formal space for main-group triatomic molecules showing selected cubical subspaces. R_1, R_2, and R_3 are the period numbers for the atoms on the "right" ends, centers, and "left" ends of the molecules in the cubical subspaces. Within each cube, the group numbers C_1, C_2, and C_3 (1 to 8) are defined similarly. This figure includes triatomics formed from atoms $(R_1,R_2,R_3) = (2,2,2)$ to $(5,5,5)$; species from $R_2 = 6$ and 7 lie higher above the figure; hydrogen and helium ($R_i = 1$) are not included in this study.

3.2. Computations

Two *ab initio* molecular modelling programs, QChem and PSI4[3], and the WebMo graphical user interface, were employed. Several basis sets were used: aug-cc-pVDZ, -pVTZ infrequently, -PVQZ; 6-311+G** and +G(d,p); and 6-31G(d) infrequently.

4. Results

As data for a given (R_1,R_2,R_3) were acquired, they were placed in bins of constant total electron count n_e, and then in order of increasing central-atom group number. The data for each molecule having more than one entry were averaged and assigned absolute and percent errors based on half the difference (for two entries), or based on the sample standard deviation. Table 1 contains the results.

Column 6 contains a code for the method(s) used to obtain each datum. Computer basis sets are **A,** aug-cc-pVDZ; **C**, aug-cc-pVQZ; **D**, 6-311+G**; **L**, 6-31G(d); and **M**, 6-311+G(d,p). The table of diatomic molecular data for use with equation (1) is **H** [2]). "**AC3G**" means computations based on basis sets **A** and **C** were each successful once and basis set **G** was successful three times. The small number of molecules listed in table 1, compared to the large count suggested by Section 2, is due to (a) only molecules *not* found in tabulations are included; (b) computations usually failed to produce meaningful molecular structures and frequencies for these unknown molecules; (c) ω_e that could be used to calculate ν_1 for unknown triatomic molecules using Equation (1) are not numerous.

Table 1. Data for molecules from $(R_1,R_2,R_3) = (2,2,2)$ in order of n_e and then C_2. The averages are derived from computed values. If two sources contributed, then the half-difference error is shown in column 4; if more contributed, then the sample standard deviation is shown. In cases when one computation gave results so discrepant from the others that it caused percent errors of over 10%, it was discarded. All values are rounded to one significant figure after the decimal.

n_e	Species	Average ν_1 (cm^{-1})	Difference error, or σ	Percent error, or σ	Basis set
11	LiBeBe	388.4			**C**
12	BeBeBe	539.5	1.5	0.3	**AD**
	LiBeB	388.4			**D**
	LiBBe	533.2	0.2	0.0	**2C**
	LiCLi	570.9			**A**
13	BBeBe	443.8			**M**
	LiCBe	626.3			**M**
	LiNLi	772.8			**A**
14	BBeB	762.2			**A**
	BeBB	635.4			**A**
15	BBeC	534.6			**A**
	BBB	535.3			**H**
	BeCB	1218.6	0.3	0.0	**CD**
	BeNBe	1290.3	2.0	0.2	**2D**
	BNLi	764.1	2.8	0.4	**AD**
	BeOLi	840.1			**C**
	LiFLi	672.2			**H**
16	CBeC	1130.0	1.3	0.1	**CD**
17	BBeO	475.4[b]			**L**
	CBeN	760.8			**A**
	BeBeF	417.3			**C**
	BeBO	646.9	31.7	4.9	**DM**
	LiBF	243.3	21.9	9.0	**AC3G**
	CCB	1526.1			**M**
	BNB	1290.8			**L**
	LiCO	652.2			**M**
	BeOB	493.3			**M**

Table 1, continued

n_e	Species	Average ν_1 (cm^{-1})	Difference error, or σ	Percent error or σ	Basis set
18	NBeN	520.0			**A**
	BBO	617.5	48.7	7.9	**CM**
	BCN	806.5	64.2	8.0	**ACM**
	BOB	1113.0			**M**
	BeFB	19.4			**M**
	LiFC	605.0			**M**
19	LiOF	804.1			**M**
	BeON	1439.5			**M**
	BOC	1141.5			**M**
20	OBeO	1060.5			**M**
	COC	1096.1			**M**
21	OBeF	660.1			**A**
	NBF	980.4			**A**
	NCO	1398.7			**A**
	BNF	980.3			**M**
	CNO	1103.2			**A**
	BeOF	128.6	5.5	4.3	**2A**
22	NCF	1205.0			**C**
	BOF	1059.8			**A**
	NON	1076.8			**A**
24	FNO	945.4	0.6	0.1	**D,M**

Acknowledgement

The School of Chemistry and Biochemistry of the Georgia Institute of Technology has our profound thanks for having allowed access to Q-Chem, PSI4 [3], and the WebMo graphical user interface.

References

[1] Lide D R 2003 *Handbook of Chemisty and Physics* (Boca Raton, FL: CRC Press)

[2] Huber K P and Herzberg G *Contants of Diatomic Molecules* (New York: Van Nostrand Reinhold)

[3] Turney A A *et al* 2012 *WIREs Comput. Mol. Sci.* **2** 556. (doi: 10.1002/wcms.93).

Temperature dependence of sodium and ionized calcium resonance lines perturbed by helium

N. F. Allard[a,b], D. Homeier[c], G. Guillon[d], A. Viel[e] and J Kielkopf[f]

[a] Observatoire de Paris, GEPI, UMR8111, CNRS, 61, Avenue de l'Observatoire, F-75014, Paris, France

E-mail: nicole.allard@obspm.fr
[b] Institut d'Astrophysique, UMR7095, CNRS, Université Paris VI, 98bis Boulevard Arago, F-75014, Paris, France
[c] Centre de Recherche Astrophysique de Lyon, UMR 5574: CNRS, Université de Lyon, École Normale Supérieure de Lyon, 46 allée d'Italie, F-69364 Lyon Cedex 07, France

E-mail: derek.homeier@ens-lyon.fr
[d] Lab. Interdisciplinaire Carnot de Bourgogne, UMR 6303, CNRS, Université de Bourgogne, 21078 Dijon Cedex, France

E-mail: gregoire.guillon@u-bourgogne.fr
[e] Institut de Physique de Rennes, Université de Rennes 1, F-35042 Rennes, France

E-mail: Alexandra.Viel@univ-rennes1.fr
[f] Department of Physics and Astronomy, University of Louisville, Louisville, Kentucky 40292 USA

E-mail: kielkopf@louisville.edu

Abstract. Traces of heavy metals in cool DZ white dwarf stars may be attributed to the accretion of circumstellar dust thought to originate from tidal disruption of rocky parent bodies. Spectra of such stars therefore provide a unique opportunity to study the composition of extra-solar planetary systems. The determination of metal abundances from stellar spectra depends on stellar atmospheric parameters and an accurate prior knowledge of the collision broadening of the line profiles by the most common constituents of the stellar atmosphere. For this purpose, we present theoretical absorption spectra of Na and Ca^+ broadened by He for the conditions prevailing in cool white dwarfs.

1. Introduction

Ultracool stellar atmospheres show absorption by alkali resonance lines significantly broadened by collisions with neutral perturbers. In the coolest and densest atmospheres, such as those of T dwarfs, Na I and K I broadened by H_2 and He can come to dominate the entire optical spectrum [1]. Na I line profiles have also been observed in extremely cool, metal-rich white dwarfs [2, 3]. The studies of cool DZ white dwarfs observed in the Sloan Digital Sky Survey (SDSS) by [4] showed clearly the importance of extended wings of calcium doublets. They pointed out the need for more accurate line profile calculations than Lorentzian profiles based on approximate impact broadening assumptions. In actuality, the lines are significantly non-Lorentzian, and are broadened due to collisions with helium with detectable wings extended as much as 1000 Å either side of line core. Through their blanketing effect these lines have a dominant influence on the model structure and thus on the determination of atmospheric parameters and element abundances.

The theory of spectral line shapes, especially the unified approach we have developed, makes possible accurate models of stellar spectra that account both for the centers of spectral lines and their extreme wings in one consistent treatment. Complete details and the derivation of the theory are given by Allard et al (1999) [5].

In this continuation of previous work [6, 7] we illustrate the evolution of the absorption spectra of Na-He and Ca^+-He for the temperatures prevailing in atmosphere of cool white dwarf stars.

2. NaHe line absorption in ultra-cool white dwarf stars

Observations of the oldest and colder white dwarfs following the study of [2, 3] have revealed two stars showing very unusual wide and deep absorption at 5000-6000 Å.

Homeier *et al.* (2007) [8] investigated effects on the Na doublet of high perturber densities occurring for metal-rich white dwarfs with a helium-dominated atmosphere. They found that the density of neutral atomic helium in these two very cool white dwarfs showing very strong Na absorption could reach several 10^{21} to 10^{22} cm^{-3} (Fig. 1). Cool white dwarfs require a specific treatment for line broadening due to the high perturber density which is involved.

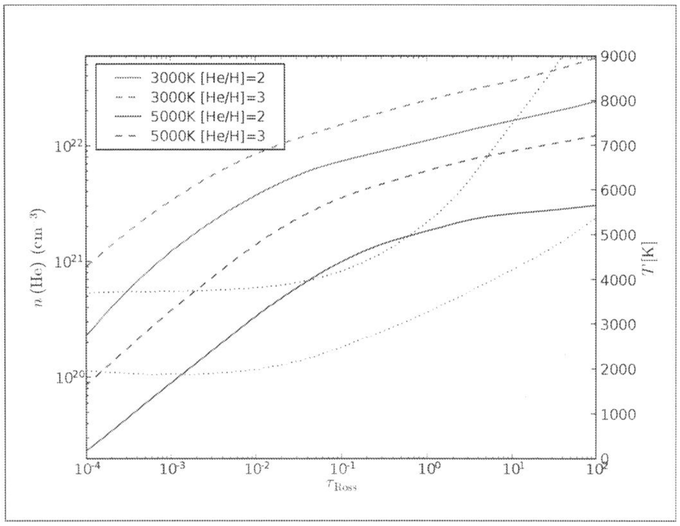

Figure 1. Number density of helium (left scale) and temperature profile (right scale) for 3000 K and 5000 K WD atmosphere models with varying [He/H] ratio (extracted from Homeier *et al.* 2007 [8]).

In dense plasmas, the possibility of several atoms interacting strongly is high, and the effects play a role in the wavelength of the line center, e.g. the shift of the line, as well as the continuum generated far from the line center. The tools the physicist uses to determine line profiles in stellar spectra are very refined, and the days when simple analytic expressions for line widths and shifts would suffice for these applications are long past.

For such high helium density the collisional effects should be treated by using the autocorrelation formalism in order to take into account simultaneous collisions with more than one perturbing atom. The line shape theory requires accurate atomic potentials and transition moments which are regrettably not yet available for many systems of astrophysical interest. For *ab initio* calculations, it is the selection of basis states and the optimization techniques that may determine the final accuracy. The current state-of-the-art in *ab initio* theories of molecular potentials (atom-atom potentials and dipole transition moments as a function of separation) yield data of spectroscopic accuracy. Our calculations reported here are based on the very recent *ab initio* Na-He potentials of [9] and the earlier radiative dipole moments of Pascale (1983) [10].

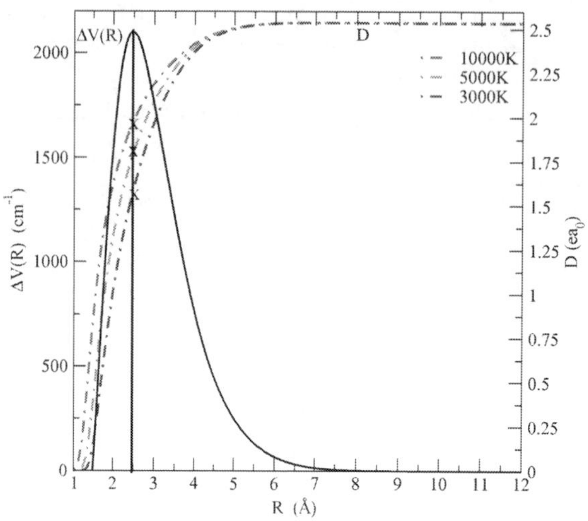

Figure 2. $\Delta V(R)$ and the temperature dependence of modulated dipole $D(R)$ corresponding to the $3s\ X \rightarrow 3p\ B\ P_{3/2}$ transition of the Na $D2$ line.

3. Temperature dependence of absorption spectra of Na and Ca^+ in dense helium

Detailed line profiles for the Na I D resonance doublets at 0.59μm calculated by Allard et al. (2003) [11] are now updated from the Na-He molecular potentials of [10] to the newer ones of [9].

The unified theory predicts that there will be line satellites centered periodically at frequencies corresponding to integer multiples of the extrema of the difference potential $\Delta V(R)$:

$$\Delta V(R) \equiv V_{e'e}[R(t)] = V_{e'}[R(t)] - V_e[R(t)] \ , \tag{1}$$

which represents the difference between the electronic energies of a quasi-molecular e- e' transition [12]. As part of a more general or unified concept of spectral line formation, satellites are the binary-collision manifestation of a ubiquitous phenomena, the many-body nature of spectral lines from dense gases.

While the position of the line satellites critically depends on the interaction potential, their strength depends on both the interaction potential *and* the radiative dipole moments, $D(R_{\rm ext})$, in the internuclear region where the line satellite is formed. In Allard *et al.* 1999 [5] we define $\tilde{d}_{ee'}(R(t))$ as a *modulated* dipole

$$D(R) \equiv \tilde{d}_{ee'}[R(t)] = d_{ee'}[R(t)] e^{-\frac{V_e[R(t)]}{2kT}} \ , \tag{2}$$

V_e is the lower state potential as we consider absorption profiles. ΔV and variation of D(R) with temperatures of Na-He modulated dipole are displayed in Fig. 2.

The presence of line satellite features is very sensitive to the temperature due to the fast variation of the modulated dipole moment with temperature in the internuclear region where the line satellite is formed (Fig 2). NaHe satellite is apparent for $T \geq 1000$ K but disappears for decreasing T when the transition moment $\tilde{d}_{ee'}[R(t)]$ becomes very small (Fig. 1 of Allard et al. (2003) [11]). In Fig. 3 we show the absorption cross section for the resonance line of Na for a He density of 10^{21} cm^{-3} and temperatures from 10000 to 3000 K. The NaHe satellite is at 0.53 μm which might be the large feature observed in the spectrum of WD2356-209 (Fig. 2B of Oppenheimer et al. (2001) [2]). There is a total blend of the line satellite in the core of the line when the density reaches 10^{22} cm^{-3} (Fig. 4).

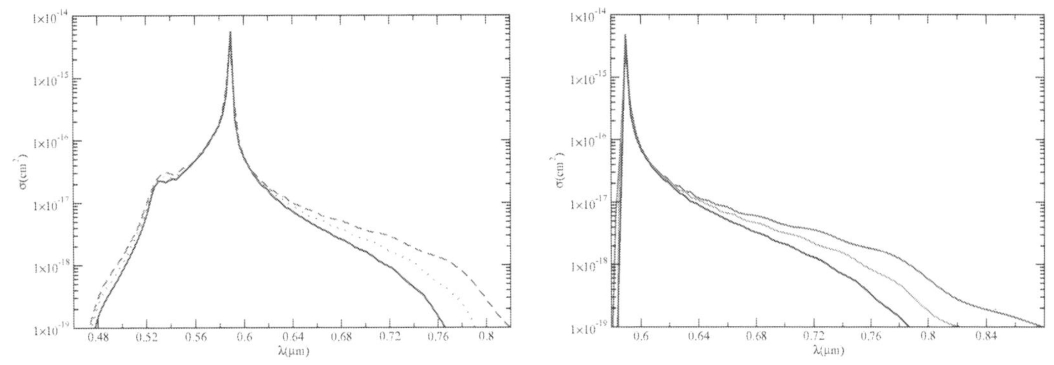

Figure 3. Variation of the absorption cross section of the Na $D2$ line (left) and the $D1$ line (right). ($n_{\rm He}$=10^{21} cm^{-3}, from top to the bottom T =10000, 5000 and 3000 K).

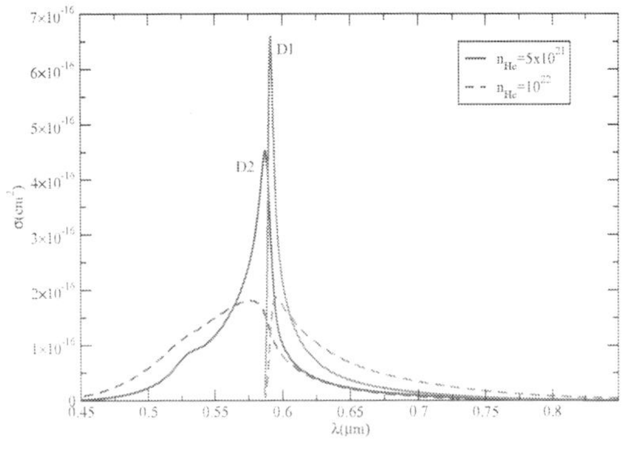

Figure 4. Variation of the absorption cross section of the Na $D2$ (blue lines) and $D1$ (red lines) with helium density (T =3000 K, from top to the bottom $n_{\rm He}$=5 $\times$ 10^{21} and 10^{22} cm^{-3}).

In Fig. 5 we show the variation of the absorption cross section of the K and H lines, of Ca^+-He at various temperatures from 10000 to 4000 K. The strength of the satellites at about 0.45 μm increases with temperature. From this dependence it is apparent that the sensitivity of the spectrum to temperature and pressure is a tool for determining basic parameters of white dwarf atmospheres, provided that the physics underlying its formation is well described.

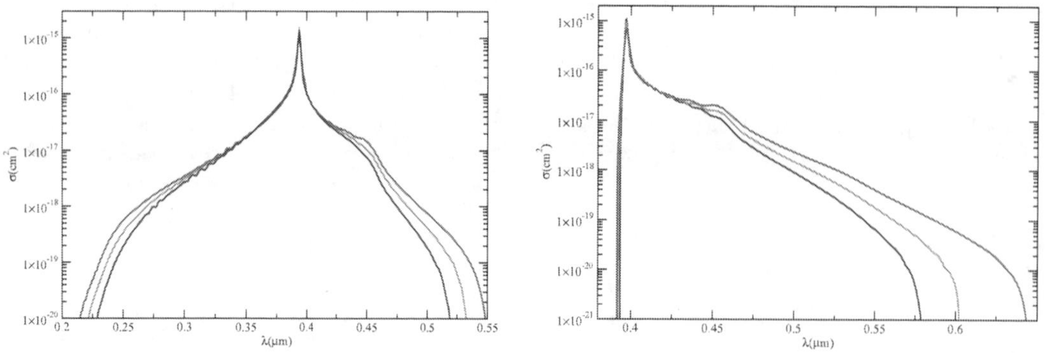

Figure 5. Variation of the absorption cross section of the Ca II K line (left) and the H line (right). ($n_{\rm He}$=5 × 10^{21} cm^{-3}, from top to the bottom T =10000, 6000 and 4000 K).

References

[1] Burrows A, M S Marley M S and Sharp C M 2000 *ApJ* **531** 438

[2] Oppenheimer B R, Saumon D, Hodgkin S T, Jameson R F, Hambly N C, Chabrier G, Filippenko A V, Coil A L and Brown M E 2001 *ApJ* **550** 448–456 (*Preprint* arXiv:astro-ph/0101258)

[3] Harris H C, Liebert J, Kleinman S J, Nitta A, Anderson S F, Knapp G R, Krzesiński J, Schmidt G, Strauss M A, Vanden Berk D, Eisenstein D, Hawley S, Margon B, Munn J A, Silvestri N M, Smith J A, Szkody P, Collinge M J, Dahn C C, Fan X, Hall P B, Schneider D P, Brinkmann J, Burles S, Gunn J E, Hennessy G S, Hindsley R, Ivezić Z, Kent S, Lamb D Q, Lupton R H, Nichol R C, Pier J R, Schlegel D J, SubbaRao M, Uomoto A, Yanny B and York D G 2003 *ApJ* **126** 1023–1040 (*Preprint* astro-ph/0305347)

[4] Koester D, Girven J, Gänsicke B T and Dufour P 2011 *A&A* **530** A114 (*Preprint* 1105.0268)

[5] Allard N F, Royer A, Kielkopf J F and Feautrier N 1999 *Phys. Rev. A* **60** 1021

[6] Allard N F and Alekseev V A 2014 *Advances in Space Research* **54** 1248–1253

[7] Allard N F 2013 Density Effects in Sodium and ionized Calcium perturbed by helium in Cool White Dwarf Stars *EAS Publications Series* (*EAS Publications Series* vol 63) pp 403–406

[8] Homeier D, Allard N, Johnas C M S, Hauschildt P H and Allard F 2007 Alkali Line Profiles in Ultra-cool White Dwarfs *15th European Workshop on White Dwarfs* (*Astronomical Society of the Pacific Conference Series* vol 372) ed R Napiwotzki & M R Burleigh p 277 (*Preprint* 0707.4138)

[9] Dell'Angelo D, Guillon G and Viel A 2012 *J. Chem. Phys.* **136** 114308

[10] Pascale J 1983 *Phys. Rev. A* **28** 632

[11] Allard N F, Allard F, Hauschildt P H, Kielkopf J F and Machin L 2003 *A&A* **411** L473

[12] Allard N F and Kielkopf J F 1982 *Rev. Mod. Phys.* **54** 1103

Journal of Physics: Conference Series **548** (2014) 012007 doi:10.1088/1742-6596/548/1/012007

Medical Applications of Laser Induced Breakdown Spectroscopy

A K Pathak[1], N K Rai[2], Ankita Singh[3], A K Rai[4], Pradeep K Rai[5] and Pramod K Rai[5]

[1]Department of Physics, Ewing Christian College, Allahabad- 211003, India
[2]Department of Physics, IISER, Bhopal, India
[3]Vinayak Hospital 30A/1A, 18 Stanly Road, Allahabad- 211 002, India
[4]Department of Physics, University of Allahabad, Allahabad- 211002, India
[5]Opal Hospital, DLW road, Varanasi-221005, India

E-mail: akpathak75@gmail.com

Abstract. Sedentary lifestyle of human beings has resulted in various diseases and in turn we require a potential tool that can be used to address various issues related to human health. Laser Induced Breakdown Spectroscopy (LIBS) is one such potential optical analytical tool that has become quite popular because of its distinctive features that include applicability to any type/phase of samples with almost no sample preparation. Several reports are available that discusses the capabilities of LIBS, suitable for various applications in different branches of science which cannot be addressed by traditional analytical methods but only few reports are available for the medical applications of LIBS. In the present work, LIBS has been implemented to understand the role of various elements in the formation of gallstones (formed under the empyema and mucocele state of gallbladder) samples along with patient history that were collected from Purvancal region of Uttar Pradesh, India. The occurrence statistics of gallstones under the present study reveal higher occurrence of gallstones in female patients. The gallstone occurrence was found more prevalent for those male patients who were having the habit of either tobacco chewing, smoking or drinking alcohols. This work further reports *in-situ* LIBS study of deciduous tooth and *in-vivo* LIBS study of human nail.

1. Introduction

Increasing incidence of gallstone formation in gallbladder is of great concern for medical practitioners which is a major cause of morbidity and mortality in developed as well as developing countries. Analysis of chemical composition of the different kinds of gallstones can provide a significant clue for the treatment and prevention of their recurrence. The presence of gallstones in the gallbladder results from changes in bile components. Bile is yellowish fluid, formed by liver, to digest and absorb the fats from food which has been broken by enzymes. These enzymes are secreted from pancreas and the intestine. Usually, gallstones form when bile stored in the gallbladder hardens into pieces of stone-like materials. Gallbladder is capable of considerable distension in some pathological conditions [1].The gallbladder mucocele or empyema occurs when a stone obstructs the neck of the gallbladder. Mucocele is a non-inflammatory condition of the gallbladder with increased mucus production from the gallbladder epithelium. The empyema of gall bladder appears to be filled with pus that may be a sequel of acute cholecystitis or the result of a mucocele being infected. As these gallstones are capable of obstructing the neck of gallbladder, thus our interest, in the present study, is to explore the elemental differences between these gallstones under these conditions. Bio-materials such as teeth and nails are excellent bio-monitors because they can show the level of deposition of essential minerals over weeks to years due to the habitat, nutrition, and other environmental conditions [2-3]. Dental related problems (like caries) are major oral health problems throughout the world. The elemental analysis of teeth samples can give information about the causes of caries in human teeth. Laser Induced Breakdown Spectroscopy (LIBS) is an excellent analytical tool suitable for the elemental analysis of any kind of materials irrespective of their phases [4-6].This manuscript also presents an excellent example of *in-situ* LIBS analysis of deciduous tooth and *in-vivo* LIBS study of human nail.

2. Materials and methods

Gallstones samples along with patient history were obtained from Opal Hospital, Varanasi, India. After the surgery of the patients the stone samples were washed with de-ionized water to remove the blood and other adherent contaminants/ impurities and brought it in the Laser Spectroscopy Laboratory, Department of Physics, University of Allahabad, Allahabad, in sealed pots. The tooth samples from patients were supplied by Vinayak Hospital, Allahabad, India. The tooth samples were washed in distilled water to remove all the contamination from its outer surface and dried at room temperature. Firstly, tooth samples were visually examined by trained examiners/Doctors and then used for the LIBS diagnosis. The nail samples obtained from different age group volunteers have been used for *in-vivo* analysis and results related to volunteer aged 32 is discussed.

The LIBS spectra of various gallstone samples were recorded using LIBS experimental set up described in detail in review by Pathak et al. [6]. The laser pulse energy and pulse repetition rate were optimized and better signal to background and signal to noise ratio were observed for gallstone and nail samples at 20 mJ energy and 10 Hz repetition rate. For tooth samples the averaged (10 laser shots) LIBS spectra having best signal-to-noise (S/N) ratio and signal-to-background (S/B) ratio was found using 30 mJ of laser energy per pulse @ 10 Hz. *In-vivo* LIBS spectra of nails and *in-situ* LIBS spectra of deciduous tooth (collected from hospital mentioned above) were recorded by directly focusing the laser beam on the surface of the samples.

3. Results and discussion

3.1. Occurrence statistics of gallstones

A total of 92 patients of gallstone disease were reported during March 2008 to July 2009 at Opal hospital, Varanasi, Uttar Pradesh, India. This section studies the risk factors for gallstone disease on the basis of database (age, gender, diet & BMI) collected to these 92 patients [7].

Table 1 shows the age and gender distribution of gallstone patients (N= 92). From the Table 1, it is observed that maximum number of gallstone patients belong to age group 35-65, irrespective of their gender. Further, it is also observed that occurrence of gallstone is almost twice in female as compared to male patients. Figure 1 clearly depicts the Gender (male & female), Diet (Veg & Non-veg) and Body Mass Index in kg/m^2 (BMI) distribution and number of gallstone occurrence for 92 patients. Though it is observed from the data of Table 1, the possibility of gallstone formation is less in male patients but on the other hand the occurrence of gallstone in male patients having habit of either chewing tobacco, smoking or drinking alcohols are more as compared to normal.

Table 1. Age and Gender distribution of Gallstone patients (N=92)

Age group (years)	**<35**	**35-65**	**Over 65**
Number of Patients	18(19.56 %)	69(75.0 %)	05(5.4 %)
Females	15(16.3 %)	49(53.3 %)	02(2.2 %)
Males	03(3.2 %)	20(21.7 %)	03(3.2 %)

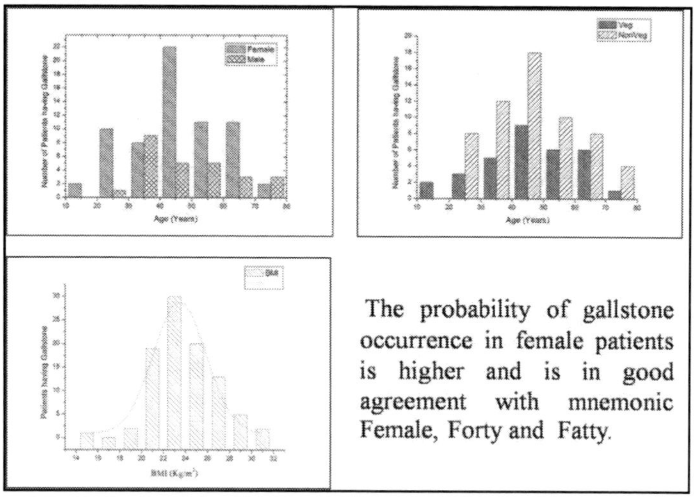

Figure 1. Bar diagram showing Age & Gender, Diet & BMI distribution of gallstone occurrence.

3.2. Study of gallstone related to empyema and mucocele state of gallbladder

We have reported studies related to different type of gallstones using LIBS [8-10] and this manuscript describes the gallstone formed under empyema and mucocele of gallbladder. Out of ninety two patients of gallbladder stones only eight (9.8%) gallstone samples belong to empyema of gallbladder and four samples (4.4%) belong to mucocele of gallbladder whose specifications are mentioned in Table 2. It is observed from Table 2, that the outer parts of the most of the gallstones belonging to empyema (E-1to E -8) and mucocele (M-1to M-4) of gallbladder are of colours yellow, white and green and the morphological classification suggests that outer parts of these may be dominated with cholesterol[6].

Table 2. Specifications of gallstones associated to empyema and mucocele (N=12)

Empyema	Color of gallstone	Patient Age(Years) Gender	Remarks
E-1	Black	56 , Female	
E-2	Outer yellow with inner brown	55, Female	Obese,Type-2 diabetes and hypertension
E-3	Outer white with light yellow inner	29, Female	
E-4	Outer green with inner brown	40, Male	Habit-Smoking
E-5	Outer green with inner brown	48, Female	
E-6	Outer brown with inner black	40, Female	
E-7	Outer yellowish white with inner brown	42, Female	
E-8	Outer yellow with inner brown	45, Female	

Mucocele	**Color of gallstone**	**Patient Age(Years) Gender**	**Remarks**
M-1	Outer white with inner black	36, Female	
M-2	Outer white with inner brown	40, Female	
M-3	Outer greenish brown with inner brownish white	60, Male	Habit-Alcohol
M-4	White	70, Female	Type-2 diabetes and hypertension with cardiovascular disease

As these gallstone samples are capable of obstructing the neck of gallbladder, thus we were interested to investigate how these gallstones are different. For this, we have focussed laser beam on outer and inner parts of these gallstones to record the LIBS spectra. Typical LIBS spectra of gallstones belonging to empyema and mucocele are shown in Figure 2. The LIBS spectra have been analyzed and spectral lines of the elements like C, Ca, Mg along with C_2 bands are observed in both types of samples. The analysed LIBS spectra of outer parts associated with empyema and mucocele gallbladder stone have prominence presence of atomic spectral signature of carbon C lines at 247.8 nm and 229.6 nm and molecular signatures of C_2 bands at 469.7 nm, 471.4 nm, 473.6 nm, as compared to inner parts of these gallstone. Thus the outer parts of these gallstone samples resemble cholesterol gallstone.

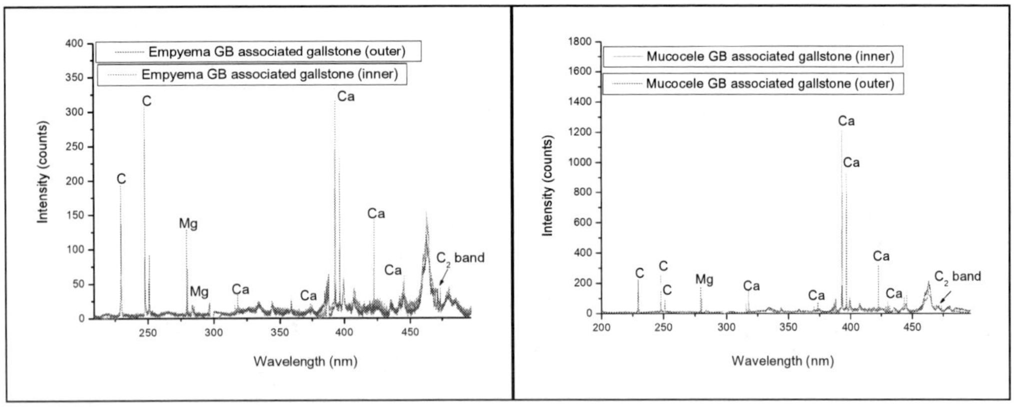

Figure 2. Typical LIBS spectra of gallstone.

In order to get the idea of hardness, we have calculated intensity ratio of ionic line of C (229.2 nm) to the corresponding neutral atomic line C (247.2 nm) for outer and inner parts of gallstones related to empyema and mucocele of gallbladder. The ratio[C(229.2)/C(247.2)] for outer parts are calculated as 1.26±0.06 and 1.55±0.13 but this ratio for inner parts are calculated as 0.75±0.05 and 0.45±0.0.24, for empyema and mucocele gallstones, respectively. Thus we found that the outer parts of both the gallstones related with empyema and mucocele state of gallbladder are having higher ionic to neutral spectral intensity ratio as compared to inner part. Based on the report by Tsuyuki et al. [11] in which he described that the intensity ratio of ionic to neutral atomic line of calcium is proportional to the compressive strength of target sample because in hard target, speed of shock wave produced is more in comparison to the soft target which results in enhancement of the ionization. Therefore from LIBS study we can conclude that outer surface of the gallstone is harder than inner part which may be the probable cause why these gallstones obstruct the neck of gallbladder.

3.3. Study of healthy and caries affected parts of deciduous tooth

Early childhood caries are becoming more common problem because of frequent intake of fast foods and sugars. The health of deciduous teeth (Primary dentition) should be also taken care because the health of permanent teeth also depends on them [4]. Therefore we have studied the caries and healthy parts of enamel of tooth sample of a child (aged 5 years) affected with early childhood caries. LIBS spectra of caries and healthy parts of the deciduous tooth sample were recorded in the spectral range 200-900 nm. LIBS spectra of the healthy part and caries affected part of the enamel of deciduous tooth sample in spectral range 225-330 nm & spectral range 575-800 nm are shown in Figure 3. Since the major constituents of the tooth's crystalline enamel structure is hydroxyapatite, $Ca_{10}(PO_4)_6(OH)_2$ and we observed the persistent emission lines of the elements Ca, P ,H and O which are the matrix elements of the tooth. Our result reveals that the intensities of the atomic lines of Ca and P are higher in healthy parts than that of caries affected parts of the tooth samples (Figure 3) which indicates that the concentrations of Ca & P are decreased in caries affected part of the tooth enamel. In contrast to the above, the intensity of atomic lines of the trace elements i.e. Mg, K, and C are higher in the caries than that of healthy part of the tooth sample and thus the concentration of these elements are increased in caries affected part in comparison to healthy part (Figures 3). We have calculated intensity ratios of different analyte emission lines of C at 247.8 nm, Mg at 285.2 nm, and K at 766.4 nm to the background were estimated using LIBS spectra of caries affected part and healthy part of the tooth sample. It is clear from Fig. 4 that the normalized intensity of C (247.8 nm), Mg (285.2 nm), and K (766.4 nm) are higher in caries affected parts of the tooth sample than that of the healthy parts which reveals that the concentration of the non-matrix elements like C, Mg, and K increases. The changes in the concentration of matrix elements may be due to fact that calcium bound to the hydroxyapatite are ionized and subsequently washed out from the tooth and are replaced by other elements like Mg, C and K. Demineralization of main matrix elements (Ca & P) occurs for several reasons, but one of the most important causes is the ingestion of sugars. In addition to, foods high in starches, such as potato chips, crackers, breads, and cakes may also be a cause to damage teeth. Our result related to caries and healthy parts of deciduous tooth are in agreement with the results reported for permanent tooth [5].

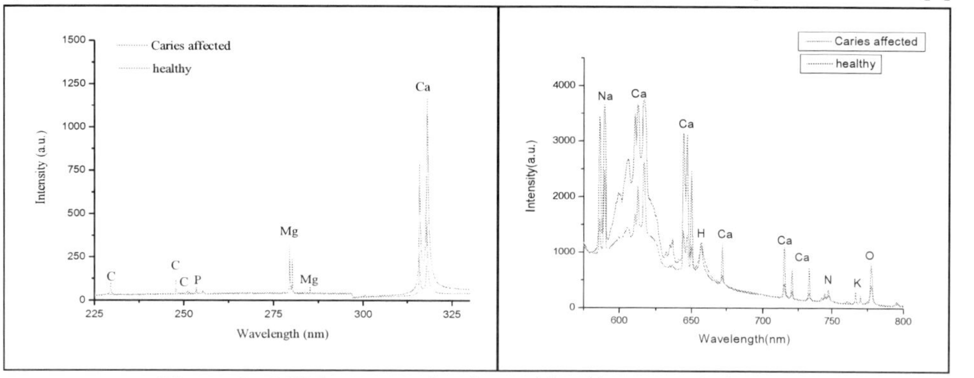

Figure 3. LIBS spectra of caries and healthy enamel of tooth.

We have also compared the hardness of healthy and caries affected part of the tooth samples. For this purpose we have used spectral line of Calcium and we have calculated intensity ratio of ionic lines Ca II (373.6 nm), to neutral lines Ca I (422.6 nm) from the LIBS spectra of the caries enamel and healthy enamel. For deciduous tooth this ratio is calculated and found as 1.11 ± 0.20 for healthy enamel but 0.63 ± 0.10 for caries enamel. This indicates that hardness of the tooth enamel decreases in the case of caries. Thus it is observed that the ionic to atomic spectral intensity ratio is higher for healthy enamel as compared to caries enamel. We have repeated this measurement for the permanent tooth sample belonging to different age group of patients and also found similar trends. Therefore the ionic to atomic LIBS spectral line intensity ratio is also an important indicator to distinguish healthy part and caries affected part of tooth.

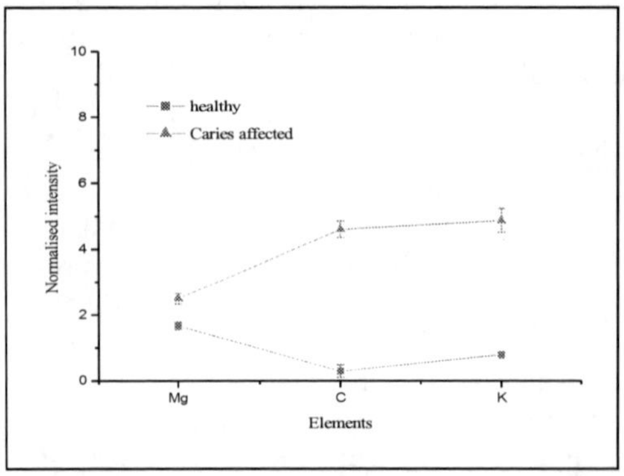

Figure 4. Variation of concentration of non-matrix elements.

3.4. *In-vivo human nail study*

In-situ LIBS study of the finger nail clippings belonging to normal nail sample and the healed injured nail sample is reported [12] and it is shown that main matrix of the nail are not affected much but mineral elements like Ca, Na & K changed much in injured/ reshaped nail. In this study we present the *in-vivo* LIBS analysis of human nails. In order to know the variations and similarities between left-hand and right-hand nails, the LIBS spectra of mineral profiles of left-hand and right-hand nails were recorded. The LIBS spectra were recorded by focusing the laser beam directly on each finger nail (Thumb finger nail:TFN, Index finger nail:IFN, Middle finger nail:MFN, Ring finger nail:RFN, Smallest finger nail:SFN) of left hand and right hand belonging to different volunteers.

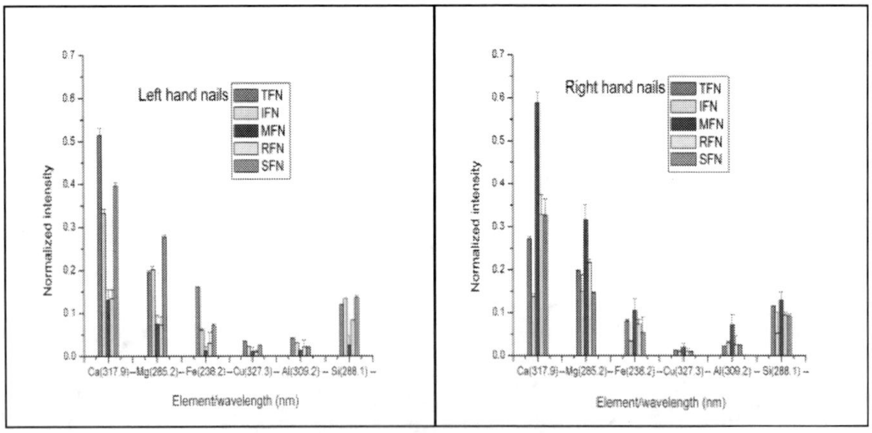

Figure 5. Bar diagram showing intensities of the spectral lines of elements in LIBS spectra.

The normalized intensity of selected spectral lines of Ca (317.9 nm), Mg (285.2 nm), Fe (238.2 nm), Cu (327.3 nm), Al (309.2 nm) and Si (288.1 nm) were evaluated by dividing intensity with corresponding background signal for each finger nails belonging to Left hand and Right hand. The *in-vivo* analysis belonging to left hand and right hand nails of volunteer aged 32 years is shown in Figure 5. It is observed from Fig. 5 that even the left hand and right hand mineral profiles are not exactly same. We have also found similar pattern of nail profiles for other volunteers. Thus *in-vivo* nail study can give instant information regarding changes in elemental profile of nail. We are also continuing our study for large sample data to find further correlation in different nail samples and the study is in progress.

4. Conclusion

The present study demonstrated how laser induced breakdown spectroscopy has been implemented to understand the role of various elements in the formation of gallstones (formed under the empyema and mucocele state of gallbladder).On the basis of database from patient history, the occurrence statistics of gallstones revealed higher occurrence of gallstones for female patients. Though the gallstone occurrence was less in male patients but cases were more prevalent for those male patients who were having the habit of either tobacco chewing, smoking or drinking alcohols. The work also reported *in-situ* study of deciduous tooth and *in-vivo* study of human nail.

Acknowledgements

A K Pathak duly acknowledges, Department of Science and Technology for providing international travel grant to attend and deliver an invited talk in 22nd ICSLS, UTSI at Tullahoma, Tennessee, USA.

References

[1] Zubair M, Habib L, Mirza M R, Channa M A, Yousuf M 2010 *Mymensingh Med J.* **19** 422
[2] Singh J P, Thakur S N 2007 *Laser-Induced Breakdown Spectroscopy* (Amsterdam: Elsevier)
[3] Rehse S J, Salimnia H and Miziolek AW 2012 *J Med Eng Technol.* **36** 77
[4] Losee F L, Custress T W, Brown R 1974 *Caries Res.* **8** 123
[5] Samek O, Telle H H and Beddows D C S, 2001 *BMC Oral Health* **1** 1
[6] Pathak A K, Kumar R, Singh V K, Agrawal R, Rai S, Rai A K 2012 *Appl. Spectrosc. Rev.* **47** 14
[7] Cuevas A, Miquel J F, Reyes M S, Zanlungo S, Nervi F 2004 *J Am Coll Nutr.* **23** 187
[8] Pathak A K, Singh V K, Rai N K, Rai A K, Rai P K, Rai P K, Rai S, Baruah G D 2011 *Lasers Med. Sci.* **26** 531
[9] Singh V K, Singh V, Rai A K, Thakur S N, Rai P K and Singh J P 2008 *Appl. Opt.* **47** G38
[10] Singh V K, Rai V, Rai A K 2009 *Lasers Med. Sci.* **24** 27
[11] Tsuyuki K, Miura S, Idris N, Hendrik K, Jie T, Kagawa K 2006 *Appl. Spectrosc.***60** 61
[12] Pathak A K and Rai A K 2010 *Asian Journal of Spectroscopy Special Issue* 147

Ultrashort Phenomena in Biochemistry and Biological Signaling

Robert Splinter[1,2]

[1]Splinter Consultants; Graham, NC 27253; USA;
[2]Physics and Optical Science, University of North Carolina at Charlotte,
Charlotte, NC 28223, USA

Email: splinter@consultant.com or rsplinter@gmail.com

Abstract. In biological phenomena there are indications that within the long pulse-length of the action potential on millisecond scale, there is additional ultrashort perturbation encoding that provides the brain with detailed information about the origin (location) and physiological characteristics. The objective is to identify the mechanism-of-action providing the potential for encoding in biological signal propagation. The actual molecular processes involved in the initiation of the action potential have been identified to be in the femto-second and pico-second scale. The depolarization process of the cellular membrane itself, leading to the onset of the actionpotential that is transmitted to the brain, however is in the millisecond timeframe. One example of the femtosecond chemical interaction is the photo-response of bacteriorhodopsin. No clear indication for the spatial encoding has so far been verified. Further research will be required on a cellular signal analysis level to confirm or deny the spatial and physiological encoding in the signal wave-trains of intercellular communications and sensory stimuli. The pathological encoding process for cardiac depolarization is however very pronounced and validated, however this electro-chemical process is in the millisecond amplitude and frequency modulation spectrum.

1. Introduction

In biology there is still insufficient information regarding the encoding of signals with respect to their location of origin. The cellular action potential is fairly similar for all cell, with variability for the specific cells, in particular the cardiac muscle cell [1]. The representative action potential is outlined in Figure 1, the time scale will vary based on a range of physiological factors.

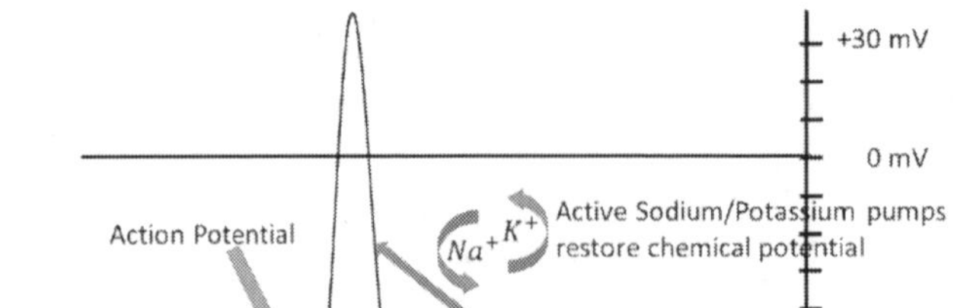

Figure 1. General outline of the human action potential, based on electrochemical mechanisms [1].

Various phenomena and diseases in humans are providing the indication that the central nervous system and the brain have a mechanism of action to identify the location from which a signal originates and what the function of this signal is. When analyzed on a pure macroscopic level the signals are transmitted either by myelinated or unmyelinated nerve cells with primarily the frequency of the pulse train offering the encoding for the magnitude of the signal [1, 2]. The frequency of the pulse-train phenomenon is outlined in Figure 2.

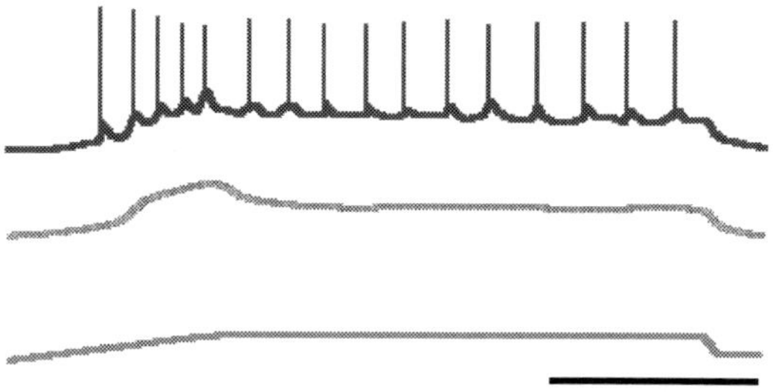

Figure 2. Frequency profile of the nerve action potential for various applications. The top pulse-train indicates the changes in magnitude of a single pressure sensor cell stimulated by pressure from the tip of a needle [1, 2].

The nerve cells are linked together to transmit the signal from the sensor or the chemical environment of the organ through a chain of command with divergence to various locations in the spine or the brain for processing. The divergence links the axon of one nerve cell to several dendrites to provide signal amplification by means of greater number of signal paths, hence a stronger signal resulting from superposition. On a practical note, there are several clinical reports that may indicate that there is a location specificity associated with the propagation of nervous impulses. For instance people with reported damage to certain segments of their nervous system have been documented to "learn" to regain control over muscles or process sensory signals through parts of the brain that were not originally associated with those tasks [1]. The fact that the brain can be retrained to overcome physiological and biological interruption in signal transduction can also be mediated by the fact that other sensor information provides the spatial information required for the learning process which is communicated on a chemical level through intra-cellular communications. Other sensors will yield feedback, such as hearing, touch, sight and for instance balance.

In order to describe the potential for high-frequency signal encoding there are documented ultra-short phenomena in physiology that can provide the mechanism-of-action to support such modulation principles [1, 2]. Two important spectroscopic phenomena with respect to biological applications are found in photosynthesis and vision [2, 3, 4, 5, 6, 7, 8]. For vision specifically, the light interaction in the rods of the eye (Rhodopsin) involving the conversion of light into changes in the chemical bond structures under ultra-fast conditions will be described to illustrate the potential for spatial encoding.

One specific example of a long time-scale physiological encoding is found in the depolarization pulse-shape of the cardiac muscle. The depolarization pulse-width for the heart muscle cells are in the order of tens of milliseconds as illustrated in figure 3.

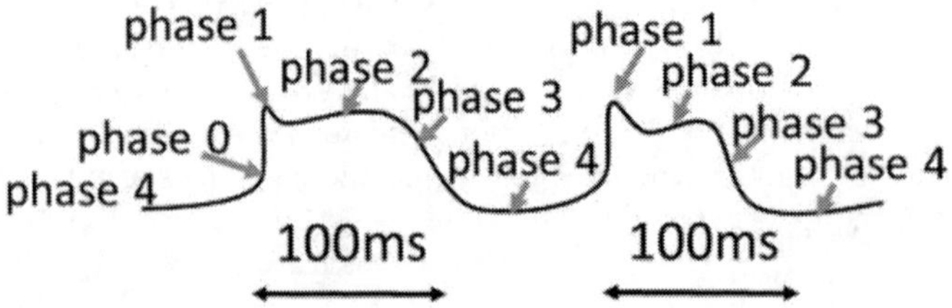

Figure 3. Depolarization pulse duration and wave front variability for the cardiac muscle cell [1].

The pulse-shape, pulse-duration, rise- and fall time for the cardiac muscle are influenced by chemical conditions. These chemical conditions include, but are not limited to: biochemical influences from hormones, next to prescription drug as well as recreational drugs, in particular cocaine. Most of these physiological chemical conditions are based on for instance: physiological factors, food or exercise (hormones, pheromones, etc.). For the cardiac muscle the pulse rise and fall time form a mechanism of changing the chemical exchange process, specifically the oxygen supply and absorption and exchange against carbon-dioxide.

2. Background

For vision specifically, the light interaction in the rods of the eye (Rhodopsin) next to the conversion of light into chemical energy and oxygen release in the mammalian skin as well as in the leafs of plants take place on a femtosecond scale.

Vision is one biological phenomenon that has very surprising capabilities [4, 5, 7, 8, 9]. The fluence detection range of the eye covers better than 14 orders of magnitude. The minimum detectable fluence by a single rod is in the order of a single photon. A single molecule of photoactivated rhodopsin (R*) in the cell membrane of the rod catalyzes the activation of up to 1000 transductin molecules, and represents the initial stage in the signal amplification process. The transduction molecules provide various stages of the activation process for depolarization and the formation of the action-potential. The entire cell membrane of the rod has multiple bacteriorhodopsin molecules spread within the membrane.

The cones are more particular about the trigger stimulus and require a significant amount more photons. However, the brain and neural communication mechanisms have filters in place (threshold action-potential, chemical transduction in synapses etc.) to trigger a conscious response when at least, depending on personal acuity, about five to nine action-potentials arrive within less than 100 ms. In the case that single photon detection were possible there would be a continuous stream of visual "noise" under dim illumination, hence this filter is a necessity, not a weakness.

On experimental level, a single rod photoreceptor cell from the eye of a frog has proven to operate as an extremely sensitive detector that can respond to individual photons and provide an electro-chemical reaction that can be registered [10, 11, 12, 13]. The frog rod can also be used establish the coherence of extremely weak pulses of light. The sensitivity of rhodopsin, the photosensitive chemical of the rods (and cones; rod as radiance indicators only, no spectral resolution required), is highest at 568-nm.

Neural filtering provides the biological means of implementing a threshold before the brain will register a sensory input. Neural convergence is the physical joining of several axons or dendrites leading to and from a multitude of nerve cells involved in the signal transduction for a single sensor cell. In the retina the ganglion cells are connected on a one to one order to the rod cells as well as for the cone cells in the fovea. Neural convergence is responsible for increasing sensitivity of the rod (radiance vision; "intensity"), however reduces the cone vision sensitivity.

2.1. Theoretical Background

The mechanism of the chemical response of bateriorhodopsin to light is described.

In biological cells there are vibrational mode transitions, specifically in amino-acids, that take place on femtosecond or shorter scale [4, 14]. Two important spectroscopic phenomena in biological functionality in the ultrashort timeframe are found in photosynthesis and vision [4, 5]. The light interaction in the cones of the eye (receptor protein: Rhodopsin, chromophore derived from vitamin-A) take place on a femtosecond scale. Rhodopsin is present in both rods and also found in certain cones for vision. In vision the quantum states of the excited molecules yields information on the energy transfer. One significant detectable interaction is $C_{11} = C_{12} isomerization.$

2.2. Mechanism

The activation of bateriorhodopsin forms the basis of the action-potential leading to neural recognition.

The energy of the photon response is provided by the cis-trans C11 torsion potential of rhodopsin. By reducing the 'twisting' force in the configuration of the C11 position in rhodopsin the twisting angle of the C11=C12 bond is reduced in the negative direction to about -80 degrees [15, 16]. The photoisomerization relaxation energy obtained by this change amounts to an order of 10 kcal/mol. The energy storage base for bathorhodopsin consists approximately 40% of charge separation, with the remaining 60% of conformational distortion, the primary energy carrier mechanism. From rhodopsin, bathorhodopsin can be experimentally observed within a 2.3 picoseconds [7, 16] with a quantum yield of 0.67 [7, 11].

3. Theoretical Analysis

The light interaction in the rods and cones of the eye (receptor protein: Rhodopsin, chromophore derived from vitamin-A) take place on a femtosecond scale. Rhodopsin is present in both rods and cones for vision. In vision the quantum states of the excited molecules yields information on the energy transfer. One significant detectable interaction is $C_{11} = C_{12} isomerization$, which reveals itself through torsional vibrations; specifically the low $C_{12} - H$ vibration frequency in bathorhodopsin that shows anomalous behavior. The first chemical process is associated with a transition defined by the Franck-Condon principle, which has an approximate lifetime of $200\,fs$. [1, 5, 6, 7, 8]. The Frank Condon principle is defined by the Franck-Condon factor (FCF) that relies on the energy eigenfunctions overlap for the two eigenconditions (associated with the Franck-Condon states, energy status based on minimal variability in coordinates within the system) with their respective waveform: $FCF = \left|\int_{-\infty}^{\infty} \Psi_n^{1*}(x)\Psi_m^2(x)dx\right|$. Initially, there is no change in the molecular geometry associated with the electronic transition, due to the short transaction time, only a vertical transition in the electronic potential. The primary chemical event in rhodopsin is an ll-cis to ll- trans photoisomerization, a conformational change that involves rearrangement of a large fraction of the retinyl polyene [7, 8]. The photon interaction process and time-line with bacteriorhodopsin is illustrated in Figure 4.

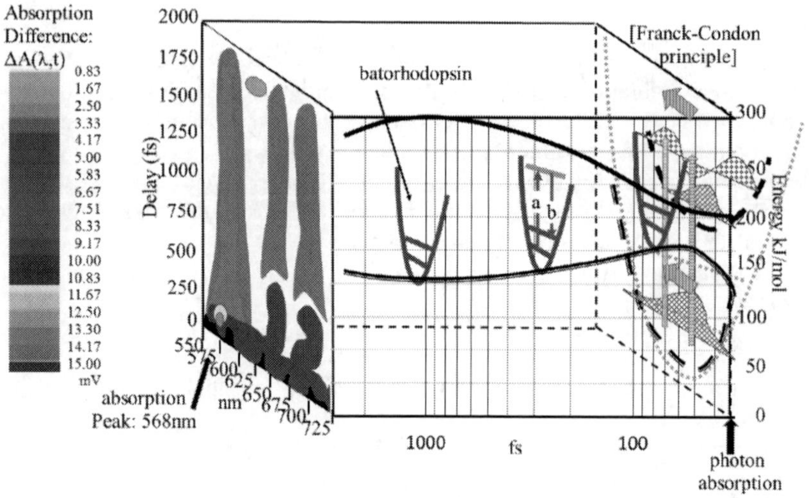

Figure 4. Temporal and spectral profile of energy transfer processes in Rhodopsin - Bathorhodopsin. The photon interaction takes place in the first 50 fs (at the far right of the graph; Franck-Condon event); 'a' is Raman emission and 'b' a broad-band energetic restructuring fs event. Bathorhodopsin is formed $\sim 1-2\ ps$ into the process. The vibrational molecular interaction is most pronounced for 568nm [5, 6], data from Yan et al 1991.

The spectrophotometric interaction for Rhodopsin is illustrated in Figure 5. The predominant absorption of 568 nm is illustrated, representing the Cis-Trans absorption spectrum for the energy transfer in the torsional state of the C11=C12 bond.

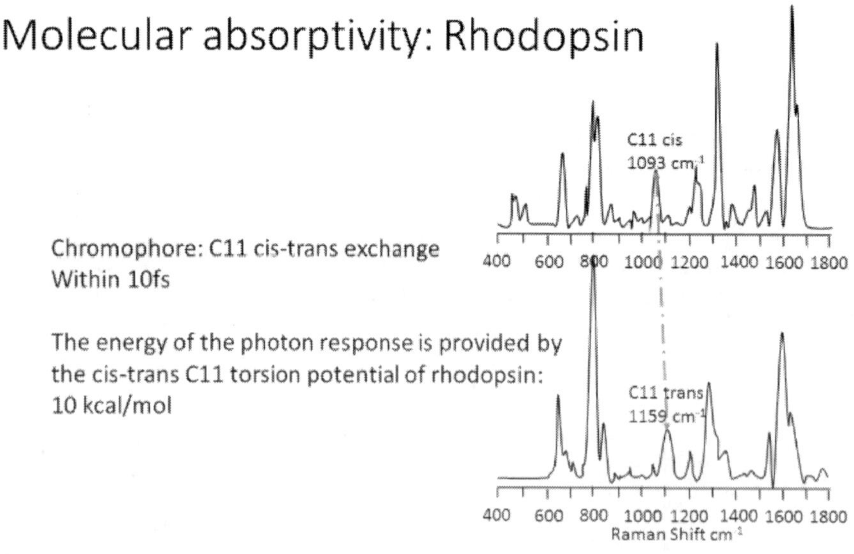

Figure 5. Spectral absorption profile for Rhodopsin, identifying the photo-chemical response of a single photon within the 10 fs interaction time frame [8, 9, 10].

After the interaction with the first photon the absorption band of the rhodopsin derivative in the rod of the eye switches its spectral profile as outlined in Figure 6.

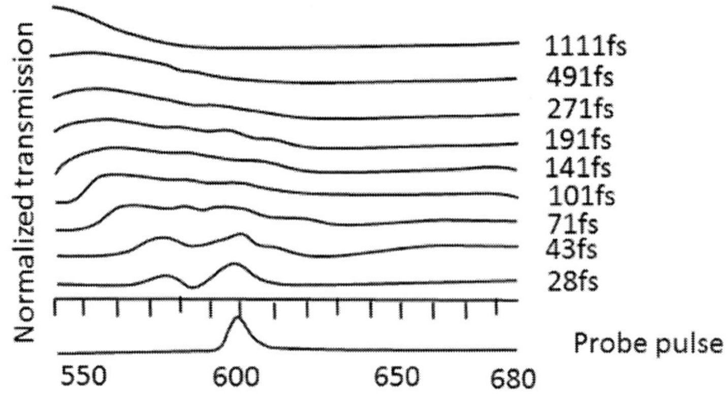

Figure 6. Time-resolved spectral attenuation profile (x-axis; wavelength: nm) for bacteriorhodopsin as a function of time (y-axis) after the seed photon absorption 1000 fs into the process. Batorhodopsin formed ~1ps after excitation [4, 5, 9, 10, 11], data from Birge 1990.

3.1. Comments

Additional research is required on a cellular level to determine the shape of the propagating exact depolarization wavefront along the path of the transported signal. This process will require in-situ non-invasive high temporal resolution measurement of the electro-chemical processes. External electrodes are insufficient to provide this kind of detail since they average the biphasic and triphasic aspects of the depolarization process based on volume conduction and superposition.

4. Conclusions

No conclusive evidence for signal encoding has so far been established. No supporting information is currently available to validate the hypothesis of location specific biological communications based on encoding. However on an electro-chemical level there are mechanisms available that can potentially support encoding at the initial stage of pulse generation providing a modulation process.

References

[1] Splinter R 2010 *Handbook of Physics in Medicine and Biology*, (Boca-Raton, FL: CRC-Press).
[2] Najarian K and Splinter R 2012 *Biomedical Signal and Image Processing, Second Edition* (Boca-Raton, FL: CRC-Press)
[3] Splinter R and Hooper B A 2006 *An Introduction to Biomedical Optics* (Boca-Raton, FL: CRC-Press)
[4] Martin J L et al 1980 *Proc. Opt.Photon. Applied to Medicine*, **128** SPIE 0211 128
[5] Yan M *et al* 1991 *Proc. Nat. Acad. Sci.* USA **88** 9809
[6] Yabushita A L, Kobayashi T and Tsuda M 2012 *J. Phys. Che. B* **116** 1920
[7] Birge R R 1990 *Biochim. Biophys. Acta* **1016** 293

[8] Sim N et al 2012 *Phys. Rev. Lett.* **109** 113601
[9] Schnapf J and Baylor D 1987 Scientific American **256** 32
[10] Baylor D A *et al* 1979 *J. Physiol. Lond.* **288** 613
[11] Birge R R and Cooper T M 1983 *Biophys. J.* **42** 61
[12] Weingart O 2007 *J. Amer. Chem. Soc.* **129** 10618
[13] Lacey C J et al 2007 *J. Neurosci.* **27** 4374
[14] Diels J-C and Rudolp W 2006 *Ultrashort Laser Pulse Phenomena* (Dordrecht, Netherlands: Elsevier).
[15] Yamada A L *et al* 2004 *Photochem. Photobiol.* **79** 476
[16] Fitter J *et al* 1997 *Biophys. J.* **73** (4) 2126

Opacity profiles in inertial confinement fusion plasmas

D Benredjem[1], J C Pain[2], F Gilleron[2], S Ferri[3] and A Calisti[3]

[1] Laboratoire Aimé Cotton, Université Paris-Sud, 91405 Orsay, France
[2] CEA, DAM, DIF, F-91297 Arpajon Cedex, France
[3] Aix-Marseille Université, CNRS, PIIM, Centre Saint-Jérôme, 13397 Marseille, France

E-mail: `djamel.benredjem@lac.u-psud.fr`

Abstract. The opacity is an important issue in the knowledge of radiative properties of ICF and astrophysical plasmas. In this work we present the opacity of dopants embedded in the ablator of some ICF capsules. The silicon is used as dopant and we are interested in C+Si mixtures. We have used two methods to calculate the opacity of C+Si. The first one involves a detailed line shape calculation in which the atomic database is provided by a MCDF code. The lineshape code PPP is then adapted to the calculation of opacity profiles. Almost all spectral broadening effects, including Zeeman splitting and Stark effect, are taken into account. This method is able to provide accurate opacity spectra but becomes rapidly prohibitive when the number of lines is large. To account for many ionic stages and thousands of lines, a second method −*hybrid* method− is prefered. This method combines detailed-line and statistical calculations. In the spectral regions where the lines are sufficiently separated and the number of radiative transitions is moderate, the hybrid method performs detailed calculations. When the number of transitions is very large and most of them merge in broad structures due to line broadening, the hybrid method performs statistical calculations.

1. Introduction

Indirect-drive inertial confinement fusion (ICF) uses a high-Z Hohlraum to convert precisely shaped laser pulse into thermal X-rays. The thermal radiation is absorbed due to the ablator opacity, setting up an ablation front that propagates into the ablator. Low density mass is ejected from the capsule surface, compressing the capsule in reaction (the DT fuel is compressed and a central hot spot is created at sufficiently high temperature and areal density for ignition). This spherical rocket compression process is dependent on the Hohlraum radiation temperature and on the opacity/emissivity of the ablator through the mass flux and the exhaust velocity.

The high quality of the X-ray radiation source is very important. Hard X-rays above 1.8 keV can propagate ahead of ablation front and preheat the inner layer of ablator next to the DT fuel, which increases the density difference between the fuel and ablator and results in the growth of hydrodynamic instabilities. Therefore, optimized designs of ignition capsule usually use the graded doped ablators such as germanium- or silicon-doped plastic. The role of the dopant is to absorb the hard X-ray radiation and to prevent preheating of the inner undoped plastic layer next to the DT fuel. The doped design limits Richtmyer-Meshkov and Rayleigh-Taylor instabilities as well as mixing at the fuel-ablator interface. The dopant concentration must be optimized: a too small quantity of dopant cannot efficiently prevent the preheating, and an excess of dopant tends to reduce the implosion velocity. The knowledge of opacity is then very important.

For low- or mid-Z dopants the number of lines is sometimes tractable. In such cases, we are able to perform a detailed line calculation to obtain the opacity of silicon. This calculation involves the MCDF [1] (atomic data) and PPP [2] (line profile) codes. When the number of transitions is large this approach becomes prohibitive. An alternative approach -hybrid approach- based on a detailed description whenever it is required and a statistical one when a large number of lines overlap, is proposed. The hybrid approach for opacity is the best compromise between precision and calculation time. The statistical part is calculated by the SCO (Super-Configuration Opacity) code [3] which relies on the concept of "super-configuration".

In the next section we show results obtained by the code PPP for C+Si. In the last section we present the hybrid approach and we investigate the variation of the opacity with mass density and fraction of silicon. All ionic stages of carbon and silicon are taken into account.

2. Detailed line calculations

In a magnetized plasma, the radiators interact with a perturbing bath formed by ions in various ionic stages and free electrons. The Hamiltonian of a radiator is written as

$$H = H_0 - \mathbf{d} \cdot \mathbf{E} - \frac{\mu_B}{\hbar}(\mathbf{L} + 2\mathbf{S}) \cdot \mathbf{B}, \tag{1}$$

where H_0 is the Hamilonian of the free radiator. The second term in the right-hand side is the interaction of the radiator with the charged particles (free electrons and ions) that create the electric field $\mathbf{E}$. The third term represents the interaction of the magnetic moment of the radiator with the magnetic field which is due to the Rayleigh-Taylor instabilities. $\mathbf{L}$ and $\mathbf{S}$ are the orbital and spin angular momenta, respectively. μ_B is the Bohr magneton, $\mu_B = -e\hbar/(2m_e)$.

The perturbing ions are assumed static and their effect on the radiator spectrum involves a static microfield distribution function [4]. The free electrons are treated in the impact approximation and their effect on line shape is represented by a collision operator [5].

The atomic transitions occur between upper $|(\alpha)JM\rangle$ and lower $|(\alpha')J'M'\rangle$ sublevels. The polarization state is determined by the q value ($q = M - M'$). The selection rules for E1 radiation are

$$\Delta J = 0,\ \pm 1 (0 \to 0 \text{ not allowed}) \qquad q = 0,\ \pm 1.$$

Transitions with $q = \pm 1$ (0) are responsible for circularly- (linearly-) polarized radiation.

The magnetic field is assumed to be fixed along the z axis. The direction of observation forms and angle α with $\mathbf{B}$. The line profile is then given by [6]

$$I(\nu, \alpha) = I_{\parallel} \cos^2 \alpha + I_{\perp} \sin^2 \alpha, \tag{2}$$

where the parallel and the transverse profiles, $I_{\parallel}(\nu)$ and $I_{\perp}(\nu)$, respectively, read

$$I_{\parallel}(\nu) = I_{+}(\nu) + I_{-}(\nu), \tag{3}$$

$$I_{\perp}(\nu) = \frac{1}{4}\left[I_{+}(\nu) + I_{-}(\nu) + 2I_0(\nu)\right]. \tag{4}$$

$I_{+/-}$ (I_0) is the intensity of the right/left circularly (linearly) polarized radiation.

In the formalism of PPP, if one neglects dynamic effects the expression of each polarized line shape is written as a sum of rational fractions:

$$I_q(\nu) = \sum_p \frac{c_{q,p}(\nu - \nu_{q,p}) + a_{q,p}\gamma_{q,p}}{(\nu - \nu_{q,p})^2 + \gamma_{q,p}^2}, \tag{5}$$

Journal of Physics: Conference Series **548** (2014) 012009 doi:10.1088/1742-6596/548/1/012009

where the $a_{q,p} + ic_{q,p}$ are the generalized intensities and $\nu_{q,p} + i\gamma_{q,p}$ the generalized frequencies (for more details, see Refs. [2] and [6]). Here, the line shape can be seen as the contribution of a set of Stark-Zeeman dressed transitions (SZDTs) defined by those generalized intensities and frequencies. The contribution of each SZDT to the opacity is given by:

$$k_{q,p}(\nu) = N_g \frac{\pi e^2}{mc} \frac{2m\nu_{q,p}}{3\hbar} a_{q,p} \psi_{q,p}(\nu), \tag{6}$$

where $\psi_{q,p}$ is the line shape of the SZDT and N_g the population of the ground levels. The polarized opacity is then the sum of all individual SZDTs contributions:

$$k_q(\nu) = \sum_p k_{q,p}(\nu). \tag{7}$$

In Fig. 1 we show the opacity of a mixture composed of carbon and silicon at the electron density $N_e = 6 \times 10^{23}$ cm^{-3} and temperature T_e=1000 eV. At these plasma parameter values, the fraction of non-hydrogenic ions is negligible. The gray curve accounts for electron and Doppler broadening. The black curve also accounts for the Stark effect due to perturbing ions. The comparison of the two curves shows that the ionic Stark effect is significant. In the range 250−500 eV we have carbon and silicon lines while above 600 eV we only have silicon lines.

We have checked, for several B values (from 10 MG up to 100 MG), that the Zeeman effect is negligible at these density and temperature.

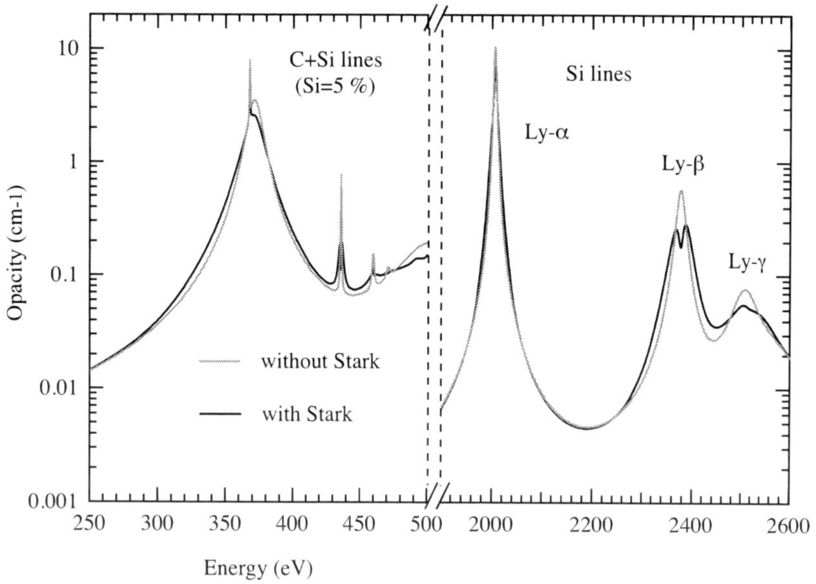

Figure 1. Opacity of the C+Si mixture. Electron density $N_e = 6 \times 10^{21}$ cm^{-3}, electron temperature T_e=1000 eV. The fraction of silicon is 5 %.

3. Hybrid calculations

Let us define some quantities which are useful in atomic spectroscopy. A non-relativistic configuration C reads $(n_1l_1)^{N_1}(n_2l_2)^{N_2}(n_3l_3)^{N_3}\cdots$, where n and l are the principal and orbital quantum numbers, respectively. N is the number of electrons occupying the sub-shell nl. The $N's$ fulfill the condition $\sum_i N_i = N_b$, where N_b is the number of bound electrons. A super-configuration (SC) reads $(\sigma_1)^{N_1}(\sigma_2)^{N_2}(\sigma_3)^{N_3}\cdots$, where the super-shell σ is an ensemble of sub-shells close in energy. Now N is the number of electrons occupying the super-shell σ. For instance $(n_1)^{N_1}(n_2)^{N_2}(n_3)^{N_3}\cdots$ is a particular case of super-configuration where all the sub-shells having the same principal quantum number are gathered. In that case, each super-shell is an ordinary shell. Note that $(n)^N$ is sometimes called a Layzer complex, but this is not true since a Layzer complex stricto sensu has a fixed parity ; therefore $(n)^N$ represents in fact two Layzer complexes. A transition array is the ensemble of transitions between two configurations of the same ion :

$$\{(n_1l_1)^{N_1}(n_2l_2)^{N_2}(n_3l_3)^{N_3}\cdots - (n_1l_1)^{N_1-1}(n_2l_2)^{N_2+1}(n_3l_3)^{N_3}\cdots\}$$

Electrons in the subshells $(n_3l_3)\cdots$ are spectators. A configuration is determined by a set of N_i: $C(N_1, N_2, N_3\cdots)$. A supertransition array (STA) [7] is the ensemble of transitions arrays $\{C(N_1, N_2, N_3\cdots) - C(N_1-1, N_2+1, N_3\cdots)\}$. The spectral broadening is often responsible for a coalescence of the lines pertaining to a transition array. In such cases the structure is named unresolved transition array (UTA) [8].

In ICF plasmas one is often faced with spectra involving a large number of lines. When most of them overlap, a detailed –line-by-line– calculation is not tractable. A statistical calculation involving the first moments of the opacity, i.e. μ_n [1], represents a satisfactory alternative. In a same spectrum we may also have separated lines which can be treated precisely, line-by-line. Then, there is a need to develop a hybrid approach which combines detailed and statistical calculations and as such would be able to treat STA and UTA structures as well as fine-structure. The hybrid code SCO-RCG [9, 10] is appropriate for such problems.

The STA approach enables one to deal with the large number of configurations but it does not provide a precise description of the individual electric-dipolar lines inside a given transition array. It can therefore be improved if several transition arrays are treated in a detailed way (line-by-line). The data required for the calculation of transition arrays (Slater integrals, spin-orbit and dipolar matrix elements) are evaluated and transmitted by SCO (Cowan's code [11] was indeed designed in such a way that the radial and dipolar integrals can be entered as parameters). Therefore, the plasma environment effects (screening) on the wave-functions are taken into account. Then, the energies of the levels and lines (positions and strengths) are calculated by an adapted RCG subroutine of Cowan's code, which performs the diagonalization of the Hamiltonian matrix (intermediate coupling).

In SCO-RCG, configuration interaction is limited to electrostatic interaction between relativistic subconfigurations (nlj orbitals) of a non-relativistic configuration (nl orbitals), namely relativistic configuration interaction (RCI). For the detailed part of the spectrum, RCI is calculated exactly, through the diagonalization of the Hamiltonian matrix. For the statistical part of the spectrum, RCI is described using an approximate model, which relies on the definition, in a transition array, of two sets of lines by considering only one relativistic orbital (the one with the largest spin-orbit integral in the transition). Corrections to the energy and strength of both structures are derived in the first-order perturbation theory, in order to include electrostatic effects between levels of the sub-arrays. Each transition array is represented by two distributions which characteristics (mean energy, strength, and number of lines) change in a continuous way from LS (where one sub-array vanishes) to jj coupling [12].

SCO-RCG involves definition of criteria in order to decide whether a detailed treatment of the lines is necessary and to determine the validity of the statistical methods. The estimate

[1] $\mu_n = \sum_{l-u} \Delta E_{lu}^n g_l f_{lu} / \sum_{l-u} g_l f_{lu}$, where ΔE_{lu}, g and f represent respectively the transition energy, the statistical weight of the lower level of the transition and the oscillator strength.

of the number of lines within a transition array is an important ingredient, in order to decide whether a detailed calculation is suitable or not for the transition array, as well as the number of levels with fixed angular momentum J which represents the size of a block of the Hamiltonian matrix. This led us to develop efficient algorithms for the numbering of LS terms and J levels [13]. We have also extended the statistical modelling of the distribution of quantum states in order to better describe configurations having open sub-shells with a large angular momentum.

Spectral broadening is taken into account. The Doppler and ionic Stark effect are modeled by a Gaussian profile. Inelastic collisions are modeled by a Lorentzian profile. The convolution of Gaussian and Lorentzian profiles yields the well-known Voigt profile. The electron collisions are described in the impact approximation within an approach proposed by Dimitrijević and Konjević [14]. Such a formulation relies on Baranger's expression for the width of an isolated line [15, 16]. The treatment of ionic Stark effect in SCO-RCG (assuming the quasi-static approximation), is very close to the one described in Refs. [17, 18]. The Zeeman line splitting is also taken into account. An approximate approach providing a fast and quite accurate estimate of the effect of an intense magnetic field on opacity was recently implemented in SCO-RCG. The formalism requires the moments of the Zeeman components of a line, which can be obtained analytically in terms of the quantum numbers and Landé factors. A good representation of the Zeeman profile is obtained using, for each component, the fourth-order Gram-Charlier expansion series [19].

In Table 1 we give for each element of the mixture, the average ion charge, the number of ionic stages involved in the hybrid calculation and the most abundant ion species.

Table 1. Average ion charge, number of ionic stages, most abundant ion. The mass density ρ and T_e are fixed to 0.01 g/cm^3 and 50 eV, respectively.

Element	$\bar{Z}$	Number of species	Most abundant ions
C	4.48	4	C^{4+}, C^{5+}
Si	8.04	8	Si^{7+}, Si^{8+}, Si^{9+}

In Fig. 2 we show the opacity of C+Si. The following spectral broadening mechanisms are included in the opacity calculation: electron impacts, ion microfield and Doppler effect. The opacity involves bound-bound, bound-free and free-free transitions. Three silicon fractions are considered: 1, 2 and 5 %. The large structure between 160 and 300 eV is composed of silicon lines. The spectral broadening due to free electrons and perturbing ions plays an important role.

Rayleigh-Taylor instabilities are expected to generate magnetic fields (see for example Ref. [20]) whose evolution depends on $\nabla N_e \times \nabla T_e$. As the magnetic field can affect the opacity, we have introduced the Zeeman effect in our opacity calculations. In Fig. 3 we compare the opacity represented in Fig. 2 and the opacity calculated with a magnetic field of 10 MG. We can see that the Zeeman effect plays a small role. A higher resolution is necessary to see the effect of the magnetic field. Then in Figure 4 we focus on a small energy interval. It is clear that the Zeeman effect plays a small role.

Figure 5 shows the variation of the opacity with density. When the density increases, the Stark broadening (electrons+ions) increases yielding a line overlapping. Moreover the number of lines decreases due to a continuum lowering. For example, the carbon peaks around 450 eV vanish at a mass density of 0.1 g/cm^3. At the highest mass density, below 150 eV the opacity is dominated by free-free transitions.

Journal of Physics: Conference Series **548** (2014) 012009 doi:10.1088/1742-6596/548/1/012009

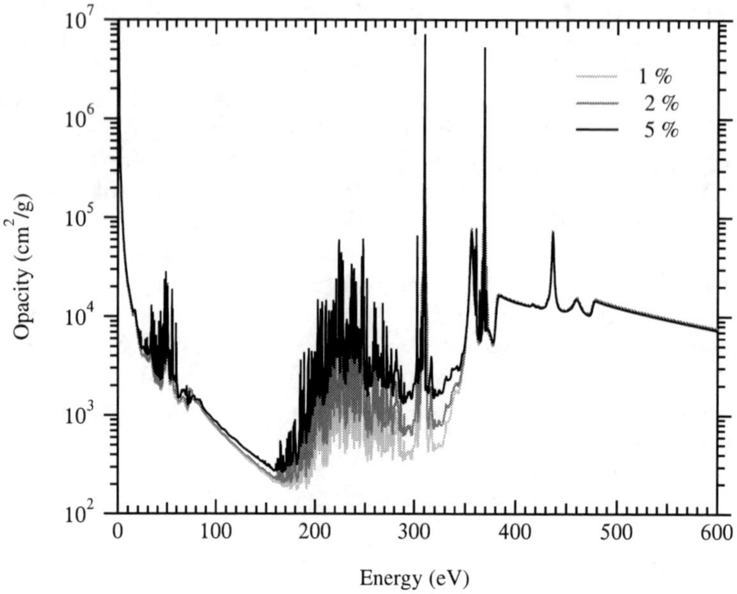

Figure 2. Opacity of C+Si for three fractions of silicon. $\rho = 0.01$ g/cm^3 and $T_e = 50$ eV.

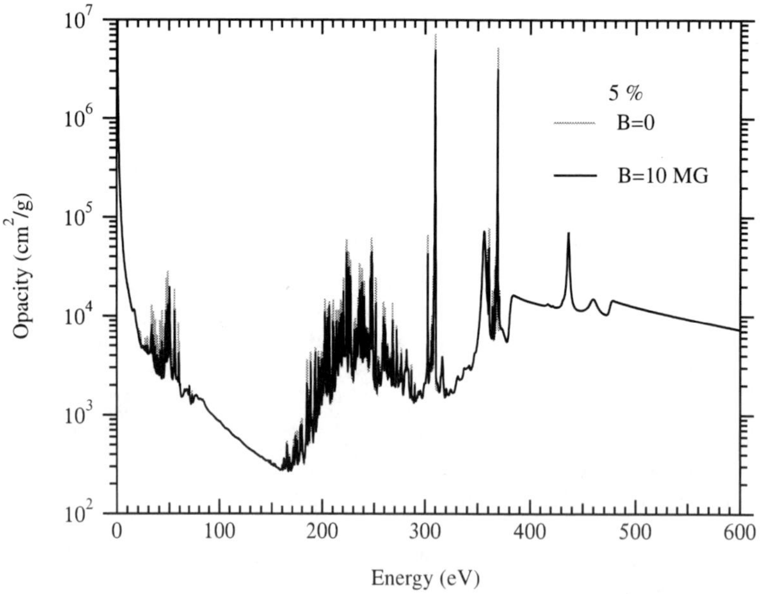

Figure 3. Opacity of C+Si. ρ and T_e, as in Fig. 2.

Finally, in Table 2 we show the variation of Rosseland and Planck mean opacities with the silicon fractions. The Planck opacity varies linearly with the silicon fraction.

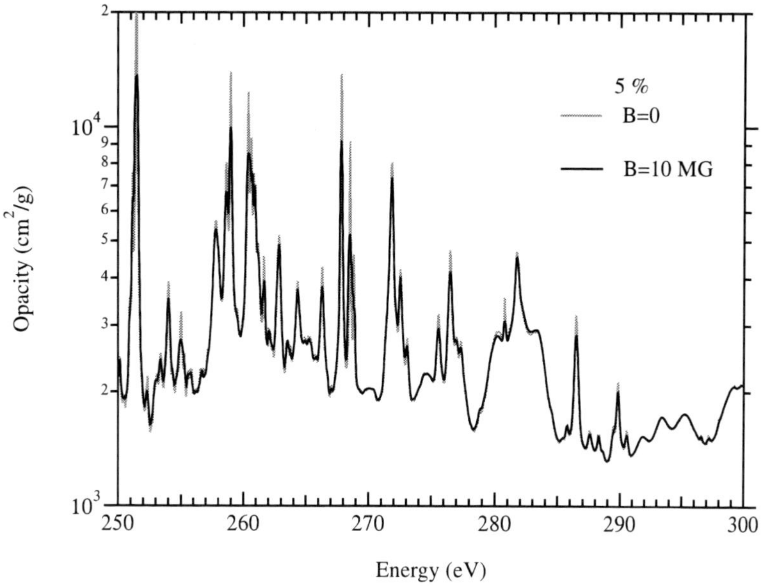

Figure 4. Same as Fig. 3 in a smaller energy interval.

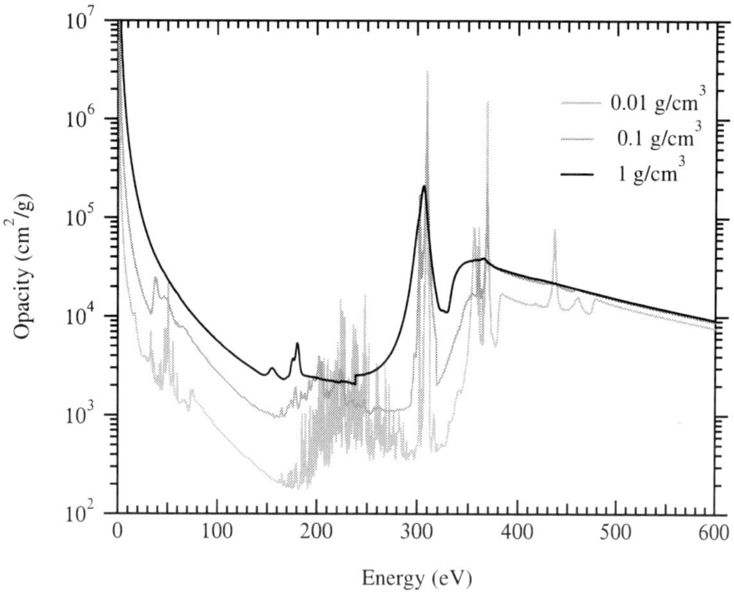

Figure 5. Opacity of C+Si for three values of ρ. T_e=50 eV and the fraction of silicon is fixed to 1%.

4. Concluding remarks

The opacity of a mixture composed of carbon and silicon is calculated for various densities and fractions of silicon. The opacity is very sensitive to both parameters. It involves a large

Table 2. Rosseland and Planck mean opacities for various silicon fractions. $\rho = 0.01$ g/cm^3 and kT_e=50 eV.

Si fraction	1%	2%	3%	4%	5%
$\kappa_R(\times 10^2)$	5.517	7.256	8.531	9.590	10.52
$\kappa_P(\times 10^3)$	4.949	5.196	5.437	5.671	5.900

number of lines which, owing to spectral broadening principally due to Stark effect, merge together yielding large structures. The treatment of such structures is performed with the help of statistical methods. Our calculations show that the ionic Stark effect plays an important role while the Zeeman effect is small in opacity calculations. In a future work we will investigate another dopant (Ge) and calculate the opacity of the mixture C+Ge.

References

[1] Grant I P, McKenzie B J, Norrington P H, Mayers D F and Pyper N C 1980 *Comput. Phys. Comm.* **21** 207
[2] Calisti A, Khelfaoui F, Stamm R, Talin B and Lee R W 1990 *Phys. Rev.* A **42** 5433
[3] Blenski T, Grimaldi A and Perrot F 2000 *J. Quant. Spectrosc. Radiat. Transfer* **65** 91
[4] Iglesias C A, DeWitt H E, Lebowitz J L, MacGowan D and Hubbard W B 1985 *Phys. Rev.* A **31** 1698
[5] Griem H R, Blaha M and Kepple P C 1979 *Phys. Rev.* A **19** 2421
[6] Ferri S, Calisti A, Mossé C, Mouret L and Talin B 2011 *Phys. Rev.* A **84** 026407
[7] Bar-Shalom A, Oreg J, Goldstein W H, Shvarts D and Zigler A1989 *Phys. Rev.* A **40** 3183
[8] Bauche J, Bauche-Arnoult C and Klapisch M 1979 *Phys. Rev.* A **20** 2024
[9] Porcherot Q, Pain J C, Gilleron F and Blenski T 2011 *High Energy Density Phys.* **7** 234
[10] Pain J C, Gilleron F, Porcherot Q and Blenski T 2013 *Proc. of the 40th EPS Conference on Plasma Physics* P4.403
[11] Cowan R D 1981 The Theory of Atomic Structure and Spectra, University of California Press, Berkeley
[12] Gilleron F, Bauche J and Bauche-Arnoult C 2007 *J. Phys. B: At. Mol. Opt. Phys.* **40** 3057
[13] Gilleron F and Pain J C 2009 *High Energy Density Phys.* **5** 320
[14] Dimitrijević M S and Konjević N 1987 *Astron. Astrophys.* **172** 345
[15] Baranger M 1958 *Phys. Rev.* **111** 494
[16] Baranger M 1962 in D. R. Bates, *Atomic and molecular processes* (Academic Press, New-York)
[17] Rozsnyai 1977 *J. Quant. Spectrosc. Radiat. Transfer* **17** 77
[18] Iglesias C A, Lebowitz J L and MacGowan D 1983 *Phys. Rev.* A **28** 1667
[19] Pain J C and Gilleron F 2012 *Phys. Rev.* A **85**, 033409
[20] Srinivasan B, Dimonte G and Tang X Z 2012 *Phys. Rev. Lett.* **108** 165002

Journal of Physics: Conference Series **548** (2014) 012010 doi:10.1088/1742-6596/548/1/012010

Atomic Collision Processes for Astrophysical and Laboratory Plasmas

M S Pindzola[1], S D Loch[1], C P Ballance[1], Sh A Abdel-Naby[1,2], T G Lee[1], J Colgan[3], C J Fontes[3], N R Badnell[4] and M G O'Mullane[4]

[1] Department of Physics, Auburn University, Auburn, AL 36849, USA
[2] Department of Physics, Beni Suef University, Beni Suef, Egypt
[3] Theoretical Division, Los Alamos National Laboratory, Los Alamos, New Mexico 87545, USA
[4] University of Strathclyde, Glasgow, UK

E-mail: pindzola@physics.auburn.edu

Abstract. An accurate knowledge of atomic collision processes is important for a better understanding of many astrophysical and laboratory plasmas. Collision databases which contain electron-impact excitation, ionization, and recombination cross sections and temperature dependent rate coefficients have been constructed using perturbative distorted-wave methods and non-perturbative R-matrix pseudo-states and time-dependent close-coupling methods. We present recent atomic collision results.

1. Introduction

Accurate atomic and molecular databases underpin current research efforts in a variety of scientific and engineering areas: including controlled fusion energy, astrophysics, radiation biophysics, fluorescent lamps, and atmospheric pollutant removal. For example, all light elements (H-O) are of interest for fusion experiments; in particular Li and B as wall coating materials and Be and C as primary wall materials.

Over the years both theory and experiment have provided increasingly more accurate cross sections for the electron-impact excitation, ionization, and recombination of atoms and their ions. In this paper we review perturbative distorted-wave methods and non-perturbative R-matrix with pseudo-states and time-dependent close-coupling methods in both their non-relativistic and fully-relativistic versions.

The rest of the review paper is structured as follows: in Section II we describe distorted-wave, R-matrix, and time-dependent close-coupling methods and give an electron-impact ionization cross section example, and in Section III we give a brief review of current projects.

2. Basic Collision Theory

2.1. Configuration-Average Distorted-Wave (CADW) Method

For ionization a general transition between configurations has the form:

$$(n_0 l_0)^{w_0} k_i l_i \rightarrow (n_0 l_0)^{w_0 - 1} k_e l_e k_f l_f \ , \tag{1}$$

where w_0 is a subshell occupation number, $n_0 l_0$ are quantum numbers of the bound electron, and $k_i l_i$, $k_e l_e$, and $k_f l_f$ are quantum numbers of the initial, ejected, and final continuum electrons.

The configuration-average ionization cross section is given by[1]:

$$\sigma_{ion} = \frac{32w_0}{k_i^3} \int_0^{E/2} \frac{d\epsilon_e}{k_e k_f} \sum_{l_i,l_e,l_f} (2l_i + 1)(2l_e + 1)(2l_f + 1)\ \mathcal{S}(n_0 l_0 k_i l_i \to k_e l_e k_f l_f)\ , \tag{2}$$

where $k = \sqrt{2\epsilon}$ and the continuum normalization is one times a sine function.

For excitation a general transition between configurations has the form:

$$(n_1 l_1)^{w_1+1} (n_2 l_2)^{w_2-1} k_i l_i \to (n_1 l_1)^{w_1} (n_2 l_2)^{w_2} k_f l_f\ , \tag{3}$$

where w_1 and w_2 are subshell occupation numbers, $n_1 l_1$ and $n_2 l_2$ are quantum numbers of the bound electrons, and $k_i l_i$ and $k_f l_f$ are quantum numbers of the initial and final continuum electrons. The configuration-average excitation cross section is given by:

$$\sigma_{exc} = \frac{8\pi}{k_i^3 k_f} (w_1 + 1)(4l_2 + 3 - w_2) \sum_{l_i,l_f} (2l_i + 1)(2l_f + 1)\ \mathcal{S}(n_1 l_1 k_i l_i \to n_2 l_2 k_f l_f)\ . \tag{4}$$

For dense plasmas the initial and final distorted-waves and the Coulomb matrix elements found in $\mathcal{S}(n_0 l_0 k_i l_i \to k_e l_e k_f l_f)$ of Eq.(2) and $\mathcal{S}(n_1 l_1 k_i l_i \to n_1 l_2 k_f l_f)$ of Eq.(4) are modified to include an exponential screening factor:

$$SF(r) = e^{-r/\Lambda}\ , \tag{5}$$

where the Debye-Huckel screening radius $\Lambda = \sqrt{T_e / 4\pi N_e}$, T_e is the electron temperature, and N_e is the electron density. Fully-relativistic subconfiguration-average distorted-wave ionization and excitation cross sections may also be calculated[2].

2.2. *R-Matrix Pseudo-States (RMPS) Method*

The R-Matrix method[3] splits the scattering process into two regions. The total wavefunction in the inner region is given by:

$$\Psi_k^{N+1} = \mathcal{A} \sum_{i,j} a_{ijk} \psi_i^{N+1} \frac{u_{ij}(r_{N+1})}{r_{N+1}} + \sum_i b_{ik} \chi_i^{N+1}\ , \tag{6}$$

where $\mathcal{A}$ is an antisymmetrization operator, ψ_i^{N+1} are channel functions obtained by coupling N - electron target states with the angular and spin functions of the scattered electron, $u_{ij}(r)$ are radial continuum basis functions, and χ_i^{N+1} are bound functions which ensure the completeness of the total wavefunction. The coefficients a_{ijk} and b_{ik} are determined by diagonalization of the total $(N+1)$ electron Hamiltonian. The total wavefunction in the outer region is given by:

$$\Psi_k^{N+1} = \sum_i \psi_i^{N+1} \frac{v_i(r_{N+1})}{r_{N+1}}\ , \tag{7}$$

where the radial wavefunctions $v_i(r)$ are obtained by solving the coupled differential equations given by:

$$[T_i(r) + V_{ij}(r)] v_i(r) = 0 \tag{8}$$

where $T_i(r)$ is a kinetic and nuclear energy operator and $V_{ij}(r)$ is an asymptotic coupling operator. The inner and outer region solutions are matched at the boundary and the K-matrix is extracted. Excitation cross sections are obtained by relating the K-matrix to the S-matrix.

Over the years, a non-relativistic RMLS suite of codes has been developed for low Z atoms and ions, a semi-relativistic RMLSJ suite of codes has been developed for medium Z atoms and ions, and a fully-relativistic RMjjJ suite of codes has been developed for high Z atoms and ions. With the addition of pseudo-states, the R-matrix method becomes more accurate for excitation to high-lying excited states. The RMPS method can also be used to calculate electron-impact ionization cross sections, including both direct ionization and excitation-autoionization.

2.3. Time-Dependent Close-Coupling (TDCC) Method

The time-dependent Schrodinger equation for two-active electron atomic systems is given by:

$$i\frac{\partial \Psi(\vec{r}_1,\vec{r}_2,t)}{\partial t} = H(\vec{r}_1,\vec{r}_2)\Psi(\vec{r}_1,\vec{r}_2,t) \ , \tag{9}$$

where

$$H(\vec{r}_1,\vec{r}_2) = -\frac{1}{2}\nabla_1^2 + V(r_1) - \frac{1}{2}\nabla_2^2 + V(r_2) + \frac{1}{|\vec{r}_1-\vec{r}_2|} \ . \tag{10}$$

Expanding in coupled spherical harmonics:

$$\Psi(\vec{r}_1,\vec{r}_2,t) = \sum_{l_1,l_2} \frac{P^{LS}_{l_1l_2}(r_1,r_2,t)}{r_1r_2} \sum_{m_1,m_2} C^{l_1l_2L}_{m_1m_2M}\ Y_{l_1m_1}(\hat{r}_1)\ Y_{l_2m_2}(\hat{r}_2) \tag{11}$$

yields the time-dependent close-coupled equations[4]:

$$i\frac{\partial P^{LS}_{l_1l_2}(r_1,r_2,t)}{\partial t} = [T_{l_1}(r_1)+T_{l_2}(r_2)]\ P^{LS}_{l_1l_2}(r_1,r_2,t) + \sum_{l'_1,l'_2} V^L_{l_1l_2,l'_1l'_2}(r_1,r_2)\ P^{LS}_{l'_1l'_2}(r_1,r_2,t) \ , \tag{12}$$

where

$$T_l(r) = -\frac{1}{2}\frac{\partial^2}{\partial r^2} + \frac{l(l+1)}{2r^2} + V(r) \tag{13}$$

and $V^L_{l_1l_2,l'_1l'_2}(r_1,r_2)$ is a two-body coupling operator.

Fully-relativistic time-dependent close-coupling equations[5] may also be derived based on the time-dependent Dirac equation. After time propagation of the non-relativistic or fully-relativistic TDCC equations, scattering probabilities and cross sections are obtained by projecting the radial wavefunctions onto antisymmetrized products of bound and continuum radial orbitals.

2.4. Comparison of the CADW, RMPS, and TDCC Methods

As an example, we calculated[6] the electron-impact ionization of Al and compared with experiment[7] in Figures 1-2. CADW direct ionization cross sections were included using Eq.(2) for the $3p$ and $3s$ subshells in Figure 1. CADW excitation-autoionization cross sections were included using Eq.(4) for the $3s \to 3p, 3d$ and $3s \to 4l(l = 0-3)$ excitations in Figure 1. The CADW total ionization cross section is about a factor of 2 above experiment. TDCC direct ionization cross sections were included for the $3p$ and $3s$ subshells in Figure 2. RMPS total ionization cross sections are also shown in Figure 2. Good agreement is found between the RMPS total ionization calculations and experiment.

2.5. Isonuclear Sequence Databases

Over the last decade light atom isonuclear sequence databases have been assembled based on the most accurate basic collision cross section calculations. Accurate databases are now available for H[8], He[8], Li[9], Be[10], and B[11], while work is in progress for C, N, O, and Ne. Electronic data is available at IAEA/ALADDIN[12]. We also note that previous CADW calculations for the electron-impact ionization of the Fe[13], Ni[14], Kr[15], Sn[16], Xe[16], and W[17] isonuclear sequences are being updated with RMPS and TDCC calculations for the atoms and low-charged ions.

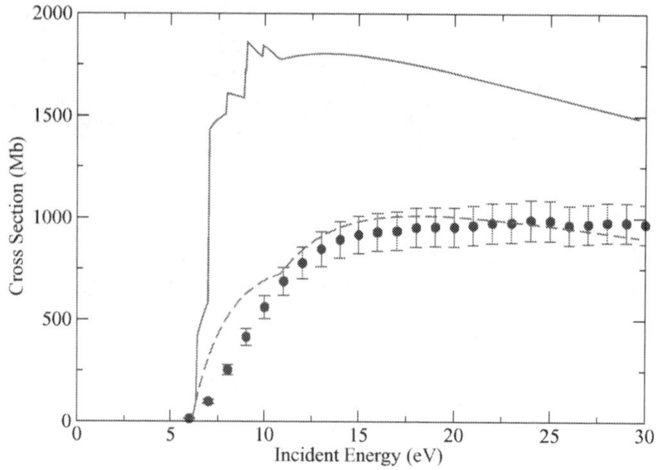

Figure 1. Electron-impact ionization of Aluminum. Solid line (red): CADW total ionization, dashed line (red): CADW direct ionization, circles (blue) with error bars: experiment [7] (1.0 Mb = 1.0 × 10^{-18} cm^2).

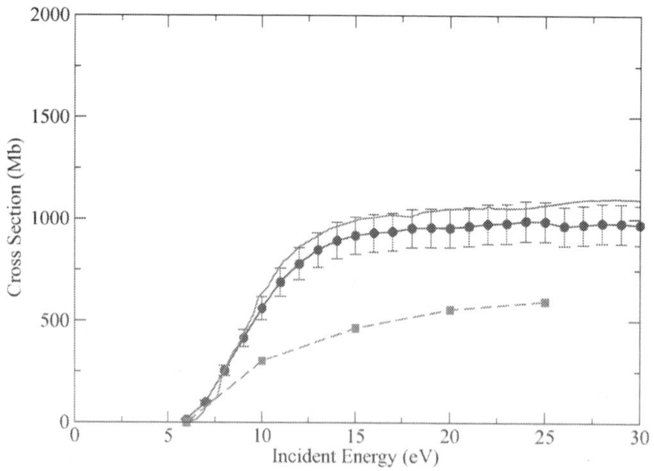

Figure 2. Electron-impact ionization of Aluminum. Solid line (red): RMPS total ionization, dashed line (red): TDCC direct ionization, circles (blue) with error bars: experiment [7] (1.0 Mb = 1.0 × 10^{-18} cm^2).

3. Summary

We are currently generating accurate electron-impact excitation, ionization, and recombination cross sections for the N isonuclear sequence. In the future we plan to calculate accurate electron-impact excitation and ionization cross sections for atoms and low-charged ions in the Mn, Fe, Co, and Ni isonuclear sequences, as well as in the heavier Mo, Xe, La, W, Au, and U isonuclear sequences.

Acknowledgments

This work was supported in part by grants from the US Department of Energy and the US National Science Foundation. Computational work was carried out at the National Energy Research Scientific Computing Center in Oakland, California and the National Center for Computational Sciences in Oak Ridge, Tennessee.

References

[1] Pindzola M S, Griffin D C and Bottcher C 1986 *NATO Advanced Study Institute, Series B: Physics* **145** 75
[2] Pindzola M S and Buie M J 1988 *Phys. Rev.* A **37** 3232
[3] Burke P G 2011 *R-Matrix Theory of Atomic Collisions*, Springer Series on Atomic, Optical, and Plasma Physics (Heidelberg ; New York : Springer-Heidelberg) **61**
[4] Pindzola M S *et al* 2007 *J. Phys.* B **40** R39
[5] Pindzola M S, Ludlow J A, Robicheaux F, Colgan J and Fontes C J 2009 *Phys. Rev.* A **80** 052706
[6] Loch S D, Ballance C P, Wu D, Abdel-Naby Sh A and Pindzola M S 2012 *J. Phys.* B **46** 065201
[7] Freund R S, Wetzel R C, Randy J S and Hayes T R 1990 *Phys. Rev.* A **41** 3575
[8] Loch S D, Ballance C P, Pindzola M S and D. P. Stotler D P 2009 *Plasma Phys. Control. Fusion* **51** 105006
[9] Loch S D, Colgan J, Witthoeft M C, Pindzola M S, Ballance C P, Mitnik D M, Griffin D C, O'Mullane M G and Summers H P 2006 *ADNDT* **92** 813
[10] Loch S D, Pindzola M S, Ballance C P, Griffin D C, Colgan J, Badnell N R, O'Mullane M G and Summers H P 2008 *ADNDT* **94** 257
[11] Loch S D, Ballance C P, Pindzola M S, Griffin D C, Colgan J, Badnell N R and O'Mullane M G 2013 *J. Phys. Conf. Ser.*
[12] IAEA/ALADDIN *www-amdis.iaea.org/ALADDIN.*
[13] Pindzola M S, Griffin D C, Bottcher C, Younger S M and Hunter H T 1987 *Nucl. Fusion Suppl.* Issue **1** 21
[14] Pindzola M S, Griffin D C, Bottcher C, Buie M J and Gregory D C 1991 *Physica Scripta* T **37** 35 (1991).
[15] Loch S D, Pindzola M S, Ballance C P, Griffin D C, Mitnik D M, Badnell N R, O'Mullane M G, Summers H P and Whiteford A D 2002 *Phys. Rev.* A **66** 052708
[16] S. D. Loch, M. S. Pindzola and D. C. Griffin, *Int. J. Mass. Spec.* **271**, 68 (2008).
[17] Loch S D, Ludlow A J, Pindzola M S, Whiteford A D and Griffin D C 2005 *Phys. Rev.* A **72** 052716

Journal of Physics: Conference Series **548** (2014) 012011 doi:10.1088/1742-6596/548/1/012011

Stark broadening by Lorentz fields in tokamak edge lasmas

J Rosato, G Seguineaud, Y Marandet, and R Stamm

Aix-Marseille Université, CNRS, PIIM UMR 7345, F-13397 Marseille Cedex 20, France

E-mail: joel.rosato@univ-amu.fr

Abstract.
We report calculations of hydrogen line shapes in tokamak edge plasma conditions. A special emphasis is put on the motional Stark effect that arises due to the electric field $\vec{v} \times \vec{B}$ present in the frame of reference of an atom moving at a velocity $\vec{v}$. The statistical repartition of the velocities results in a broadening of the lines, which can be significant if the upper level of the transition under consideration is high. Applications to Balmer lines are shown here.

1. Introduction

Line shape models are routinely used in magnetic fusion for diagnostic purposes, given the dependence of spectra on the plasma parameters. Hydrogen Balmer line series with a high principal quantum number n (up to 10 and higher) have been observed in divertor plasmas at conditions such that a large amount of excited atoms is present (recombining "detached plasma" regime) [1, 2, 3]. It is widely recognized that an analysis of the width of such lines yields information on the electron density, given their sensitivity to Stark broadening. However, as already noticed in [1], the electron density may not be sufficiently high so that other mechanisms contribute significantly to the line broadening. The effect discussed in this work concerns the Stark broadening due to the Lorentz field $\vec{F}_L = \vec{v} \times \vec{B}$ present in the emitters' frame of reference (motional Stark effect or MSE). Several works have reported on this effect in magnetized plasmas. Analytical expressions have been obtained for Lyman and Balmer hydrogen lines with a low upper principal quantum number n, in magnetic fusion conditions [4, 5], in high-density plasmas [6], and also in astrophysics [7, 8]. A rough estimate indicates that this broadening scales as $n^2 B\sqrt{T_{\text{at}}}$ (with T_{at} being a characteristic atomic temperature), meaning that this additional broadening can enter in competition with the other line broadening mechanisms for lines with a sufficiently high n, provided the magnetic field and the atomic temperature are sufficiently high. Note, this effect is to be distinguished from the MSE splitting involved in active spectroscopic diagnostics based on highly energetic neutral beams with a strongly shifted velocity distribution function (e.g. [9]). We report here on recent investigations of high-n Balmer line shapes [10] and perform new calculations of MSE broadening.

2. Line shapes in magnetized plasmas

The development hereafter follows the lines of [10]. Consider a set of atoms immersed in a magnetized plasma. The spectral profile of a line, $I(\omega, \vec{n})$, is given in terms of the line profile in

Journal of Physics: Conference Series **548** (2014) 012011 doi:10.1088/1742-6596/548/1/012011

the atom's frame of reference $I_0(\omega, \vec{n})$ by the following integral (Doppler convolution)

$$I(\omega, \vec{n}) = \int d^3v f(\vec{v}) I_0(\omega - \omega_0 \vec{v} \cdot \vec{n}/c, \vec{n}). \tag{1}$$

Here, ω_0 is the central angular frequency of the line under consideration, ω and $\vec{n}$ denote the frequency and the observation direction, and $f(\vec{v})$ is the atomic velocity distribution function (VDF). For Maxwellian VDF, the line shape is a Gaussian function if there is no broadening mechanism in the atom's frame of reference [i.e. when $I_0(\omega, \vec{n}) \equiv \delta(\omega - \omega_0)$] and a Voigt function if natural broadening is retained. In the general case, $I_0(\omega, \vec{n})$ is proportional to the Fourier transform of the atomic dipole autocorrelation function $C_{\vec{n}}(t)$

$$I_0(\omega, \vec{n}) \propto \frac{1}{\pi} \mathrm{Re} \int_0^\infty dt C_{\vec{n}}(t) e^{i\omega t}, \tag{2}$$

$$C_{\vec{n}}(t) = \mathrm{Tr}\left\{\vec{d}_\perp(0) \cdot \vec{d}_\perp(t) \rho\right\}. \tag{3}$$

Here, the trace is performed over the atomic states and denotes a statistical average, ρ is the projection of the density operator to the atom's Hilbert space evaluated at initial time, the brackets $\{...\}$ denote an average over the perturber trajectories (classical path approximation), and $\vec{d}_\perp = \vec{d} - (\vec{d} \cdot \vec{n})\vec{n}$ denotes the projection of the atomic dipole operator onto the polarization plane, in the Heisenberg picture. In Eq. (3), restrictions of this operator between the upper and lower levels of the transition are implied (no-quenching approximation). Formally, this means that a projector to the final states has to be included between $\vec{d}_\perp(0)$ and $\vec{d}_\perp(t)$. We do not write it here for the sake of simplicity. It is customary to write the autocorrelation function in terms of the evolution operator $U(t)$. Because of the identity $\vec{d}_\perp(t) \equiv U^\dagger(t)\vec{d}_\perp(0)U(t)$, a calculation of the line shape requires the knowledge of the matrix elements of $\vec{d}$ and an evaluation of $U(t)$. The evolution operator obeys the time-dependent Schrödinger equation

$$i\hbar \frac{dU}{dt}(t) = [H_0 + V(t)]U(t). \tag{4}$$

Here H_0 is the Hamiltonian including both the atomic energy level structure (with a non-Hermitian part accounting for natural broadening) and the eeman effect, and $V(t) = -\vec{d} \cdot \vec{F}(t)$ is the time-dependent Stark effect term (Schrödinger picture) resulting from the action of the microscopic electric field $\vec{F}(t)$. When this term is neglected, the Schrödinger equation has the trivial solution $U(t) = \exp(-iH_0 t/\hbar)$, which shows, using Eqs. (3) and (2), that $I_0(\omega, \vec{n})$ reduces to a set of delta functions (or Lorentzian functions if the natural broadening is retained). By contrast, the case where $\vec{F}(t)$ is significant is much more intricate because there is no general exact analytical solution. Several models, based on suitable approximations, have been developed in such a way to provide an analytical expression for the line shape (e.g., the impact and static approximations; the model microfield method; etc.). Fully numerical simulations can also be used.

In the presence of a magnetic field, the atom "feels" an electric field $\vec{F}_L = \vec{v} \times \vec{B}$ present in its frame of reference (Lorentz field), which results in an additional Stark effect term $-\vec{d} \cdot \vec{F}_L$ that must be retained in the Hamiltonian (motional Stark effect or MSE). The evolution operator, the autocorrelation function and thus the line profile I_0 depend on $\vec{v}$ [i.e. $I_0(\omega, \vec{n}) \equiv I_0(\omega, \vec{n}; \vec{v})$], so that the integral in Eq. (1) is no longer a convolution. A common approximation done in the case of a strongly shifted velocity distribution function (e.g., for an energetic quasi-monokinetic neutral beam) consists in evaluating the Lorentz field at the average velocity $\vec{v}_0$. This approximation has been used for diagnostic purposes in the interpretation of MSE spectra

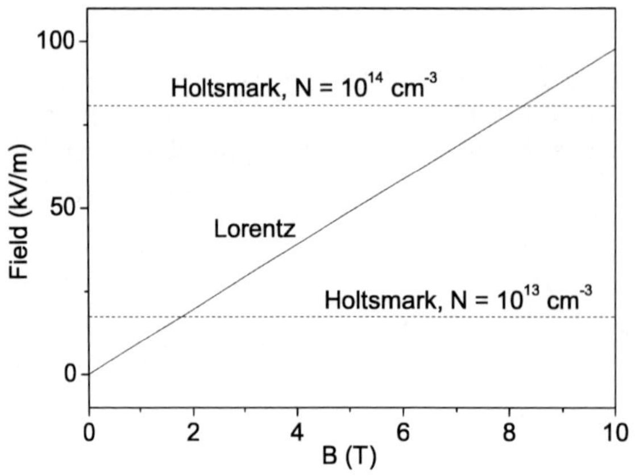

Figure 1. Plot of the thermal Lorentz field in terms of the magnetic field, assuming a Maxwellian gas of deuterium atoms at the temperature $T_{\text{at}} = 1$ eV. At low densities, the Lorentz field can be larger than the Holtsmark field, indicating that the TMSE can exceed the usual (microfield) Stark broadening.

in tokamaks [9]; such experiments consist in injecting a beam of neutrals, sufficiently energetic so that the MSE yields a splitting of the hydrogen lines that exceeds the other line broadening mechanisms. In a plasma with a thermal velocity distribution function (e.g., a Maxwellian function with $\vec{v}_0 = \vec{0}$), the MSE splitting due to the thermal motion of the emitters (which will be referred to as thermal MSE, or TMSE in the following) is not strong enough and the corresponding effect on spectra is a broadening of the line under consideration. This broadening can be significant with respect to other line broadening mechanisms (see Fig. 1).

3. Calculation of spectra

We have calculated profiles of Balmer lines with a high upper principal quantum number using the simulation method. In the plasma-microfield-Stark broadening model, the ions are described as particles moving along straight lines around the emitter, in a cube with periodic boundary conditions; the initial conditions (on the positions and velocities of the perturbers) are generated randomly; correlations are retained through the use of a Debye electric field model; and the broadening due to the electrons is described with a collision operator. The eeman effect and the TMSE are retained through their corresponding Hamiltonians. Fine structure corrections are not retained because they are negligible on lines with a high principal quantum number. The average over the atom's velocities Eq. (1), required for the Doppler and TMSE broadenings, is formed by random sampling using a Maxwellian VDF. Figure 2 shows a profile of the D_{12} line (Balmer 12 of deuterium) obtained assuming a plasma density of 10^{13} cm^{-3}, a value of 1 eV for the temperatures T_i, T_e, T_{at}, and a magnetic field of 8 T. Such plasma conditions were inferred from spectra analysis in a low-density discharge in Alcator C-Mod [1]. A line-of-sight perpendicular to the magnetic field has been assumed. The TMSE yields a significant additional broadening and more structure. In contrast, the Doppler broadening contribution is weak. This result suggests the TMSE should be accounted for in the interpretation of high-n line shapes.

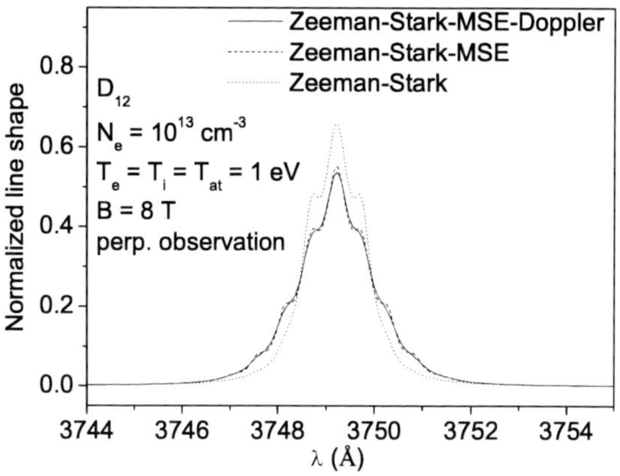

Figure 2. Profile of the D_{12} line calculated at conditions relevant to a plasma discharge reported in [1], retaining various line broadening mechanisms. The thermal MSE yields a significant additional broadening and more structure. The Doppler broadening contribution is weak.

4. Revisiting the Inglis-Teller formula in magnetized plasmas

The Inglis-Teller limit is the upper principal quantum number n of the last distinguishable line in a series before the lines merge into the continuum. A simple formula is provided by equating the half separation between two consecutive energy levels $\Delta E \simeq E_I/n^3$ (with E_I being the Rydberg ionization energy) to the Stark width $(3/2)n^2 ea_0 F_0$ associated with the plasma microfield F_0 (see details in [11]). sing the Holtsmark formula for the microfield yields the following relation

$$Nn^{15/2} \simeq 0.027a_0^{-3}, \tag{5}$$

which provides an estimate of the density N in terms of the upper principal quantum number n of the last line. Such a procedure, with a relation similar to Eq. (5) (apart from a multiplicative factor in the right-hand side [12]), was followed by the authors in [1] to estimate the plasma density. The Inglis-Teller formula should be reconsidered in the case where a magnetic field is present, because the TMSE can dominate the plasma-microfield-Stark broadening. Replacing F_0 by the Lorentz field F_L estimated at the thermal velocity yields a relation that does not contain the density, but the magnetic field-strength and the atomic temperature instead

$$n^{10}B^2T_{\rm at} \simeq 1.5 \times 10^{14}\frac{m_{\rm at}}{m_p}. \tag{6}$$

Here the temperature is expressed in eV and the other quantities are in SI units. This expression could serve as a diagnostic of $T_{\rm at}$. For example, applying it to the low-density discharge reported in [1] with $n > 16$ and $B = 8$ T yields a temperature smaller than 5 eV, which is in the range expected from the Doppler analysis performed in [1]. Note, the eeman effect, not considered here, can be significant with respect to the TMSE. The derivation of a more accurate formula, which accounts for this effect, could be done along the lines of the analytical model reported in [10].

5. Conclusion

We have presented an investigation of atomic line shapes broadened due to the thermal motional Stark effect (TMSE) at conditions relevant to magnetic fusion experiments, following a recent work [10]. If the magnetic field is sufficiently strong, the TMSE broadening enters in competition with the usual (Doppler, eeman, and plasma-microfield-Stark) broadening mechanisms considered in diagnostic models. We have confirmed this by numerical simulations of Balmer lines with a high principal quantum number. An adaptation of the Inglis-Teller formula to the Lorentz field has also been considered and it has been shown that it also provides the correct order of magnitude for the temperature. The analysis of high-n hydrogen spectra can be used to reinforce the usual diagnostics for the atomic temperature that utilize Doppler-broadened lines. Our work has focused on the plasma discharges in Alcator C-Mod reported in [1] because the high-n Balmer lines observed in this machine and the strong magnetic field present therein are also expected in the ITER divertor (detached plasma regime). The diamagnetic quadratic eeman effect, not considered in this paper, may be significant on high-n lines (e.g. [13]) and should be investigated in details in a future work.

Acknowledgments

This work was carried out within the framework of the European Fusion Development Agreement and the French Research Federation for Fusion Studies. It is supported by the European Communities under the contract of Association between Euratom and CEA. The views and opinions expressed herein do not necessarily re ect those of the European Commission. Financial support was also received from the Agence Nationale de la Recherche, project ANR-11-BS09-023 (SEDIBA).

References

[1] Welch B L, Griem H R, Terry J, Kurz C, LaBombard B, Lipschultz B, Marmar E and McCracken G 1995 *Phys. Plasmas* **5** 4246
[2] Wenzel U, Behringer K, Carlson A, Gafert J, Napiontek B and Thoma A 1999 *Nucl. Fusion* **39** 873
[3] Koubiti M, Loch S, Capes H, Godbert-Mouret L, Marandet Y, Meigs A, Stamm R and Summers H 2003 *J. Quant. Spectrosc. Radiat. Transfer* **81** 265
[4] Isler R C 1976 *Phys. Rev. A* **14** 1015
[5] Breton C, Michelis C D, Finkenthal M and Mattioli M 1980 *J. Phys. B: At. Mol. Phys.* **13** 1703
[6] Hoe N, Grumberg J, Caby M, Leboucher E and Coulaud G 1981 *Phys. Rev. A* **24** 438
[7] Brillant S, Mathys G and Stehlé C 1998 *Astron. Astrophys.* **339** 286
[8] Stehlé C, Brillant S and Mathys G 2000 *Eur. Phys. J. D* **11** 491
[9] Donné A J H *et al.* 2007 *Nucl. Fusion* **47** S337
[10] Rosato J, Marandet Y and Stamm R 2014 *J. Phys. B: At. Mol. Opt. Phys.* **47** 105702
[11] Inglis D R and Teller E 1939 *Astrophys. J.* **90** 439
[12] Griem H R 1964 *Plasma Spectroscopy* (McGraw-Hill)
[13] Hey J D, Chu C C and Mertens P 2002 *AIP Conf. Proc.* **645** 26

Parameterization of Balmer-alpha asymmetric line shape in tokamak SOL plasmas

AB Kukushkin[1,2], VS Neverov[1], MB Kadomtsev[1], V Kotov[3], AS Kukushkin[4], MG Levashova[1], SW Lisgo[4], VS Lisitsa[1], VA Shurygin[1], AG Alekseev[1]

[1] NRC "Kurchatov Institute", Moscow, 123182, Russia
[2] National Research Nuclear University "MEPhI", Moscow, 115409, Russia
[3] Forschungszentrum Jülich, Euratom Association, Jülich, 52425, Germany
[4] ITER Organization, Route de Vinon sur Verdon, 13115 St Paul Lez Durance, France

E-mail: Kukushkin_AB@nrcki.ru

Abstract. A parameterization of the Balmer-alpha spectral line shape asymmetry in the tokamak scrape-off layer (SOL) plasmas is suggested, which describes the contribution of non-Maxwellian components of the neutral atom velocity distribution function. Parameterization is needed for a fast-routine interpretation of high-resolution spectroscopy data and should be incorporated into the algorithms for the recovery of hydrogen neutral atom parameters in the SOL. We illustrate the efficiency of the parameterization on the example of spectral data calculated using the predictive modeling of the International Thermonuclear Experimental Reactor (ITER) tokamak operation.

1. Introduction

The asymmetry of the Balmer-alpha spectral line shape [1] is observed in tokamak experiments and simulated with kinetic codes (see, e.g., [2-6] and references therein). The asymmetry is manifested in the broad wings of the spectral lines and is thought to be caused by the charge-exchange neutrals and reflection of fast neutral atoms from the first wall of the vacuum chamber. These processes produce a strong non-Maxwellian component of the velocity distribution function (VDF) of hydrogen neutral atoms. A semi-analytic one-dimensional model [7,8] for neutral atom VDF in the scrape-off layer (SOL), based on the dominance of ballistic flights of neutral atoms, has shown good agreement with the EIRENE code stand-alone (Monte-Carlo) simulations of neutral deuterium VDF, applied on the divertor&SOL plasma background calculated for the International Thermonuclear Experimental Reactor (ITER) tokamak by the SOLPS4.3 (B2-EIRENE) code [9-11].

An example of a strong asymmetry may be found in the spectroscopic data from the recent experiments on the JET tokamak with the ITER-like wall (JET-ILW). An enhanced blue wing of the line shape of the central Zeeman component of the deuterium Balmer-alpha line was observed at the line-of-sight directed along major toroidal radius at the limiter stage of discharge when the toroidal plasma is attached to the inner wall limiters and the Balmer-alpha signal comes from the inner-wall SOL (see Figure 1(left) in Ref. [12]). It appears that such a wing of the line shape cannot be fitted with a set of exponential functions (Gaussians), normally used to fit the Doppler-broadened lines in the SOL plasmas, because the wing has a power-law slope.

Here we suggest a parameterization of the Balmer-alpha spectral line shape asymmetry in the tokamak SOL plasmas, which is based on the success of the model [7] and describes the contribution of non-Maxwellian components of the VDF in terms of major kinetic processes responsible for this effect. Such a parameterization is needed for a fast-routine interpretation of the Balmer-alpha high-resolution spectroscopy (HRS) data and should be incorporated into the algorithms for the recovery of hydrogen neutral atom parameters in the SOL. The parameterization has been used for analyzing the HRS D-alpha data in JET-ILW [12]. Here we show the efficiency of the parameterization on the example of spectral data [13] calculated using the SOLPS4.3 (B2-EIRENE) code predictive modeling of ITER operation, with account of the poloidally resolved plasma recycling from the first wall in the frame of the "extended grid" [14].

2. A model of neutral atom velocity distribution function

The asymmetry of the Balmer-alpha spectral line shape in the tokamak SOL plasmas originates from the substantial non-locality of particle transport. In contrast to divertor, here the mean-free-path of neutral hydrogen atom is large (see the Ballistic Model [7]) because of (i) the high rate of the resonance charge exchange of relatively cold first-generation atoms (of few-several eV temperature near the wall) with hot ions of background plasma in the SOL (of temperature up to few-several hundreds of eV in the region of the essential observed emissivity of Balmer-alpha line) and (ii) the low plasma density in the SOL. In the tokamak SOL in a strong magnetic field, the Balmer-alpha line shape is determined by the Zeeman splitting and the Doppler broadening of the Zeeman components. Under these conditions, the asymmetry of the line shape of each Zeeman component is caused by the asymmetry of inward-outward fluxes of neutrals, which is dominated by the moderate- and high-energy fractions of the neutral atom VDF. The asymmetry of the VDF is substantially enhanced by the reflection of fast atoms from the wall. In this section we suggest a simple model for the VDF and test it on the example of the data from the EIRENE code stand-alone (Monte-Carlo) simulations of neutral deuterium VDF, applied on the divertor&SOL plasma background calculated for tokamak ITER by the SOLPS4.3 (B2-EIRENE) code.

We consider the following model for the one-dimensional VDF in velocity v and space coordinate x in the direction perpendicular to the wall:

$$F_{VDF}(x,v) = \sum_{m=1}^{M1} n_M^m(x) F_{Gauss}(v, T_M^m(x)) + \sum_{n=1}^{M2} n_N^n(x) F_{Gauss}(v, T_N^n(x)) D_v(v,x)\eta(v), \quad (1)$$

where F_{Gauss} is the Maxwellian distribution (normalized in the velocity space); $M1$ and $M2$ are the number of, respectively, Maxwellian and non-Maxwellian fractions of the VDF; n_M^m and n_N^n are the particle density, respectively, of the m-th Maxwellian and the n-th non-Maxwellian fractions of the VDF,

$$D_v(v,x) = \exp\left(-v_{damp}(x)/v\right), \quad (2)$$

where v_{damp} is the characteristic velocity in the damping factor D_v which describes the attenuation of the inward flux, $\eta(x)$ is the Heaviside function, the x-axis is directed inward to plasma from the wall. The characteristic velocity should allow for the fact that in the SOL the inward flux of neutral atoms is attenuated mainly due to (i) ionization of neutral atoms by electron impact and (ii) their charge exchange with ions [7]:

$$v_{damp}(x) \sim \int_0^x \left[n_e(\tilde{x}) \langle \sigma_{ai} v_e \rangle (\tilde{x}) + n_i(\tilde{x}) \langle \sigma_{cx} v_i \rangle (\tilde{x}) \right] d\tilde{x}, \quad (3)$$

where n_e and n_i are the densities of, respectively, electrons and ions; σ_{ai} is the cross section of ionization; σ_{cx} is the cross-section of the charge exchange; v_e and v_i are velocities of, respectively, electrons and ions; brackets denote averaging over electron or ion Maxwellian distribution.

A comparison of the characteristic velocity v_{damp} (3) with the respective value recovered from the fitting of the EIRENE code simulation data [15] with the model VDF (1) is given in Figure 1, for two cases, namely low (the case “d” = “Low density far-SOL, L-mode pedestal” in notations of Ref. [15]) and high density of neutral deuterium (the case “i" = “High density far-SOL, H-mode pedestal”), within the case ITER SOL&divertor#1514 for background plasma (cf. [9]).

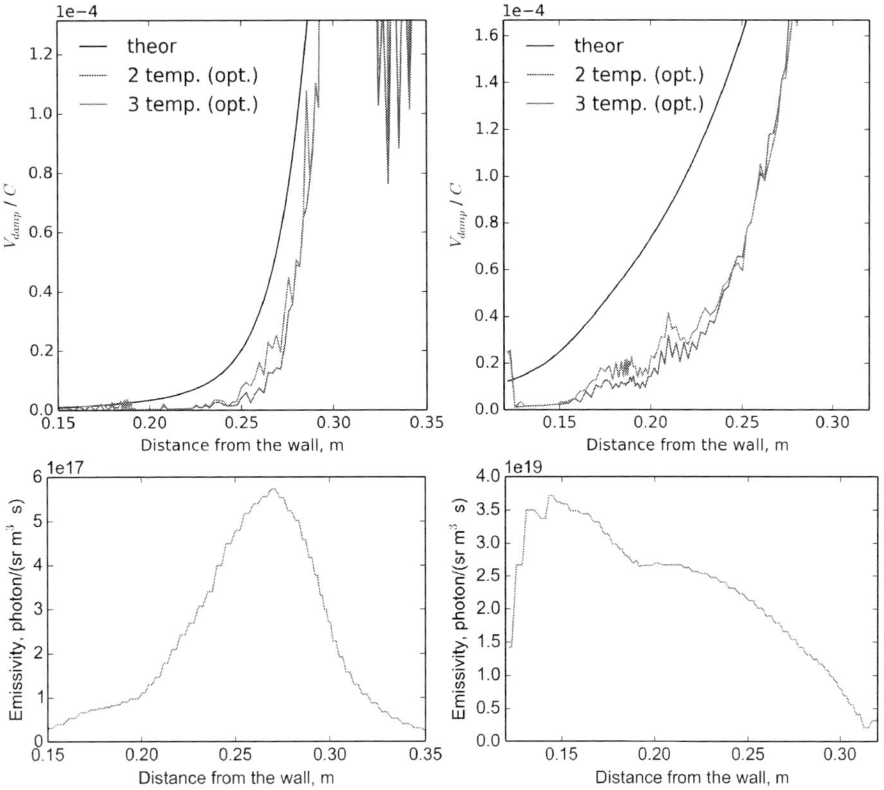

Figure 1. A comparison of v_{dump} (3), divided by the speed of light, (black curve) with the value recovered from the fitting of the EIRENE code simulation data with the model VDF in Eq. (1) in the case of low (upper left) and high density in the SOL (upper right) The results of fitting are given for the model (1) with two-temperature Maxwellian and one-temperature non-Maxwellian, $M1$=2, $M2$=1 (blue) and for the model (1) with $M1$=3, $M2$=2 (green). The respective profiles of Balmer-alpha line emissivity, in photons/(sr m^3 s), are shown in the bottom figures.

Note that the recovery is not good in the region of low emissivity: here the contribution to emissivity comes from fast particles whose Monte-Carlo modeling statistics is poor. Figure 1 suggests that the simple model (1), (2) with the v_{damp} taken as a free parameter may be used in the fitting of the VDF obtained by a numerical modeling of such a complicated kinetic problem. The success of the respective fitting is shown in Ref. [16].

3. A model of Balmer-alpha spectral line shape

The results of the above comparison enable us to suggest a simple model for the line shape of the observed signal because the latter integrates the local VDF data over a line-of-sight. This should give a reasonable parameterization of an asymmetric line shape in the SOL. We suggest the following model for the spectrum of the central (π-) Zeeman component:

$$\begin{aligned} S^{\pi}_{SOL}(\lambda) &= \sum_{m=1}^{M1} S^{m}_{M} F_{Gauss}\left(\lambda-\lambda_{D\alpha}, T^{m}_{M}\right) \\ &+ \sum_{n=1}^{M2} S^{n}_{N} F_{Gauss}\left(\lambda-\lambda_{D\alpha}, T^{n}_{N}\right) D_{\lambda}\left(\lambda-\lambda_{D\alpha}, \Lambda^{n}\right) \eta\left(\left(\lambda-\lambda_{D\alpha}\right)(\mathbf{k},\mathbf{l})\right), \end{aligned} \tag{4}$$

where F_{Gauss} is the Gaussian line shape (normalized in the wavelength space)); S^{m}_{M} and S^{n}_{N} are statistical weights of the contribution, respectively, from the *m*-th Maxwellian and the *n*-th non-Maxwellian fractions of the VDF. The damping factor D_λ in the line shape is derived from the damping factor (2) in the VDF (1):

$$D_{\lambda}(\Delta\lambda, \Lambda) = \exp\left(-\frac{\Lambda}{|\Delta\lambda|}\right). \tag{5}$$

The values Λ^{n} in Eq. (4) are the free parameters which have to be determined by the fitting the experimental line shape. The argument of the Heaviside function allows for mutual direction of the line-of-sight, defined by the unit vector **l** from pupil to plasma, and the direction of atom inward flux in the SOL, defined by the unit vector **k** either at inner or outer part of the inner wall of the toroidal vacuum chamber. The total contribution of all Zeeman components takes the form:

$$S_{SOL}(\lambda) = C_{\pi} S^{\pi}_{SOL}(\lambda) + \frac{1-C_{\pi}}{2}\left(S^{\pi}_{SOL}\left(\lambda-\Delta\lambda^{Zeem}\right) + S^{\pi}_{SOL}\left(\lambda+\Delta\lambda^{Zeem}\right)\right), \tag{6}$$

where $\Delta\lambda^{Zeem}$ is the Zeeman splitting for magnetic field in a rather thin layer of essential emissivity in the SOL (see Figure 1, bottom figures). The factor $C_{\pi} = ½ \sin^{2}\theta$ is determined by the angle θ between the line-of-sight and the above-mentioned magnetic field.

The first application of the above model has been made to interpret the data from tokamak JET-ILW experiments at the limiter stage of discharge (see Figure 1(left) in Ref. [12]). Here we illustrate the fitting of the line shapes for the case of ITER predictive modeling data [15], [9] used in Figure 1.

The test of the above model can be made only in the frame of a "synthetic" diagnostic. Such a diagnostic generates "phantom" experimental data from the results of predictive numerical modeling of plasma parameters and allows direct comparison of the pristine (i.e. taken as known) and the recovered values of main parameters. The estimation of accuracy cannot be made using experimental data only because no diagnostic can directly measure the distribution of neutral atoms in velocity and space coordinates. The data [15], [9] enabled us to compare the results for temperatures (for non-Maxwellian fractions of the VDF, the effective ones, see the second term in the right hand side of Eq. (1)) averaged over the line-of-sight in the local kinetic model with those recovered from an algorithm of interpretation the observed signal, which originally operates with parameters averaged over the line-of-sight. The respective results are presented in Ref. [12], and in more detail in Ref. [16], they show good enough agreement of assumed and recovered values of temperatures.

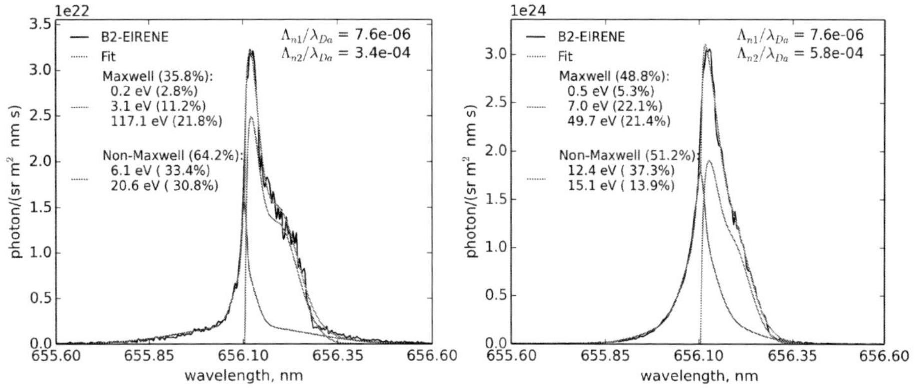

Figure 2. The results of fitting the deuterium Balmer-alpha line shape with the model (4) for the cases defined in Figure 1. The temperatures and relative particle densities of various fractions of neutral atoms, as well as the values of Λ in the damping factor (5), are indicated.

4. Conclusions

The models (1) and (2) for neutral hydrogen velocity distribution function (VDF) with the ν_{damp} taken as a free parameter may be used for fitting the VDF obtained with the numerical modeling (the success of such a fitting is shown in Ref. [16]). The models (4) and (5) for the Balmer-alpha line shape may be effective in fitting the spectral intensity of Balmer-alpha emission in the tokamak SOL plasmas under condition of a strong spectral asymmetry not treatable with the standard (set-of-Gaussians) model. The parameterization of the line shape is tested on the data of predictive modeling of ITER operation.

Acknowledgements

The authors are grateful to K.Yu. Vukolov, E. Veshchev, M. von Hellermann, M.F. Stamp, and S. Brezinsek for collaborations in the research associated with the ITER H-alpha (and visible light) spectroscopic diagnostic. The present research effort is carried out to support the work under the contract with the RF State Corporation ROSATOM (№ H.4k.52.9B.14.1002 of 31.12.2013).

The views and opinions expressed herein do not necessarily reflect those of the ITER Organization.

References

[1] Mirnov S V, Semenov I B 1970 *At. Energ.* **28** 129
[2] Reiter D, *et al.* 1992 *J. Nucl. Mat.* **196-198** 1059
[3] Stotler D P, *et al.* 1996 *Phys. Plas.* **3** 4084
[4] Koubiti M, *et al.* 2002 *PPCF* **44** 261
[5] Marandet Y, *et al.* 2004 *Nucl. Fusion* **44** S118
[6] Brezinsek S, *et al.* 2005 *PPCF* **47** 615
[7] Kadomtsev M B, *et al.* 2012 *Proc. 39th EPS Conf. (Stockholm, Sweden)* P4.093
[8] Kadomtsev M B, *et al.* 2013 *Proc. 40th EPS Conf. (Espoo, Finland)* P1.135
[9] Kukushkin A S, Pacher H D, Loarte A, Kotov V, *et al.* 2009 *Nucl. Fusion* **49** 075008
[10] Braams B J 1986 *PhD thesis* (Utrecht: Rijksuniversitet)
[11] Reiter D, Baelmans M, Boerner P 2005 *Fusion Sci. Tech.* **47** 172
[12] Kukushkin A B, Neverov V S, Stamp M F, *et al.* 2013 *JET Preprint* EFDA–JET–CP(13)04/05
[13] Kukushkin A B, *et al.* 2012 *Proc. 24th IAEA FEC (San Diego, USA)* ITR/P5-44
[14] Lisgo S W, Börner P, *et al.* 2011 *J. Nucl. Mater.* **415** S965
[15] Lisgo S W 2012 *Private communication*
[16] Neverov V S *et al.* 2014 Asymmetry of Balmer-alpha line shapes and reconstruction of effective temperatures of hydrogen in tokamak edge plasmas *Plasma Phys. Reports* (Accepted)

Journal of Physics: Conference Series **548** (2014) 012013 doi:10.1088/1742-6596/548/1/012013

Pressure Broadening of the Cadmium 326.1 nm Line

G D Roston and M S Helmi

Department of Physics, Faculty of Science, Alexandria University, P.O. Box 21511, Alexandria, Egypt

Dr.gamal_daniel@yahoo.com

Abstract. The temperature dependence of the Cd line absorption profile at 326.1nm perturbed by inert gases (Xe, Kr, Ar, Ne and He) has been carefully studied over a wide spectral range in both blue and red wings using a high-resolution double-beam spectrometer. The atomic densities of inert gases (N_{gas}) and cadmium (N_{Cd}) was sufficient to study the wing of the Cd line at 326.1nm. The temperature dependence of the studied line profile was analyzed in the framework of the quasi-static theory. The van der Waals coefficient differences (ΔC_6^0 and ΔC_6^1) between the ground $X0^+$ state and the two excited states A^30^+ and B^31 were obtained from the near red wing profile using Kuhn's law. All the results of the well depths with their positions for the ground ($X0^+$), and the excited (^{31}, $^30^+$) were determined. The obtained results are compared with the corresponding theoretical and experimental molecular beam experiments results.

1. Introduction

The Cd – Cd and Cd – inert gas systems are considered to be a class of excimer molecules having shallow ground states and several much more strongly bound excited states. Recently, there have been numerous theoretical and experimental spectroscopic determinations of interatomic potentials for the ground X^10^+ and the two excited states A^30^+ and B^31 of these systems [1-38]. This paper is mainly devoted to obtain the interaction potentials of the ground state (XO^+) and both excited states A^30^+ and B^31 from the temperature dependence of the profile of Cd line 326.1 nm broadened by inert gases (Xe, Kr, Ar , Ne and He) pressure. These profiles are assigned to the transition ($A \rightarrow X0^+$ and $B \rightarrow X0^+$) this was illustrated in Refs. [1-7].

2. Theoretical aspects

2.1 The quasi-static theory

The Unified Frank-Condon (UFC) treatment of pressure broadening of spectral lines developed by Szudy and Baylis [39] reduces in the appropriate limits to the quasi- static theory in the far wings of the line. Using this theory, the reduced absorption coefficient (K_n^m ($\Delta\nu$, T) cm^5) in the line wings as a consequence of two particle interactions normalized to the perturbing (N_{gas}) and the radiating (N_{cd}) atom densities is given by

$$K_n^m(\Delta\nu,T)=\frac{k^m(\Delta\nu,T)(cm^{-1})}{N_{Cd}N_{gas}}=4\pi C_m R^2\left|\frac{dR}{d(\Delta V_m)}\right|\exp\frac{-\nu_g(R)}{KT}, \qquad (1)$$

where the symbol m = 0 and 1 are assigned to the $^3 0$ and $^3 1$ excited states respectively, R is the internuclear separation, ΔV_m (R) is the potential difference between the excited $V_m(R)$ and the ground $V_g(R)$ state potentials, given by $\Delta V_m (R) = V_m (R) - V_g (R) = h\,\Delta\nu$ and $\Delta\nu = \nu_o - \nu$ is the frequency separation from the unperturbed line frequency ν_o . C_m is the constant which relates the absorption coefficient to the normalized emission intensity. It is given by:

$$C_m = (\pi e^2 /m\, c)\, (g_m/g_j)\, f_{eg}$$

where e and m are the electron charge and mass, c is speed of light, gj and gm are the atomic and molecular statistical weights (g_m = 1 or 2 for m = 0 or 1) and f_{eg} is the oscillator strength for this atomic transition.

2.2 Determination of the van der Waals Coefficients.

The part of wings near to the line center is arising as a result of electronic transition between the states which are only slightly perturbed by the long-range van der Waals interactions of the form $\Delta V(R) = \Delta C_6 R^{-6}$. In this range of interaction , $V_g(R) \ll kT$, then Eq.(1) leads to Kuhn`s law as:

$$K_n^m(\Delta\nu) = \frac{2\pi C_m}{3} h^{-3/2} (\Delta C_6^m)^{1/2} \Delta\nu^{-3/2} = A_m (\Delta\nu)^{-\frac{3}{2}} \qquad (2)$$

If log K_n^m ($\Delta\nu$) is plotted versus log ($\Delta\nu$), a straight line results with a slope of (-3/2). The ΔC_6^m constant may then be obtained from the calculation of the coefficient Am. Generally for Cd – rare gases such a linear dependence on log K_n^m ($\Delta\nu$) is observed in two regions of the near red wing profile for more detailed see [2-6, 12].

2.3 Determination of potentials from line shape data

A more general method of inversion for the relation between the spectral line profile and $\Delta V_m(R)$ has been developed by Behmenburg [40] and Gallagher [41]. In this method the condition that $R(\Delta\nu)$ is a single-valued function and no overlapping bands ought to be fulfilled, then the spectral line profile can be interpreted in the framework of the quasi- static theory. From the temperature dependence spectra K_n^m ($\Delta\nu$, T) of Eq.(1), it is possible to deduce both $V_g(R)$ and $\Delta V_m(R)$ and hence the excited state potentials from line shape data. The plot between log K_n^m ($\Delta\nu$,T) and 1000/T for a fixed $\Delta\nu$ gives a straight line. The slope of this plot gives $V_g(\Delta\nu)$ and extrapolation to the infinite temperature gives K_n^m ($\Delta\nu$,T $\to\infty$). for more detailed see [2-6, 12].

3. Experimental aspects

The apparatus for this experiment is identical to that described elsewhere [1-8, 12]. A high pressure XBO 150 xenon lamp was used as a background source for absorption. In this case two identical absorption quartz cells were used, a sample cell filled with Cd + inert gas mixture and an empty reference cell. They were placed in a special double cell oven which could operate at a temperature up to 1300 K. The contributions due to the optical system were canceled in view of the double beam method being used. The atomic densities of inert gas N_{gas} and cadmium N_{Cd} were taken as those in [1-9]. The vapor pressures were calculated using Bousquet`s formula [11].

4. Results and discussion

4.1 Blue wing profile of the Cd line 326.1 nm.

Due to large numbers of graphs of Cd –inert gases, sometimes the Cd – Ar graphs are presented as examples. Fig.(1) illustrates the experimental blue wing profile resulting from Cd –inert gases and the (Cd-Cd) interactions [1-8] at the same cadmium density. It is seen from this figure that the contributions from Cd-Cd is very small in the range of frequencies $20 \leq \Delta\nu \leq 200$ cm^{-1}. For frequencies $\Delta\nu \geq 200$ cm^{-1} the absorption is mainly attributed to the Cd-Cd system. Consequently, the absorption profile of the experimental Cd-inert gas was corrected for Cd-Cd contributions in the range of frequencies $20 \leq \Delta v \leq 200$ cm^{-1} and was tentatively analyzed for pure Cd-gas up to 200 cm^{-1} [2-8].

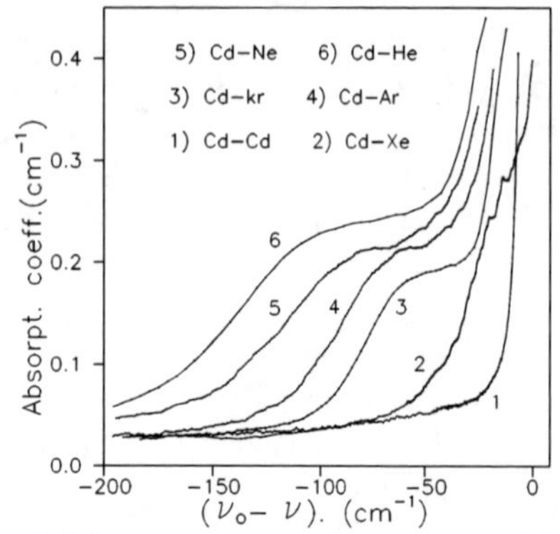

Figure 1: The experimental blue wing profile for pure Cd – Cd and Cd -inert gases interactions.

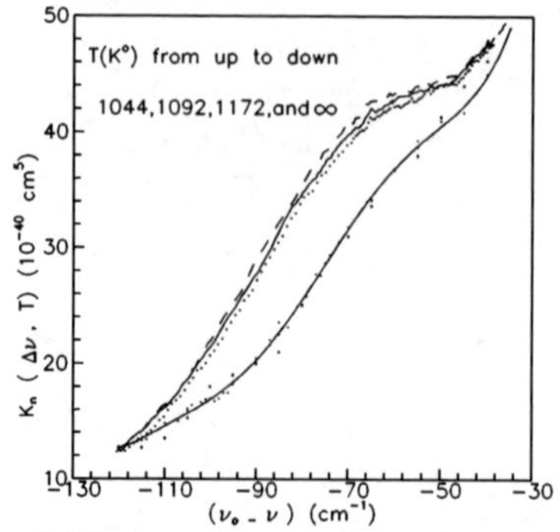

Figure 2:The temperature dependence for the blue wing of the pure Cd-Ar profile.

It is also seen from this figure that the profile has a quadratic form in the near wing which emphasizes that the profile is Lorentzian in this region. The existence of a broad maximum (blue satellite) is attributed to a maximum in the potential difference ΔV1 = V(B31) –V(X0+). The temperature dependence of the pure Cd-Ar profile (as an example of Cd–inert gases) is illustrated in Fig.(2). According to the UFC theory developed by Szudy and Baylis [39], the profile in the region of the satellite can be interpreted on the frame work of the quasi- static theory [Eq.(1)] if it is multiplied by (T)1/6 [2-8]. It is seen from Fig. (2), that the ground state potential Vg = 0, is at the intersection point of the profiles (Δv ≈ 125 cm-1) for Cd – Ar interaction.

Figure (3) shows the ln Kmn (Δv,T) values plotted against 1000/T for some values of Δv. Within the error bounds, the linear relation is well observed. The slopes of the calculated least- squares lines give the ground state potential Vg(Δv) as a function of Δv. This is shown in Fig.(4). The infinite temperature intercepts determine the Kmn (Δv,T → ∞) values [see Fig.(2)].

The relation between Δv1 and the internuclear separation R can be obtained using the integration (6). For Cd-Ar, the starting point of this integration Ra was taken as the position of the ground state potential well depth deduced by [8] (Ra = 4.3 Å). The corresponding frequency separation, Δva ≈72 cm-1, is taken from Fig.(4).

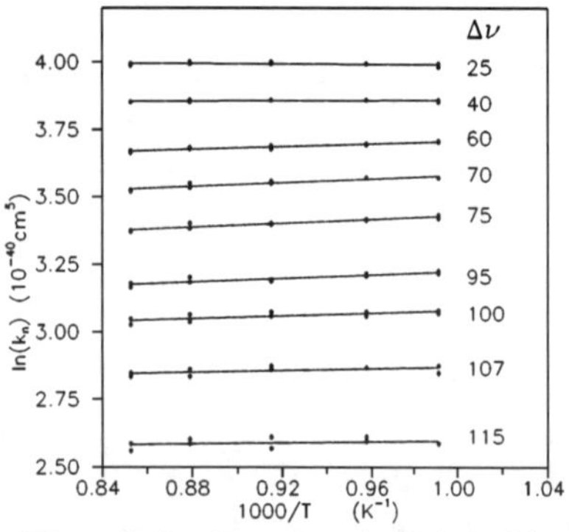

Fig. 3: The relation between the Cd-Ar blue wing profile lnk(Δv) against 1000/T for different values of Δv (cm-1).

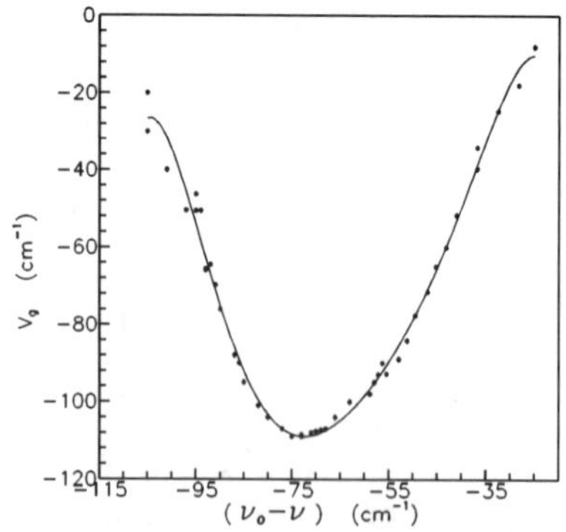

Fig. 4: The Vg potential plotted against Δv, experimental results(**) with its fitting curve (—).

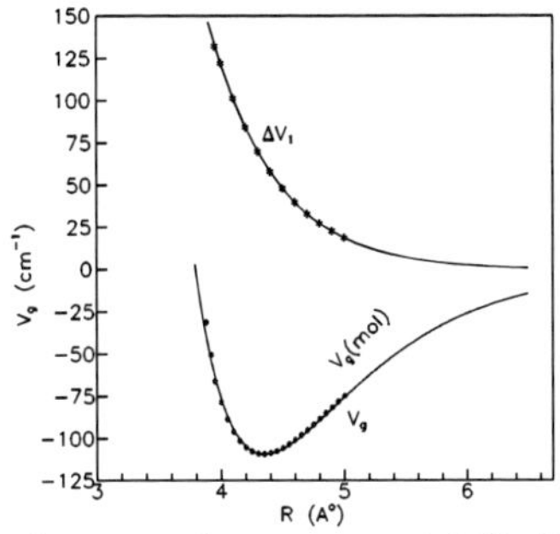

Figure 5: The ground state potential Vg(R), and the potential difference ΔV1(R) for cd-Ar system.

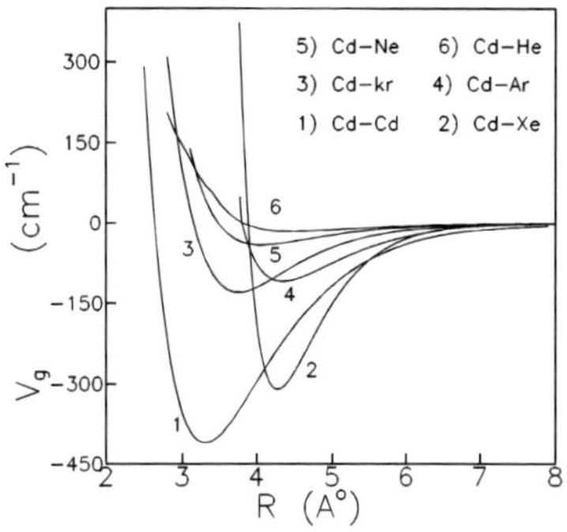

Figure 6: The ground state potential Vg(R), for pure Cd – Cd and Cd inert gases system.

Using these values (R_a = 4.3 Å), $\Delta v_a \approx 72$ cm^{-1}) and the absorption profile at T$\rightarrow \infty$ (K_m^n (Δv, T$\rightarrow \infty$)) from Fig.(2), then the potential difference $\Delta V_1(R)$ can be obtained, then it is fitted to the polynomial function with a proper parameters as shown in Fig. (5). As $\Delta V_1(R)$ is obtained, the ground state V_g as a function the internuclear separation R can be recalculated, then $v_g(R)$ is fitted to the Morse potential as showing in Fig. (5). Then the potential curve of the excited state $V_1(R)$ [$V_1(R) = V_g(R) + \Delta V_1(R)$] can be calculated. Fig. (6, 7) shows ground V_g (R) and the excited $V_1(R)$ states for pure Cd – Cd and Cd inert gases.

4.2 Red wing profile of the Cd line 326.1 nm.

The red wing profile of the Cd 326.1 nm line broadened by inert gases [2, 8], compared with the self broadened line profile obtained from [1], in logarithmic scale is illustrated in Fig.(8). In this figure one can see a single satellite at about 10 cm^{-1}. In the vicinity of this band two linear regions with slope (- 3/2) are observed for the (Cd-gas) system. The first region lies between 2 and 10 cm^{-1} from the line center may be interpreted as the quasi-static line wing which is formed by transitions between the ground state $^10^+$ and excited states $^30^+$ as well as ^{31}, while the second region lies between 18 and 100 cm^{-1} is attributed to the transition between the ground X^10^+ and the excited ($^30^+$) states only. The values of van der Waal's coefficient difference ΔC_6^o and ΔC_6^1 for the two transition ($A^30^+ \rightarrow X^10^+$) and ($B^31 \rightarrow X^10^+$) are obtained respectively. These results are illustrated in Table I.

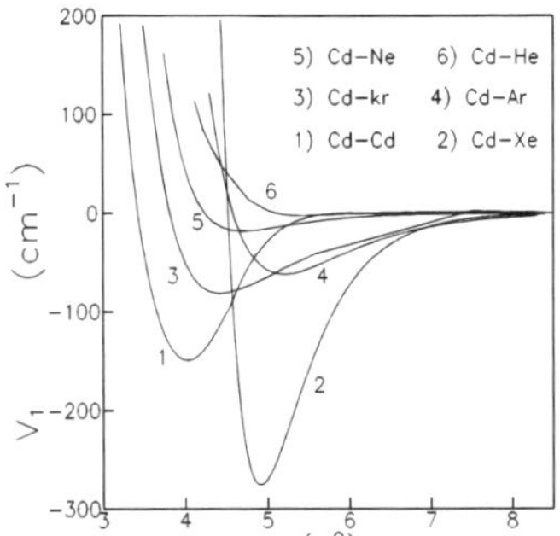

Figure 7: The excited state potential V1(R), for pure Cd – Cd and Cd inert gases.

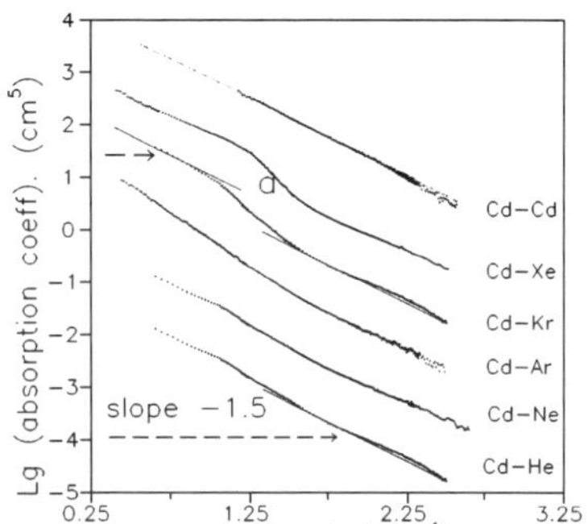

Figure 8: Red wing profiles for Cd-Cd and Cd-inert gases in logarithmic scale (10-40 cm5). The Cd-Xe, Cd-Kr, Cd-Ar, Cd-Ne and Cd-He profiles are divided by 10, 102, 103, 104, 105. respectively

Table 1. The van der Waals coefficients in eV Å6 for the Cd-inert gases system

Molecule	Cd – Cd	Cd - Xe	Cd – Kr	Cd - Ar	Cd - Ne	Cd - He
ΔC_6^o	-----	43.0 ± 3	37.8 ± 2	28.7 ± 2	23.5 ± 2	14.8 ± 2
ΔC_6^1	106.8 ± 2.3	86.0 ± 4	58.5 ± 3	48.0 ± 4	39.3 ± 3	28.5 ± 3

The potential $V_o(R)$ for the state ($^30^+$) can be obtained from the temperature dependence of the red wing profile. As the temperature dependence of this wing is very weak, then the ground state potential $V_g(R)$ can't be determined from the red wing profile. Nevertheless in this case the excited state potential $V_o(R)$ can be obtained with sufficient accuracy, if $V_g(R)$ is known from other data. By using the ground state potential $V_g(R)$ obtained from the blue wing and the red wing profile at any temperature T (K_m^n (Δv,T)), it was possible to obtain the potential difference $\Delta V_o(R)$ if the starting point R_a (Δv_a) in the integration (8) is established. This point was obtained due to the fact that Kuhn's law is observed in the near red wing at $20 < \Delta v < 100$ cm^{-1} and the van der Waals potential difference $\Delta V_a = \Delta C_6^o R_a^{-6}$ can be applied. $\Delta V_o(R)$ with the used ground state are presented in Fig.(9). Fig. (10) shows the determined $V_o(R)$ for pure Cd – Cd and Cd -inert gases interactions.

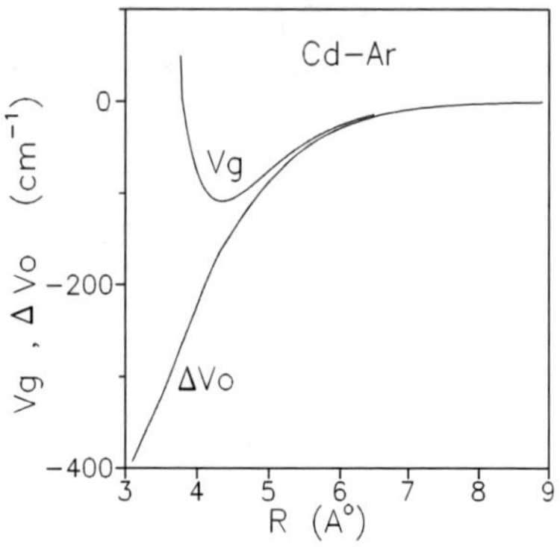

Figure 9: The ground state potential $V_g(R)$, and the potential difference $\Delta V_0(R)$ for Cd-Ar.

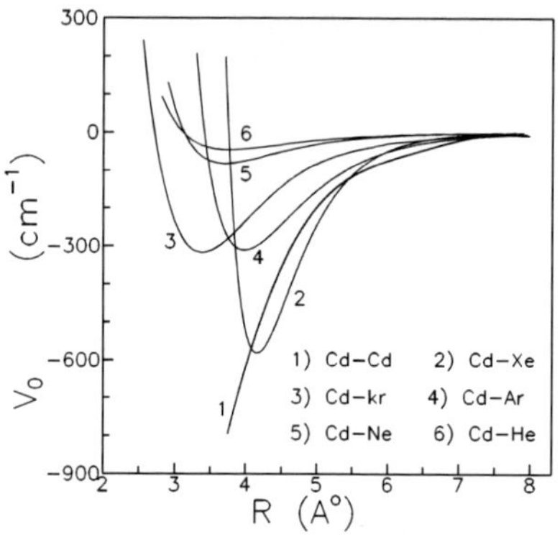

Figure10: The excited state potential $V_0(R)$, for pure Cd – Cd and Cd inert gases.

The potential parameters , well depth ε (cm^{-1}) and its positions R (Å) of the ground X^10^+ and the two excited ^{31} and $^30^+$ states for pure Cd - Cd and Cd – Inert gases (Xe, Kr, Ar, Ne and He) are illustrated in tables II and III.

Table II: Comparison of our data for the ground X10g+ and the two excited states $A^30_u^+$ and B^31_u potentials parameters (ε (cm-1), R (Å)) for pure Cd – Cd.

Molecule	$X^10_g^+$		$A^30_u^+$		B^31_u		
	ε_g (cm^{-1})	R_g (Å)	ε_0 (cm^{-1})	R_0 (Å)	ε_1 (cm^{-1})	R_1 (Å)	Ref.
	386±27	3.3±0.1	293±18	3.5	>1400	≤3	1
	386±27	4.82	290±18	4.95	-	-	1,8
	410±30	3.3±0.1	-	-	-	-	11
Cd-Cd	380	4.82	-	-	-	-	12
	150	4.76	-	-	5650		13
	323	4.26	252	4.52	-	-	14
	-	-	235	-	≥315	-	15
	328 ± 3	3.76, 3.78	259.9 ± 2.9	4.02±0.03	-	-	16,17

Table III: Comparison of our data for the ground $X^1 0^+$and the two excited states $A^3 0^+$ and $B^3 1$ potentials parameters (ε (cm^{-1}), R (Å)) for Cd – inert gases.

Molecule	$X\,^0 1^+$		$A\,^3 0^+$		$B\,^3 1$		
	ε_g (cm^{-1})	Rg (Å)	ε_0 (cm^{-1})	R_0 (Å)	ε_1 (cm^{-1})	R_1 (Å)	Ref.
Cd-Xe	192±18	4.28	988±41	3.93	152±15	4.41	2
	230.4	4.28	-	-	-	-	9
	370	4.73	-	-	-	-	18
	185	-	-	-	-	-	19
	-	-	1086±60	-	-	-	20
	198	4.45	-	-	677±4	4.26±0.05	21,22
Cd-Kr	134±7	3.99	389±10	3.25	73±4	4.45	3, 8
	130	3.75	-	-	-	-	11
	-	-	-	-	109.7	4.73	22
	131	-	373	-	72	-	23
	114.5	4.5	498.9	3.5	60.2	5.07	24
	310	3.63	407.2	3.44	-	3.5	18
	165, 129	4.27	-	-	-	-	25,19
	-	-	555.0 ± 0.5	3.3 ±0.03	-	-	26
Cd-Ar	109.7±3	4.3	310.315	3.95	63.2+2	4.77±0.15	4, 8
	-	-	326	3.38	-	-	18
	-	-	325	3.45	-	-	20
	-	-	-	-	55.0±0.8	5.01±0.05	22,23
	106.5	4.3	-	-	59.7	5.03±0.03	24
	106	4.31	322	3.5	56	-	28
	106.8	4.31	314.9 ± 6.1	3.51±0.03	57.0 ±2.9	5.07 ± 0.03	29,26
	112	4.29	-	-	-	-	30
	106.7	4.31	-	-	-	-	31
Cd-Ne	41.4±3	4.37±0.2	82.5±4	3.7±0.2	18.03±2	4.25±0.3	6, 8
	-	-	-	-	5.01±0.05	5.29±0.05	22
	39	-	75	-	21		24
	38	4.1±0.2	90.6	3.1±0.2	19.3	4.8	28
	-	4.26	-	3.62±0.05	-	-	20
	27	4.28	75	3.61	19.3	4.96	32
	28.3	4.32	70.5	-	19.6	4.98	33,34
Cd-He	16.3±2	4.5±0.2	45.8±4	3.8±0.2	2.55±0.5	4.94±0.3	7,8, 9
	-	-	-	-	7.8±0.2	7.8±0.2	22
	14.2	4.42	41.2	2.92	6.1	4.54	35
	-	-	-	-	2.5	5.715	36
	14.2	4.6	41.2	3.7	6.1	4.8	37
	15.1	4.5	36	3.57	7.8	5.55	30
	16.8	4.2	27.5	3.57	-	-	38

It is seen from Tables II and III, that the obtained potential parameters of the ground X10+ and the two excited states $^3 0^+$ and $^3 1$ are in good agreement with the corresponding theoretical and experimental molecular beams.

References

[1] Helmi MS, Grycuk T, Roston GD 1996 *Spectrochim. Acta B* **51** 633
[2] Helmi MS, Grycuk T, Roston GD 1996 *Chem. Phys.* **209** 53
[3] Roston GD, Ghatass ZF 2008 *J. Quant. Spectrosc. Radiat. Transf.* **109** 2427
[4] Roston GD, Grycuk T, Helmi MS 1996 *Chem. Phys.* **213** 365
[5] Roston GD, Helmi MS 2000 *Chem. Phys.* **258** 55
[6] Roston GD 2004 *J. Quant. Spectrosc. Radiat. Transf.* **87** 83
[7] Roston GD 2005 *Phys. Scr.* **72** 31
[8] Roston GD, Helmi MS 2006 *J. Quant. Spectrosc. Radiat. Transf.* **101** 201
[9] Roston GD, Ghatass ZF 2006 *J. Quant. Spectrosc. Radiat. Transf.* **101** 205
[10] Roston G.D, Helmi MS 2009 *J. Chem. Phys.* **358** 30
[11] Bousquet C 1986 *J. Phys. B: At. Mol. Phys.* **19** 3859
[12] Roston GD, Helmi MS 2014 *J. Quant. Spectrosc. Radiat. Transf.* **140** 30
[13] Bender CF, Rescigno TN, Schaefer H F, Orel AH 1979 *J. Chem. Phys.* **71** 1122
[14] Czajkowski M, Bobkowski R, Krause L 1989 *Phys. Rev.* **40** 4338
[15] Kowalski A, Czajkowski M, Breckenridge WH 1985 *Chem. Phys. Lett.* **119** 368
[16] Strojecki M, Ruszczak M, Łukomski M, Koperski J 2007 *Chem. Phys.* **340** 171
[17] Strojecki M, Krośnicki M, Zgoda P, Koperski J 2010 *Chem. Phys. Lett.* **489** 20
[18] Czuchaj E, Sienkiewicz J 1984 *J. Phys. B* **17** 2251
[19] Funk J D, Breckenridge WH *J. Chem. Phys.* **90** 2927
[20] Kvaran A, Funk D J, Kowalski A, Breckenridge WH 1988 *Chem. Phys.* **89** 6069
[21] Strojecki M, Krośnicki M, Łukomski M, Koperski J 2009 *Chem. Phys. Lett.* **471** 29
[22] Koperski J, Urbańczyk T, Krośnicki M, Strojecki M 2014 *Chem. Phys.* **428** 43
[23] Urbanczyk T, Koperski J 2014 *Chem. Phys. Lett.* **591** 64
[24] Kowalski A, Czajkowski M, Breckenridge W H 1985 *Chem. Phys. Lett.* **121** 217
[25] Czajkowski M, Bobkowski R, Krause L 1991 *Phys. Rev. A* **44** 5730
[26] Urbanczyk T, Strojecki M, Koperski J 2011 *Chem. Phys. Lett.* **503** 18
[27] Koperski J, Łukomski M, Czajkowski M 2002 *Spectrochim. Acta Part A* **58** 2709
[28] Bobkowski R, Czajkowski M, and Krause L 1990 *Phys. Rev A* **41** 234
[29] Koperski J, Kielbasa M, Czajkowski M 2000 *Spectrochim. Acta Part A* **56** 1613
[30] Czuchaj E, Stoll H 1999 *Chem. Phys.* **248** 1
[31] Koperski J, Czajkowski M 2002 *J. Mol. Spectrosc.* **212** 162
[32] Czajkowski M. Krause L, Bobkowski R 1994 *Phys. Rev A* **49** 775
[33] Strojeck M, Krośnicki M, Koperski J 2009 *J. Mol. Spectrosc.* **256** 128
[34] Koperski J , Czajkowski M 2000 *Eur. Phys. J. D* **10** 363
[35] Koperski J, Czajkowski M 1998 *J. Chem. Phys.* **109** 459
[36] Zagrebin A L, Lendnev M G 1993 *Opt. Spectrosc.* **75** 562
[37] Koperski J,Czajkowski M A 2001 *Chem. Phys. Lett.* **350** 367
[38] Czuchaj E, Krośnicki M, Stoll H 2001 *Theor. Chem. Acc.* **105** 219
[39] Szudy J, Baylis W 1975 *J. Quant. Spectrosc. Radiat. Transf.* **15** 641
[40] Behmenburg W 1972 *Z. Naturforsch.* **27a** 31
[41] Gallagher A 1975 *Atomic Physics,* vol. 4 (New York and London: Plenum)

Journal of Physics: Conference Series **548** (2014) 012014 doi:10.1088/1742-6596/548/1/012014

Infilling of Vector Collision–Sequence Interference Dips in Collision–Induced Spectra

John Courtenay Lewis
Memorial University of Newfoundland St. John's NL Canada A1B 3X7
E-mail: john.lewis@mun.ca

Abstract. The vector interference dip most clearly seen in the fundamental collision–induced band of H_2 and H_2 mixtures at densities upward of 20 amagat, is well known to persist at densities down to 1 amagat. However, At lower densities it is not seen. Lewis and Herman gave a theory of this infilling. In this work we approach the problem from the simple statistical models for collision–induced absorption introduced by Lewis in 2000 and subsequently much elaborated. The statistical model shows infilling, but rather than a uniformly infilling dip the model shows a small central peak forming in the dip, then broadening with increasing amplitude.

1. Introduction

The vector interference dip most clearly seen in the fundamental collision–induced band of H_2 and H_2 mixtures at densities upward of 20 amagat, is well known to persist at densities down to 1 amagat. A. R. W. McKellar noted it at 8 amagat and Bragg *et al.* [1] found it down to 1 amagat. Indeed, D.H. Rank *et al.* [2] were able to see the fundamental quadrupole line of H_2 because it sat in an interference dip.

On the other hand the studies of van der Waals complexes of H_2, measured at sub–amagat densities, do not show any such dips [3–5] (note in particular Fig. 1 of ref. [4]). A qualitative explanation appears straightforward: the relevant autocorrelation function is not

$$\langle \boldsymbol{\mu}(t_1) \cdot \boldsymbol{\mu}(t_2) \rangle \tag{1}$$

but rather

$$\left\langle \boldsymbol{\mu}(t_1) \cdot \boldsymbol{\mu}(t_2)\, e^{\iota \boldsymbol{k} \cdot [\boldsymbol{x}(t_1) - \boldsymbol{x}(t_2)]} \right\rangle \tag{2}$$

where $\boldsymbol{k}$ is the wave vector of the incident radiation. The factor

$$e^{\iota \boldsymbol{k} \cdot [\boldsymbol{x}(t_1) - \boldsymbol{x}(t_2)]} \tag{3}$$

is very close to unity for a smallish molecule, or for collision – induced spectra where the mean free path between collisions is short compared to the wavelength.

However, if the density becomes sufficiently low so that the mean free path is comparable to (actually, rather greater than) the mean free path, the factor (3) becomes in effect a number of modulus 1 and random phase, and on average destroys the negative correlation in the induced dipole moment and so leads to the disappearance of the interference dip.

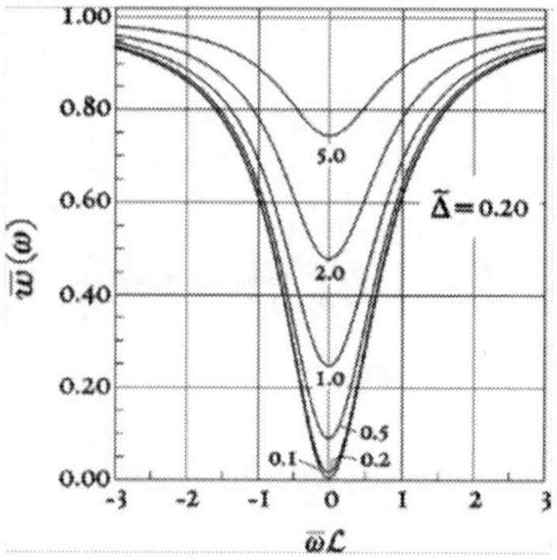

Figure 1. Theoretical results from ref. [6]. $\mathcal{L}$ is the mean free path of the H_2. The persistence of velocity is 0.20.

Lewis and Herman [6] developed a theory of the infilling effect, which predicted that the dip would fill in in a regular fashion, the dip simultaneously becoming shallower and narrower as the ratio of mean free path to wavelength increases (Fig. 1).

It is important to note that there is currently no experimental data showing the infilling of the interference of the dip.

2. The simple statistical models

The simple statistical models of Lewis *et al.* in their several variations are summarized in ref. [7]. An essential feature is that they suppose instantaneous collisions. Hence the resultant spectra are flat except around the interference dip. The Cartesian components of the velocities are taken from a Gaussian distribution, which for convenience is of zero mean and unit standard deviation. Although cases of zero persistence of velocity are interesting, in this study I usually assume a non–zero persistence, ranging up to 0.99. The velocity series are generated with an $AR(1)$ process [8]:

$$\boldsymbol{v}_{j+1} = \varphi \boldsymbol{v}_j + \boldsymbol{\epsilon}_j \tag{4}$$

where the $\boldsymbol{\epsilon}_k$ are independently distributed Gaussian random variables such that each component has zero mean and standard deviation 1. The impulses are given by

$$\boldsymbol{f}_j = \boldsymbol{v}_{j+1} - \boldsymbol{v}_j \tag{5}$$

and the integrated induced dipole moments are taken proportional to the impulses:

$$\boldsymbol{\mu}_j = \boldsymbol{f}_j \,. \tag{6}$$

The positions $\boldsymbol{x}(t_j)$ can be obtained by accumulation:

$$\boldsymbol{x}(t_j) = \boldsymbol{x}(t_{j-1}) + (t_j - t_{j-1})\,\boldsymbol{v}_j \,. \tag{7}$$

Then the total integrated induced dipole moment at time t is given by

$$\boldsymbol{M}(t) = \sum_j \boldsymbol{\mu}_j \delta(t - t_j)\, e^{\imath \boldsymbol{k}\cdot\boldsymbol{x}(t_j)} \,. \tag{8}$$

The Fourier transform of this is

$$\boldsymbol{\mathcal{M}}(\omega) = \sum_j \boldsymbol{\mu}_j e^{\iota \boldsymbol{k}\cdot\boldsymbol{x}(t_j)} e^{-\iota\omega t_j} \,. \tag{9}$$

If $\{t_j\}_j$ is the sequence of collision times lying within $(0\,,T)$ then

$$\boldsymbol{\mathcal{M}}(\omega)_T = \sum_{t_j \epsilon (0\,,T)} \boldsymbol{\mu}_j e^{\iota \boldsymbol{k}\cdot\boldsymbol{x}(t_j)} e^{-\iota\omega t_j} \tag{10}$$

is an estimate of the Fourier transform of $\boldsymbol{M}$ and the averaged periodogram is

$$S_T(\omega) = \frac{1}{T}\left\langle \boldsymbol{\mathcal{M}}(\omega)_T \cdot \boldsymbol{\mathcal{M}}(\omega)_T^* \right\rangle \,. \tag{11}$$

3. Computational approach

We take $\langle \cdots \rangle$ to mean an average over a large number of sets of collision times $\{t_j\}_j$ and velocities $\{\boldsymbol{v}_j\}_j$.

The simulations were carried out in Python 2.7, and made use of the module *random.py* for the random number generators and the module *cmath.py* for the complex exponentials. The module *random.py* uses the Mersenne twister for its basic random number generator.

A total time T input together with a number of collisions N. Then N collision times t_k are chosen by the uniform random number generator to lie within (0, T). N was always taken to be 16000 in the simulations reported here. T ranged from 8000 to 2097152000.

The calculation of S_T from eq. (11) for a given parameter set was repeated many times, and the mean taken. For fitting purposes either 5000 or 10000 repetitions were used.

The ratio of the mean free path to the wavenumber k is

$$k/\lambda = kN/T \tag{12}$$

as the mean speed is taken to be unity.

The velocity autocorrelation function for velocity in a single direction is

$$\langle v_x(t)\, v_x(t+\tau)\rangle = e^{-(1-\varphi)\tau} \tag{13}$$

and so we define an effective mean free path, the distance a particle will travel before being significantly deflected, as

$$\lambda/(1-\varphi) \,. \tag{14}$$

The ratio of the wavenumber k to this is

$$\mathcal{R} = k/\left[\lambda/(1-\varphi)\right] = kN(1-\varphi)/T \tag{15}$$

and this we use as the independent parameter in our plots.

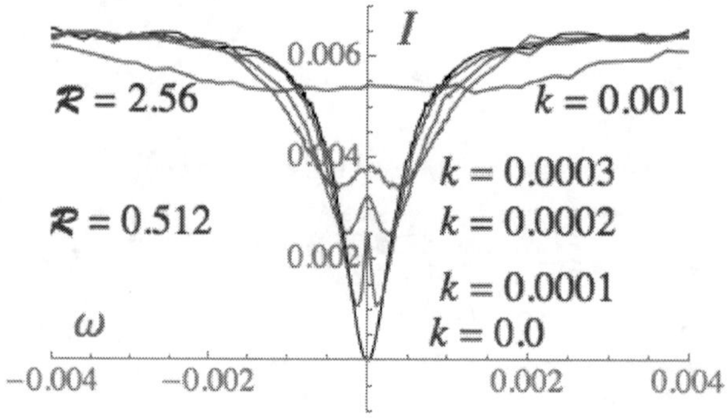

Figure 2. Infilling of the dip from the simple statistical model. This graph shows results for a series of runs with increasing k. Here $\varphi = 0.8$, $N = 16000$ and $T = 8192000$.

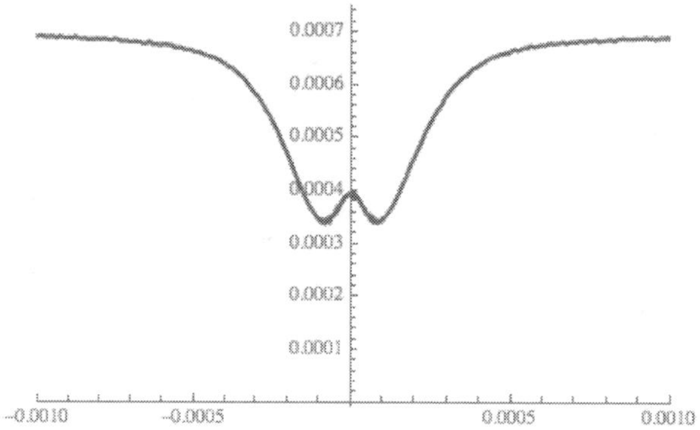

Figure 3. Spectrum for an intermediate case. 90000 repetitions were used. At each ω the mean and one standard deviation on either side of the mean are shown. Here $\varphi = 0.4$, $k = 0.0001$, $N = 16000$ and $T = 98304000$. $\mathcal{R} = 1.024$.

4. Results and analyses

We can analyze such a spectrum as the sum of a Lorentzian dip and a smaller Lorentzian peak:

$$W(\omega) = A\left[1 - \frac{1}{1 + (\omega/w_{dip})^2}\right] + \frac{B}{1 + (\omega/w_{peak})^2}. \tag{16}$$

As the dip should be a pure Lorentzian, this works well provided that the area of the peak is less than 0.1 of the area of the dip. It is not very good for a spectrum such as that of Fig. 3, which exhibits substantial infilling.

Incorporation of a quartic term in both the dip and the peak gives a much better fit:

$$W(\omega) = A\left[1 - \frac{1}{1 + (\omega/w_{dip})^2 + (\omega/w_{dQ})^4}\right] + \frac{B}{1 + (\omega/w_{peak})^2 + (\omega/w_{pQ})^4}. \tag{17}$$

The results of this analysis for the above figure are shown below. Both the dip and the peak are slightly broader than their Lorentzian counterparts.

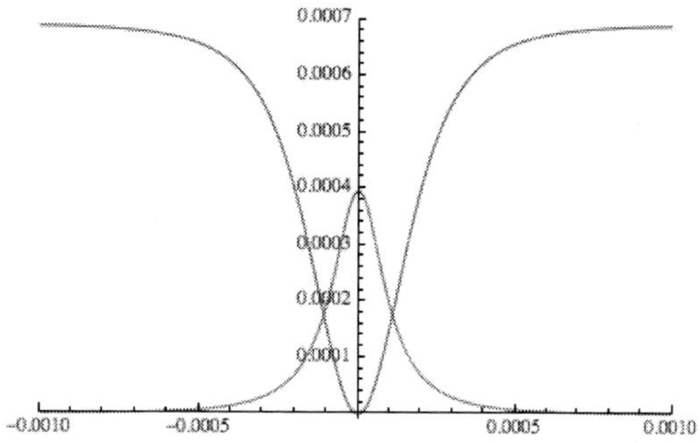

Figure 4. Decomposition of the spectrum of Fig. 3 into quartic line shapes.

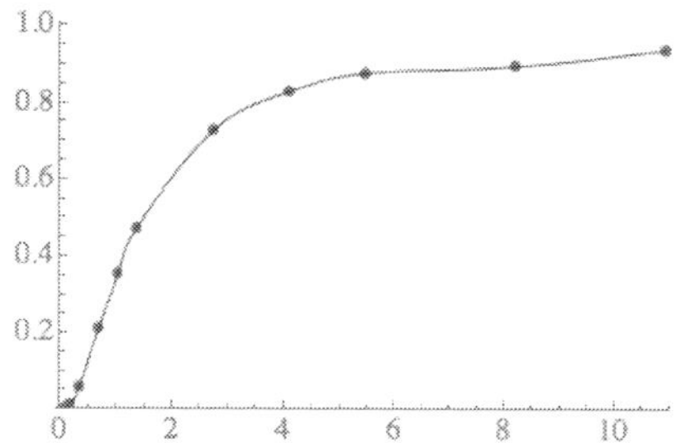

Figure 5. Ratio of area of peak to dip area as a function of the effective mean free path $\mathcal{R}$.

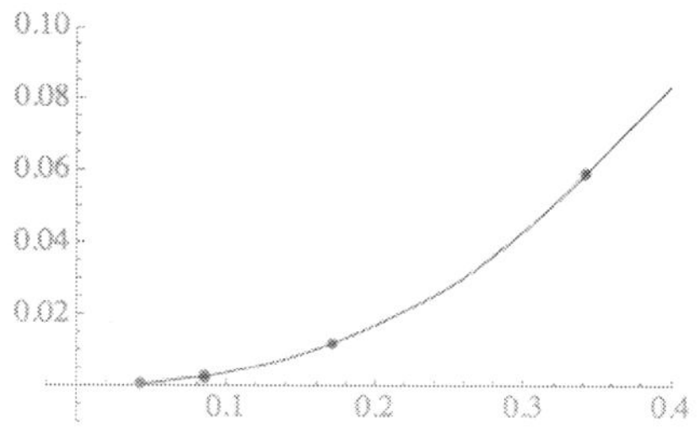

Figure 6. Detail of Fig. 5.

A good way to parametrise the infilling of the dips is by the ratio of the area of the peak to the area of the dip. This ratio lies between 0 and 1. Typical results are shown in Figs. 5 and 6 above. The ratio may depend somewhat on the method of analysis. Use of Lorentzian peaks and dips leads to somewhat smaller numbers for the ratios, but only for cases where the fit residuals indicate poor fits.

Similar graphs are obtained by changing φ or changing $\boldsymbol{k}$. However, no scaling which reduces them to a universal curve has been found.

5. Summary

(i) The interference dips do fill in in this version of the simple statistical models.

(ii) Rather than fill in uniformly, as per Lewis and Herman, the first infill is a small sharp Lorentzian peak which increases in amplitude and width until the dip is completely filled.

(iii) The reasons for this behaviour are not totally clear.

(iv) It is evident by comparing Fig. 1 of ref. [4] with Fig. 1 and Fig. 3 above that the very limited observational data are not sufficient to decide between the two descriptions of the infilling.

References

[1] Bragg S L, Brault, J W and Smith W H 1982 *Ap. J.* **263** 999
[2] Rank D H, Fink U, and Wiggins T A, 1966 *Ap. J.* **143** 980
[3] McKellar A R W and Welsh H L 1971 *J. Chem. Phys.* **55** 595
[4] McKellar A R W and Welsh H L 1972 *Can. J. Phys.* **50** 1458
[5] McKellar A R W and Welsh H L 1974 *Can. J. Phys.* **52** 1082
[6] Lewis J C and Herman R M 2003 *Phys. Rev.* A **68** 032703
[7] Lewis J C and Herman R M 2012 *J. Phys.: Conf. Series* **397** 012032
[8] Wheeler H T and Lewis J C *J. Phys.: Conf. Series* **397** 012010

Precise cavity enhanced absorption spectroscopy

A Cygan*, P Wcisło, S Wójtewicz, P Masłowski, J Domysławska, R S Trawiński, R Ciuryło, D Lisak

Institute of Physics, Faculty of Physics, Astronomy, and Informatics, Nicolaus Copernicus University, Grudziadzka 5, 87-100 Torun, Poland

E-mail: agata@fizyka.umk.pl

Abstract. A short review of recent achievements in high-precision cavity enhanced absorption spectroscopy is presented. Actual challenges and development paths in modern line shape study are indicated and discussed. The importance of identification and quantification of systematic instrumental errors affecting the measured line shape is highlighted. New, alternative measurement methods based on cavity enhanced spectroscopy are proposed.

1. Introduction

Satellite and ground-based studies of the atmosphere [1–3], control of the composition and purity of the gas mixtures used in production processes [4], determination of physical constants [5–9] as well as calibration of measurement equipment continuously increase demands on precision molecular spectroscopy. Recent progress in development of precise spectrometers revealed that the main reason of systematic errors of measurement of gas concentration, temperature and pressure is a simplified description of the spectral line shape. Particularly in the case of very weak molecular lines, important in the study of the atmosphere, the lack of a unified theoretical model and reference data inhibit the progress in these applications. Hence a strong effort is nowadays put into proper analysis of experimental spectral line shapes with the use of the complex line-shape models [10], incorporation of advanced and fast algorithms of the line shapes computation [11–13] as well as elimination of numerical correlations between fitted line-shape parameters by application of the multispectrum fitting procedure [14, 15].

Cavity enhanced absorption spectroscopy (CEAS) using high Q-factor optical cavities yields the highest sensitivity of absorption measurements and enables study of weak molecular spectra with potentially very high resolution and precision [16]. Combination of cavity ring-down spectroscopy (CRDS) with frequency stabilization of cavity modes realized at NIST in 2004 [17, 18] provided the proper tool for probing the absorption lines with unprecedented precision. Further evolution of this technique [19] including spectral narrowing and frequency stabilization of the probe laser [20–22] as well as the use of optical frequency combs [23, 24] resulted in the signal-to-noise ratio exceeding 10^5 [25, 26] and control of the frequency axis with the kilohertz level of accuracy [27]. This created favorable conditions for testing advanced line-shape models and above all it allowed to determine spectral parameters with subpercent accuracy [28–34] demanded in satellite research.

This paper is a condensed review about recent achievements in development of precise cavity enhanced absorption spectroscopy. At extremely high level of precision of spectroscopy experiments systematic instrumental errors affecting the shape of investigated line can be

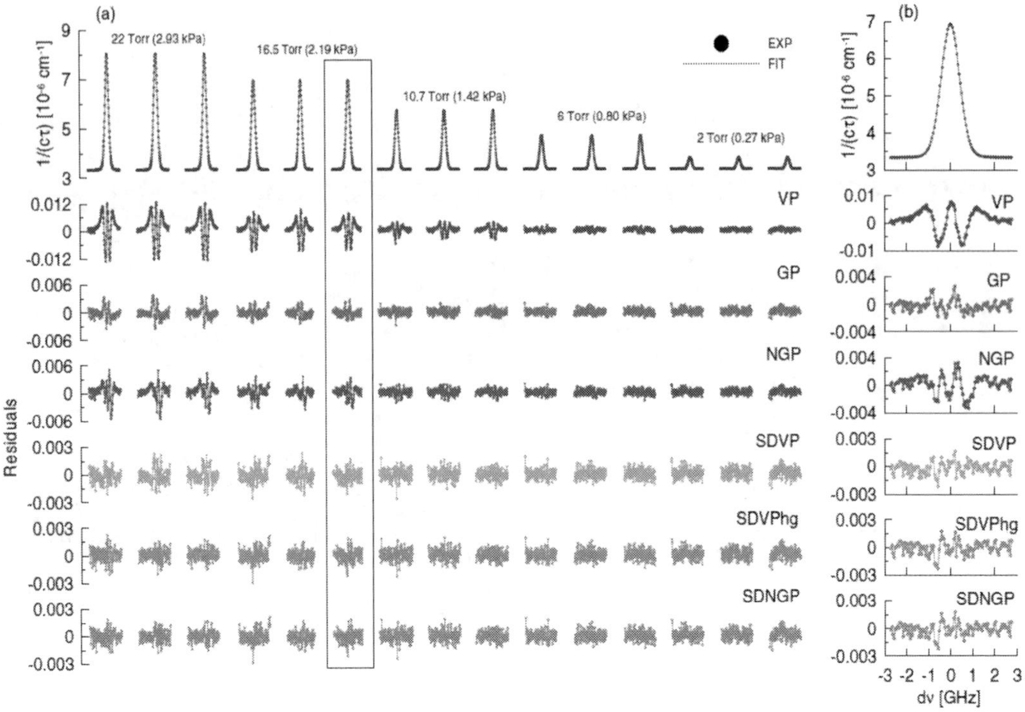

Figure 1. Experimental spectra of R1 Q2 O_2 B band lines recorded at pressures 2-22 Torr and residuals from multispectrum fits with the use of six line-shape models: VP, GP, NGP, SDVP, $SDVP_{hg}$ and SDNGP. The subscript "hg" denotes that more realistic dependence on absorber speed described by means of hipergeometric function was taken into account in SDVP. Source: [33].

exposed as well. In this paper alternatives to the CRDS method for the measurement of weak spectral lines will be presented. One of them bases on dependence of the width of the optical cavity mode on the absorption coefficient [35, 36]. This method seems to be complementary to CRDS and under certain conditions more attractive.

2. Line shape description

A proper description of the line shape is a nontrivial task. It demands careful preparation of high-precision experimental data, only slightly affected by systematic distortions, and choosing adequate theoretical model which allows to reproduce experimental line shape with the highest possible accuracy. During recent years it was demonstrated many times that the Voigt profile (VP), which is just a simple convolution of Lorentzian and Doppler shapes, cannot be valid and cannot provide subpercent accuracy of line-shape parameters even at low pressures [31]. Instead of the VP other more complex line-shape models are often considered. They take into account Dicke narrowing [37], both in case of soft (Galatry profile, GP) [38] and hard (Nelkin-Ghatak profile, NGP) [39,40] collision model, the speed dependence of collisional broadening and shifting [41] marked mostly by abbreviation SD, as well as line-mixing and correlations between phase- and velocity- changing collisions [40]. Recently, partially-correlated speed-dependent Nelkin-Ghatak profile (pCSDNGP) [42] with quadratic speed dependence of absorber speed was tested in a wide range of pressures, from single Torrs to atmospheric pressures [12]. Its high flexibility, unification and variety of physical effects accounted for in this model, as well as recent development of fast numerical computation of this model [13] based on the SDVP algorithm introduced by Boone *et al.* [43], caused that the pCSDNGP became attractive especially for atmospheric study and development of new generation of spectroscopic data bases. Recently, the pCSDNGP was recommended by IUPAC as a proper isolated line-profile for representing high-resolution spectroscopic transitions [44].

Typical thorough analysis of the line shape looks like this presented in Fig. 1. Fifteen weak self-broadened spectra of the R1 Q2 line from the oxygen B band [33] were recorded by CRDS at pressures from 2 to 22 Torr. Below are residuals from fits with VP and other more sophisticated profiles. As would be expected the VP gives the worst quality of the fit. High measurement precision as well as the wide range of investigated pressures allows to distinguish which line-narrowing mechanism, Dicke narrowing or speed dependence of collisional broadening, is dominating in case of the R1 Q2 line. As seen all profiles taking into account speed dependence of collisional broadening give better fit quality. Moreover, it is clear that the hard collision model in the NGP reproduces experimental data worse than the soft collision model provided by the GP.

A complication of line-shape model implies growth of number of fitted line-shape parameters and hence increase of degree of numerical correlations between them. Such numerical correlations may strongly affect the line-shape analysis and perturb achieving results, especially when the range and values of investigated pressures are small. To overcome this problem, the multispectrum fit procedure [14, 15] is used. It is based on analysis of spectral line shapes simultaneously for all measured pressures, see Fig. 1. It should be noted as well, that the multispectrum fit analysis can help to determine line-shape parameters even when spectra are affected by instrumental errors introduced by the measurement procedure [45, 46].

3. CEAS precision and accuracy

Systematic increase of CEAS precision observed during last 15 years is mainly caused by techniques of frequency stabilization of probing light as well as frequency modulation methods [47, 48] allowing for realization of spectroscopy near the shot noise limit. Narrowing of laser linewidth much below the width of the cavity resonance, possible thanks to Pound-Drever-Hall (PDH) [50] or optical feedback [51, 52] methods, increased significantly the number of coincidences between probing light and the cavity mode and hence improved a repetition rate of ring-down events in CRDS. A detection limit of 1×10^{-12} $cm^{-1}Hz^{-1/2}$ obtained in this way by Spence *et al.* [53] in 2000 is the lowest achieved in CRDS to this day. A table presented in Fig. 2 (c) includes the most sensitive cavity-enhanced spectrometers reported in the literature. As seen the highest precision of 1×10^{-14} $cm^{-1}Hz^{-1/2}$ provides so far the NICE-OHMS (Noise-Immune Cavity-Enhanced Optical Heterodyne Molecular Spectroscopy) technique [54] which combines two methods: PDH laser linewidth narrowing and phase modulation of laser light at frequency equal to the cavity free spectral range. It should be noted as well, that two modes of detection (absorption and dispersion) occurring in NICE-OHMS allows for quantitative comparison of differences in accuracy of determined line-shape parameters in case of both modes, as was shown recently by Wang *et al.* [49]. The third, also very sensitive technique mentioned in table in Fig. 2 (c) is a frequency-agile, rapid scanning CRDS (FARS-CRDS) developed recently by Truong *et al.* [55]. This modified CRDS method allows for fast spectra recording, for example a 2 GHz spectrum within several miliseconds.

Active stabilization of the comb of cavity modes provides a stable relative frequency axis for recorded spectra. Moreover, it gives an opportunity for effective spectra averaging leading to further improvement of measurement sensitivity. In Figs. 2 (a) and (b) two recent spectra demonstrating extremely high signal-to-noise ratio are presented. The former was obtained in

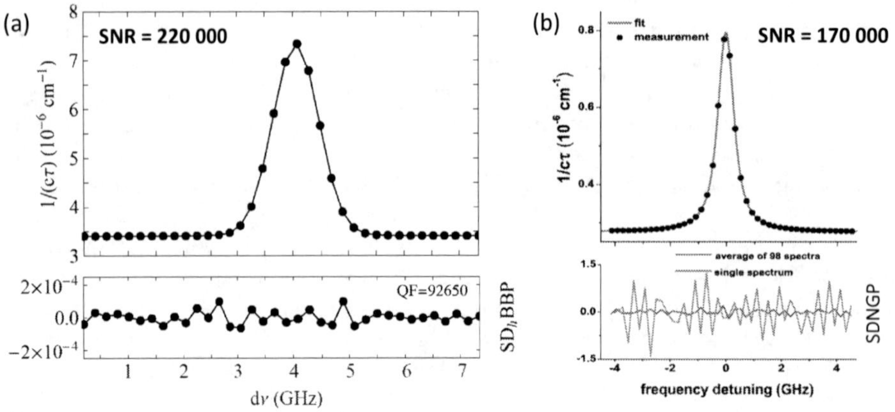

(c)

Technique	Ref.	NEA (cm^{-1} Hz$^{-1/2}$)	Laser source	Tuning range (nm)
NICE-OHMS	[6]	1×10^{-14}	cw-Nd:YAG	0.1
CRDS	[5]	1×10^{-12}	cw-Nd:YAG	0.14
FARS-CRDS	Present	1.7×10^{-12}	ECDL	60
NICE-OHMS	[7]	5.6×10^{-12}	Fiber laser	1.0
CRDS	[4]	1.43×10^{-11}	Ti:sapphire	200
FARS-CRDS	[12]	2×10^{-11}	Fiber laser	1.0
NICE-OHMS	[1]	6.9×10^{-11}	ECDL	14
NICE-OHMS	[2]	8.5×10^{-11}	ECDL	50

Figure 2. (a) - The R7 Q8 O_2 B band spectrum and residuals from fit with speed-dependent billard ball profile. Source: [11]. (b) - The R26e CO_2 spectrum and residuals from fit with SDNGP. Source: [26]. (c) - The most sensitive cavity-enhanced spectrometers reported in the literature. Source: [35].

2012 by the use of PDH-locked frequency-stabilized CRDS [25], while the later was obtained in 2013 thanks to differential frequency-stabilized cavity ring-down spectroscopy described in Ref. [26]. It should be noted that although such ultra-high precision makes possible to better test line-shape models, it also reveals problem of distinguishing systematic distortions of recorded spectra due to experimental imperfections from these caused by the wrong choice of the line-shape model.

Currently the strong effort is put into reduction of systematic instrumental errors as well as development of new, more accurate spectroscopic techniques. For example, parasitic etaloning effects appear as sine structures on line-shape residuals were firstly subtracted or fitted as artificial background [56]. Later, techniques of spectra averaging allowed for significant suppression of etalons amplitude because the etalon phase changes randomly in time during spectra measurement [25, 26]. Recently, Courtois *et al.* [26] discussed how etalons are caused by coupled-cavity interactions between primary ring-down cavity and other optical elements in the experiment. They also proposed differential technique which under conditions of carefully selected frequency interval between measurements points provided spectra free from unwanted interferences.

Standard CEAS experiments are based on measurements of attenuation of light transmitted through the cavity filled with absorbing medium. Such approach although provide fast data acquisition is very sensitive to power variations of probing light. For this reason, CRDS methods are considered as more promising for accurate measurements. However, light rapidly decaying

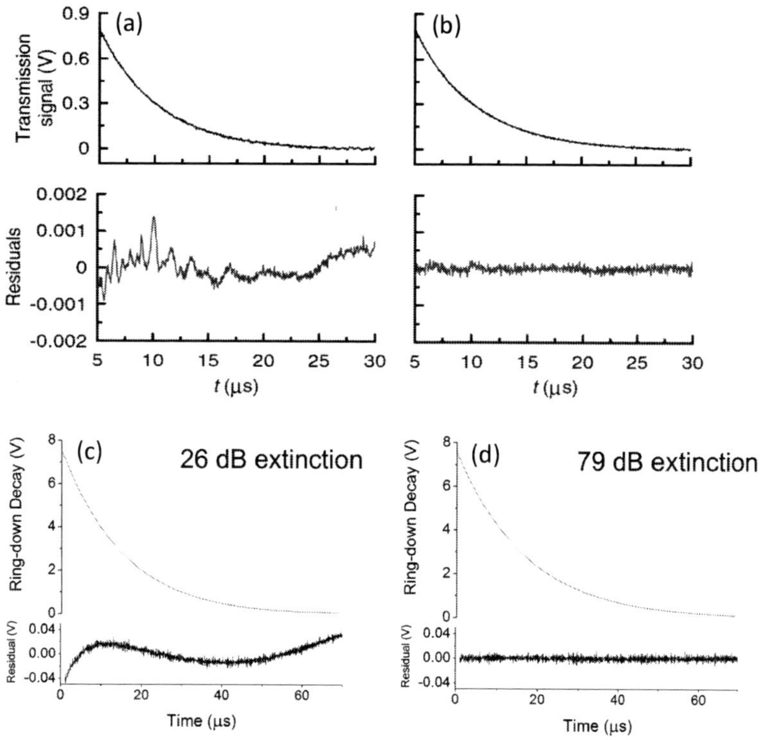

Figure 3. (a), (b) - Registered ring-down decays and fit residuals from a single-exponential fit for two different ADC measurement cards. Source: [31]. (c), (d) - Measured ring-down decays and fit residuals from a single-exponential fit for two different beam extinction ratios. Source: [58].

from the cavity needs to be recorded by a fast-response detector characterized by high linearity both in amplitude as well as gain for whole bandwidth. Any nonlinearities of detection system lead to distortions of measured ring-down events which strongly affect determined line-shape parameters. In paper [31] we demonstrated comparison of two different high-speed analog-to-digital converters (ADC) used as detectors in CRDS experiment. Their influence on recorded ring-downs is presented in Figs. 3 (a) and (b), for the worse and better ADC, respectively. Reported systematic errors of determined line intensity were 5 % and 0.5 % for those cases, respectively. Apart from the high-quality detection system, also the system of light beam interruption in CRDS has to be fast enough. In general, it consists of an acoustooptic or an electrooptic modulator driven by the RF signal which is switched off when the laser beam has to be blocked. If the shut-off of the excitation laser is incomplete, the residual optical power couples to the excited cavity mode. Since the coherence time of most lasers is less than the cavity storage time, this extra field adds with random phase to the ring-down decay creating additional interference noise on it. It is worth noting that the tight locking of the laser to the cavity, leading to laser line width narrowing, positively affects also on the reduction of unwanted interference noise. The impact of the RF signal extinction ratio on the recorded ring-down decay was discussed in recent papers [57, 58] and is presented in Figs. 3 (c) and (d). As seen to get undistorted ring-down decay, beam extinction ratio has to be at least 80 dB which, however, is

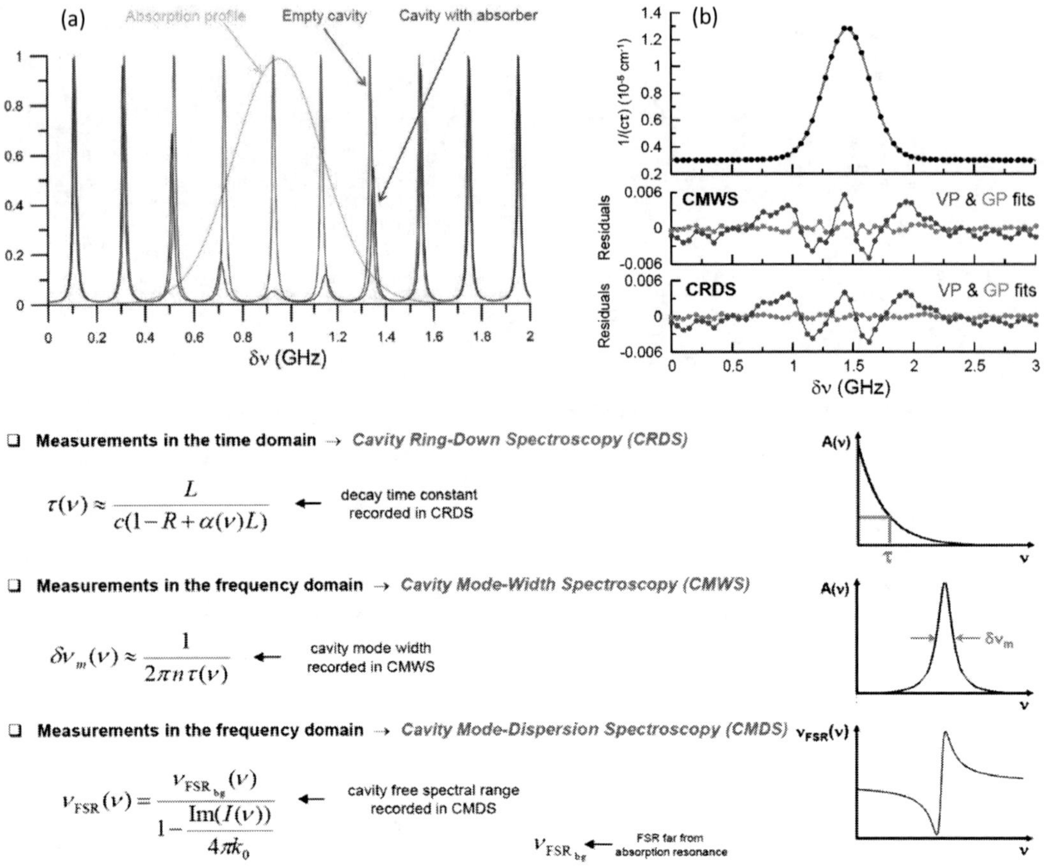

Figure 4. (a) - Spectral broadening and dispersion shift of cavity modes in the vicinity of absorber. (b) - Experimental spectrum of P3 (3 ← 0) $^{13}C^{16}O$ line recorded at pressure 4.8 Torr and residuals from VP and GP fits. Below graphs (a) and (b) - mathematical formulas for calculation of physical quantities measured in CRDS, CMWS and CMDS exepriments, respectively.

much larger than typical extinction ratios of 50 dB offered in commercial devices. A description of a home-made microwave switch with the extinction ratio variable in range 30-80 dB can be found in Ref. [57].

4. Alternative measurement methods

As was mentioned above identification and quantification of systematic instrumental errors is the most challenging task of modern line shape study. To solve this problem alternatives to the CRDS method for the measurement of weak spectral lines are now being developed.

The photon lifetime $\tau(\nu)$ in the cavity, commonly measured from the ring-down decays in CRDS, is directly related to the spectral width of the cavity resonant modes which also are shifted due to dispersion, see Fig. 4 (a). This first phenomenon inspires relatively new technique named cavity mode-width spectroscopy (CMWS) in which absorption information can be retrieved by precise measurements of widths $\delta\nu_m(\nu)$ of cavity modes. In 1994 Nakagawa *et al.* [59] reported for the first time that a higher intracavity absorption coefficient is accompanied by a larger cavity mode width. Since then no one explored this phenomenon until 2013. Very recently experimental spectrum determined from cavity mode widths measurement was demonstrated by Long *et al.* with the use of the FARS technique [35,55]. At the same time we have measured CMWS spectrum of the oxygen line with the use of ultranarrow and tunable in the range of 40 GHz external cavity diode laser [36]. The CMWS seems to be an intermediate technology between CRDS and CEAS. No need to use fast detectors in CMWS results in a wide dynamic range of absorption measurements, as was also indicated in [35]. Also possibility to measure high absorptions by the use of lower-reflectivity mirrors makes CMWS technique similar to CEAS. However, low sensitivity of CMWS to laser power variations allows to find similar features with CRDS technique. Qualitative and intuitive comparison of CRDS and CMWS methods reveals their complementarity in the sense that they achieve their best precision in different pressure ranges. For low absorptions the best precision is achieved with the CRDS technique, where the ring-downs are long and hence they can be well determined. In the opposite case, where the absorption is high and hence cavity modes become more broadened, the precision of CMWS is enhanced. In Fig. 4 (b) very recent experimental spectra of P3 ($3 \leftarrow 0$) $^{13}C^{16}O$ line and residuals from VP and GP fits are presented both in case of CMWS and CRDS experiments. We found 0.7% and 0.3% agreement between collisional widths and line areas obtained from CRDS and CMWS measurements, respectively.

It is worth noting that measurement of dispersion shift of cavity modes delivers another way for potentially accurate studies of molecular spectra. Direct use of frequency domain quantities to obtain absorption coefficient thanks to Kramers-Krönig relation [60] in here-called cavity mode-dispersion spectroscopy (CMDS) may prevent from nonlinearities of detection system and hence may minimize contribution of systematic instrumental errors in the total shape of investigated spectral lines.

Acknowledgments

The research is part of the program of the National Laboratory FAMO in Toruń, Poland, and is supported by the Polish National Science Centre Project nos. DEC-2011/01/B/ST2/00491, UMO-2012/05/N/ST2/02717 and UMO-2013/11/D/ST2/02663. The research is also supported by the Foundation for Polish Science TEAM and HOMING PLUS Projects co-financed by the EU European Regional Development Fund. A. Cygan is supported by the Foundation for Polish Science START Project.

References

[1] Russell III J M, *et al.* 1993 *J. Geophys. Res.* **98** 10777
[2] Crisp S, *et al.* 2004 *Adv. Space Res.* **34** 700
[3] Miller C E, *et al.* 2005 *C. R. Phys.* **6** 876
[4] Abe H 2009 *Synthesiology* **2** 206
[5] Yamada K M T, *et al.* 2009 *C. R. Phys.* **10** 907
[6] Lemarchand C, *et al.* 2011 *N. J. Phys.* **13** 073028
[7] Sun Y R, *et al.* 2011 *Opt. Express* **19** 19993
[8] Moretti L, *et al.* 2013 *Phys. Rev. Lett.* **111** 060803
[9] Cygan A, *et al.* 2010 *Phys. Rev. A* **82** 032515
[10] Ciuryło R 1998 *Phys. Rev. A* **58** 1029
[11] Wcisło P, *et al.* 2013 *Phys. Rev. A* **88** 012517
[12] Ngo N H, *et al.* 2013 *J. Quant. Spectrosc. Radiat. Transf.* **129** 89; 2014 **134** 105
[13] Tran H, *et al.* 2013 *J. Quant. Spectrosc. Radiat. Transf.* **129** 199
[14] Benner D C, *et al.* 1995 *J. Quant. Spectrosc. Radiat. Transf.* **53** 705
[15] Pine A S, Ciuryło R 2001 *J. Mol. Spectrosc.* **208** 180
[16] Berden G and Engeln R 2009 *Cavity Ring-Down Spectroscopy: Techniques and Applications* (Wiley-Blackwell, The Netherlands)
[17] Hodges J T, *et al.* 2004 *Rev. Sci. Instrum.* **75** 849

[18] Long D A, *et al.* 2012 *Chem. Phys. Lett.* **536** 1
[19] Cygan A, *et al.* 2013 *Eur. Phys. J.-Spec. Top.* **222** 2119
[20] Cygan A, *et al.* 2011 *Rev. Sci. Instrum.* **82** 063107
[21] Cygan A, *et al.* 2011 *Meas. Sci. Technol.* **22** 115303
[22] Martinez R Z, 2016 *J. Opt. Soc. Am. B* **23** 727
[23] Maddaloni P, 2010 *J. Chem. Phys.* **133** 154317
[24] Domysławska J, *et al.* 2012 *J. Chem. Phys.* **136** 024201
[25] Cygan A, *et al.* 2012 *Phys. Rev. A* **85** 022508
[26] Courtois J, *et al.* 2013 *J. Opt. Soc. Am. B* **30** 1486
[27] Truong G-W, *et al.* 2013 *J. Chem. Phys.* **138** 094201
[28] Hodges J T, *et al.* 2008 *J. Mol. Spectrosc.* **249** 86
[29] Long D A, *et al.* 2009 *Phys. Rev. A* **80** 042513
[30] Lisak D, *et al.* 2009 *Phys. Rev. A* **79** 052507
[31] Wójtewicz S, *et al.* 2011 *Phys. Rev. A* **84** 032511
[32] Bielska K, *et al.* 2013 *Geophys. Res. Lett.* **40** 6303
[33] Domysławska J, *et al.* 2013 *J. Chem. Phys.* **139** 194312
[34] Wójtewicz, *et al.* 2014 *J. Quant. Spectrosc. Radiat. Transf.* **114** 36
[35] Long D A, *et al.* 2014 *Appl. Phys. B* **114** 489
[36] Cygan A, *et al.* 2013 *Opt. Express* **21** 29744
[37] Dicke R H 1956 *Phys. Rev.* **89** 472
[38] Galatry L 1961 *Phys. Rev.* **122** 1218
[39] Nelkin M, Ghatak A 1964 *Phys. Rev.* **135** A4
[40] Rautian S G, Sobelman I I 1966 *Usp. Fiz. Nauk* **90** 209; 1967 *Sov. Phys. Usp.* **9** 701
[41] Berman P R 1972 *J. Quant. Spectrosc. Radiat. Transf.* **12** 1331
[42] Pine A S 1999 *J. Quant. Spectrosc. Radiat. Transf.* **62** 397
[43] Boone C D, *et al.* 2007 *J. Quant. Spectrosc. Radiat. Transf.* **105** 525
[44] Tennyson J, *et al.* 2014 *Pure. Appl. Chem.* in press.
[45] Brault J W, *et al.* 2003 *J. Mol. Spectrosc.* **222** 220
[46] Bui T Q, *et al.* 2014 *J. Chem. Phys.* submitted.
[47] Bjorklund G C 1980 *Opt. Lett.* **5** 15
[48] Ye J, *et al.* 1996 *Opt. Lett.* **21** 1000
[49] Wang J,*et al.* 2014 *J. Quant. Spectrosc. Radiat. Transf.* **136** 28
[50] Drever W P, *et al.* 1983 *Appl. Phys. B* **97** 31
[51] Morville J, *et al.* 2005 *Appl. Phys. B* **80** 1027
[52] Kassi S, Camparague A 2012 *J. Chem. Phys.* **137** 234201
[53] Spence T G, *et al.* 2000 *Rev. Sci. Instrum.* **71** 347
[54] Ye J, *et al.* 1998 *J. Opt. Soc. Am. B* **15** 6
[55] Truong G-W, *et al.* 2013 *Nature Photon.* **7** 532
[56] Hodges J T, Lisak D 2006 *Appl. Phys. B* **85** 375
[57] Huang H, Lehmann K K 2009 *Appl. Phys. B* **94** 355
[58] Long D A, *et al.* 2013 *Proc. SPIE* **8726** 8726O-1
[59] Nakagawa K, *et al.* 1994 *Opt. Commun.* **107** 369
[60] Libbrecht K G, *et al.* 2006 *Am. J. Phys.* **74** 1055

Resonant four-wave mixing spectra: A fresh look at hotodissociation dynamics

A Kouzov[1], P Radi[2], P Maksyutenko[2], and D Kozlov[3]

[1]Saint-Petersburg State University, 198504 Saint-Petersburg, Russia
[2]Paul Scherrer Insitute, 5232 Villigen, Switzerland
[3]A.M. Prokhorov General Physics Institute, 119991 Moscow, Russia

E-mail: alex@ak1197.spb.edu, peter.radi@psi.ch, pavlo.maksyutenko@gmail.com, dnk@kapella.gpi.ru

Abstract. Frequency- and polarization-resolved degenerate resonant four-wave mixing is used for the first time to study the rotational anisotropy of the OH radicals produced by photolysis of H_2O_2 by a plane-polarized laser radiation at 266 nm. For that, a flexible polarization setup is designed to directly isolate the rotationally anisotropic signals in the $A^2\Sigma^+ - X^2\Pi(0,0)$ OH vibronic band. The results are quantitatively interpreted by using the line-space approach and accounting for all possible correlations between the electric field $\mathbf{E}_D$ of the photolysis laser, the recoil velocity $\mathbf{v}$ and the angular momentum $\mathbf{J}$ of nascent particles. In so doing, the resonance amplitudes are decoupled into products of the polarization factors, the Doppler-broadened profiles and the Einstein B-coe cients. The anisotropic signals have characteristic shapes of well resolved Doppler doublets providing a direct evidence of translational helicity of the nascent OH radicals, a hitherto exprementally inaccessible key characteristic of the photodissociation path.

1. Introduction

The photodissociation reaction $AB \overset{\hbar\omega_D}{\longrightarrow} A + B$ has a paramount importance for theoretical chemistry, for its understanding can shed light on the final stage of reactive collisions [1]. Advantageously, the reaction is naturally controlled by the photolysis laser intensity I_D and, what is by far more important, by the frequency ω_D and polarization $\mathbf{e}_D$. Here, we shall consider only moderate intensities whose variation has a minute effect on the reaction dynamics and results only in a linear scale of the production rate. Conversely, the energy $(\hbar\omega_D)$ and the polarization of an absorbed photon affect the reaction path and, therefore, are strongly imprinted on the properties of the nascent particles A and B. Due to the basic inseparability principle [2], the wave functions of A and B are always unknown linear combinations of eigenvectors of the free species. In other words, the quantum states of products are correlated and this correlation is given in terms of their density matrices ρ^A and ρ^B, the most detailed observable characteristics of the photolytic reaction which can be obtained by studying the nascent particles. The strongest correlations arise between the $(2J+1)$-degenerate substates of products corresponding to different space orientations of their total angular momenta $\mathbf{J}_n$ $(n = A, B)$. To describe the effect, the so-called state multipoles ρ^n_K $(K = 0, 1, 2...)$ are introduced [2], with the monopole term $(K = 0)$ corresponding to the completely chaotic (isotropic) $\mathbf{J}_n$−orientation and the others characterizing the reaction-induced rotational anisotropy.

An impressive progress achieved so far in purely optical studies of photofragments almost entirely owes to the use of laser-induced uoresence (LIF) [1]. However, a LIF setup operates with two polarizations and, because of this, the detected signal is informative only on the K =0,2,4 state multipoles [3]. Advantageously, four polarizations are controllable if one uses the Resonant Four-Wave Mixing (RFWM) techniques. Due to this sophistication, all state multipoles with $K \leq 4$ contribute to the (coherent) response thus allowing a deeper insight into the reaction path.

Here, we provide the reader with further details omitted in our previous paper [4] on degenerate RFWM (i.e. one-color or DFWM) spectra of the OH radicals produced by laser photolysis of H_2O_2 molecules at 266 nm.

2. Outline of theory

The symbol m denotes a collection of the principle quantum numbers (including also the total angular momentum J_m) which defines the photofragment eigenstate. Omitting for brevity the photofragment label, the rotational anisotropy of the state m can be generally introduced with the help of the irreducible tensor operators

$$N_Q^{(K)}(m) = \rho_{mm}^{-1/2} \sum_{M_m M'_m} (-1)^{J_m - M_m} C^{KQ}_{J_m M_m J_m - M'_m} \mid mM_m\rangle\langle mM'_m \mid \tag{1}$$

where $C^{KQ}_{J_m M_m J_m M'_m}$ are the Clebsch-Gordan coefficients and ρ_{mm} is the total Boltzmann factor of the state m, i.e. its total population divided by $2J_m + 1$. sing Dixon's approach [5] to account for the anisotropic distribution of the translational velocity $\mathbf{v} = (v, \Omega_v)$, we obtain

$$\rho = \sum_{mKL\lambda} t^{(m)}_{KL\lambda}(\{C^{(\lambda)}(\Omega_D) \otimes C^{(L)}(\Omega_v)\}^{(K)}, N^{(K)}(m)) \tag{2}$$

where $C^{(l)}_\sigma$ $(\sigma = -l, -l+1, ..l)$ are the Racah spherical harmonics; the outer parentheses and the braces respectively denote the scalar and tensor contraction operations [6]; Ω_D is the orientation of the polarization plane $\mathbf{e}_D$ of the photolysis laser beam in the laboratory frame (X, Y, Z), with OZ assumedly coinciding with the beam wave vector. Besides, only the ranks $\lambda = 0, 2$ are allowed in (2). The coefficients $t^{(m)}_{KL\lambda}$ are proportional to the statistical weights of the velocity- and state-dependent correlations imposed on the nascent molecules by the electric field $\mathbf{E}_D (= E_D \mathbf{e}_D)$. The $\lambda = K = L = 0$ term refers to the isotropic case for which the $\mathbf{v}$- and $\mathbf{J}$−orientations are completely chaotic. The remaining terms with $\lambda = 0$ describe the mutual $\mathbf{v} - \mathbf{J}$ correlations of order one, the helicity effect for which $K = L = 1$ and $\mathbf{v}$ and $\mathbf{J}$ can be either parallel or antiparallel. At $\lambda = 0$, the alignment $K = L = 2$ terms may also contribute as well as the higher-order ones $(K = L = 3, 4, ..)$, all independent on the photolysis laser polarization. In the $\lambda = 2$ case, the effects relate to the field direction and include the double $\mathbf{E}_D - \mathbf{v}$ (for which $K = 0$ and $L = 2$) and $\mathbf{E}_D - \mathbf{J}$ $(L = 0, K = 2)$ alignment terms as well as the triple $(\mathbf{E}_D - \mathbf{v} - \mathbf{J})$ ones when both K and L are distinct from zero. Of particular importance is the field-induced helicity contribution, i.e. the term with $\lambda = 2, K = L = 1$.

The key quantity targeted by the RFWM theory is the resonance part of the third-order susceptibility, $\chi^{(res)}$. As in the isotropic case [7], the line-space formalism appears to be an efficient tool to study this quantity. Another salient feature of our approach is the velocity averaging which was analytically performed in the limit when the photon energy $\hbar\omega_D$ much exceeds the dissociation threshold ε_D. In this limit, the kinetic energy T of a nascent OH radical is almost fixed, $T = (\hbar\omega_D - \varepsilon_D)/2 = m_{OH}v^2/2$, and the role of the Doppler effect on the RFWM shapes is paramount. After tedious tensor recouplings, the final result reads

$$\chi^{(res)} = \rho_{ii} \sum_{\lambda KL} t^{(i)}_{KL\lambda} \sum_{r_1 r_2} B^2_{if} F(J_i, J_f; r_1, r_2, K)\{C^{(\lambda)}(\Omega_D) \otimes G^{(K)}(r_1, r_2)\}^{(L)}_0 \Phi_L(\Delta\Omega/kv) \tag{3}$$

where B_{if} is the Einstein B-coefficient for the pumped transition. Here, $G_\sigma^{(K)} = \{\{e_1^{(1)} \otimes e_2^{(1)*}\}^{(r_1)} \otimes \{e_3^{(1)} \otimes e_4^{(1)*}\}^{(r_2)}\}_\sigma^{(K)}$ $(\sigma = -K, -K+1, ...K)$ is the spherical tensor built on the polarization unit vectors $\mathbf{e_n} \equiv e_n^{(1)}$ $(n = 1, 2, 3, 4)$ of the DFWM beams. At detunings $|\Delta\Omega|$ from the resonance frequency ω_{fi} which do not exceed $kv = \omega_{fi}v/c$, the shape function $\Phi_L(\Delta\Omega/kv)$ coincides with the Legendre polynomial P_L of the same argument; otherwise, $\Phi_L = 0$. Structurally, Eq. (3) reproduces the result of Wasserman et al [8, 9] who treated the anisotropy induced by optical pumping; however, the $F-$coefficients derived by us contain also $9j$-symbols missing in the previously derived formuli [9].

The DFWM technique allows a handy way to separate the signal entirely induced by anisotropy from the by far dominating "isotropic" background term proportional to the state monopole $t_{000}^{(i)}$. For that, the input beams 1,2,3 are to be polarized along OY and the signal channel polarizer should be aligned with OX, that is shortly denoted as $(\mathbf{e}_1, \mathbf{e}_2, \mathbf{e}_3, \mathbf{e}_4) = YYYX$. Equation (3) can be used to analyze $\chi^{(res)}$ as a function of the angle φ between $\mathbf{e}_D$ and OX. In so doing, we obtain

$$\begin{aligned}\chi^{(res)} &= i[a_1\Phi_1(\Delta\Omega/kv) + a_3\Phi_3(\Delta\Omega/kv)] + i[b_1\Phi_1(\Delta\Omega/kv) + b_3\Phi_3(\Delta\Omega/kv)]\sin 2\varphi \\ &+ [c_0\Phi_0(\Delta\Omega/kv) + c_2\Phi_2(\Delta\Omega/kv) + c_4\Phi_4(\Delta\Omega/kv)]\cos 2\varphi \qquad (4)\end{aligned}$$

where the φ-independent terms are proportional to $t_{110}^{(i)}$ and $t_{330}^{(i)}$, correspondingly. Similarly, b_1 and b_3 correspond to the field-induced odd-rank anisotropies characterized by $t_{112}^{(i)}$ and $t_{332}^{(i)}$. Finally, the terms varying as $\cos 2\varphi$ are, respectively, proportional to $t_{202}^{(i)}, t_{222}^{(i)}$ and $t_{242}^{(i)}$ and thus refer to the binary $\mathbf{E}-\mathbf{J}$ and triple $\mathbf{E}-\mathbf{J}-\mathbf{v}$ correlations. Since all coefficients a, b and c are real quantities and the intensity I is proportional to $|\chi^{(res)}|^2$, the interference between the terms proportional to the odd- and even-order Legendre polynomials is forbidden and, therefore, the observable shapes appear to be even functions of $\Delta\Omega$. Since for odd orders $P_L(0) = 0$, the $YYYX$ profiles measured at $\varphi = \pi/4$ in a collisionless medium should have a vanishing intensity at $\Delta\Omega = 0$ provided that all nascent particles have the same recoil velocity v. In the simplest case when only the a_1- and b_1-terms are of importance, all anisotropic profiles should behave as $\Delta\Omega^2$ at $|\Delta\Omega| \leq kv$ and disappear at higher detunings.

3. Experimental detalis

The forward-box geometry of DFWM was used, for which all the laser beams, including that of the photolysis laser, were almost co-propagating along OZ. While the photolysis laser wavelength was fixed at 266 nm, the (common) frequency of the input and output beams was scanned that allowed to study the DFWM resonances in the $A^2\Sigma^+ - X^2\Pi(0,0)$ OH vibronic band with resolution of about 0.1 cm^{-1}. All spectra were recorded by applying laser pulses with durations of 6 ns. The reader is referred to our previous publication [4] where the DFWM setup is explicitly described.

Results of two series of low-pressure experiments ($p \leq 1$ mbar) are presently reported. In the first series, no delay was introduced between the dissociation and pump probe pulses and the $YYYX$ responses were measured for different angles φ. Second, the DFWM pulses were delayed by time t_d and the anisotropic signals were studied as function of $t_d \leq 100$ ns at $\varphi = \pi/4$. Estimations show that the results obtained in the first series of experiments can be interpreted assuming the absence of collisions at times not longer than the pulse durations. However, at delays longer than few tens of ns the signals appear to dramatically change with t_d demonstrating the increasing role of collisions in this time scale.

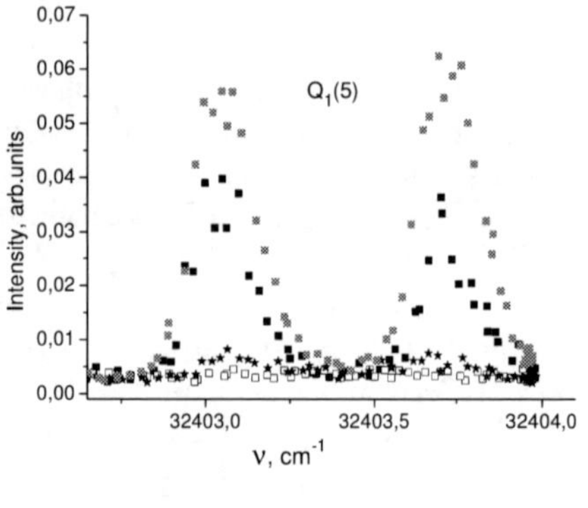

Figure 1. $Q_1(5)(0-0)$ profiles at YYY geometry for different angles φ between dissociation laser polarization and O (black squares) $\varphi=-\pi/4$; (red squares) $\varphi=\pi/4$; (empty squares) $\varphi=0$; (stars) $\varphi=\pi/2$

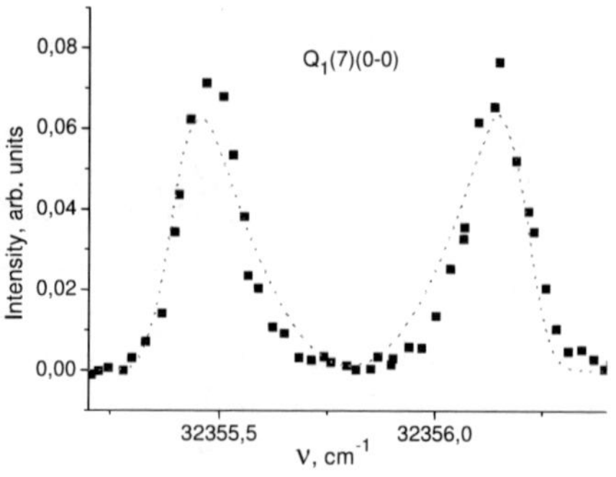

Figure 2. Experimental anisotropic YYY shape of $Q_1(7)(0-0)$ line at $\varphi=-\pi/4$ (squares) and its theoretical fit (dashed line).

4. Discussion

Four plots in Fig. 1 show how the $Q_1(5)$ shape is altered by variations of φ. Equation (4) shows that the even-order polynomials do not contribute to I at $\varphi=\pm\pi/4$. Since the intensity is tangibly changed when φ is switched between these angles, we conclude that this variation should be assigned to interference between the terms of $\chi^{(res)}$ proportional to the $a-$ and $b-$ coefficients. Besides, the intensity almost vanishes at $\varphi=0$ showing that the squares of the $a-$ and $c-$terms do not practically contribute to I. However, I is strongly enhanced when φ is set equal to $\pm\pi/4$ which implies the predominance of the b-terms while the $a-$ ones play a minor role. Furthermore, based on the fact that $I(\pi/4)>I(-\pi/4)$, we conclude the a and b coefficients to be of the same sign. Since the correlations are known to rapidly decrease with their orders, the a_1 and b_1 terms are the most important ones. Recall that the smaller a_1-term refers to the $\mathbf{v}-\mathbf{J}$ correlation not subjected to the orientation of $\mathbf{e}_D$, while the b_1-term originates because of the triple $(\mathbf{E}_D-\mathbf{v}-\mathbf{J})$ correlations. Both terms are caused by the helicity effect for which the mean value of $\cos(\mathbf{v},\mathbf{J})$ is distinct from zero. In other words, some of the OH radicals move like bullets rotating either clockwise or counterclockwise along the trajectory direction.

The existence of helicity finds further support from the analysis of the anisotropic signal shapes. The theory based on Eq. (4) predicts the measured profiles to be well-resolved doublets with an almost zero intensity in the center. A good agreement between theory and experiment is demonstrated by Figure 2 where the simulated and experimental profiles of the $Q_1(7)$ line are depicted. The parabolic theoretical shape $\Delta\Omega^2$ was cut off at $|\Delta\Omega|=kv$ and then convolved with the Gaussian line to account for the finite laser line width and for the recoil velocity dispersion. Fittings resulted in the Gaussian FWHH=0.12 cm^{-1} and kv =0.42 cm^{-1}, in a reasonable accord with estimations of these parameters.

To the best of our knowledge, the rotation-translation helicity of the photolysis products is a new effect which has not been detected so far by using purely optical means. The proof of helicity is a new step in the investigation of photolytic decomposition of H_2O_2 for which a pronounced rotation-translation alignment has been already established. However, the knowledge of alignment can not help us to distinguish between the clockwise and

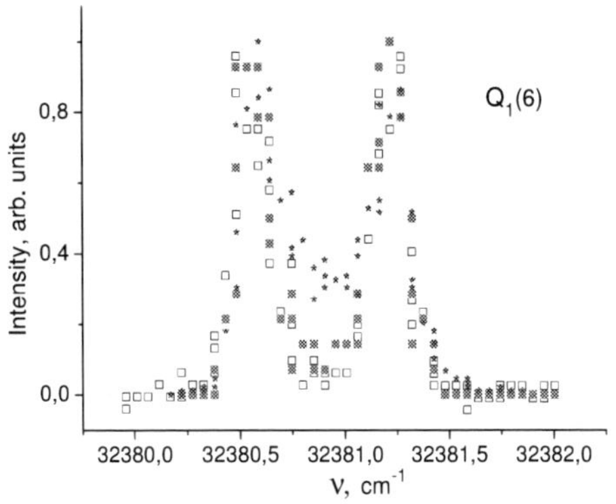

Figure 3. Transformation of the normalized YYY $Q_1(6)(0-0)$ profiles (p=0.5 mbar) with dissociation-probe delay time t (empty squares) $t=0$ ns; (red squares) $t=20$ ns; (blue stars) $t=50$ ns. Intensities at t=20 ns and t=50 ns were magnified by factors 2 and 8, correspondingly

counterclockwise rotations whereas a nonzero helicity proves that the OH radicals are rotating only in one direction. nfortunately, we are yet unable to find out which direction is realized since upon squaring of $\chi^{(res)}$ the information on the helicity sign is lost. To resolve the issue, an implementation of other RFWM techniques for which a useful signal scales linearly with $\chi^{(res)}$ is required.

Another series of experiments was performed with a varied delay time t_d between the pulses of the dissociation and DFWM lasers. Due to the delay, the rotational anisotropy is increasingly lost with t_d. Indeed, as seen from Fig. 3, the $Q_1(6)$ anisotropic signal becomes eight times weaker upon the 50 ns delay showing considerable randomization of the rotational momentum orientation. Also, smoothing of the signal shape with increasing t_d is evident from Fig. 3 that gradually fills the gap between the doublet components. This effect can be explained by velocity changing collisions which cause the spectral exchange and the intensity transfer to the doublet center. However, the overall Doppler width which correlates with v is not visibly affected. Therefore, we may conclude that the absolute value of the recoil velocity does not considerably vary at the time scale of 50 ns whereas the orientation of $\mathbf{v}$ becomes tangibly randomized by elastic collisions.

Acknowledgements

Support from the Science and Technology Cooperation Program Switzerland-Russia (GA 128546), the Swiss Federal Office of Energy, the Swiss National Science Foundation (# 200020_146387) and from the Russian Foundation for Basic Research, grants #15-02-05290 and # 15-03-04997, is gratefully acknowledged.

References

[1] Sato H 2001 *Chem. Rev.* **101** 2687
[2] Blum K 2012 *Density Matrix. Theory and Applications* (Berlin: Springer)
[3] Greene C H and Zare R N 1983 *J. Chem. Phys.* **78** 6741
[4] Maksyutenko P, Radi P P, Kozlov D N, and Kouzov A P 2013 *J. Raman Spectrosc.* **44** 1349
[5] Dixon R N 1986 *J. Chem. Phys.* **85** 1866
[6] Varshalovich D A, Moskalev A N, and Khersonskii V K 1988 *Quantum Theory of Angular Momentum* (Singapore: World Scientific)
[7] Kouzov A and Radi P 2014 *J. Chem. Phys.* **140** 194302
[8] Wasserman T A W, Vaccaro P H, and Johnson B R 1997 *J. Chem. Phys.* **106** 6314
[9] Wasserman T A W, Vaccaro P H, and Johnson B R 1997 *J. Chem. Phys.* **108** 7713

Raman Under Liquid Nitrogen (RUN)

R N Compton*[1] and N I Hammer[2]
[1] Departments of Physics and Chemistry, University of Tennessee, Knoxville, TN 37996, USA
[2] Department of Chemistry and Biochemistry, University of Mississippi, PO Box 1848, University, MS 38677, USA

E-mail: rcompton@utk.edu

Abstract. The recording of Raman spectra for many molecules in air at room temperature is difficult or impossible as a result of sample degradation which is due to a combination of laser heating and oxidation. Often nitrogen gas is applied over the sample in an attempt to reduce oxidation. Also, the samples are sometimes cooled to reduce ro-vibrational "hot bands" and enhance the spectra. We have found great utility in recording Raman spectroscopy of samples *under* liquid nitrogen, a technique we call RUN. The RUN spectra show much higher resolution as a result of ro-vibrational cooling and in some cases cooling produces only the lowest energy conformer of the molecular ensemble further simplifying the spectra. A very sharp Raman peak at 2327.0 cm^{-1}, due to liquid nitrogen, also serves as a convenient wavelength calibration. We also demonstrate the ability to clearly delineate the lattice modes for naphthalene and benzene crystals.

1. Introduction

The Raman effect was first described in 1928 by Sir Chandrasekhara Venkata Raman and Kariamanickam Krishnan which resulted in the Nobel Prize in Physics for Raman only two years later. Since that time Raman spectroscopy, when combined with computations, has become a powerful method for determining molecular structure and for examining intermolecular reactions and molecular dynamics in the condensed phase. Raman and infra-red (IR) spectroscopy provide complementary information on molecular vibrations and in cases of high molecular symmetry (e.g., I_h, O_h, etc.) the two methods are "mutually exclusive", i.e., vibrational modes are either Raman active or IR active but not both. The selection rules for the interaction of light and matter predict that asymmetric vibrations are excited in IR absorption and symmetric vibrations are active in Raman scattering. Together these two complimentary methods of molecular spectroscopy have been essential techniques for understanding the structure of inorganic and organic molecules since the 1930's. When proper attention is addressed to removing the strong Rayleigh line, as well as laser scattered light near the laser line, Raman spectroscopy has also contributed greatly to the understanding of low frequency molecular vibrations as well as lattice modes in solids.

Raman spectroscopy has also proven to be a useful tool for the analysis of biological systems such as conformational changes of peptides and proteins in aqueous solutions. Likewise nucleic acids and polynucleotides have been studied. A general comparison of Raman and IR spectroscopy can be

*To whom any correspondence should be addressed.

found in the books by Ingle and Crouch [1] and by Nakamoto [2]. An excellent introduction to Raman spectroscopy can be found in the general education article by Tobias [3]. Other sources for elementary discussions of Raman spectroscopy can be found in the books by Engel [4], Struve [5], Steinfeld [6] and Harris and Bertolucci [7] among others cited therein.

In 2004 we introduced an exceedingly simple method of recording Raman spectra at 77 K by submerging the sample under liquid nitrogen [8]. We named this method Raman Under Nitrogen or RUN. This study presented the first high resolution Raman spectra for the extremely air sensitive uranocene molecule as well as for other air sensitive metallocenes. In addition to RUN we have found that surface-enhanced Raman spectroscopy (SERS) under liquid nitrogen can be greatly enhanced as a result of the increased dielectric properties of the liquid nitrogen over that of air [9]. This technique is called SERS under nitrogen or SERSUN.

In this article we further describe RUN and present Raman spectra and analysis for the C_{60} fullerene molecule and the $C_{60}Br_{24}$ fullerene derivative. In addition we demonstrate the utility of examining the lattice modes of benzene and naphthalene crystals under liquid nitrogen.

2. Experimental details

The experimental arrangement for recording Raman spectra for samples under liquid nitrogen is exceedingly simple as illustrated in Figure 1 below. The outer container for the liquid nitrogen consists of Styrofoam packing container. The inner container is a Styrofoam coffee cup supported by three toothpicks from the outer container. A few holes are placed in the coffee cup in order for the liquid nitrogen to flow between the two Styrofoam containers. The inner Styrofoam cup reduces the amount of surface rippling due to boil off, vibrations and refilling to keep the level constant. The sample holder is a stainless steel block containing indentations to hold the liquid or solid samples. Sometimes it is necessary to place a screen grid over the solid sample to prevent the sample from floating to the surface of the liquid nitrogen. In most cases the boiling off of the liquid nitrogen prevents fogging of the lens in the Raman apparatus by "pushing" the water vapor in the room away from the lens.

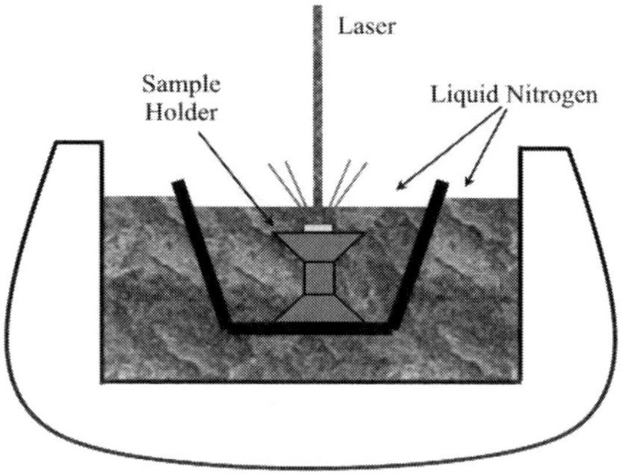

Figure 1. Set-up of the Raman Under Nitrogen (RUN) Spectroscopy experiment.

In the experiments reported herein two different Raman spectrometers were used: (1). At the University of Tennessee the Raman spectra were recorded using a Dilor XY 800 Raman spectrometer. Laser excitation was obtained using a Lexel, Inc. 3500 argon ion laser (5145 Å) and a Coherent Innova 300 krypton ion laser (6764 Å). Raman spectra were taken through Olympus confocal optics in a micro-Raman configuration utilizing a double fore-monochromator coupled to a liquid nitrogen cooled Spectrum One CCD and (2) At the University of Mississippi Raman spectra was recorded using a Labview-controlled Jobin-Yvon Ramanor HG2-S Raman spectrometer employing a 90° scattering geometry relative to the incident laser. The spectrometer is equipped with a double grating (2000 grooves/mm) monochromator and a photomultiplier tube detector. The 514.5 nm laser line of a Spectra Physics Stabilite 2018 Ar^+/Kr^+ laser was used as the Raman excitation source.

The surrounding liquid nitrogen provides a narrow wavelength calibration. For example, the Raman spectrum of the liquid nitrogen shows two narrow features one at 2328.4 cm^{-1} and a weaker peak at 2290.2 cm^{-1} which are due to $^{14}N_2$ and $^{14}N^{15}N$, respectively as shown in Figure 2. The Raman spectra for liquid nitrogen have been well characterized and more accurate values for the N_2 Raman lines are 2327.0 cm^{-1} and 2288.8 cm^{-1} (see reference [10]). The liquid nitrogen vibration is only red-shifted by ~ 3.4 cm^{-1} from the gas-phase value [11]. These peaks are always present and serve to provide a convenient wavelength calibration for the spectrum. The width of the $^{14}N_2$ peak can also be used to provide an estimate the resolution of the apparatus. These lines are very narrow and the positions of the peaks are sensitive to the atmospheric pressure. The presence of liquid oxygen is also seen as a peak at ~1552 cm^{-1}.

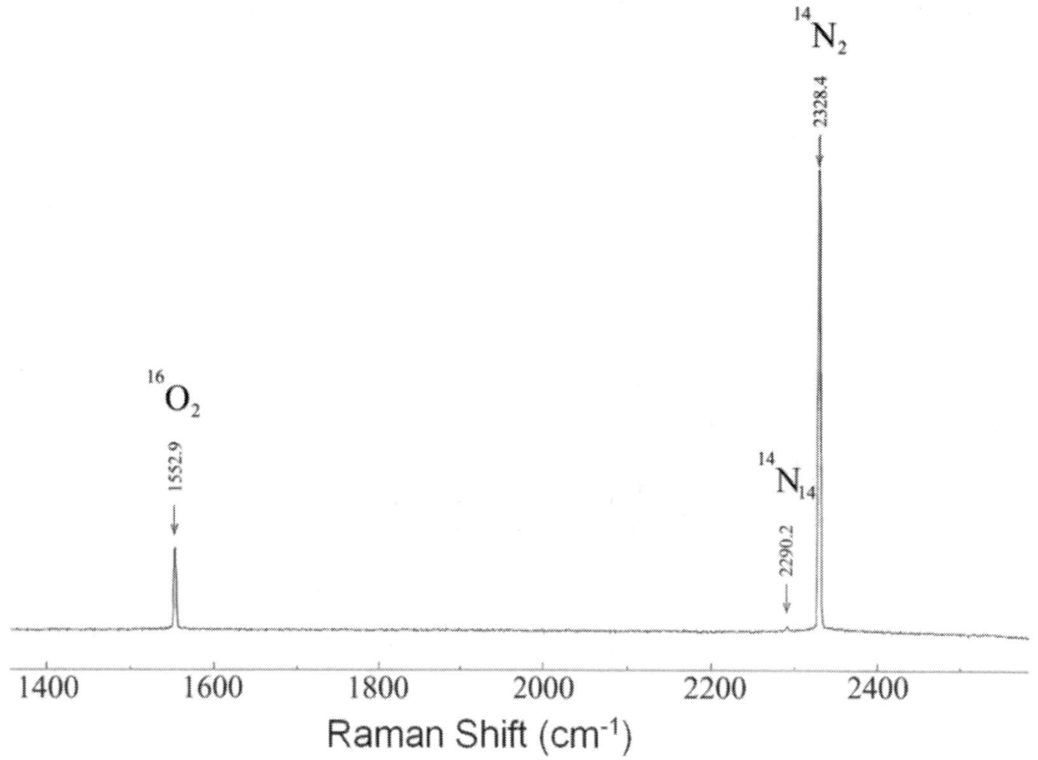

Figure 2. Raman spectrum of liquid nitrogen with the presence of some dissolved oxygen. Oxygen is generally not present in "fresh" samples of liquid nitrogen.

Journal of Physics: Conference Series **548** (2014) 012017 doi:10.1088/1742-6596/548/1/012017

3. Examples of RUN spectra

The spectra to be presented are representative of that for over 150 samples studied over the years. We will discuss results for a few of these below. One of the first molecules to be studied by Sir Chandrasekhara Raman was tetrachlorobenzene, CCl_4. Figures 3 shows a room temperature spectrum for CCl_4 take at UM together with a simulated calculation of the symmetric stretch ν_1 vibration.

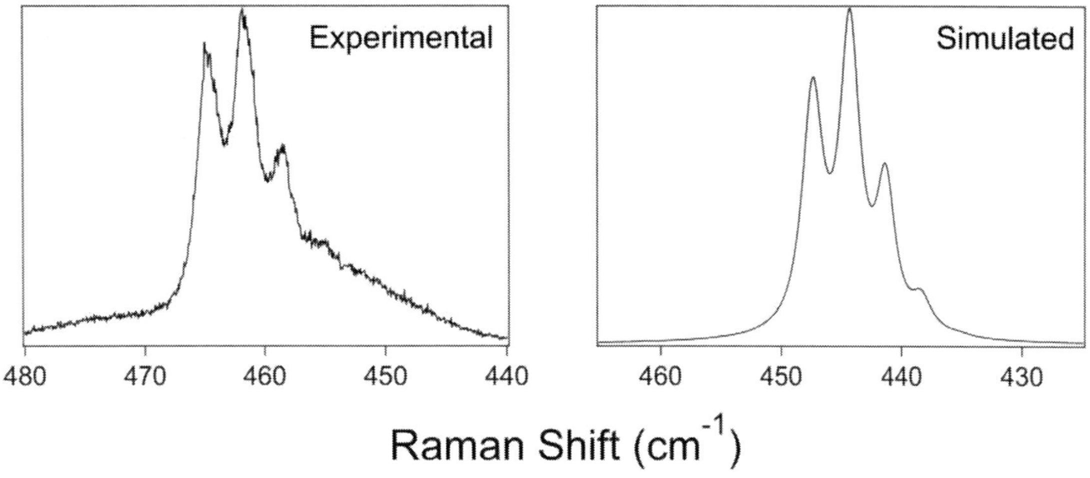

Figure 3. Room temperature Raman spectrum of carbon tetrachloride's symmetric stretch, ν_1 compared to an *ab initio* simulation.

This spectra is clearly degraded by ro-vibrational contributions along with "hot bands". Figure 4 below shows the same vibrational frequency, ν_1, under RUN conditions. In this example, ro-vibrational cooling allows for the individual isotopic contributions of the symmetric stretch, ν_1, to be resolved. Comparison of the results from RUN and room temperature is dramatic. Other spectra of CCl_4 have been able to resolve the ^{13}C contributions to CCl_4 showing up between the Cl vibrations.

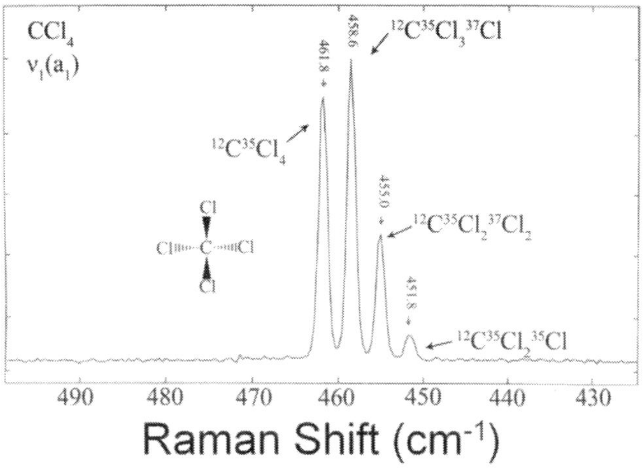

Figure 4. High resolution spectrum of carbon tetrachloride.

A full Raman spectrum of the C_{60} fullerene molecule is shown in Figure 5 below. In this study the C_{60} is dissolved in CS_2. Figure 5shows the ten first order Raman lines for C_{60} are seen to be in reasonable accord (+/- 3 cm^{-1}) with the two A_g and eight H_g allowed Raman vibrational energies. A complete compilation of the vibrational frequencies for C_{60} has been tabulated by Dresselhaus et al. [12]. The spectra in Figure 6 agree with previous results however, the spectra seem to be more narrow than previous data. Run spectra for C_{70} and C_{84} show similar resolution and simplicity. Below we discuss RUN spectra for a fullerene derivative, $C_{60}Br_{24}$.

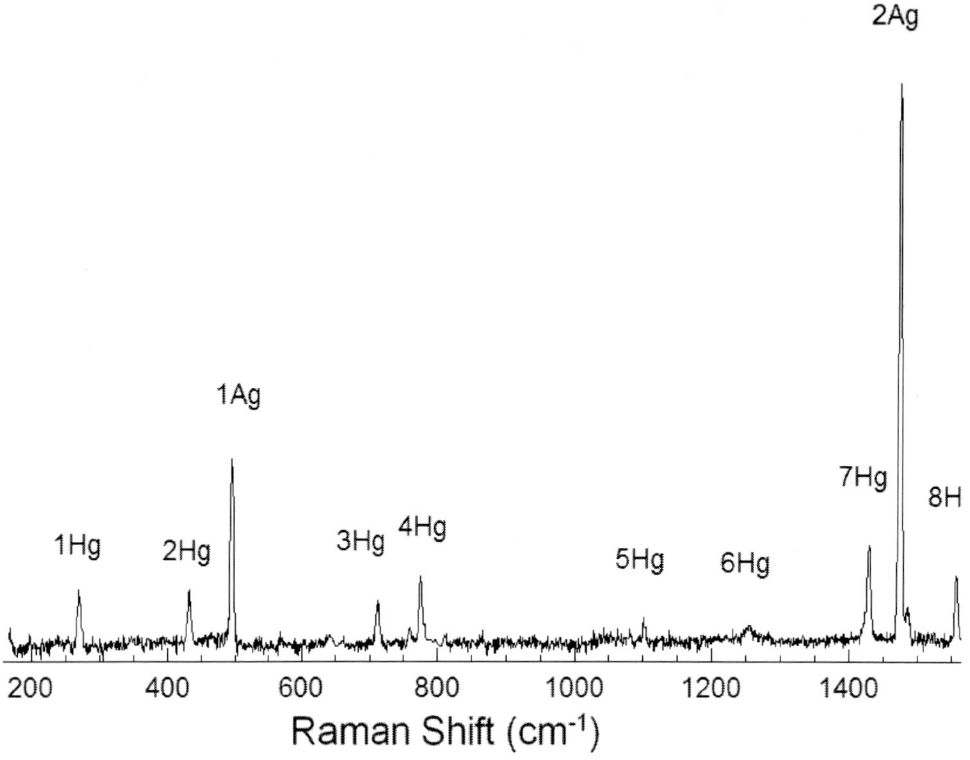

Figure 5. RUN spectrum of C_{60} disolved in carbon disulfide.
The CS_2 Raman spectrum has been removed.

In 1992, Tebbe et al.[13] reported the synthesis and single-crystal X-ray structure of $C_{60}Br_{24}$ exhibiting T_h symmetry. Raman spectroscopy is also used to examine the structure of solid $C_{60}Br_{24}$ prepared by reactions between C_{60} and liquid bromine in the presence of an ion wire. Raman spectra recorded for $C_{60}Br_{24}$ under liquid nitrogen (RUN) reveals all of the allowed Raman active bands to be in good agreement with those calculated from density functional theory. An unexplained peak at 311 cm^{-1} is attributed to interstitial bromine molecules as previously suggested. The fundamental vibration for the bromine molecule is 323 cm^{-1} (see Reference [14]).

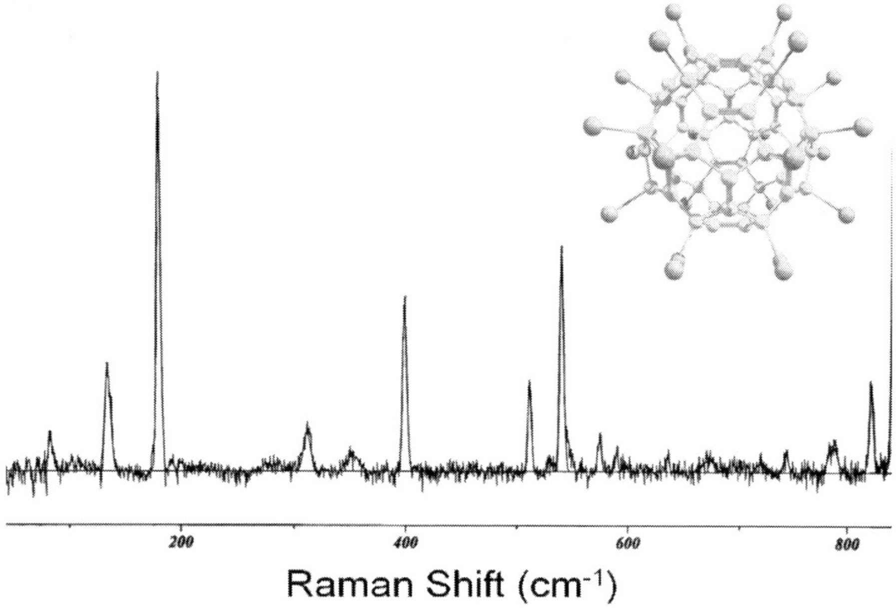

Figure 6. RUN spectrum of $C_{60}Br_{24}$.

Raman Under Nitrogen can also provide another method to examine the lattice modes of molecular crystals at liquid nitrogen temperature. We use RUN of benzene and naphthalene as examples of this application when Raman scattering near the Rayleigh line is possible. Figures 7 and 8 show RUN spectra of benzene and naphthalene, respectively. In both cases, low energy modes are sharpened by the reduction in temperature and collapse into the lowest rotational states. Fig. 9 shows a comparison of the low energy modes of naphthalene at room temperature and under liquid nitrogen.

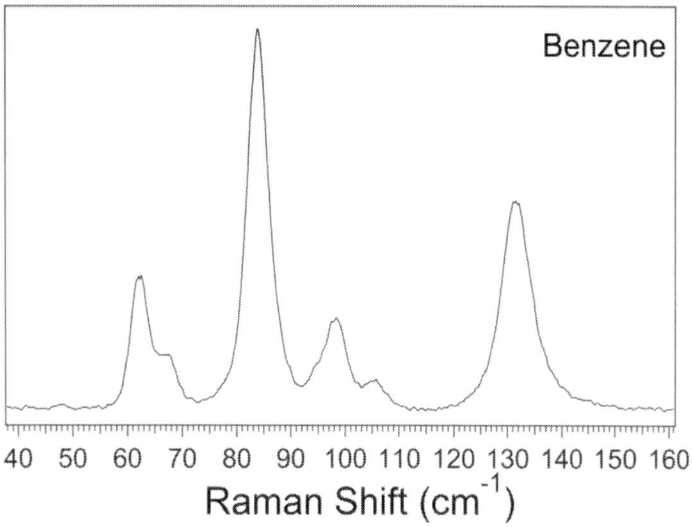

Figure 7. RUN spectrum of benzene in the low energy region of the spectrum recorded at UM. Lattice modes involving motions between benzene molecules are evident.

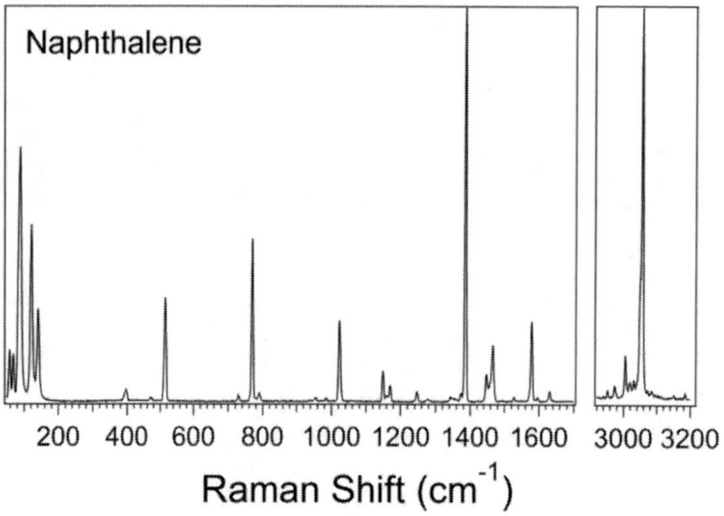

Figure 8. RUN spectrum of naphthalene recorded at UM.

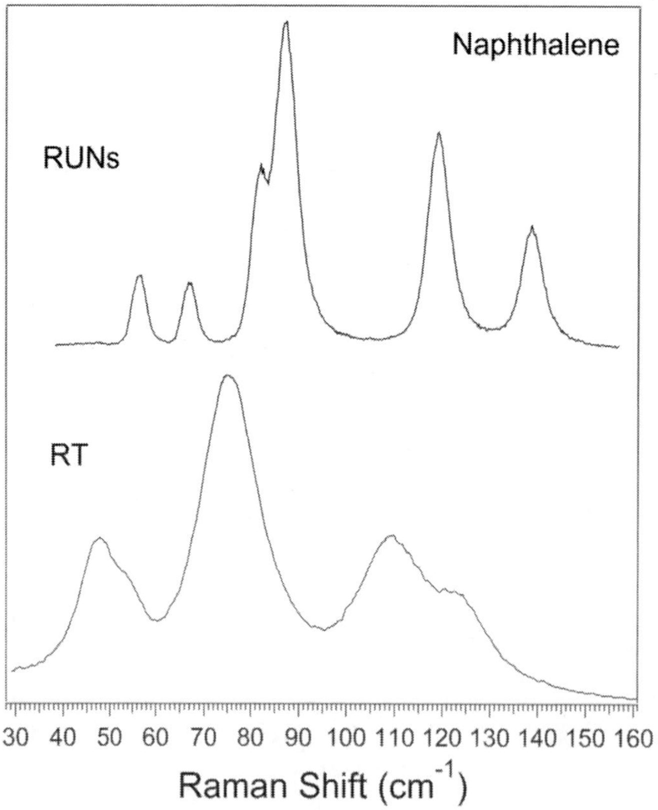

Figure 9. Room temperature Raman spectrum vs RUN spectrum of naphthalene in the low energy region. Normal modes are observed to blue shift and sharpen under liquid nitrogen.

The low energy modes of benzene [15] and naphthalene [16] have been studied in detail previously. Normal modes in this region are observed to blue shift and sharpen under liquid nitrogen. The use of RUN offers key advantages in systems such as these, including polycyclic aromatic hydrocarbons (PAHs). These include: (1) the ability to exclude atmospheric oxygen and water from the sample as the molecules start to interact in the crystalline phase and (2) the ability to study small shifts in vibrational frequencies due to the effects of noncovalent interactions. The latter feature is afforded by the sharper linewidth observed in RUN spectra.

Acknowledgements

This work has been supported by the National Science Foundation (EPS-0903787, CHE-0955550, and CHE-1156713). The authors thank the many students and colleagues who have contributed to this research over the years.

References

[1] Ingle JD Jr. and Crouch SR 1988 *Spectrochemical Analysis* (Englewood Cliffs: Prentice Hall) pp 494-524
[2] Nakamoto K 1997 *Infrared and Raman Spectra of Inorganic and Coordination Compounds* (New York: John Wiley & Sons)
[3] Tobias RS 1967 *J. Chem. Educ.* **44** 2
[4] Engel T 2006 *Quantum Chemistry and Spectroscopy* (San Francisco: Pearson Benjamin Cummings)
[5] Struve WS 1989 *Fundamentals of Molecular Spectroscopy* (New York: John Wiley & Sons)
[6] Steinfeld JI 1985 *Molecules and Radiation: An Introduction to Modern Molecular Spectroscopy* (Cambridge: The MIT Press)
[7] Harris DC and Bertolucci MD 1978 *Symmetry and Spectroscopy: An Introduction to Vibrational and Electronic Spectroscopy* (New York: Dover Publications)
[8] Hager JS, J. Zahardis, J Pagni RM, Compton RN, Li J *J. Chem. Phys.* **120** 2708
[9] Hinde RJ, Sepaniak MJ, Compton RN, Lavrik NV 2001 *Chem. Phys. Letts.* **339** 167
[10] Everitt KF, Skinner L, Ladanyi BM 2002 *J. Chem. Phys.* **116** 179
[11] Everitt KF and Skinner JL 2002 *J. Chem. Phys.* **116** 8015
[12] Dresselhaus MS, Dresselhaus G, and Eklund PC 1996 *Science of Fullerenes and Carbon Nanotubes* (San Diego: Academic Press)
[13] Tebbe FN, Harlow C , Thorn DL, Campbell GC Jr., Calabrese JC, Herron N, Young RJ Jr. and Wasserman E 1992 *Science* **256** 822
[14] Huber KP and Herzberg G. 1979 *Constants of Diatomic Molecules* (New York: Van nostrand Reinhold)
[15] Kan K-C, Huang T-L, Tin C-H 1945 *Chin. J. Phys.* **13** 6
[16] Nicol M, Vernon M, and Woo JT 1975 *J. Chem. Phys.* **63** 1992

Radiative transitions under resonance charge exchange. $H + H^-$ collisions

A Devdariani[1,2], A Dadonova[2] and J Shevtsova[1]

[1] Department of Optics and Spectroscopy, St.Petersburg University, St.Petersburg 198904, Russia
[2] Department of Theoretical Physics and Astronomy, A.I.Herzen University, St.Petersburg 191186, Russia

Email: snbrn2@yandex.ru

Abstract. The $H + H^- \rightarrow H + H + \hbar\omega, H + H^- + \hbar\omega \rightarrow H + H + e$ reactions give examples of bound-bound and bound-free electron transitions in the $H + H^-$ quasimolecules temporally formed during collisions. The present work is aimed at the calculations of spectral profiles produced in the reactions.

1. Introduction

We present our theoretical study of radiating transitions produced in slow collisions of stable negative ions with neutral parent atoms, $A + A^- \rightarrow A^- + A + \hbar\omega$, $A + A^- + \hbar\omega \rightarrow A + A + e$.

In atomic collision physics the process of resonant charge exchange $A + A^- \rightarrow A^- + A$ is well studied. It can be regarded either in terms of transitions of an electron between two identical atoms – diabatic approach – or in terms of the evolution of an atomic system along *gerade* and *ungerade* potential energy curves of the adiabatic states of a quasimolecule formed during collisions – adiabatic approach. Accordingly, the radiative transitions produced during collisions can also be regarded in the frames of two approaches. In the diabatic approach the jumps of the electron from one atom to another produce emission of photons. The starting point for the calculation of spectral profiles is the well-known formula of classical electrodynamics for dipole radiation, see Ref. [1]. In the adiabatic approach emission is treated as a consequence of the dipole transition between two quasi-molecular states involved [2].

The radiating transitions have been supposed to take place between two discrete quasi-molecular states. However, another type of transitions can also be treated, namely $A + A^- + \hbar\omega \rightarrow A + A + e$. The process can be attributed to bound-free electron transitions in quasimolecules or to photo detachment of quasimolecules. The main characteristics of the process are dipole moments of bound-free transitions, which have been calculated in the frame of the LCAO approximation.

2. Bound-bound transitions

We study the specific example of $H + H^-$ collisions in the frame of the adiabatic approach in order to clarify the main features of spectral profiles produced in the collision processes of charge exchange.

2.1. Potential energy terms

As the first step in accordance with the approach one needs to provide quantum-chemical information on potential energy curves and a dipole transition moment [2]. It is well known that a negative ion of hydrogen can be described with reasonable accuracy by the Zero Range Potential Model [3]. Thus we use the three-dimensional model to calculate the energy terms and the dipole moment [4,5] that depend on the interatomic distance.

It is interesting to note that the one-dimensional zero range potential model (ZRPM) was also used to clarify questions about the calculation of the exchange energy of a homonuclear positive molecular ion H_2^+ [6]. It was noted that the exchange energy in this model problem can be expressed via the Lambert function. In our problem the potential energy functions can also be expressed in the closed forms through the Lambert function $W(z)$ which has recently attracted much attention [7] and is a solution of the equation

$$z = W(z) exp W(z) . \tag{1}$$

For example, the potential energies are

$$U_{g,u}(R) = -\frac{1}{2}\chi_{g,u}^2 \tag{2}$$

where $\chi_{g,u}$ are connected with the Lambert function through

$$\chi_{g,u}(R) = \alpha + \frac{W[\pm \exp(-\alpha R)]}{R} , \tag{3}$$

and $-\alpha^2/2$ is the bound energy of an additional electron in the atomic negative ion.

Below the difference potential energy $\Delta U(R) = U_u - U_g$ is used. Applying Taylor series to the Lambert function leads to

$$\Delta U = \frac{2\alpha}{R} e^{-\alpha R} - \frac{2}{R^2} e^{-3\alpha R} \tag{4}$$

at $\alpha R \geq 1$. Interestingly, that the Lambert function was recently used in the theory of blackbody radiation functions [8].

2.2. Dipole transition moment

For the first time the dipole transition moment for radiating transitions between *gerade-ungerade* states produced in charge exchange has been calculated by Mulliken [9] in the linear combination of atomic orbitals (LCAO) approximation

$$D(R) \approx \frac{R}{2} . \tag{5}$$

It is the approximation that has been used in [4,5] to calculate the Einstein coefficient for spontaneous emission of the quasimolecule

$$A(R) = \frac{4}{3g}\left(\frac{\omega}{137}\right)^3 D(R)^2 . \tag{6}$$

In the present work, in contrast [4, 5] the dipole moment has been calculated directly in the frame of the ZRPM without any further simplifications.

2.3. Spectral profiles
The spectral profile in the quasi-static approximation is

$$I(\omega) = 4\pi R_c^2 \left[\frac{A(R_c)}{d\Delta U/dR_c}\right] exp\left[-\frac{U_u(R_c)}{kT}\right], \tag{7}$$

where R_c is the position of the Condon point that is the root of the equation

$$\Delta U(R_{c)} = \omega \quad , \tag{8}$$

which has the solution at $\alpha R \geq 1$ as

$$R_c = \frac{1}{\alpha} W\left(2\frac{\alpha^2}{\omega}\right) . \tag{9}$$

The results of the profile calculations are shown in figures 1 and 2 for two temperatures versus frequency of the radiation and the interatomic distance, respectively.

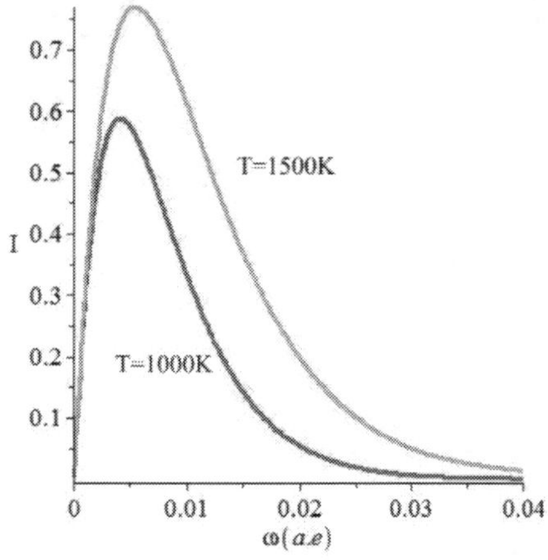

Figure 1. Spectral profile produced in the $H + H^-$ collisions versus frequency of the radiation for two temperatures.

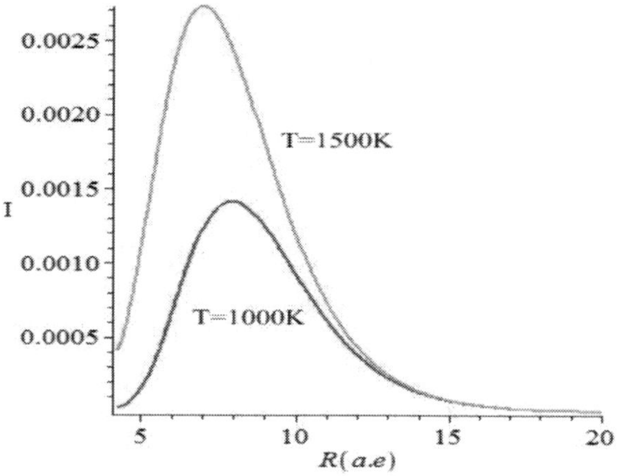

Figure 2. Spectral profile produced in the $H + H^-$ collisions versus the interatomic distance for two temperatures.

The main part of the transitions are concentrated in the region of comparatively large interatomic distances 8-10a.u. or 10^3cm^{-1}. It should be noted the temperature values on similar graphs in [4] based on LCAO should be changed, that is $1000 \leftrightarrow 1500$.

3. Free-bound transitions
In the previous Section the radiating processes produced by charge exchange have been considered as an example of bound-bound transitions in quasimolecules. Here, we consider the reaction $A + A^- + \hbar\omega \rightarrow A + A + e$ as an example of bound-free transitions in quasimolecules or, in other words, photo detachment of quasimolecules.

The cross-section of photo detachment is proportional to the density of oscillator strengths which is the main characteristic of the dipole transitions

$$\frac{df_{bc}}{d\omega} = 2\omega_{bc}|x_{bc}|^2, \tag{10}$$

where the matrix element is

$$x_{bc} = \int_{-\infty}^{\infty} \langle \Psi_b | x | \Psi_c \rangle dx \,. \tag{11}$$

For simplicity, we discuss the one-dimensional case. Equation (9) indicates that continuous wave functions ψ_c of the quasimolecule are normalized in accordance with $\langle \Psi_c(E) | \Psi_c(E') \rangle = \delta(E - E')$. In the frame of the LCAO approximation

$$\psi_b^{g,u} = \frac{1}{\sqrt{2}}(\varphi_b^+ \pm \varphi_b^-),\ \psi_c^{g,u} = \frac{1}{\sqrt{2}}(\varphi_c^+ \mp \varphi_c^-)\,, \tag{12}$$

where $\varphi_{b,c}^{\pm}$ mean one-electron wave functions for bound and continuous states centered at the atoms that is at $\mp\frac{R}{2}$. The symmetry of the quasi-molecular functions in equation (11) accounts for the symmetry of ionic functions $\varphi_b(-x) = \varphi_b(x), \varphi_c(-x) = -\varphi_c(x)$.

Substitution of Eq. (12) into Eq. (11) leads to the following formulae for the matrix elements of bound-free transitions between g and u quasi-molecular states

$$x_{b=g,c=u} = \frac{1}{2}\langle \varphi_b | x | \varphi_c \rangle + \frac{1}{2}\langle \varphi_b^+ | x | \varphi_c^- \rangle,$$

$$x_{b=u,c=g} = \frac{1}{2}\langle \varphi_b | x | \varphi_c \rangle - \frac{1}{2}\langle \varphi_b^+ | x | \varphi_c^- \rangle\,. \tag{13}$$

The functions $\varphi_{b,c}$ are centered at the origin, and the factor ½ is due to the existence of two initial and two final states.

Let us take the one-dimensional ZRPM to describe the negative ion of hydrogen quasimolecule, then

$$\varphi_b^{\pm} = \sqrt{\alpha} exp\left(-\alpha\left[x \pm \frac{R}{2}\right]\right),\ \varphi_c^{\pm} = \frac{1}{\sqrt{\pi k}} sink\left|x \pm \frac{R}{2}\right|, E = \frac{k^2}{2}\,, \tag{14}$$

and

$$x_{b=g,c=u} = x_{bc}^0 \cos^2\frac{kR}{2} + \frac{1}{4}\sqrt{\frac{\alpha}{\pi k}}\frac{\alpha R}{\omega}\sin kR, \tag{15}$$

$$x_{b=u,c=g} = x_{bc}^0\, sin^2\frac{kR}{2} - \frac{1}{4}\sqrt{\frac{\alpha}{\pi k}}\frac{\alpha R}{\omega}\sin kR, \tag{16}$$

where $x_{bc}^0 = \alpha\sqrt{\frac{\alpha k}{\pi}\frac{1}{\omega^2}}$ is the matrix element for a single ion and $\omega = \frac{\alpha^2+k^2}{2}$. Although the quasi-molecular matrix elements increase with increasing R it does not lead to difficulties at large R. In this case $U_u = U_g$, and it is obligatory to use the sum of amplitudes of two different ways of detachment, which leads to $x_{b=g,c=u} + x_{b=u,c=g} = x_{bc}^0$.

4. Conclusions

Bound-bound and bound-free transitions in quasimolecules formed by colliding atoms and stable ions have been considered in the frame of the ZRPM which is suitable to describe the negative ion of hydrogen. The model used is one-electron, thus one-electron states involved must be brought in line with the real H_2^- states. It can be shown qualitatively that g and u states discussed above are connected with the $\Sigma_{g,u}^2$ states of the quasi-molecular ion, but the boundary of the continuum is related with the Σ_u^3 state of the hydrogen molecule.

Three main results should be emphasized. First, it is the collision $A + A^-$ that produces emission in the absence of any initial electronic excitation in colliding particles. Second, quasi-molecular radiative transitions lead to the formation of ro-vibrational states of the H_2^- molecules at comparatively large interatomic distances about 8-10 a.u. Third, collisions result in a lowering of the threshold of the electron detachment compered with the single ion because u bound quasi-molecular term is repulsive. To calculate the cross-section of the detachment near the threshold, only the square of the matrix element, $x_{b=u,c=g}^2$, equation (16), is necessary without any contributions from the $g \to u$ transitions which are forbidden.

Acknowledgments

We are grateful to the reviewers for their valuable comments and the suggestion of the Ref. [8].

References

[1] Landau L D and Lifshitz E M 1971 *The classical Theory of Fields* (Oxford: Pergamon Press)
[2] Devdariani A Z 1999 *Opt. Spectr.* **86** 853
[3] Demkov N and Ostrovsky V N 1988 *Zero-range Potentials and their Applications in Atomic Physics* (New York: Plenum)
[4] Dadonova A V, Devdariani A Z 2012 *J. Phys. : Conf. Series.* **397** 012040
[5] Dadonova A V, Devdariani A Z 2013 *Vestnik St. Petersburg University* **4** 155
[6] Scott T C, Babb J F, Dalgarno A and Morgan III J D 1993 *J. Chem. Phys.* **99** 2841
[7] Corless R M, Connet G H, Hare D E G, Jeffrey D J and Knuth D E 1996 *Adv. Comput. Math.* **5** 329
[8] Stewart S M 2012 *J. Quant. Spectrosc. Radiat. Transfer* **113** 232
[9] Mulliken R S 1939 *J. Chem. Phys.* **7** 20

Journal of Physics: Conference Series **548** (2014) 012019 doi:10.1088/1742-6596/548/1/012019

Measurements of Line Strength and Line Shape Parameters of the HCl R3 (0-2) 5739.263 cm^{-1} and the HF R1 (0-2) 7823.820 cm^{-1} Transitions

D H Plemmons,[1] N A Galyen,[2] W J Phillips,[1] and B C Winkleman[1]

[1]Senior Scientist, Aerospace Testing Alliance, Arnold Air Force Base, USA
[2]Engineer, Aerospace Testing Alliance, Arnold Air Force Base, TN 37389

E-mail:david.plemmons@arnold.af.mil

Abstract. Line strengths and line shape parameters have been measured for a single absorption transition of both hydrogen chloride and hydrogen fluoride gases. These parameters were determined from spectral absorbance measurements over a range of temperatures and pressures using single-mode laser diodes as the laser sources. The measured parameters are compared to values reported in scientific literature. Significant differences are observed that will impact spectroscopic measurements of species concentrations, temperature, and pressure.

1. Introduction

Laser diode absorption spectroscopy is a powerful nonintrusive diagnostic technique for gas phase systems. It is often used to measure species concentrations, temperatures, and velocities in high-temperature, harsh, or otherwise inaccessible environments [1-4]. In order to minimize the uncertainties in the measured quantities, accurate line shape parameters are required. For this paper, the interest is in measuring concentrations of hydrogen chloride (HCl) and hydrogen fluoride (HF). This paper presents a methodology to determine the line shape parameters for these two molecular species.

Single-mode laser diodes typically have a spectral line width that is several orders of magnitude smaller than the width of a measured absorption line. Hence, the measured absorption profile is an accurate representation of the actual line shape with no need to deconvolve the laser line width from the spectral line shape. The line shapes are modeled as Voigt profiles. Voigt profiles contain both Doppler (temperature) and Lorentzian (pressure) broadening effects. The absorption model assumes that the line strength and pressure broadening parameters are well known. These parameters are available in the HITRAN database [5]. However, the uncertainties of the tabulated parameters are usually not well known.

This paper describes the research efforts to validate the HITRAN parameters for single absorption transitions of HCl [R3 (0-2), 5739.263 cm^{-1}] and HF [R1 (0-2), 7823.820 cm^{-1}] and presents the results of the measured values of line strength and pressure broadened line shape parameters. Low concentration HCl and HF gas, on the order of parts per million (ppm), was introduced into a temperature and pressure controlled absorption cell and interrogated using collimated laser diode beams. The resulting absorption profiles were collected and analyzed to determine the targeted line shape parameters. The measured line shape parameters were then compared to the values listed in the HITRAN database.

2. Laser absorption diagnostics

The fundamental equation for absorption spectroscopy is the Beer-Lambert Law [6]. The Beer-Lambert relation gives the transmitted radiation intensity, *I(ω),* of a collimated beam of light after passing through some homogeneous absorptive medium as

$$I(\omega) = I_0(\omega) \cdot e^{-\kappa(\omega,P,T,\chi)}, \tag{1}$$

where $I_0(\omega)$ is the incident radiation intensity, $\kappa(\omega,P,T,\chi)$ is the spectral absorbance, ω is the optical frequency, P is the pressure, T is the temperature, and χ is the mole fraction of the absorbing species. Rearranging equation (1), the spectral absorbance is expressed as:

$$\kappa(\omega,P,T,\chi) = -\ln\left(\frac{I(\omega)}{I_0(\omega)}\right). \tag{2}$$

Figure 1 (a) shows a notional absorption profile, Eq. (1), and (b) an absorbance profile, Eq. (2).

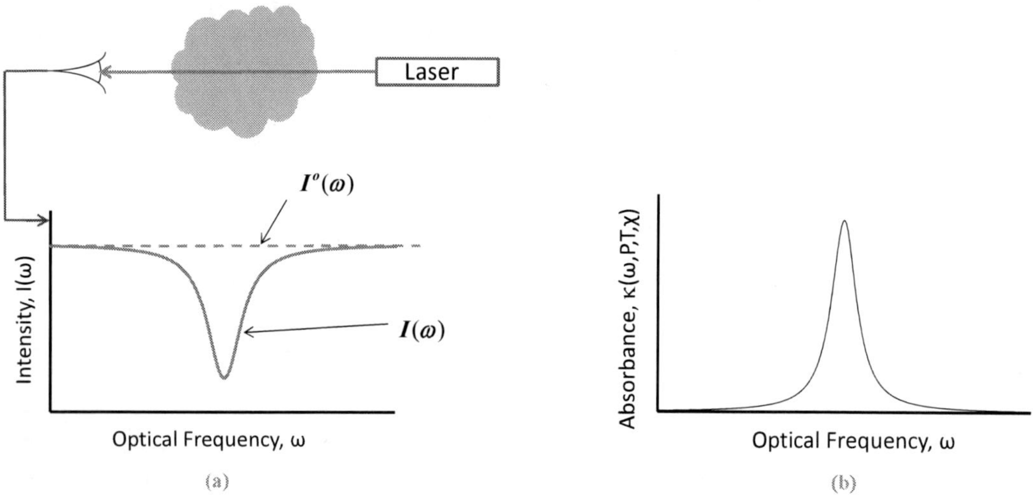

Figure 1. (a) Absorption profile, Eq. (1), observed in a homogenous medium and (b) absorbance profile, Eq. (2).

For a single molecular transition, the absorbance can be expressed as

$$\kappa(\omega,P,T,\chi) = S^0(T)\chi N(T,P)L\Phi(T,P), \tag{3}$$

where $S^0(T)$ is the line strength, $N(T,P)$ is the total number density, L is the optical path length, and $\Phi(T,P)$ is the line shape function. A Voigt profile is used as the line shape function in this work. The Voigt profile is the convolution of a Doppler-broadened line shape and a pressure (or Lorentz)-broadened line shape and is given as

$$\Phi(T,P) = \frac{a\sqrt{\ln 2}}{\pi^{3/2}\gamma_D}\int_{-\infty}^{\infty}\frac{e^{-y^2}}{a^2+(x-y)^2}dy, \tag{4}$$

where

$$a = \frac{\gamma_L}{\gamma_D}\sqrt{\ln 2} \text{ and } x = \frac{\omega - \omega_0}{\gamma_D}\sqrt{\ln 2}\,.$$

Here, γ_D is the Doppler-broadened line half-width at half maximum (HWHM), γ_L is the Lorentz-broadened line HWHM, and ω_0 is the optical frequency at the absorption line center or peak. The Doppler HWHM is given as

$$\gamma_D(T) = \frac{\omega_0}{c}\sqrt{\ln 2 \frac{2kT}{m}} = 3.581\times 10^{-7}\,\omega_0\sqrt{\frac{T}{M}}\,,$$

where c is the speed of light, k is the Boltzmann constant, m is the molecular mass, and M is the molecular weight.

The Lorentz HWHM is an empirical value given by

$$\gamma_L(P,T,\chi) = \gamma_{air}^0(T_0)\left(\frac{T_0}{T}\right)^{\eta} P(1-\chi) + \gamma_{self}^0(T_0)\left(\frac{T_0}{T}\right)^{\eta} P\chi\,, \tag{5}$$

where $\gamma_{air}^0(T_0)$ is the air-broadened HWHM in cm^{-1}/atm. at the reference temperature T_0 and $\gamma_{self}^0(T_0)$ is the self-broadened HWHM cm^{-1}/atm. at the reference temperature T_0. γ_{air}^0, γ_{self}^0, and η are constants and are available in the HITRAN database. The reference temperature used for the HITRAN database is 296 K. For this work, χ is on the order of parts per million and equation (5) is approximated as

$$\gamma_L(P,T) \approx \gamma_{air}^0(T_0)\left(\frac{T_0}{T}\right)^{\eta} P\,. \tag{6}$$

For analysis purposes, it is convenient to calculate the integrated absorbance (A) using equation (3)

$$A = \int_{-\infty}^{\infty} \kappa(\omega,P,T,\chi)\,d\nu = S^0(T)\chi N(T,P)L \tag{7}$$

since by definition the integrated Voigt profile is equal to unity.

Equation (7) can also be written as

$$A = S^0(T)\Theta(T,P), \tag{8}$$

where $\Theta(T,P)$ is referred to as the column density. The line strength is a function of temperature only and can be expressed as

$$S^0(T) = S^0(T_0)f(T), \tag{9}$$

where the constant $S^0(T_0)$ is the line strength at the reference temperature T_0. The temperature correction factor is given by

$$f(T) = \frac{Q(T_0)}{Q(T)} \frac{1-\exp\left(-\frac{hc\omega_0}{kT}\right)}{1-\exp\left(-\frac{hc\omega_0}{kT_0}\right)} \exp\left(-\frac{hc}{k}E''\left(\frac{1}{T}-\frac{1}{T_0}\right)\right),$$

where *E"* is the lower energy state, *Q(T)* is the molecular partition function at temperature *T*, and *h* is the Plank constant. Combining Eqs. (8) and (9) yields

$$A = S^0(T_0) f(T) \Theta(T,P). \tag{10}$$

3. Experimental approach

The quantities tabulated in the HITRAN database that are measured during this effort are as follows:

- $S^0(T_0)$, the line strength at T_0;
- $\gamma_{air}^0(T_0)$, the air-broadened HWHM at one atmosphere pressure and T_0;
- η, the coefficient of temperature dependence of the air-broadened HWHM.

Line strength is determined from the integrated absorbance, equation (10). $\gamma_{air}^0(T_0)$ and η are determined from a linearized version of equation (6). A detailed description of the analysis process is given in Section 4.

Examination of equations (6) and (10) reveals that absorption data are needed over a range of temperatures and pressures in order to accurately determine these three parameters. An absorption cell was constructed that allowed independent control over the temperature and pressure. A schematic of the equipment setup is shown in Figure 2.

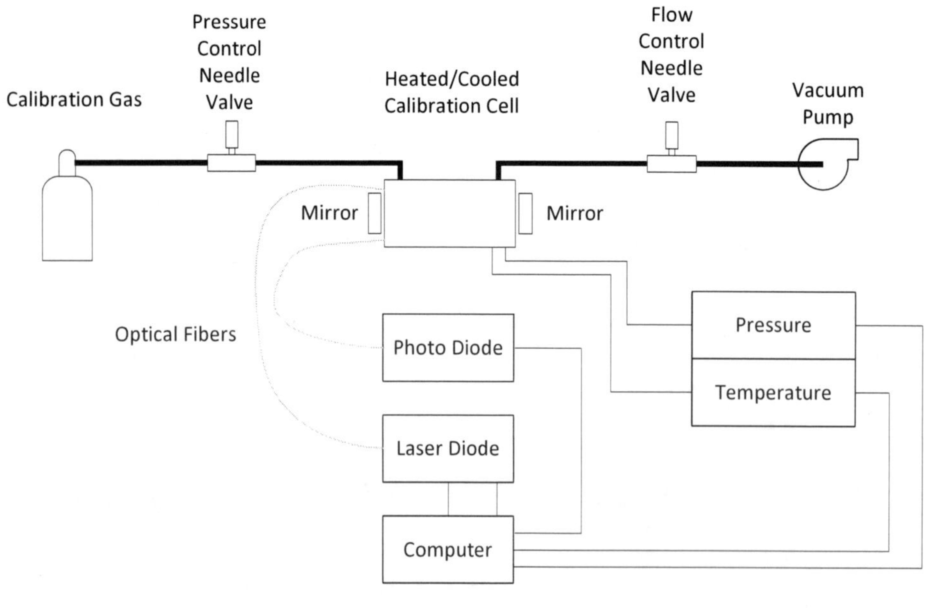

Figure 2. Schematic of the experimental arrangement.

The calibration gases were introduced into a temperature-controlled absorption cell. A vacuum pump and a needle value on the downstream side of the absorption cell were used to control the flow rate of the calibration gas. A needle valve on the upstream side of the absorption cell was used to control the pressure by creating a critical orifice and choking the flow. The absorption cell consisted of a stainless-steel tube sealed with windows on both ends. The fiber-coupled beam from the laser diode was routed to the absorption cell and directed through the calibration gas sample using a collimator. A pair of mirrors was used to create a multipass optical arrangement. The laser beam was then coupled into a second optical fiber and directed to a photodiode. The photodiode output signal was recorded on the computer hard disk along with the temperature and pressure of the calibration gas measured inside the calibration cell.

At each temperature set point, the pressure was adjusted to the desired value and the flow rate was minimized to increase the calibration gas dwell time in the absorption cell. Each data point was recorded when the temperature of the calibration gas inside the absorption cell had stabilized.

4. Data analysis and results

Absorption data were acquired at various concentrations, path lengths, temperatures, and pressures for both HCl and HF. The spectral data were curve fit with the Voigt profile and a polynomial background for $I_0(\omega)$ to obtain the integrated absorbance, the Lorentz HWHM, and the column density.

4.1. Line strength determination

From Equation (10), the integrated absorbance results can be expressed as

$$A_i = S^0(T_0)f(T_i)\Theta(T_i, P_i) \tag{11}$$

where the subscript i indicates each unique temperature, pressure, and mole fraction data point in the test matrix. When A_i is plotted against $f(T_i)\Theta(T_i, P_i)$ the resulting slope is equal to the reference line strength $S^0(T_0)$.

Figure 3 shows a plot of Equation (11) for the HF data acquired during this effort along with the linear fit to the data. The line resulting from the line strength in the HITRAN 2012 database is also shown for comparison. The line strength value from our measurements is 6.79×10^{-20} as compared to the HITRAN value of 7.62×10^{-20}. The measured value is 10% lower than the HITRAN value.

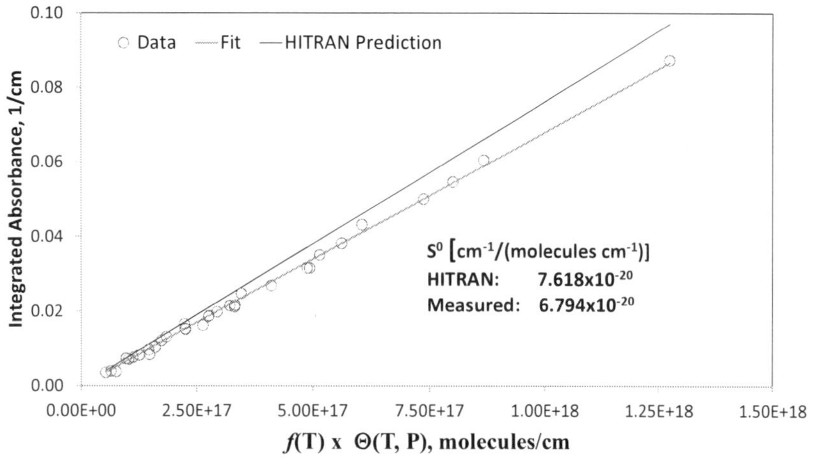

Figure 3. Plot of HF integrated absorbance vs. *f*(T)Θ(T, P).

4.2. Lorentz HWHM scaling parameters determination

From Equation (6), the Lorentz HWHM can be expressed as

$$\gamma_i(P_i,T_i)=\gamma_{air}^0(T_0)\left(\frac{T_0}{T_i}\right)^{\eta}P_i. \tag{12}$$

Taking the logarithm of equation (12) yields

$$\log(\gamma_i)-\log(P_i)=\log(\gamma_{air}^0)+\eta\log\left(\frac{T_0}{T_i}\right), \tag{13}$$

which is a linear equation where η is the slope and $\ln(\gamma_{air}^0)$ is the intercept.

Figure 4 shows a plot of equation (13) for the HF data acquired during this effort and the resulting straight line fit. The inferred Lorentz line parameters are shown in Figures 5 and 6. Figure 5 shows the HF temperature-corrected Lorentz HWHM from equation (12) plotted as a function of pressure. Also shown in Figure 5 is the temperature-corrected Lorentz HWHM predicted using HITRAN tabulated values for η and γ_{air}^0. The measured value of γ_{air}^0 is 0.098 compared to the HITRAN value of 0.085. The measured value is 15% higher than the HITRAN value.

Figure 6 shows the pressure-corrected Lorentz HWHM vs. temperature ratio. Also shown in Figure 6 is the predicted value using HITRAN parameters. The measured value for η is 0.65. The HITRAN value is 0.50. The measured value is 30% higher than the HITRAN value.

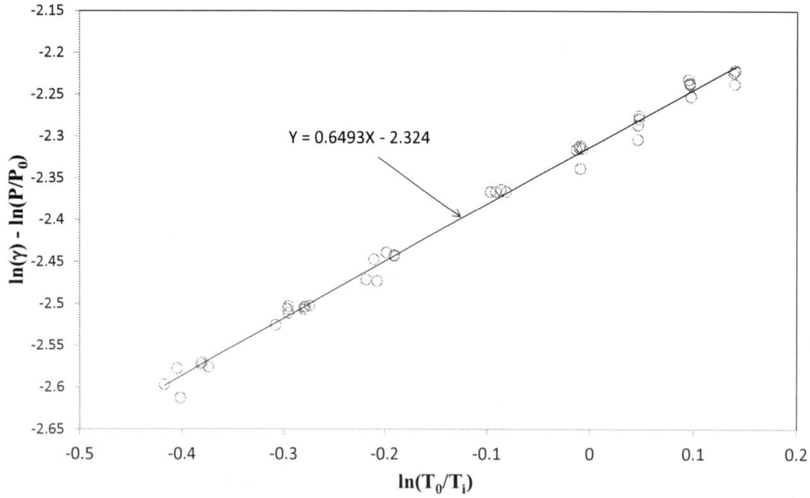

Figure 4. Plot of equation (13) for the HF absorption data analyzed during this effort.

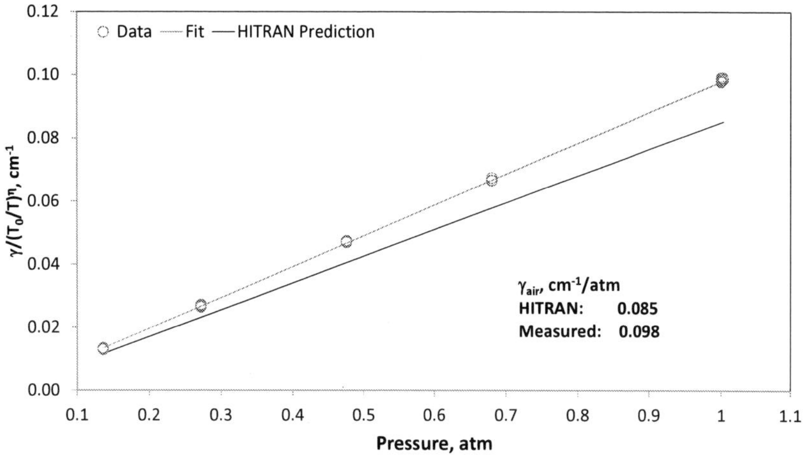

Figure 5. Temperature-corrected fitted Lorentz HWHM vs. pressure.

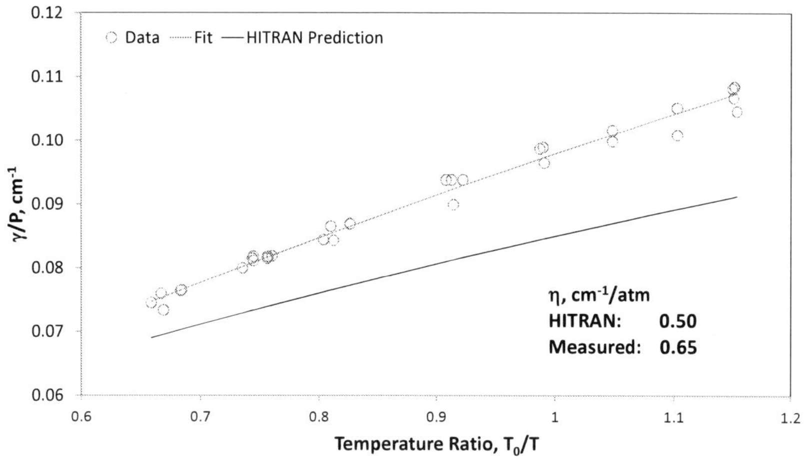

Figure 6. Pressure-corrected air-broadened HWHM vs. temperature ratio.

4.3. Results summary

This analysis process was repeated for HCl with similar results. A summary of the results for both gases is shown in

Table 1. The differences between our measured values and the values retrieved from the HITRAN database varies from 10 to 45%. These difference have a significant impact on the measured mole fractions and on the properties inferred from line shape, such as temperature and pressure.

Table 1. Comparison of measure line shape parameters and values listed in the HITRAN database.

Species	Parameter	Units	Measured Value	HITRAN Value	Difference
HF	$S^0(T_0)$	$\frac{molecules \cdot cm^{-1}}{cm^{-2}}$	6.81×10^{-20}	7.62×10^{-20}	11%
HF	$\gamma_{air}^0(T_0)$	$\frac{cm^{-1}}{atm}$	0.098	0.085	15%
HF	η	–	0.65	0.5	30%
HCl	$S^0(T_0)$	$\frac{molecules \cdot cm^{-1}}{cm^{-2}}$	1.37×10^{-20}	1.25×10^{-20}	10%
HCl	$\gamma_{air}^0(T_0)$	$\frac{cm^{-1}}{atm}$	0.0833	0.0574	45%
HCl	η	–	0.622	0.566	10%

5. Summary and conclusion

HF and HCl line strengths and line shape parameters have been experimentally measured over a range of temperatures and pressures. Significant differences between the measured parameters and the parameters listed in the HITRAN database have been noted. The differences between our measured values and the values retrieved from the HITRAN database vary from 10 to 45%. These differences have a significant impact on the measured mole fractions and on the properties inferred from line shapes, such as temperature and pressure.

6. Acknowledgments

Distribution A: Approved for public release; distribution is unlimited.

References

[1] Plemmons D H, Galyen N A, Mills M L 2014 *AIAA SciTech Conference (National Harbor, MD, 13-17 January 2014)* AIAA-2014-1243
[2] Hanson R K 2011 *Proc. Combustion Inst.* **33** 1
[3] Galyen N A, Plemmons D H 2010 *27th AIAA AMT/GT Conference (Chicago, IL, 28 June – 2 Juy 2010)* AIAA-2010-4910
[4] Allen, M G 1998 *Measurement Science and Technology* **9** 545
[5] Rothman, L S *et al.* 2013 *J. Quant. Spectrosc. Rad. Transfer* **130** 4
[6] Thorne, A P 1988 *Spectrophysics* (New York: Chapman and Hall)

Journal of Physics: Conference Series **548** (2014) 012020 doi:10.1088/1742-6596/548/1/012020

Spectroscopy of atom and nucleus in a strong laser field: Stark effect and multiphoton resonances

A V Glushkov

Odessa State Environmental University, L'vovskaya str. 15, Odessa-16, 65016, Ukraine

E-mail: dirac13@mail.ru

Abstract. The consistent relativistic energy approach to atoms in a strong realistic laser field, based on the Gell-Mann and Low S-matrix formalism, is applied in the study of resonant multiphoton ionization of krypton by intense uv laser radiation and for the computation of the resonance shift and width in krypton. The approach to the treatment of the multiphoton resonances in nuclei is outlined for the ^{57}Fe nucleus.

1. Introduction

At the present time, the physics of multiphoton phenomena in atoms and molecules shows great progress that is stimulated by development of new laser technologies (see Refs. [1-10]). The availability of powerful laser sources allows one to obtain a radiation field amplitude on the order of atomic fields in the wide range of wavelengths. Consequently, systematic investigations of the nonlinear interaction of radiation with atomic and molecular systems can are possible [1-14]. At the same time, direct laser-nucleus interactions traditionally have been dismissed because of the well known effect of small interaction matrix elements [9-11]. Some exceptions such as an interaction of X-ray laser fields with nuclei in relation to alpha, beta-decay and X-ray-driven gamma emission of nuclei have been considered earlier. With the advent of new coherent X-ray laser sources in the near future, however, these conclusions have to be reconsidered. From the design report (see Table II in Ref.[10]) for SASE 1 at TESLA XFEL and parameters for current and future ion beam sources, the signal rate due to spontaneous emission after real excitations of the nuclei can be estimated. For nuclei accelerated with an energy resolution of 0.1%, 12.4 keV photons produced by SASE 1 become resonant with the $E1$ transition in a whole number of nuclei (for example, ^{153}Sm, ^{181}Ta, ^{223}Ra, ^{225}Ac , ^{227}Th etc). In other words, the resonance condition ($\omega \sim \Delta\varepsilon$, where $\Delta\varepsilon$ is a typical level spacing, ω is a laser frequency) can be fulfilled [10]. The coherence of the laser light expected from new sources (TESLA XFEL at DESY) may allow one to access the extended coherence or interference phenomena. In particular, in conjunction with moderate acceleration of the target nuclei it allows principally to achieve realization of multi-photon phenomena, nuclear Rabi oscillations or more advanced quantum optical schemes in nuclei.

The interaction of atoms with the external alternating fields, in particular, laser fields, has been the subject of intensive experimental and theoretical studies (see, for example, Refs. [1-8, 12-24]). A definition of the k-photon emission and absorption probabilities and atomic levels shifts, study of dynamical stabilization and field ionization *etc.* are the most current problems to be solved. At present time, a progress is achieved in the description of the processes of atoms interacting with the harmonic emission field [1, 12-14]. But in a realistic laser field the according processes differ significantly from the ones in the harmonic field. A substantial role is played by the photon-correlation effects and the influence of the multi-mode character of laser pulse. Notably, a number of different theoretical approaches have been developed in order to give adequate description of the atoms in a strong laser field. Methods of interest include standard perturbation theory (surely for low laser filed intensities), Green function method, the density-matrix formalism, time-dependent density functional formalism, direct numerical solution of the Schrödinger and/or Dirac equation, multi-body multi-photon approach, the time-independent Floquet formalism *etc.* (see [1-8,12-24] and references therein). The effects of the different laser line shape on the intensity and spectrum of resonance fluorescence from a two-level atom are studied in Refs.[1-5,15-17,19-23]. Earlier, the relativistic energy approach to studying the interaction of atoms with a realistic strong laser field, based on the Gell-Mann and Low S-matrix formalisms, has been developed. Originally, Ivanov has proposed an idea to describe quantitatively the behavior of an atom in a realistic laser field by means of studying the radiation emission and absorption lines and, further, the theory of interaction of an atom with a Lorentzian laser pulse and calculation of the corresponding line moments has been developed in detail in Ref. [19-25]. Multi-photon resonances shifts and widths in hydrogen and caesium have been confirmed in numerical simulations. Theory of interaction of an atom with the Gaussian and soliton-like laser pulses and calculations of the corresponding lines moments have been presented in detail in Refs. [23, 26, 27]. Here we apply this approach to the study of the resonant multiphoton ionization of krypton by intense uv laser radiation and compute the multiphoton resonances shift and width. We also outline the corresponding scheme to treating the multiphoton resonances in nuclei on example of ^{57}Fe nucleus.

2. Relativistic energy approach to atom in a strong laser field: Multiphoton resonances

The relativistic energy approach in the different realizations and the radiation lines moments technique are presented in detail in Refs. [19-30]. Here we only present the essential elements. In the theory of the non-relativistic atom, a convenient field procedure is known for calculating the energy shifts δE of degenerate states. This procedure is connected with the secular matrix M diagonalization. In constructing M, the Gell-Mann and Low adiabatic formula for δE is used [20-23, 31]. In relativistic theory, the Gell-Mann and Low formula δE is connected with electrodynamic scattering matrices, which include interactions of a laser field with a photon vacuum field. The case of interaction with photon vacuum corresponds to standard theory of radiative decay of excited atomic states. In relativistic theory the secular matrix elements are already complex in the second perturbation theory (PT) order. Their imaginary parts are connected with radiation decay possibility. The total energy shift is usually presented in the form [23]:

$$\delta E = \mathrm{Re}\delta E + \mathrm{i}\ \mathrm{Im}\delta E\ ,\ \ \mathrm{Im}\ \delta E = -\mathrm{P}/2, \tag{1}$$

where P is the level width (decay possibility). Spectroscopy of an atom in a laser field is fully defined by position and shape of the radiation emission and absorption lines. The lines moments μ_n are strongly dependent upon the laser pulse quality, namely intensity and mode composition [15-23]. Let us describe the interaction "atom-laser field" by the Ivanov potential [21,23]:

$$V(r,t)= V(r) \int d\omega\, f(\omega - \omega_0)\ \sum_{n=-\infty}^{\infty} \cos\ [\ \omega_0 t + \omega_0\, n\tau]. \tag{2}$$

Here, ω_0 is the central laser radiation frequency, n is an integer number. The potential V represents the duration of laser pulses, τ, with known frequency. The function $f(\omega)$ is a Fourier component of the laser pulse. The condition $\int d\omega f^2(\omega)=1$ normalizes the potential $V(rt)$ on the definite energy in the pulse. Let us consider the pulses with Lorentzian shape (coherent 1-mode pulse): $f(\omega) = \beta/(\omega^2+\Delta^2)$, Gaussian one (multi-mode chaotic pulse): $f(\omega) = \beta \exp[\ln 2(\omega^2/\Delta^2)]$ and the soliton-like pulse: $f(t) = \beta\, ch^{-1}[t/D]$ (β -normalizing multiplier). The case of the Lorentzian shape has been earlier studied [20-23]. The case of the Gauss and soliton-like shape is considered in Refs.[23,26,27]. The master program results in calculating an imaginary part of energy shift $\mathrm{Im}\delta E_\alpha\ (\omega_0)$ for any atomic level as the function of the central laser frequency ω_0. An according function has the shape of the resonance, which is connected with the transition α-p (α, p-discrete levels) with absorption (or emission) of the "k" number of photons. For the resonance we calculate the following values [20-23]:

$$\delta\omega(p\alpha|k) = \int' d\omega \ \mathrm{Im}\ \delta E_\alpha(\omega)\ (\ \omega - \omega_{p\alpha}/k)\ /\ N, \tag{3}$$

$$\mu_m = \int' d\omega \ \mathrm{Im}\ \delta E_\alpha(\omega)\ (\varpi - \omega_{p\alpha}/k)^m\ /\ N,$$

where $\int' d\omega \mathrm{Im} E_\alpha$ is the normalizing multiplier; $\omega_{p\alpha}$ is position of the non-shifted line for transition α-p, $\delta\omega(pa|k)$ is the line shift under k-photon absorption; $\varpi_{p\alpha}=\omega_{p\alpha}+k\cdot\delta\omega(p\alpha|k)$. The first moments μ_1, μ_2 and μ_3 determine the atomic line centre shift, its dispersion and the asymmetry. To find μ_m, we need to obtain an expansion of E_α to PT series: $E_\alpha = \sum E_\alpha^{(2k)}(\omega_0)$. One may use here the Gell-Mann and Low adiabatic formula for δE_α :

$$\delta E_\alpha = \lim_{\gamma\to 0} \mathrm{i}\gamma g \ln\langle\Phi_\alpha|S_\gamma(0,-\infty|g)|\Phi_\alpha\rangle|_{g=1} \tag{4}$$

The representation of the S-matrix in the form of the PT series induces the expansion for δE_α:

$$\delta E_\alpha(\omega_0) = \lim_{\gamma\to 0} \gamma \sum_{k_1 k_2 \ldots k_n} a\,(k_1, k_2,\ldots,k_n), \tag{5}$$

$$I_\gamma(k_1, k_2,\ldots,k_n) = \prod_{j=1} S_\gamma^{(kj)}, \tag{6}$$

$$S_\gamma^{(m)} = (-1)^m \int_{-\infty}^{0} d\,t_1 \ldots \int_{-\infty}^{t_m-1} d\,t_m \langle\Phi_\alpha\,|\,V_1V_2\ldots V_m\,|\,\Phi_\alpha\rangle, \tag{7}$$

$$V_j = \exp(\mathrm{i}H_0\,t_j)\ V(rt_j)\ \exp(-\mathrm{i}H_0\,t_j)\ \exp(\gamma t_j). \tag{8}$$

Here, H is the atomic Hamiltonian, $a\,(k_1, k_2,\ldots,k_n)$ are the numerical coefficients. The structure of matrix elements $S_\gamma^{(m)}$ is further described in [19-23]. Here we only note that one may simplify the considerations by accounting for the k-photon absorption contribution in the first two PT orders. Moreover, summation on laser pulse is exchanged with the integration. The corresponding (l+2k+1)-times integral on (l+2k) temporal variables and ρ ($l=0,2$) (integral I_γ) are calculated [19-23]. Finally, and only following some tedious transformations, one can obtain the expressions for the line moments. The corresponding expressions for the Gaussian laser pulse are as follows:

$$\delta\omega(p\alpha\mid k) = \{\pi\Delta/(k+1)\mathrm{k}\}\ [\,E(p, \omega_{p\alpha}/k) - E(\alpha, \omega_{p\alpha}/k)], \tag{9}$$

$$\mu_2 = \Delta^2/\mathrm{k}$$

$$\mu_3 = \{4\pi\Delta^3/[k\,(k+1)]\}\ [\,E(p, \omega_{p\alpha}/k) - E(\alpha, \omega_{p\alpha}/k)],$$

where

$$E(j, \omega_{p\alpha}/k) = 0{,}5 \sum_{p_i} V_{jpi} V_{pij} \left[\frac{1}{\omega_{jp_i} + \omega_{p\alpha}/k} + \frac{1}{\omega_{jp_i} - \omega_{p\alpha}/k} \right] \tag{10}$$

The summation in Equation (10) is over all atomic states. Note that these formulas for the Gaussian pulse differ from the expressions for Lorentzian-shape laser pulses [21-23]. For the soliton-like pulse, it is necessary to carry out the numerical calculation or use some approximations to simplify the expressions [27]. To evaluate Eq. (10), we use an effective Ivanov-Ivanova technique [22,28] of calculating sums of QED PT second order, which has been applied earlier in calculations of atomic and meso-atomic parameters [26,27,30-32]. Finally, the computational procedure includes solving an ordinary differential equations system for above described functions and integrals. In specific numerical calculations the PC "Superatom-ISAN" package is used. The construction of the operator wave functions basis within the QED PT, the technique of calculating the matrix elements in Eqs. (9, 10) and other details are presented in Refs. [19-30]. Special features of treating the multiphoton resonances in a nucleus within the outlined approach are obviously connected with estimating the corresponding matrix elements in the basis of the nuclear wave functions and some other details. In a modern theory of a nucleus, there are a sufficiently great number of the different models for generating the proton and neutron wave functions bases. At present time, it is acceptable that a quite adequate description of the nuclear density is provided by the relativistic mean-field (RMF) and other models of the nucleus [32-35]. As alternative approach one could use the advanced RMF or shell models based on the effective Dirac-Wood-Saxon type Hamiltonian [32].

3. Results and conclusions

Here we present the results of the numerical simulation for the three-photon resonant, four-photon ionization profile of atomic krypton (the $4p \to 5d[1/2]_1$ and $4p \to 4d[3/2]_1$ three photon Kr resonances are considered). In Ref. [18], an experimental study has been presented of the resonant multi-photon ionization of krypton by intense uv (285-310 nm) laser radiation for the irradiance range of 3×10^{12}-10^{14} W/cm^2. The experiment consisted of the measurement of the number of singly charged Kr and Xe ions produced under collisionless conditions as a function of laser frequency and intensity. The output of a dye-laser system operating at 2.5 Hz is frequency doubled in a 1-cm potassium dihydrogen phosphate (KDP) crystal to give a 0.5-mJ, 1.3-ps, transform-limited 0.1-nm-bandwidth beam, tunable between 285 and 310 nm. In these experiments, the corresponding parameters of the following transions have been determined: (i) $4p \to 5d[1/2]_1$ and (ii) $4p \to 4d[3/2]_1$ three photon Kr resonances. The resonance shift is proportional to intensity with a width dominated by lifetime broadening of the excited state. The corresponding shift and width have been found as follows: (i) the shift $\delta\omega_0(p\alpha|3)=aI$, $a_{exp}=3.9$ meV/(Tw·cm^{-2}); width $b_{exp}=$ 1.4 meV/(Tw·cm^{-2}); (ii) shift $\delta\omega_0(p\alpha|3)=aI$, $a_{exp}=8.0$ meV/(Tw·cm^{-2}); width $b_{exp}=4$ meV/(Tw· cm^{-2}). The authors have applied [18] a quite simple model for an effective two-level atom with the assumption of a rate limiting three-photon excitation step followed by rapid one-photon ionization from the excited state. As expected, the three-photon resonances broaden and shift further as the laser pulse intensity is increased. The important feature of the corresponding profiles is linked to asymmetry [18]. Naturally, it is easy to understand that the asymmetric profile is typical of realistic laser pulses with the spatially and temporally varying intensity. Besides, the authors of Ref. [18] have noted that while all resonances are "blue" shifted, ac Stark shift calculations are difficult to perform for excited states leading to both "blue" and "red" shifts. Our numerical simulation results for the (i) $4p \to 5d[1/2]_1$ and (ii) $4p \to 4d[3/2]_1$ three-photon Kr resonances are as follows: (i) the shift $\delta\omega_0(p\alpha|3)=aI$, $a_{exp}=3.95$ meV/(Tw·cm^{-2}) and width $b_{exp}=$ 1.5 meV/(Tw·cm^{-2}); (ii) shift $\delta\omega_0(p\alpha|3)=aI$, $a_{exp}=8.1$ meV/(Tw·cm^{-2}) and width $b_{exp}=4.2$ meV/(Tw·cm^{-2}). One could conclude that there is a physically reasonable agreement of the theoretical and experimental data. Analysis shows that the shift and width of the multi-photon resonance line for the interaction "atom- multimode laser pulse" is greater than the corresponding shift and width for a specific case of the "atom- single-mode pulse" (the Lorentzian pulse model) interaction. From the physical point of view, it is obviously provided by the action of the photon-correlation effects and influence of the multi-mode laser pulse. Of great interest is the possibility of quantitative construction of the corresponding resonance profiles with explanation of the asymmetric nature by means of calculating sufficiently "large" number of the multiphoton transition line moments. Such an approach may easily explain the qualitative features of the multiphoton resonances lines in the ^{57}Fe nucleus. According to Ref. [34], the nuclear multiphoton transitions are taking a place in ^{57}Fe nucleus subjected to radio-frequency field ω_0=30MHz. This scenario was experimentally observed in the Mössbauer spectra of ^{57}Fe nuclei in Permalloy by Tittonen *et al.* [35]. Really, the eight transitions are possible between the four hyperfine sub-states of the 14.4 keV excited level e and the two sub-states of the ground state g in the radio-frequency magnetic field [34]. If the static magnetic hyperfine splitting of the ground and excited states are respectively $\omega_g>0$ and $\omega_e>0$, the transition frequencies corresponding to forbidden γ-ray transitions are $(E_e-E_g)/h\pm3\omega_g/2\pm\omega_g/2$, where E_e and E_g are respectively the energies of the 14.4-keV and ground states of the ^{57}Fe nucleus in an absence of any external field.

Acknowledgements
The author would like to thank Dr. Christian G. Parigger for invitation to make contributions to ICSLS-22 (Tullahoma, Tennessee, USA, 2014), useful advice and help in preparing the manuscript. The comments of the anonymous referees are acknowledged too.

References

[1] Batani D, Joachain C, eds 2006 *Matter in super-intense laser fields* (New York: AIP Series)
[2] Ivanova E P 2011 *Phys. Rev. A* **84** 043829
[3] Khetselius O Yu, Florko T A, Svinarenko A A and Tkach T B 2013 *Phys.Scr.* **T153** 014037
[4] Florko T , Ambrosov S, Svinarenko A and Tkach T 2012 *J. of Phys.: Conf. Ser.* **397** 012037
[5] van der Hart H W, Lysaght M A and Burke P G 2007 *Phys. Rev.* A **76** 043405
[6] Mercouris T and Nikolaides C A 2003 *Phys. Rev. A* **67** 063403
[7] Brandas E and Floelich P 1977 *Phys. Rev. A* **16** 2207
[8] Wangt K, Hot T-S and Chu S-I 2006 *Phys. Rev. A* **73** 023403
[9] Glushkov A V, Ivanov L N and Letokhov V S Nuclear quantum optics 1991 *Preprint of Institute for Spectroscopy, USSR Academy of Sciences*, Troitsk-N4
[10] Bürvenich T, Evers J and Keitel C 2006 *Phys.Rev.Lett.* **96** 142501; 2006 Phys.*Rev.C* **74** 044601
[11] Khetselius O Yu 2009 *Int. J. Quant. Chem.* **109** 3330
[12] Fedorov M V 1999 *Phys.-Uspekhi* **169** 66
[13] Delone N B and Kraynov V P 1998 *Phys.-Uspekhi* **168** 531
[14] Belyaev V S, Kraynov V P, Lisitsa V S and Matafonov A P *Phys.-Uspekhi* **178** 823
[15] Zoller P. 1982 *J.Phys. B: At. Mol. Opt. Phys.* **15** 2911
[16] Lompre L, Mainfrau G, Manus C and Marinier J 1981 *J.Phys. B: At. Mol. Opt. Phys.* **14** 4307
[17] Khetselius O Yu 2012 *J. of Phys.: Conf. Ser.* **397** 012012; 2009 *Phys. Scr.* **T.135** 014023
[18] Landen O L, Perry M D and Campbell E M 1987 *Phys. Rev. Lett.* **59** 2558
[19] Ivanov L N and Letokhov V S 1975 *JETP* **68** 1748; 1985 *Com. Mod. Phys.* D4 169
[20] Ivanov L N, Ivanova E P and Aglitsky E V 1988 *Phys. Rep.* **166** 315
[21] Glushkov A V and Ivanov L N 1985 *Preprint of ISAN, USSR Academy of Sci.*, Troitsk, N3
[22] Glushkov A V, Ivanov L N and Ivanova E P 1986 *Autoionization Phenomena in Atoms*, (Moscow: University Press) p 58
[23] Glushkov A V and Ivanov L N 1992 *Phys. Lett. A* **170** 33
[24] Glushkov A V and Ivanov L N 1992 *Proc. of 3rd Seminar on Atomic Spectroscopy* (Chernogolovka, Moscow reg.) p 113
[25] Glushkov A and Ivanov L N 1993 *J.Phys. B: At. Mol. Opt. Phys.* **26** L179
[26] Glushkov A V, Loboda A, Gurnitskaya E and Svinarenko A 2009 *Phys. Scr.* **T135** 014022
[27] Glushkov A V, Khetselius O Yu, Loboda A V and Svinarenko A A 2008 *Frontiers in Quantum Systems in Chem. and Phys., Ser. Progress in Theoretical Chemistry and Physics* vol. 18 ed. S Wilson, P J Grout, J Maruani, G Delgado-Barrio, P Piecuch (Berlin: Springer) p 543
[28] Ivanov L N and Ivanova E P 1974 *Theor. Math. Phys.* **20** 282
[29] Khetselius O Yu 2008 *Spectral Line* Shapes, Vol. 15, AIP Conf. Proceedings. **1058** 363
[30] Glushkov A V, Malinovskaya S V, Sukharev D E, Khetselius O Yu, Loboda A V and Lovett L 2009 *Int. J. Quant. Chem.* **109** 1717
[31] Glushkov A V, Khetselius O Yu and Malinovskaya S V 2008 *Europ. Phys. Journ ST* **160** 195
[32] Glushkov A V 2012 *J. of Phys.: Conf. Ser.* **397** 012011
[33] Serot B and Walecka J 1986 *Adv. Nucl. Phys.* **16** 1
[34] Olariu S, Sinor T W and Collins C B 1994 *Phys. Rev. B* **50** 616
[35] Tittonen I, Lippmaa M, Ikonen E, Linden J and Katila T 1992 *Phys. Rev. Lett.* **69** 281

Journal of Physics: Conference Series **548** (2014) 012021 doi:10.1088/1742-6596/548/1/012021

Line Narrowing in Mössbauer Spectra of Superparamagnetic Fe_3O_4 Nanoparticles

H-Y Hah[1,2], S Gray[1,2], C E Johnson[1], J A Johnson[1,2], V Kolesnichenko[3], P Kucheryavy[3] and G Goloverda[3]

[1]Center for Laser Applications, University of Tennessee Space Institute, Tullahoma, TN, 37388

[2]Department of Mechanical, Aerospace and Biomedical Engineering, University of Tennessee Space Institute, Tullahoma, TN, 37388

[3]Department of Chemistry, Xavier University, New Orleans, LA, 70125

Email: hhah@utsi.edu

Abstract. Nanoparticles of magnetic crystals below a critical size are single domain and exhibit superparamagnetism. If there are N atoms or molecules with magnetic moment μ in each particle, the magnetic moment of the particle is $N\mu$. At high temperatures the thermal fluctuations of the magnetic moments give an ensemble average moment of zero and the Mössbauer spectrum is a single line. As the temperature, T, is lowered the fluctuations slow down and the sample acquires a magnetization and the Mössbauer line broadens and eventually shows magnetic hyperfine splitting. We have observed ^{57}Fe line broadening in nanoparticles of ferrimagnetic Fe_3O_4 with diameters of 5.3 and 10.6 nm. The results have been analyzed using the motional narrowing equation familiar in nuclear magnetic resonance to determine the superparamagnetic fluctuation time and magnetic anisotropy.

1. Introduction

Nanoparticles of ferrimagnetic iron oxide Fe_3O_4 have applications in medicine and in information technology. They are potential contrast agents in Magnetic Resonance Imaging (MRI), and in targeting and destroying cancer cells. The sharpness of an MRI image is determined by the proton (usually) spin relaxation times which can be decreased by interactions with electronic magnetic moments. Currently this is achieved by injecting gadolinium compounds, since gadolinium atoms have a large magnetic moment of 7 μ_B but magnetic nanoparticles can have magnetic moments exceeding 10^4 μ_B so are potentially more powerful. Targeting of diseased organs may be achieved either chemically (with specific molecules) or physically (with magnets) and destruction of cells is done as by heating in an rf field (hypothermia). In this paper measurements of the Mössbauer spectra of Fe_3O_4 nanoparticles of different sizes and at varying temperatures are described and used to study their magnetic properties. Mössbauer spectra have a special value in characterizing these materials since Fe_3O_4 readily oxidizes to γ- Fe_2O_3 and shows a different spectrum but as it has the same crystal structure it cannot easily distinguished by x-ray diffraction.

Small particles of magnetic crystals below a critical size are single domain and exhibit superparamagnetism. The magnetic moment of each particle is $N\mu$ where there are N atoms with magnetic moment μ. According to Néel's theory [1] of superparamagnetism, the fluctuation time τ of particles of volume V and magnetic anisotropy K is

$$\tau = \tau_0 \exp\left(\frac{KV}{kT}\right) \tag{1}$$

where τ_0 is a characteristic time usually estimated to be between 10^{-9} and 10^{-10} s.

At high temperatures the fluctuations are fast and the macroscopic magnetic moment averages to zero and the hyperfine field measured by the Mössbauer spectrum is also zero giving a single line. As the temperature is lowered the fluctuations slow down and the line becomes broadened. Eventually at low temperatures the fluctuations are blocked and a fully split hyperfine pattern is observed. The temperature at which this happens is called the blocking temperature T_B.

This line broadening with decreasing temperature is a similar phenomenon to the motional narrowing observed in nuclear magnetic resonance (NMR) for increasing temperature. Borrowing the notation from NMR [2] the linewidth, Γ, is broadened approximately by an amount

$$\Delta\Gamma = \omega^2\tau, \tag{2}$$

where ω is the hyperfine splitting expressed as an angular frequency. This expression has the advantage of simplicity and transparency and is valid for $\omega\tau << 1$. Motional narrowing has long been studied in NMR [2] and is discussed in detail in Chapter X of Abragam's book [3] and has been applied to Mössbauer spectra by Blume and Tjon [4].

In this paper we use Mössbauer spectroscopy to characterize Fe_3O_4 nanoparticles and measure their magnetic anisotropy.

2. Experimental details

We have measured the Mössbauer spectra of ^{57}Fe in two samples of nanoparticles of Fe_3O_4 between 6 K and 300 K. The samples were prepared from a magnetite colloid and passivated by adsorbed diethylene glycol. TEM images showed that their diameters were 10.6 and 5.3 nm. The Mössbauer spectrometer was manufactured by SEECO, Edina, MN and low temperatures were obtained with a cryogen-free continuous-flow cryostat (Janis/SEECO). The measurements were made in the region of their superparamagnetic blocking temperatures T_B, i.e., where the line becomes broadened.

3. Results

The spectra show a single line at room temperature. As the temperature is lowered the line broadens and this becomes further broadened at lower temperatures and eventually shows magnetic hyperfine splitting. In general the results are complicated and the spectra are difficult to fit but they become considerably simpler in the extreme cases of high and low temperatures. Figures 1(a) and 1(b) show Mössbauer spectra of the 10.6 nm and 5.3 nm particles, respectively, at temperatures in the region of T_B. Note that for the smaller (5.3 nm) particles the transition takes place at a lower temperature (Figure 1(b)) as expected from Equation (1).

In contrast with the usual way of plotting with descending temperature from top to bottom, the spectra are plotted with temperatures in ascending order so as to emphasize the motional narrowing. At the lowest temperatures, resolved hyperfine splitting into six lines is evident, even though the lines are broad. With increasing temperature the lines coalesce into a single broad line which narrows with further increase in temperature.

4. Analysis and interpretation

The single-line spectra were fitted to a Lorentzian profile to yield values for the linewidth Γ. The broadening is $\Delta\Gamma = \Gamma - \Gamma_0$ where Γ_0 is the linewidth of 0.3 mms^{-1} observed at room temperature. There is an instrumental increase over the natural lifetime value of 0.21 mms^{-1}. The fluctuation times may be derived from Equation (2). ω is taken as the full hyperfine in. eraction which gives a total splitting of about 16 mms^{-1} in the Mössbauer spectra, corresponding to a magnetic field of about 50 T and a value $\omega = 1.105 \times 10^{-9}$ rad/s. The resulting fluctuation times are given in Table 1. Since $\omega\tau \sim 0.3$, the approximation indicated in Equ. (2) is valid.

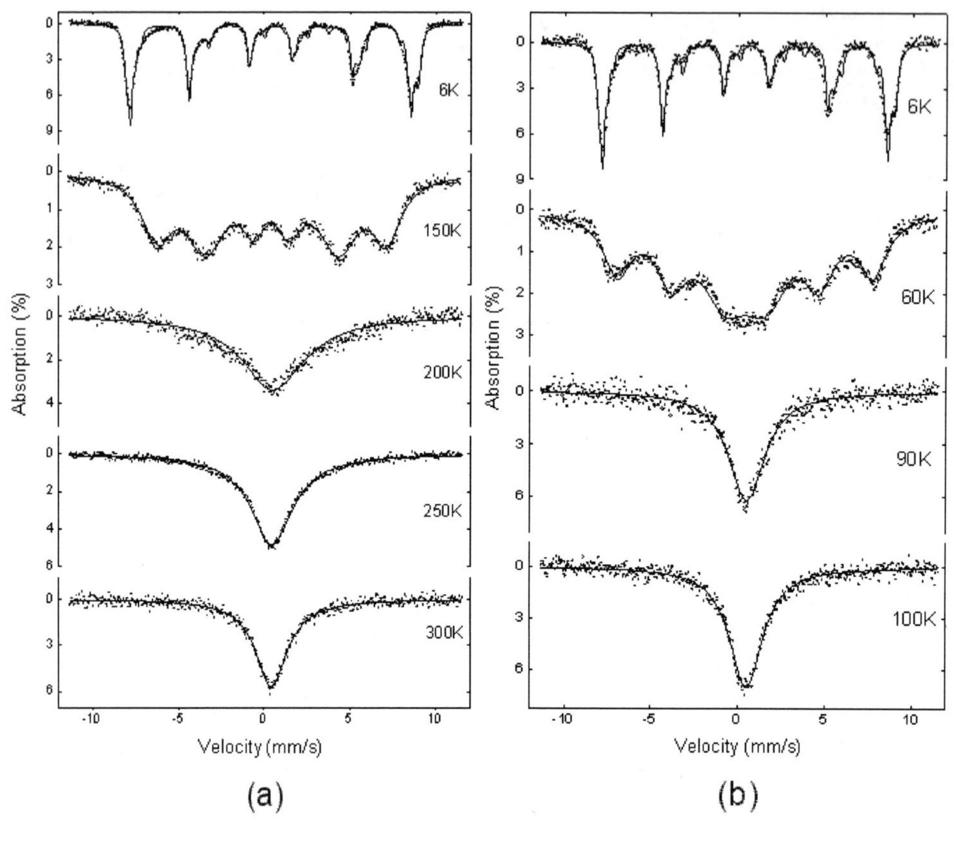

Figure 1. ^{57}Fe Mössbauer spectra of Fe_3O_4 at temperatures in the region of the blocking temperatures T_B for (a) 10.6 nm and (b) 5.3 nm, particles.

Table 1. Nanoparticle fluctuation times.

Particle size (nm)	Temperature (K)	Linewidth Γ (mms^{-1})	fluctuation time τ (s)
10.6	200	6.55	3.35×10^{-10}
	250	3.12	1.67×10^{-10}
	300	2.37	1.11×10^{-10}
5.3	90	2.57	1.23×10^{-10}
	100	2.52	1.10×10^{-10}

5. Superparamagnetism

From Equation (1) it follows that

$$ln\,\tau = \frac{KV}{kT} + ln\,\tau_0 \qquad (3)$$

Hence a plot of *ln* τ vs. 1/*T* yields a straight line with slope *KV/k* and intercept $ln\,\tau_0$. From the value of KV the anisotropy can be deduced knowing the particle volume. Such a plot is shown in Figure 2 and the values of K are given in Table 2.

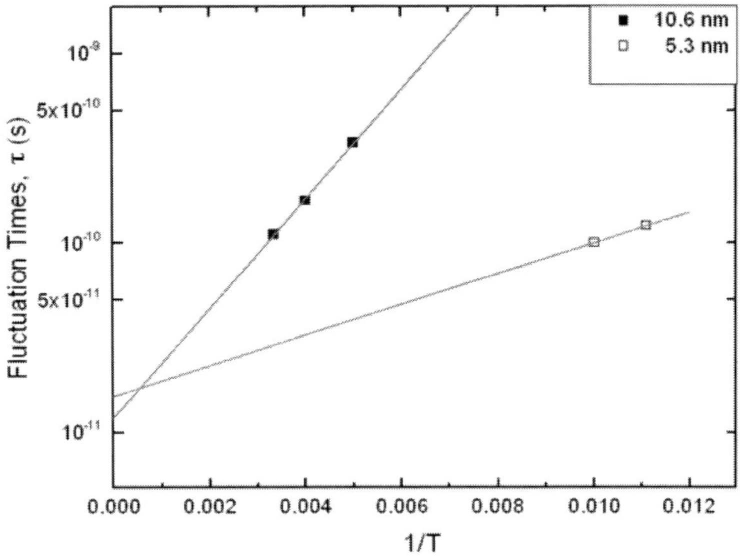

Figure 2. Plot of *ln τ* vs. *1/T* for 10.6 nm (■) and 5.3 nm (□) particles.

Table 2. Magnetic anisotropy values.

Particle size (nm)	***KV/k* (K)**	***KV* (J)**	***K* (Jm^{-3})**
10.6	665	9.18×10^{-21}	1.47×10^{4}
5.3	186	2.57×10^{-21}	3.30×10^{4}

Values of *K* have been previously determined for the same specimens from observation of collective magnetic excitations obtained from low temperature Mössbauer spectra [5]. Their values are 2.30×10^4 Jm^{-3} (10.6 nm) and 7.77×10^4 Jm^{-3} (5.3 nm). These are considerably larger (about double) than those which were measured at higher temperature. From ferromagnetic resonance measurements Bickford [6] also finds an increase in *K* as the temperature is lowered. The discrepancy probably partly arises from the simplifying assumptions of the fluctuation model we have used. It has been assumed that all the unresolved hyperfine transitions have the same linewidth whereas those from the inner lines of the hyperfine spectra will be broader than the outer lines. Other assumptions are that K is uniaxial and does not depend upon temperature (known to be wrong) or volume. K may also depend upon the shape of the particles through different surface contributions. Interaction between the particles has been neglected. Also the Mössbauer lines have not been corrected for thickness broadening.

6. The blocking temperature T_B

T_B is not a transition temperature in the thermodynamic sense as it depends upon the time, τ_m, of the measurement. For susceptibility measurements, τ_m is usually taken to be 100 s. For Mössbauer spectra Wickman [8] quotes a smaller value of $\tau_m = 5 \times 10^{-9}$ s. So,

$$T_B = aKV/k \tag{4}$$

where $a = 1/\ln(\tau_m/\tau_0)$.

Combined with the value of the anisotropy (Table 2) a value of about 0.2 is found for a. This gives T_B = 133 K (10.6 nm) and 37 K (5.3 nm). It is difficult to get precise values of T_B from the Mössbauer spectra because the spread in particles sizes broadens the transition region, but inspection of Figure 2 shows that these are of the right order of magnitude.

7. The characteristic time τ_0

The intercept with the time axis in Fig. 2 gives a value of about 10^{-11} s for Néel's characteristic time, τ_0. Previous estimates [7] on different materials range from 3.57×10^{-10} for α-Fe, and 1.05×10^{-11} for ferritin to 10^{-13} for Fe films.

8. Conclusions

The paper shows that the change in linewidth of the Mössbauer line in the region of the blocking temperature of magnetic nanoparticles can be used to deduce values for the magnetic anisotropy consistent with values deduced from low temperature measurements and consistent with the values of the blocking temperatures. This observation contributes to laying a foundation for the control over magnetic properties using nanoscience.

Acknowledgements

This publication was supported by Grant Number LA-SIGMA EPS-10038971 from the National Science Foundation to Xavier,University, New Orleans. We acknowledge support from the Center for Laser Applications and The University of Tennessee Space Institute.

References

[1] Néel L 1949 *Ann. Geophys.* **5** 99
[2] Pines D and Slichter C P 1955 *Phys. Rev.* **100** 1014
[3] Abragam A 1961 *The Principals of Nuclear Magnetism* (Oxford: Oxford University Press)
[4] Blume M and Tjon J A 1968 *Phys. Rev.* **165** 446
[5] Hah H-Y, Cole M, Gray S, Johnson C E, Johnson J A, Kolesnichenko V, Kucheryavy P and Goleverda G 2012 *Bull. Amer. Phys. Soc.* **57** SES KA60
[6] Bickford L E 1950 *Phys. Rev.* **78** 449
[7] Dickson D P E, Reid N M K, Hunt C, Williams H D, El-Hilo M and O'Grady K 1993 *J. Mag. Mag. Mat.* **125** 345
[8] Wickman H H 1966 *Mössbauer Effect Methodology* vol 2 (New York: Plenum Press) p 39

Line-shapes analysis with ultra-high accuracy

Piotr Wcisło, Agata Cygan, Daniel Lisak, Roman Ciuryło

Institute of Physics, Faculty of Physics, Astronomy and Informatics, Nicolaus Copernicus University, Grudziadzka 5, 87-100 Torun, Poland

E-mail: rciurylo@fizyka.umk.pl

Abstract. We present analysis of the R7 Q8 O_2 B-band rovibronic transition measured with ultra-high signal-to-noise ratio by Pound-Drever-Hall-locked frequency-stabilized cavity-ring-down spectroscopy. For line-shape calculations *ab intio* in spirt approach was used based on numerical solution of the proper transport/relaxation equation. Consequences for spectroscopic determination of the Boltzmann constant as well as precise determination of the line position in the Doppler limited spectroscopy are indicated.

1. Introduction

Dynamic development of experimental techniques of high resolution molecular spectroscopy allows to obtain molecular spectra with an extremely high signal-to-noise ratio. Even relatively weak molecular transition can be measured with a signal-to-noise ratio exceeding 2×10^5 [1]. To analyze so accurate experimental data it is necessary to use theory allowing line shape calculation with a similar accuracy. Such theory [2] must take into account a number of effects such as Doppler and collisional broadening and shift of the spectral line, Dicke narrowing [3], the speed dependence of collisional broadening and shifting [4]. The description of molecular collisions should be based on realistic potential energy surface.

2. Line-shape analysis

Starting from the first principles simple analytical expressions [5, 6, 7] describing the shape of the spectral line can not be obtained in general case. Realistic spectral line-shape calculations require to solve a proper transport/relaxation kinetic equation [2, 8, 9]. Numerical methods usually used for this purpose fail when the collisional effects are small comparing to the Doppler broadening [10, 11].

We present here application of an efficient iterative approach to this problem allowing the line-shape calculations [11] in a wide range of pressure and temperature of an investigated gas sample. In this approach we decompose the operators and functions from the transport/relaxation equation in the Burnett basis, what allows to convert the integral equation into the algebraic problem. Then we used the iterative technique to solve it [11].

It was shown that use of a collisional operator obtained from even very simple description of interaction between colliding molecules can lead to a very good fit quality of the calculated profile to the measured spectra. Residuals obtained from such a fit correspond to signal-to-noise ratio approaching 10^5. In our study we assumed that speed dependence of collisional broadening and shifting is described by a hypergeometric function [4]. We analyse several line-shape models including speed-dependent Blackmore profile ($SD_h BP$) [12], speed-dependent billiard ball profile (SD_hBBP) [9], exact speed-dependent Galatry profile (eSD_hGP) [9], and speed-dependent Nelkin-Ghatak profile (SD_hNGP) [13].

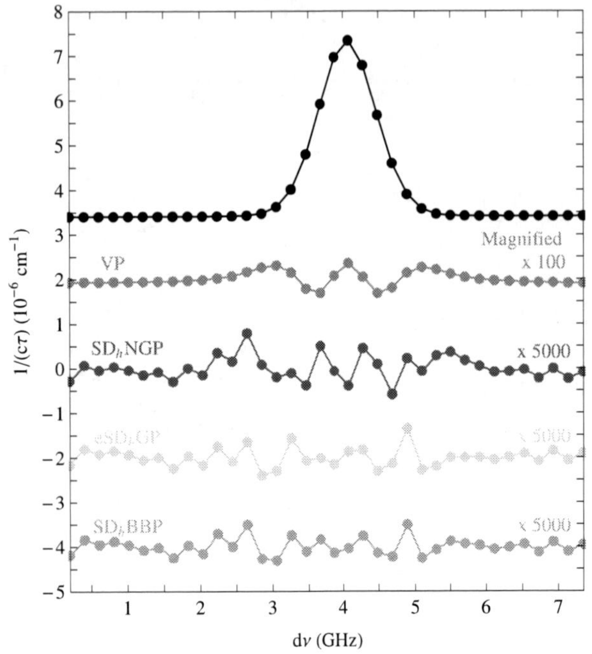

Figure 1. The R7 Q8 O_2 B-band rovibronic spectral line measured in pure oxygen at 933 Pa. Below residuals from fitting with the Voigt profile (VP), SD_hNGP, eSD_hGP, and SD_hBBP, respeectively.

In Fig. 1 we present residuals obtained from the fits with SD_hNGP, eSD_hGP, SD_hBBP. All three profiles have the same number of fitted parameters. The best quality of the fit was obtained for SD_hBBP. Noticeably worse quality of the fit can be seen in case of SD_hNGP. Using SD_hBBP we were able to reach 3 kHz precision of determination of the line position in the Doppler limited spectroscopy. Such precision is crucial to test quantum calculations of molecular energy levels and search a new physics beyond the Standard Model [14]. We also found that when SD_hBBP and SD_hNGP were used for Doppler width determination the results differed by 166 kHz what corresponds to relative difference about 3×10^{-4}. These results demonstrates that more advanced models, like SD_hBBP, need to be used to eliminate additional systematic errors originating from oversimplified description of molecular collisions. At the moment, these systematic errors give the largest contribution to the error budget for the spectroscopic determination of the Boltzmann constant [15, 16]. More details about this research can be found in Ref. [11].

3. Conclusion

We showed that experimental spectra can be fitted with high quality corresponding to signal-to-noise ratio approaching 10^5 level with SD_hBBP. Presented approach can be used to minimize systematic errors in spectroscopic determination of the Boltzmann constant. As was indicated in Ref. [17] both speed-dependent effects and velocity changing collision can affect this kind of experiments. At present for data analysis in the Doppler width thermometry, leading experimental groups [15, 16] use profile [18] equivalent to SD_hNGP or its simplifications, what can lead to some errors. Finally, improvement in line position determination is important for tests of the fundamental theories of interactions in molecular systems [14].

Acknowledgments
The research is part of the program of the National Laboratory FAMO in Torun, Poland, and is supported by the Polish National Science Centre Project no. DEC-2011/01/B/ST2/00491. The research is also supported by the Foundation for Polish Science TEAM Project co-financed by the EU European Regional Development Fund. A. Cygan is supported by the Foundation for Polish Science START Project.

References
[1] Cygan A, Lisak D, Wójtewicz S, Domysławska J,Hodges J T, Trawiński R S, Ciuryło R 2012 *Phys. Rev. A* **85** 022508
[2] May A D, Liu W K, McCourt F R W, Ciuryło R, Sanchez-Fortún Stoker J, Shapiro D, Wehr R 2013 *Can. J. Phys.* **91** 879
[3] Dicke R H 1953 *Phys. Rev.* **89** 472
[4] Berman P R 1972 *J. Quant. Spectrosc. Radiat. Transfer* **12** 1331
[5] Ciuryło R, Pine A S, Szudy J 2001 *J. Quant. Spectrosc. Radiat. Transfer* **68** 257
[6] Priem D, Rohart F, Colmont J M, Wlodarczak G, Bouanich J P 2000 *J. Mol. Struct.* **517-518** 435
[7] Ngo N H, Lisak D, Tran H, Hartmann J M, *J. Quant. Spectrosc. Radiat. Transfer* **129** 89
[8] Blackmore R 1987 *J. Chem. Phys.* **87** 791
[9] Ciuryło R, Shapiro D A, Drummond J R, May A D 2002 *Phys. Rev. A* **65** 012502
[10] Tran H, Ngo N H, Hartmann J M, Gamache R R, Mondelain D, Kassi S, Campargue A, Gianfrani L, Castrillo A, Fasci E, Rohart F 2013 it J. Chem. Phys. **138** 034302
[11] Wcisło P, Cygan A, Lisak D, Ciuryło R 2013 *Phys. Rev. A* **88** 012517
[12] Wcisło P, Ciuryło R 2013 *J. Quant. Spectrosc. Radiat. Transfer* **120** 36
[13] Lance B, Blanquet G, Walrand J, Bouanich J P 1997 *J. Mol. Spectrosc.* **185** 262
[14] Salumbides E J, Koelemeij J C J, Komasa J, Pachucki K, Eikema K S E, Ubachs W 2013 *Phys. Rev. D* **87** 112008
[15] Lemarchand C, Mejri S, Sow P L T, Triki M, Tokunaga S K, Briaudeau S, Chardonnet C, Darquié B, Daussy C 2013 *Metrologia* **50** 623
[16] Moretti L, Castrillo A, Fasci E, De Vizia M D, Casa G, Galzerano G, Merlone A, Laporta P, Gianfrani L 2013 *Phys. Rev. Lett.* **111** 060803
[17] Cygan A, Lisak D, Trawiński R S, Ciuryło 2010 *Phys. Rev. A* **82** 032515
[18] Pine A S 1999 *J. Quant. Spectrosc. Radiat. Transfer* **62** 397

Journal of Physics: Conference Series **548** (2014) 012023 doi:10.1088/1742-6596/548/1/012023

Line shape measurements of rubidium 5S-7S two-photon transition

P. Morzyński[1], P. Wcisło[1], P. Ablewski[1], R. Gartman[1], W. Gawlik[2], P. Masłowski[1], B. Nagórny[1], F. Ozimek[3], Cz. Radzewicz[3], M. Witkowski[1,4], R. Ciuryło[1] and M. Zawada[1]

[1] Institute of Physics, Faculty of Physics, Astronomy and Informatics, Nicolaus Copernicus University, Grudziądzka 5, PL-87-100 Toruń, Poland

[2] Institute of Physics, Faculty of Physics, Astronomy and Informatics, Jagiellonian University, Reymonta 4, PL-30-059 Kraków, Poland

[3] Institute of Experimental Physics, Faculty of Physics, University of Warsaw, Hoża 69, PL-00-681 Warsaw, Poland

[4] Institute of Physics, University of Opole, Oleska 48, PL-45-052 Opole, Poland

E-mail: rciurylo@fizyka.umk.pl

Abstract. We report the use of a digital lock to measure the line profile and center frequency of rubidium 5S-7S two-photon transitions with a cw laser referenced to an optical frequency comb. The narrow, two-photon transition, 5S-7S (760 nm), insensitive to first-order in a magnetic field, is a promising candidate for frequency reference.

1. Introduction

The International Committee for Weights and Measures recommended several radiations for the practical realization of the metre. At the border of the visible and near-infrared ranges, in particular, the International Bureau of Weights and Measures (BIPM) recommends the $5S_{1/2}$(F=3)-$5D_{5/2}$(F=5) two-photon transition in ^{85}Rb with a standard uncertainty of 5 kHz [1]. Recent development in phase-stabilized optical frequency combs based on mode-locked femtosecond lasers allowed determination of the absolute frequency of a similar transition in Rb, $5S_{1/2}$-$7S_{1/2}$, which is 100 times weaker than 5S-5D, yet less sensitive to stray magnetic fields. At the 5S-5D transition the rubidium atoms must be carefully shielded against the magnetic field to avoid any linear Zeeman shifts. On the other hand, the 5S and 7S levels have the same Landé g factors which cancels the linear Zeeman shifts in the 5S-7S transition. The ac-Stark effect in the 5S-7S transition is also smaller than in the 5S-5D case. All these factors explain the growing interest in the $5S_{1/2}$-$7S_{1/2}$ transition in rubidium [2, 3, 4, 5].

2. Experimental arrangement

The details of the experimental set-up and apparatus are described elsewhere [6]. Here, we focus only on the description of the digital lock system and its application to the measurements of the line profile.

The idea of the digital lock is depicted in Fig. 1. An acousto-optic modulator (AOM) driven by a direct digital synthesizer (DDS) square-wave modulates the light frequency with

Figure 1. The scheme of the digital lock.

the step $2\Delta f$ equal to the half-width of the line. The AOM carrier frequency f_{AOM} is chosen such that the AOM efficiency with $f_+ = f_{AOM} + \Delta f$ and $f_- = f_{AOM} - \Delta f$ is the same. The microcontroller (Atmel AT91SAM7S), which controls the DDS, counts the photomultiplier pulses from the fluorescence signal of the 6P-5S transition. The error signal for the laser lock is calculated from the difference of the counts corresponding to fluorescence for f_+ and f_-. The software PI regulator in the microcontroller calculates the correction Δf_L and applies it to the TiSa laser. Switching between the $f_\pm$ frequencies in the DDS is completed in 150 ns and the acoustic wave needs few μs to completely propagate the frequency changes through the AOM $PbMo0_4$ crystal. The following acquisition of the photomultiplier pulses takes 38 ms which is long enough to ignore chirping effects from switching the $f_\pm$ frequencies. The frequency of the laser f_{LASER}, before shifting by the AOM, is measured in the real-time by an optical frequency comb.

The digital lock can be easily used to measure the line profile (Fig. 2) by changing Δf in the digital lock algorithm and recording accordingly the fluorescence intensity. Furthermore, recording the frequency measured by the optical frequency comb for different Δf detects a potential asymmetry of the line profile.

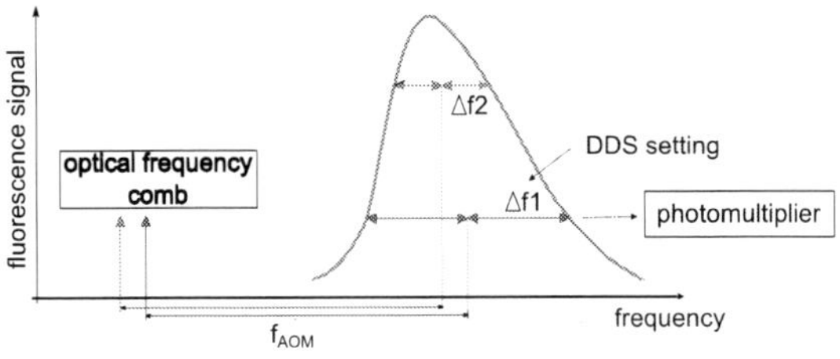

Figure 2. The measurement of the line profile.

3. Results

The examples of the measured profiles are presented in Fig. 3. The two series of graphs correspond to the measurements of the ^{87}Rb F=2-F'=2 and ^{85}Rb F=3-F'=3 transitions for different intensities of the probing light. In the two-photon spectroscopy the counter-propagating beams are not focused, the transition is characterized by a power-broadened Lorentzian profile. The measurement accuracy may be reduced at the line center where the digital lock is less precise because of small value of the fluorescence derivative with respect to the laser frequency leading to underestimation of measured fluorescence at given frequency.

In Table 1 we present the measured frequencies of four $5S_{1/2}$-$7S_{1/2}$ transitions in ^{87}Rb and ^{85}Rb. With the knowledge of the hyperfine splitting of the $5S_{1/2}$ state [7, 8] the hyperfine A constants of the $7S_{1/2}$ state were derived (94678.4(2.3) kHz for ^{85}Rb 7S and 319747.9(2.3) kHz for ^{87}Rb 7S) as well as the $5S_{1/2}$-$7S_{1/2}$ transition isotope shift (131529.6(6.6) kHz) [6].

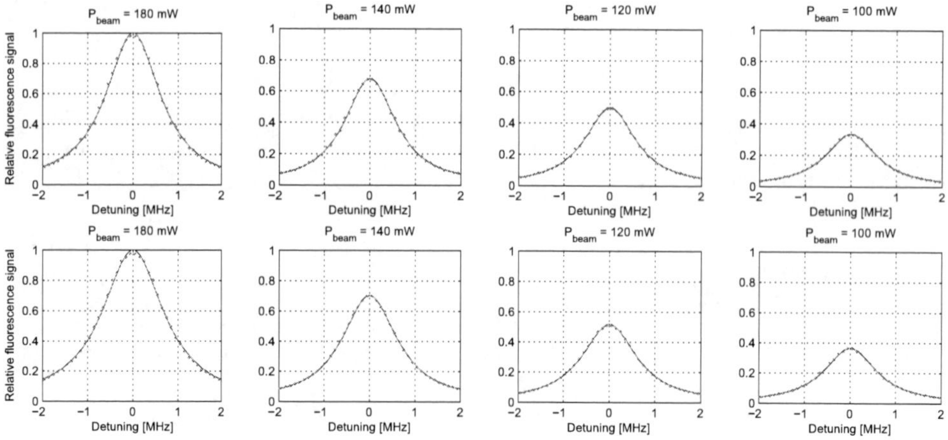

Figure 3. The measurements of the ^{87}Rb F=2-F'=2 (top) and ^{85}Rb F=3-F'=3 (bottom) transitions for different intensities of the probing light.

Table 1. Measured absolute frequencies of the two-photon $5S_{1/2}$-$7S_{1/2}$ transitions [6].

Transitions	Frequency [kHz]
^{85}Rb F=2-F'=2	394399282854.2(2.4)
^{85}Rb F=3-F'=3	394397907005.6(2.8)
^{87}Rb F=1-F'=1	394400482036.5(3.8)
^{87}Rb F=2-F'=2	394397384443.1(2.6)

4. Conclusion

We present the use of a digital lock to measure the line profile and center frequency of rubidium $5S_{1/2}$-$7S_{1/2}$ two-photon transitions. We have performed a series of measurements of the line profile and the absolute frequency of the $5S_{1/2}$-$7S_{1/2}$ transitions in rubidium vapor with a cw laser referenced to an optical frequency comb. We have also estimated the A constants of the hyperfine splitting of the $7S_{1/2}$ state and the isotope shift between ^{85}Rb and ^{87}Rb of the $5S_{1/2}$-$7S_{1/2}$ transition.

Acknowledgments

This work has been performed in the National Laboratory FAMO in Toruń and supported by the Polish National Science Centre Projects No. 2012/07/B/ST2/00235 and 2013/11/NST2/00846 and the TEAM Project of the FNP, co-financed by the EU within the European Regional Development Fund.

References

[1] Bureau International des Poids et Mesures (BIPM), in *Report of the 86th Meeting of the Comité International des Poids et Mesures*, (BIPM, Sèvres, France, 1997)
[2] Chui H C, Ko M S, Liu Y W, Shy J T, Peng J L and Ahn H 2005 *Opt. Lett.* **30** 842
[3] Marian A, Stowe M C, Felinto D and Ye J 2005 *Phys. Rev. Lett.* **95** 023001
[4] Pandey K, Kiran Kumar P V, Suryanarayana M V and Natarajan V 2008 *Opt. Lett.* **33** 1675

[5] Barmes I, Witte S and Eikema K S E 2013 *Phys. Rev. Lett.* **111** 023007
[6] Morzynski P, Wcislo P, Ablewski P, Gartman R, Gawlik W, Maslowski P, Nagorny B, Ozimek F, Radzewicz C, Witkowski M, Ciurylo R and Zawada M 2003 *Opt. Lett.* **22** 4581
[7] Bize S, Sortais Y, Santos M S, Mandache C, Clairon A and Salomon C 1999 *Europhys. Lett.* **45** 558
[8] Arimondo E, Inguscio M and Violino P, 1971 *Rev. Mod. Phys.* **49** 31

Journal of Physics: Conference Series **548** (2014) 012024 doi:10.1088/1742-6596/548/1/012024

Alternative approaches to cavity enhanced absorption spectroscopy

A Cygan*, **P Wcisło, S Wójtewicz, P Masłowski, R S Trawiński, R Ciuryło, D Lisak**

Institute of Physics, Faculty of Physics, Astronomy, and Informatics, Nicolaus Copernicus University, Grudziadzka 5, 87-100 Torun, Poland

E-mail: agata@fizyka.umk.pl

Abstract. New approaches to cavity enhanced absorption spectroscopy are presented. They are based on precise measurements of shape of cavity modes broadened and shifted in the vicinity of the absorber placed inside the high-finesse cavity. The usefulness of both techniques to accurate line-shape study is demonstrated on an example of a weak $(3 \leftarrow 0)$ $^{13}C^{16}O$ transition.

1. Introduction

A great potential of cavity-enhanced absorption spectroscopy (CEAS) techniques in application to ultrasensitive measurements of absorption grew up from significant elongation of optical path of light inside the high-finesse cavity. One of the well known variants of CEAS is the cavity ring-down spectroscopy (CRDS) which is based on variation of lifetime of the laser light in the optical cavity with absorbing medium. This lifetime can be measured from the ring-down decay time constant and is insensitive to the laser power uctuations. It is also directly related to the spectral broadening of the cavity resonant modes. This phenomenon was originally observed by Nakagawa *et al.* [1] in 1994, however until recently it was not exploited for spectroscopy. Development of laser stabilization technique [2–4] allows measurement of the width of the single mode of the ultra stable cavity with high accuracy [5]. Very recently experimental spectrum based on cavity mode widths determination was demonstrated by Long *et al.* with the use of the FARS technique [6, 7] and by Cygan *et al.* [8] who used an ultranarrow and tunable in the range of 40 GHz external cavity diode laser to measure a weak oxygen transition. Such approach to precise measurements of cavity mode widths inspired a development of a new CEAS technique called cavity mode-width spectroscopy (CMWS) [7, 8]. Qualitative and intuitive comparison of CRDS and CMWS methods presented in [8] reveals theirs complementarity in the sense that they achieve their best precision in different pressure ranges. For low absorptions the best precision is achieved with the CRDS technique, where the ring-downs are long and decay constants can be well determined. In the opposite case, where the absorption is high and cavity modes become broader, the precision of CMWS is enhanced.

In this paper we present precise CMWS spectrum measured for a P3 $(3 \leftarrow 0)$ $^{13}C^{16}O$ transition. Contrary to our previous measurement technique of an experimental setup described in [8] here we measured cavity mode widths relative to the neighboring cavity mode, to which the probe laser is locked. We also present dispersive spectrum of $^{13}C^{16}O$ line which is a result of measurement of cavity modes pulling caused by a dispersive shift of light near the $^{13}C^{16}O$

absorption line. In contrast to the NICE-OHMS (Noise-Immune Cavity-Enhanced Optical Heterodyne Molecular Spectroscopy) technique [9] which is based on amplitude measurements of dispersion profiles, in this work to obtain dispersive spectrum only frequency domain quantities were used.

2. Absorptive broadening and dispersive pulling of cavity modes

Near the resonance transition of the absorbing medium placed inside the high-finesse cavity one may observe two characteristic phenomenon - absorptive broadening and dispersive shift of cavity modes. Both these effects can be described by an extended formula for electromagnetic field power transmitted through the cavity given in the form [8]

$$I_T(\nu) = I_0 \frac{(1-R)^2 e^{-\alpha(\nu)L}}{(1-Re^{-\alpha(\nu)L})^2 + 4Re^{-\alpha(\nu)L}\sin^2(\frac{\pi\nu}{\nu_{\mathrm{FSR}}(\nu)})}, \tag{1}$$

where I_0 is the incident field power, R is the intensity re ectivity of the cavity mirrors and L is the cavity length. Quantities $\alpha(\nu)$ and $\nu_{\mathrm{FSR}}(\nu)$ are the absorption coefficient and free spectral range of the cavity, respectively, and they are related to imaginary $\kappa(\nu)$ and real $n'(\nu)$ part of the complex refractive index, given generally in the form $n_a(\nu) = n'(\nu) - i\kappa(\nu)$, by equations $\alpha(\nu) = 2k_0\kappa(\nu)$ and $\nu_{\mathrm{FSR}}(\nu) = c/(2n'(\nu)L)$, where k_0 is the value of a wave vector for the resonance transition and c is the speed of light. In case when the absorption coefficient $\alpha(\nu)$ is described by a Lorentzian line shape, formulas for $n'(\nu)$ and $\kappa(\nu)$ are simply given by ramers- rönig relations [10]. sing ramers- rönig relations we can write that

$$n'(\nu) = n + cNS\,\frac{\mathrm{Im}[\mathcal{I}(\nu)]}{2k_0}, \tag{2}$$

$$\kappa(\nu) = cNS\,\frac{\mathrm{Re}[\mathcal{I}(\nu)]}{2k_0} \tag{3}$$

in which n describes the value of refractive index far from the resonance transition, N is the number density of the absorber expressed in units molecule/cm^3, S is the line intensity in units cm^{-1}/(molecule/cm^2) and $\mathcal{I}(\nu)$ is a normalized complex line-shape function, i.e. $\int \mathcal{I}(\nu)d\nu = 1$. According to above relations Eq. (2) describes pulling of the cavity modes and hence the dispersive shape of the investigated transition, whereas Eq. (3) is related to broadening of cavity modes and an absorption line shape usually measured in CEAS experiments. Since presence of absorbing medium inside the cavity shortens the lifetime of laser light inside the cavity, it is clear that also widths of cavity modes have to become broader according to the Fourier transform relations. In this way frequency domain picture used in CMWS and time domain picture used in CRDS can be simply related by the formula $\delta\nu_m = 1/(2\pi n\tau)$ where $\delta\nu_m$ and τ are the main quantities measured in both these techniques, i.e. the cavity mode width and time constant of the ring-down decay, respectively.

3. Experiment and results

In order to precisely measure absorptive (CMWS) and dispersive spectra of P3 $(3 \leftarrow 0)$ $^{13}C^{16}O$ line presented in Fig. 1. we designed an experiment independent of any cavity length changes which were the main source of noise in our previous measurements of an oxygen line [8]. In our new system the probe laser is tightly locked by the Pound-Drever-Hall [11] method to the cavity. In this way the slow drift of the cavity length as well as cavity vibrations are transferred to the probe laser, but the overlap between laser and cavity mode frequencies is maintained constantly during cavity mode widths and shifts measurements. Our spectrometer uses two-beams configuration system in which the probe laser beam is divided into two beams one

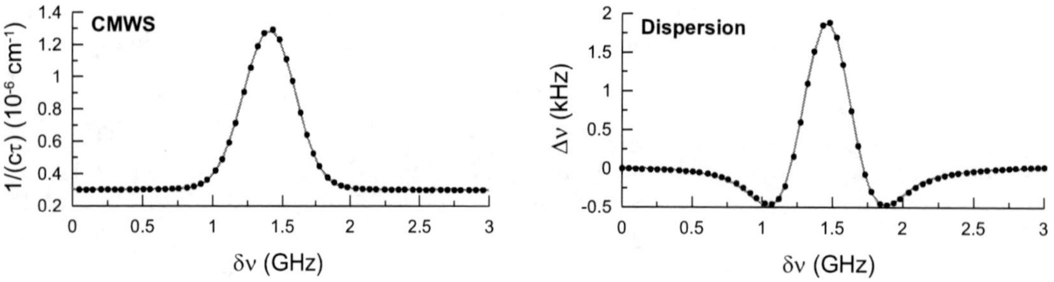

Figure 1. Absorptive (CMWS) and dispersive spectrum of the $(3 \leftarrow 0)$ $^{13}C^{16}O$ line recorded at pressure 4.8 Torr by the CMWS and dispersive techniques, respectively. Both spectra were compared with the Voigt profile.

,,lock beam" used for PDH locking of the probe laser to the cavity and the second one ,,probe beam" used for probing cavity mode widths and shifts. Both beams are mode matched to two consecutive longitudinal modes of the probe cavity. The relative scan of the ,,probe beam" around the cavity mode is realized by frequency tuning of the RF signal driving an acoustooptic modulator inserted into the optical path of the ,,probe beam". The Voigt profile commonly used in spectroscopic databases was used for comparison with absorptive (CMWS) and dispersive line shapes presented in Fig. 1. Closer look at experimental results in case of both methods suggests that the Voigt profile is not good enough to reproduce experimental data with high accuracy.

4. Conclusions

In this study two alternative approaches to cavity-enhanced absorption spectroscopy, namely CMWS and dispersive method, were proposed. We tested these new methods on the CO rovibrational transition from the $(3 \leftarrow 0)$ band, which is located in the spectral region around 6201 cm^{-1}. Both methods indicated that the Voigt profile is too simple model to describe the investigated line shape with high accuracy. Comparison of various CEAS techniques is presently of great importance for the line-shape analysis of experimental spectra, where one of the most challenging task is to distinguish the systematic errors due to experimental imperfections from these caused by the wrong choice of the line-shape model, see e.g. [12].

Acknowledgments

The research is part of the program of the National Laboratory FAMO in Toruń, Poland, and is supported by the Polish National Science Centre Project nos. DEC-2011 01 B ST2 00491, MO-2012 05 N ST2 02717 and MO-2013 11 D ST2 02663. The research is also supported by the Foundation for Polish Science TEAM, HOMING PL S and START Projects co-financed by the E European Regional Development Fund.

References

[1] Nakagawa K, *et al.* 1994 *Opt. Commun.* **107** 369
[2] Drever W P, *et al.* 1983 *Appl. Phys. B* **97** 31
[3] Stoehr H, *et al.* 2006 *Opt. Lett.* **31** 736
[4] Ludlow A D, *et al.* 2007 *Opt. Lett.* **32** 641
[5] Lisak D, *et al.* 2012 *Acta Phys. Pol. A* **121** 614
[6] Truong G-W, *et al.* 2013 *Nature Photon.* **7** 532
[7] Long D A, *et al.* 2014 *Appl. Phys. B* **114** 489
[8] Cygan A, *et al.* 2013 *Opt. Express* **21** 29744

[9] Ye J, *et al.* 1998 *J. Opt. Soc. Am. B* **15** 6
[10] Libbrecht K G, *et al.* 2006 *Am. J. Phys.* **74** 1055
[11] Drever R W P, *et al.* 1983 *Appl. Phys. B* **31** 97
[12] Wójtewicz S, *et al.* 2011 *Phys. Rev. A* **84** 032511

Journal of Physics: Conference Series **548** (2014) 012025 doi:10.1088/1742-6596/548/1/012025

Spectroscopy of cooperative laser electron-γ-nuclear processes in polyatomic molecules

A V Glushkov, P A Kondratenko, V V Buyadgi, A S Kvasikova, T N Sakun and A N Shakhman

Odessa State Environmental University, L'vovskaya str.15, Odessa-16, 65016, Ukraine

E-mail: dirac13@mail.ru

Abstract. The consistent quantum approach to calculating the electron-nuclear γ transition spectra (a set of the vibration-rotational satellites in a molecule) of a nucleus in polyatomic molecules is used to determine accurate data of the vibration-nuclear transition probabilities. We present results for emission and absorption γ-spectra of nucleus ^{191}Ir ($E^{(0)}_{\gamma}$= 82 keV) in the molecule of IrO_4.

1. Introduction

Any alteration of the molecular state must be manifested in the quantum transitions, for example, in a spectrum of the γ-radiation of a nucleus. It is well known that it is possible to transfer part of nuclear energy to atoms or molecules by emitting or absorbing γ quanta (c.f.[1-12]). A spectrum contains a set of the electron-vibration-rotation satellites, which are due to an alteration of the state of system interacting with photons. A mechanism of forming satellites in the molecule is connected with a shaking of the electron shell resulting from the interaction between a nucleus and γ quantum. [1, 3, 13-20]. Appearance of the molecular nuclear lines is of great interest as one can possibly modify the γ-radiation spectrum by means of changing the vibration-rotational state of a molecule by a laser light [1,12,16]. The well known example is the Szilard-Chalmers effect (molecular dissociation because of the recoil during radiating gamma quantum with large energy; c.f. [1, 2]).

This work addresses our study of the co-operative dynamical phenomena (c.f.[14-19]) due the interaction between atoms, ions, molecule electron shells and nuclei nucleons. We present an advanced approach to description of a new class of dynamical laser-electron-nuclear effects in molecular spectroscopy, in particular, we study the nuclear gamma-emission or absorption spectrum of a molecule. A consistent quantum- mechanical approach to calculation of the electron-nuclear γ transition spectra (set of vibration-rotational satellites in molecule) of a nucleus in the polyatomic molecules has been proposed earlier [14-16]. It generalizes the well known Letokhov-Minogin model [3]. Estimates of the vibration-nuclear transition probabilities in a case of the emission and absorption spectrum of nucleus ^{188}Os ($E^{(0)}_{\gamma}$= 155 keV) in the molecule of OsO_4 and other molecules were listed. Here we report accurate data on the vibration-nuclear transition probabilities in a case of the emission and absorption spectrum of nucleus nucleus ^{191}Ir ($E^{(0)}_{\gamma}$= 82 keV) in the molecule of IrO_4.

2.The electron-nuclear γ transition spectra of nucleus in multi-atomic molecules

As the method of computing is presented in detail earlier [2,3,14-16], here we consider only the key aspects. The main purpose is calculating a structure of the gamma transitions (a probability of transition) or spectrum of the gamma satellites because of changes to the electron-vibration-rotational states of the polyatomic molecules under gamma quantum radiation (absorption). Let us note that obviously the γ-spectrum component intensities because of the electron transitions are very small (the smallness parameter m_e/M, where m_e is an electron mass, M is mass of a molecule). So, it is of an especial interest studying nuclear γ-spectra in molecules, because of the vibration-rotational transitions. Comparing the averaged energies for excitation of the molecule vibration and rotation, one could remind that for polyatomic molecules it is typical $B/\hbar\omega \sim 10^{-4}$-$10^{-2}$. If the γ-active nucleus is located in the centre of mass of the molecule, then it is possible to consider only the vibration transitions of the molecule and neglect of the rotational transitions [2,5]. This approximation is particularly important for those molecules in which the position of the γ-active nucleus in the centre of mass is due to the symmetry of the molecule and is not random. If the γ-active nucleus in a molecule is out of the centre of mass, then the situation becomes much more complicated. Formal foundations of the general theory of the γ-emission and absorption in the nuclei of symmetric polyatomic molecules have been presented in [2,3], where the authors considered the case of molecules of the XY_2 type.

Here we limit our study to five-atomic molecules (of XY_4 type; T_d). The interaction Hamiltonian of the γ-radiation with a system of nucleons for the first nucleus can be expressed through the co-ordinates of nucleons r_n' in a system of the mass centre of the one nucleus [3,15]:

$$H(r_n) = H(r_n')\exp(-ik_\gamma u) \tag{1}$$

where k_γ is a wave vector of the γ-quantum; u is the shift vector from equality state (coinciding with molecule mass centre) in system of co-ordinates in the space. The matrix element for transition from the initial state "a" to the final state "b" is presented as usual:

$$< \Psi_b^* \,|\, H \,|\, \Psi_a > \bullet < \Psi_b^* \,|\, e^{-ik_\gamma u} \,|\, \Psi_a > \tag{2}$$

where a and b is a set of quantum numbers, which define the vibrational and rotational states before and after interaction (with gamma- quantum). The first multiplier in Eq. (2) is defined by the gamma transition of nucleus and is not dependent upon the internal structure of molecule in good approximation. The second multiplier is the matrix element of transition of the molecule from the initial state "a" to the final state "b":

$$M_{ba} =< \Psi_b^*(r_e) \,|\, \Psi_a(r_e) > \bullet < \Psi_b^*(R_1, R_2) \,|\, e^{-ik_\gamma R_1} \,|\, \Psi_a(R_1, R_2) > \tag{3}$$

The expression in Equation (3) gives a general formula for calculating the probability of changing the internal state of molecule during absorption or emission of a γ-quantum by a nucleus. It determines an intensity of the corresponding γ-satellites. Their positions are fully determined as follows:

$$E_\gamma = E_\gamma^0 \pm R + \hbar k_\gamma v \pm (E_b - E_a) \tag{4}$$

Here, M is the molecule mass, v is a velocity of molecule before interaction of nucleus with γ quantum; E_a and E_b are the energies of the molecule before and after interaction; E_γ is an energy of nuclear transition; R_{om} is an energy of recoil: $R_{om}= [(E_\gamma^{(o)}]^2/2Mc^2$.

Further we miss the molecule rotations and consider γ-spectrum of a nucleus in the molecule mass centre as a spectrum of the vibration-nuclear transitions (see details in Refs. [3,15]). Further one can suppose that only single non-generated normal vibration (vibration quantum $\hbar\omega$) is excited and initially a molecule is on the vibrational level $v_a =0$. If we denote a probability of the corresponding excitation as $P(v_b, v_a)$ and use expression for shift u of the γ-active nucleus through the normal co-ordinates, then an averaged energy for excitation of the single normal vibration is as follows [3]:

$$\overline{E}_{\text{vib}}=\sum_{v=0}^{\infty}\hbar\omega(v+\tfrac{1}{2})\overline{P}(v,0)-\hbar\infty/2=\sum_{v=0}^{\infty}\hbar\omega(v+\tfrac{1}{2})P(v,0)-\hbar\omega/2=$$

$$=\sum_{v=0}\hbar\omega(v+\tfrac{1}{2})\frac{z^{v}}{v!}e^{-z}-\frac{\hbar\omega}{2}=\frac{1}{2}R\left(\frac{M-m}{m}\right), \quad (5)$$

where

$$z=(R/\hbar\omega)[M-m/m]\cos^{2}\vartheta,$$

and m is the mass of γ-active nucleus, ϑ is an angle between nucleus shift vector and wave vector of γ-quantum and line in $\overline{E}_{\text{vib}}$ means averaging on orientations of molecule (or on angles ϑ).

A shift u of the γ-active nucleus can be expressed through the normal co-ordinates $Q_{s\sigma}$ of a molecule:

$$u=\frac{1}{\sqrt{m}}\sum_{s\sigma}b_{s\sigma}Q_{s\sigma} \quad (6)$$

where m is a mass of the γ- active nucleus; components of the vector $b_{s\Phi}$ of nucleus shift due to the Φ-component of "s" normal vibration of a molecule are the elements of matrix b [2]. It realizes the orthogonal transformation of the normal co-ordinates matrix Q to matrix of masses of the weighted Cartesian components of the molecule nuclei shifts q. According to Equation (1), the matrix element can be written as multiplying the matrix elements on molecule normal vibration, which takes contribution to a shift of the γ-active nucleus:

$$M(b,a)=\prod_{s}\left\langle v_{s}^{\ b}\,|\prod_{\sigma}\exp(-ik_{\gamma}b_{s\sigma}Q_{s\sigma}/\sqrt{m})v_{s}^{\ a}|\right\rangle. \quad (7)$$

It is obvious that missing molecular rotations means missing the rotations which are connected with the degenerate vibrations. The wave functions of a molecule should be written for non-degenerate vibration as: $|\,v_{s}\ \rangle=\Phi_{vs}(Q_{s})$, for double degenerate vibration in the form:

$$|v_{s}\rangle=(v_{s}+1)^{-\frac{1}{2}}\sum_{vs\sigma_{1},vs\sigma_{2},vs\sigma_{3}}\Phi_{v_{s\sigma_{1}}}(Q_{s\sigma_{1}})\Phi_{v_{s\sigma_{2}}}(Q_{s\sigma_{2}}) \quad (8)$$

(where $v_{s\sigma_{1}}+v_{s\sigma_{2}}=v_{s}$) and for triple degenerate vibration as follows:

$$|v_{s}\rangle=\left(\frac{2}{(v_{s}+1)(v_{s}+2)}\right)^{\frac{1}{2}}\sum_{vs\sigma_{1},vs\sigma_{2},vs\sigma_{3}}\Phi_{v_{s\sigma1}}(Q_{s\sigma_{1}})\Phi_{v_{s\sigma_{2}}}(Q_{s\sigma_{2}})\Phi_{v_{s\sigma_{3}}}(Q_{s\sigma_{3}}) \quad (9)$$

where
$$v_{s\sigma_{1}}+v_{s\sigma_{2}}+v_{s\sigma_{3}}=v_{s}.$$

In the simple approximation function $\Phi_{v_{s\sigma}}(Q_{s\sigma})$ can be chosen in a form of the linear harmonic oscillator one. Taking into account the explicit expressions for the wave functions, the calculation of the matrix element (7) is reduced to the computation of the matrix elements for each component of the normal vibration.

3. Results and conclusions

Below we present the advanced data on the vibration-nuclear transition probabilities in a case of the emission and absorption γ-spectrum of nucleus ^{191}Ir ($E^{(0)}_{\gamma}$= 82 keV) in the molecule IrO_4. The main difficulty during calculating (7) is connected with definition of the values $b_{s\sigma}$- of the normalized shifts of γ-active decay. It is known that if a molecule has the only normal vibration of the given symmetry type, then the corresponding values of $b_{s\sigma}$ can be found from the well-known Eckart conditions. For several normal vibrations of one symmetry type, a definition of $b_{s\sigma}$ requires solving the secular equation for molecule $|GF\text{-}\lambda E|=0$ [21-24]. We have used the results of advanced theoretical calculation of electron structure of the studied system within an advanced scheme of the X_{∇}- scattered waves method [22,24]. Other details can be found in Refs. [3,15]. In Table 1 we list the results of calculating probabilities of the first several vibration-nuclear transitions in a case of the emission and absorption γ-spectrum of nucleus ^{191}Ir ($E^{(0)}_{\gamma}$= 82 keV) in IrO_4 and compare with available data in Ref.[2,15], where more simple approximation has been used. Analysis shows that more sophisticated calculation gives higher values for probabilities.

Table 1. Probabilitites of the vibrational-nuclear transitions in spectrum of IrO_4.

Vibration transition $v_3^a, v_4^a - v_3^b, v_4^b$	$\overline{P}(v_3^a, v_4^a - v_3^b, v_4^b)$ [15]	$\overline{P}(v_3^a, v_4^a - v_3^b, v_4^b)$ This work
0,0 – 0,0	0.863	0.953
1,0 – 0,0	0.025	0.032
0,1 – 0,0	0.097	0.104
1,0 – 1,0	0.812	0.901
0,1 – 0,1	0.731	0.813

References

[1] Letokhov V S 1977 *Laser Spectroscopy* (New York: Academic Press)
[2] Letokhov V S 1974 *Phys. Lett. A* **46** 257
[3] Letokhov V S and Minogin V G 1985 *JETP.* **69**(5) 1568
[4] Ivanov L N and Letokhov V S 1975 *JETP.* **68** 1748
[5] Glushkov A V, Loboda A, Gurnitskaya E and Svinarenko A 2009 *Phys. Scr.* **T135** 014022
[6] Glushkov A V 2012 *J. of Phys.: Conf. Ser.* 397 012011
[7] Glushkov A V and Ivanov L N 1992 *Phys. Lett.* A **170** 33
[8] Law J and Campbell J L 1982 *Phys. Rev. C* **25** 514
[9] Khetselius O Yu 2008 *Spectral Line* Shapes, Vol. 15, AIP Conf. Proceedings. **1058** 363
[10] Wauters L, Vaeck N, Godefroid M, var der Hart H and Demeur M 1997 *J. Phys. B.* **30** 4569
[11] Shahbaz A, Müller C, Staudt A, Bürvenich T and Keitel C 2007 *Phys. Rev .Lett.* **98** 263901
[12] Fedorov M V 1999 *Phys.-Uspekhi* **169** 66
[13] Ivanov L N and Letokhov V S *Com. Mod. Phys.* D **4** 169
[14] Glushkov A V, O Yu Khetselius, Malinovskaya S V 2008 *Europ. Phys. J. ST* **160** 195
[15] Glushkov A V, Khetselius O Yu and Malinovskaya S V 2008 *Molec. Phys.* **106** 1257
[16] Glushkov A V, Khetselius O Yu and Malinovskaya S V 2008 *Frontiers in Quantum Systems in Chem. and Phys., Ser. Progress in Theoretical Chemistry and Physics*, vol. 18. ed. S Wilson, P J Grout, J Maruani, G; Delgado-Barrio, P Piecuch (Berlin: Springer) p 523
[17] Glushkov A V, Malinovskaya S V, Svinarenko A A and Chernyakova Yu G 2004 *Int. J. Quant. Chem.* **99** 889.
[18] Glushkov A V, Ivanov L N and Ivanova E P 1986 *Autoionization Phenomena in Atoms*, (Moscow: University Press) p 58
[19] Khetselius O Yu 2012 *J. of Phys.: Conf. Ser.* **397** 012012
[20] Khetselius O Yu 2009 *Phys. Scr.* **T.135** 014023
[21] Glushkov A V 1993 *J. Struct. Chem.* **34**(1) 3
[22] Glushkov A V 2008 *Quantum mechanics of atomic systems* (Odessa: Astroprint)
[23] Wilson E B, Decius J C and Cross P C 1995 *Molecular Vibrations* (New York: Dover)
[24] Gedasimov V, Zelenkov A, Kulakov V, Pchelin V, Sokolovskaya M, Soldatov A and Chistyakov L 1984 *JETP.* **86** 1169; Soldatov A 1983 Preprint of I.V.Kurchatov Institute for Atomic Energy IAE-3916, Moscow

Collisional shift of hyperfine line for rubidium in an atmosphere of the buffer inert gas

A V Glushkov, O Yu Khetselius, Y M Lopatkin, T A Florko, O A Kovalenko and V F Mansarliysky

Odessa State Environmental University, L'vovskaya str. 15, Odessa-16, 65016, Ukraine

E-mail: dirac13@mail.ru

Abstract. New relativistic approach, based on the relativistic many-body perturbation theory using optimized wave functions sets, is applied to calculate the hyper fine structure collision shift for rubidium atom in atmosphere of the helium inert gas. Data for the collisional shifts of the Rb-He system are presented and compared with data available in the literature.

1. Introduction

This paper presents our recent progress in the theoretical study of hyperfine lines shift and broadening for heavy atoms, namely, alkali and others, in atoms in an atmosphere of the buffer gases (see, for example, Refs. [1-5]). The study of the collisional shifts and broadening of the hyperfine structure lines for heavy elements (alkali, alkali-earth, lanthanides, actinides and others) in an atmosphere of inert gases is one of the important and current topics of modern quantum chemistry, atomic optics and spectroscopy [5-8]. Furthermore, heavy atoms are interesting from the point of view of studying the role of weak interactions in atomic optics and heavy-elements chemistry. Besides, calculation of the hyperfine structure line shift and broadening allows one to check the quality of wave functions and study a contribution of relativistic effects in two-center (multi-center) atomic systems. From the applied science point of view, the mentioned physical effects form a basis for creating an atomic quantum measure of frequency. For a long time, the corresponding phenomenon for thallium atoms attracted a special attention because of possibility to create the thallium quantum frequency measure. Naturally, the inert buffer gases (He, Ar *etc.*) were used. The detailed non-relativistic theory of the collisional shift and broadening of the hyperfine structure lines for simple elements (light alkali elements *etc.*) has been developed by many authors (see discussions in refs. [5-9]). However, consideration of heavy elements faces serious difficulties in accounting for the relativistic and correlation corrections.

It is noteworthy that until now a consistent, accurate quantum mechanical approach for calculating main characteristics of the collisional processes was not developed though many different simplified models have been proposed (see, for example [5-9]). The most widespread approach is based on the calculation of the corresponding collision cross-section, in particular, in a case of the van der Waals interaction between colliding particles. However, such an approach does not factually define any difference between the Penning process and resonant collisional one and gives often non-correct results for cross-sections. More consistent method requires data on the process probability $G(R)$ as a function of inter nuclear distance. It should be noted that these data are practically absent at present time.

In this paper a new relativistic optimized approach, based on the gauge-invariant perturbation theory (PT) with using the optimized wave functions basis's, is applied to calculating the hyperfine structure collision shift for rubidium atom in atmosphere of the helium inert gas. The basic expressions for the collision shift and broadening of the hyperfine structure spectral lines are taken from the kinetic theory of spectral lines [5-7]. The exchange perturbation theory (the modified version EL-HAV) has been used for calculating the corresponding potentials (see details in [5,6])..

2. Hyperfine lines shift for system "alkali atom- inert gas"

As the method of computing hyperfine lines shift is presented in detail earlier[1-5], here we highlight key elements. To calculate the collision shift of hyperfine structure spectral lines one can use the following expression known from kinetic theory of spectral line form (see [5-7]):

$$f_p = \frac{\Delta}{p} = \frac{4\pi N_b}{p}\int_0^\infty e^{-U(R)/kT}\left[1+g(R)\right]\delta\omega(R)R^2 dR, \tag{1}$$

where $U(R)$ is the effective potential of the interatomic interaction, which has a central symmetry in a case of the systems A-B *(*in our case, for example, B=He; A=Rb*)*; T is the temperature, w_0 is a frequency of the hyperfine structure transition in the isolated active atom; $d\omega\,(R)=Dw\,(R)/w_0$ is the relative local shift of the hyperfine structure lines, which is arisen due to the disposition of the active atoms (say, atom of Rb and He) on a distance R, $\{1+g(R)\}$ is a temperature factor. To calculate an effective potential of the interatomic interaction, the method of the exchange perturbation theory (the modified version EL-HAV) (see details in [1,7] has been used. Accurate to second order terms in the potential of the Coulomb interaction of the valent electrons and atomic cores one can write:

$$\delta\omega(R) = \frac{S_0}{1-S_0} + \Omega_1 + \Omega_2 - \frac{C_6}{R^6}\left(\frac{2}{\overline{E}_a} + \frac{1}{\overline{E}_a + \overline{E}_B}\right), \tag{2}$$

where C_6 is the van der Waals constant for interaction A-B*;* I, $E_{1a,b}$ are the ionization potential and excitation energy on the first level for atoms A, B correspondingly; S_0 is the overlap integral. The value of $\overline{E}_{a,b}$ can be simply defined as follows:

$$\overline{E}_{a,b} = \left(I_{a,\,b} + E_{1a,\,b}\right)/2,$$

The values Ω_1, Ω_2 in the expression (2) are the non-exchange and exchange non-perturbation sums of the first order correspondingly, which are defined as follows:

$$\Omega_1 = \frac{2}{N(1-S_0)\rho_0}\sum_k{}' \frac{\langle \Phi_0'(1)|H_{CT}'|\Phi_k'(1)\rangle V_{k0}}{E_0 - E_k},$$

$$\Omega_2 = \frac{2}{N(1-S_0)\rho_0}\sum_k{}' \frac{\langle \Phi_0'(1)|H_{CT}'|\Phi_k'(1)\rangle U_{k0}}{E_0 - E_k}$$

$$\rho_0 = \langle \Phi_0^1(1)|H_{CT}'|\Phi_0'(1)\rangle / \langle \Phi_0^1(1)|\Phi_0'(1)\rangle$$

where H'_{CT} is the operator of hyperfine interaction, N is the total number of electrons taken into account in the calculation; E_{k}, $\Phi'_k(1) = F'_{k_a}(1)\varphi_{k_b}(2...N)$ — energy and non-symmetrized wave function of state $k = \{k_a, k_b\}$ for isolated atoms A and B.

The non-exchange matrix element of the Coulomb interatomic interaction is as follows:

$$V_{ko} = < \Phi'_k(1) \mid V(1) \mid \Phi'_0(1) >.$$

and, correspondingly, the exchange matrix element is as follows:

$$U_{k0} = \sum_{i=2}^{N} \langle \Phi'_k(1) | V(i) | \Phi'_0(i) \rangle .$$

The operator $V(i)$ is, for example, in a case of system Yb-He, as follows:

$$V(i) = U_{SCF}(r_{a3}) + U_{SCF}(r_{a4}) - 2U_{SCF}(R) + \frac{1}{r_{bi}}, \tag{3}$$

where $U_{SCF}(r)$ is the self-conjunctive field, created by the ytterbium atomic core.

Let us return to consideration of the van der Waals constant C_6 for the interatomic A-B interaction. As a rule, one could use the approximate values for the van der Waals constant C_6 *etc*. Often the sufficiently large errors in determination of the van der Waals constants leads to low accuracy of the inter-atomic potentials calculation and may introduce further inaccuracies. The van der Waals constant may be written as follows [10]:

$$C_6(L,M) = C_{6,0}(L) - \frac{3M^2 - L(L+1)}{(2L-1)(2L+3)} \cdot C_{6,2}(1) \ , \tag{4}$$

where $C_{6,0}(L)$ is the isotropic component of the interaction and $C_{6,2}(L)$ is the component corresponding to the $P_2(\cos\theta)$ term in the expansion of the interaction in Legendre polynomials, where the angle specifies the orientation in the space-fixed frame.

The dispersion coefficients $C_{6,0}(L)$ and $C_{6,2}(L)$ may be expressed in terms of the scalar and tensor polarizabilities $\alpha_0(L;iw)$ and $\alpha_2(L;iw)$ evaluated at imaginary frequencies [5,8]. In particular, for the helium case one may write:

$$C_{6,0}(L) = \frac{3}{\pi} \cdot \int_0^{\infty} \alpha_0(L;iw)\overline{\alpha}_{He}(iw)dw, \tag{5}$$

where $\overline{\alpha}_{He}$ is the dynamical polarizability of helium. The polarizabilities at imaginary frequencies are given in Ref. [3].

Usually (see [6]) the non-relativistic Hartree-Fock sets of the wave functions are used. A more sophisticated approach is based on using the relativistic Dirac-Fock wave functions (first variant) [5-7]. Another variant is using the relativistic wave functions as the solutions of the Dirac equations with different model potentials (of the Ivanov-Ivanova type) and effective potentials [6,10-15]. In this paper we have used the set of the relativistic functions, generated by the Dirac equation with the optimized Dirac-Kohn-Sham exchange-correlation functionals [16]. The detailed treatment of this approach in studying spectra and radiative characteristics of different atomic systems is given in Refs. [1-3, 16].

3. Results and conclusions

In Table 1 we present our theoretical results for the line shift f_p (1/Torr) for the Rb-He pair. The observed value of the line shift (T=323K) and other theoretical results for f_p are also included in Table 1. Other theoretical data are obtained on the basis of the exchange PT [6] with using the wave functions sets in Clementi et al [17] and Hartree-Fock approximations. The important feature of our scheme is a correct account of the correlation and polarization effects with using special effective functionals from [12,14,16]. A difference between our theoretical data and other calculation data is explained by using the different sets of wave functions and different PT schemes. For other temperatures there are no quite precise data. It is obvious that using the gauge-invariant optimized sets of the wave functions and correct version of the exchange PT will be necessary in a case of the alkali elements in an atmosphere of more heavy inert gases.

Table 1. Observed and theoretical f_p (10^{-9} 1/Torr) shift data for pair of Rb-He.

T, K	Experiment	This work	Theory [1]	Theory [6]	Theory [6]
223	-	116	113	81	79
323	105	103	101	75	73
423	-	91	89	64	62
523	-	82	80	56	55
623	-	75	73	50	50
723	-	73	71	47	47
823	-	70	69	-	-

References

[1] Glushkov A V, Khetselius O Yu, Gurnitskaya E P, Loboda A V and Mischenko E V 2009 *Theory and Applications of Comp. Chem.* (AIP) **1102** 172
[2] Khetselius O Yu, Florko T A., Svinarenko A A and Tkach T B 2013 *Phys.Scr.* **T153** 014037
[3] Florko T, Ambrosov S, Svinarenko A and Tkach T 2012 *J. of Phys.: Conf. Ser.* **397** 012037
[4] Malinovskaya S V, Glushkov A V, Khetselius O Yu, Svinarenko A A, Mischenko E V and Florko T A 2009 *Int. J. Quant. Chem.* **109** 3325
[5] Khetselius O Yu, Glushkov A V, Gurnitskaya E P, Loboda A V, Mischenko E V, Florko T A and Sukharev D E 2008 *Spectral Line* Shapes, Vol. 15, AIP Conf. Proceedings. **1058** 231
[6] Batygin V V, Gorny M B and Gurevich B M 1978 *J. Techn. Phys.* **48** 1097
[7] Kaplan I G 1985 *Theory of intermolecular interactions* (Moscow: Nauka)
[8] Chi X, Dalgarno A and Groenenborn G 2007 *Phys. Rev. A.* **75** 032723
[9] Khetselius O Yu 2009 *Phys.Scr.* **T135** 014023
[10] Glushkov A V, Khetselius O Yu, Gurnitskaya E P, Loboda A V, Florko T A, Sukharev D E and Lovett L 2008 *Frontiers in Quantum Systems in Chemistry and Physics, Series: Progress in Theoretical Chemistry and Physics* vol 18, ed S Wilson, P J Grout, J Maruani, G Delgado-Barrio and P Piecuch (Berlin: Springer) p 505
[11] Malinovskaya S V, Glushkov A V, Khetselius O Yu, Loboda A V, Lopatkin Yu M, Svinarenko A A, Nikola L V and Perelygina T B 2011 *Int. J. Quant. Chem.* **111** 288
[12] Ivanova E P, Ivanov L N, Glushkov A V and Kramida A E 1985 *Phys. Scripta* **32** 512
[13] Khetselius O Yu 2012 *J. of Phys.: Conf. Ser.* **397** 012012
[14] Glushkov A V and Ivanov L N 1992 *Phys. Lett.* A **170** 33
[15] Khetselius O Yu 2008 *Spectral Line* Shapes, Vol. 15, AIP Conf. Proceedings. **1058** 363
[16] Glushkov A V, Khetselius O Yu and Lovett L 2009 *Advances in the Theory of Atomic and Molecular Systems Dynamics, Spectroscopy, Clusters, and Nanostructures, Series: Progress in Theoretical Chemistry and Physics* vol 20, ed P Piecuch, J Maruani, G Delgado-Barrio, S Wilson (Berlin: Springer) p 125
[17] Clementi E, Roothaan C C J and Yoshimine M 1962 *Phys. Rev.* **127** 1618

Calculated isotropic Raman spectra from interacting H_2–rare-gas pairs

M. Gustafsson[1], W. Głaz[2], T. Bancewicz[2], J-L. Godet[3], G. Maroulis[4], and A. Haskapoulos[4]

[1]Applied Physics, Division of Materials Science, Department of Engineering Science and Mathematics, Luleå University of Technology, SE-97187 Luleå, Sweden
[2]Nonlinear Optics Division, Adam Mickiewicz University, Poznań, Poland
[3]Laboratoire de Photonique d'Angers, Université d'Angers, France
[4]Department of Chemistry, University of Patras, Greece

E-mail: [1]magnus.gustafsson@ltu.se

Abstract. We report on a theoretical study of the H_2–He and H_2–Ar pair trace-polarizability and the corresponding isotropic Raman spectra. The conventional quantum mechanical approach for calculations of interaction-induced spectra, which is based on an isotropic interaction potential, is employed. This is compared with a close-coupling approach, which allows for inclusion of the full, anisotropic potential. It is established that the anisotropy of the potential plays a minor role for these spectra. The computed isotropic collision-induced Raman intensity, which is due to dissimilar pairs in H_2–He and H_2–Ar gas mixtures, is comparable to the intensities due to similar pairs (H_2–H_2, He–He, and Ar–Ar), which have been studied previously.

1. Introduction

Light propagating through a molecular medium can be attenuated by absorption, reflection and scattering. In light scattering the polarizability tensor of the molecule determines the response to the incident radiation and thereby the intensity and polarization of scattered radiation. Mathematically the polarizability tensor can be decomposed into its isotropic (trace) part and the traceless (anisotropy) part. Following this partition the light scattering intensity is a sum of a polarized (isotropic) and a depolarized part stemming from the trace and the anisotropy of the polarizability, respectively [1].

The polarizability of an interacting pair is different from the sum of the unperturbed monomers' polarizabilities. This results in so-called collision-induced, or interaction-induced, light scattering which is density-dependent and relatively weak at atmospheric pressures. Modern techniques for measurement of light scattering are, however, so refined that the signals due to interacting pairs can be detected [2]. Theoretical and experimental studies of such spectra give insight into the fundamental interactions and properties of molecular systems.

In this work we report on the interaction-induced trace polarizability and the isotropic light scattering spectra for H_2–He and H_2–Ar complexes. A more complete presentation, which includes three more H_2–rare-gas systems, is available in Ref. [3]. In that work we also compared a classical approach with the quantum mechanical methods used here for the calculations of spectra. We will not elaborate on this further, but note that for the systems we studied

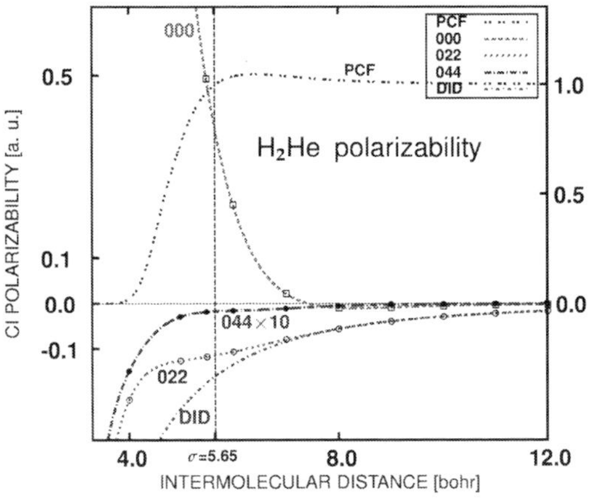

Figure 1. H_2–He trace polarizability components $\Delta\alpha^{(0)}_{\lambda\lambda}(R)$. Reprinted with permission from [3]. Copyright (2014), American Institute of Physics.

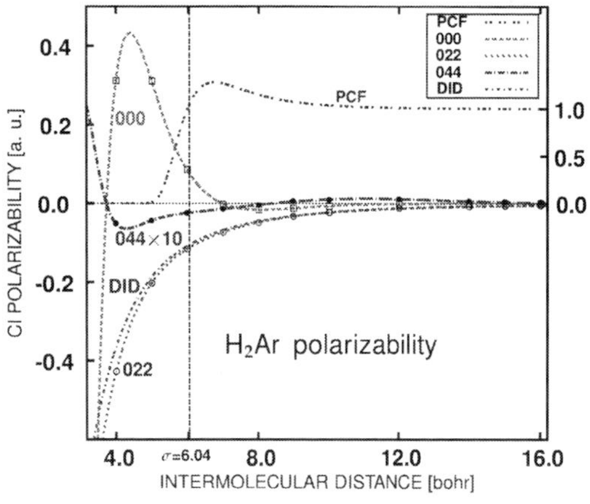

Figure 2. H_2–Ar trace polarizability components $\Delta\alpha^{(0)}_{\lambda\lambda}(R)$. Reprinted with permission from [3]. Copyright (2014), American Institute of Physics.

calculations based on classical trajectories may be used if the desired level of accuracy is less than about 6 percent.

2. Calculations

The interaction-induced polarizability may be computed with *ab initio* methods and expanded according to

$$\Delta\alpha^{(K)}_{\mu}(\hat{\mathbf{r}},\mathbf{R}) = \frac{4\pi}{\sqrt{2K+1}} \sum_{\lambda L} \Delta\alpha^{(K)}_{\lambda L}(R)\; Y^{K\mu}_{\lambda L}(\hat{\mathbf{r}},\hat{\mathbf{R}}) \tag{1}$$

where the angular functions are defined as

$$Y^{K\mu}_{\lambda L}(\hat{\mathbf{r}},\hat{\mathbf{R}}) = \sum_{m_\lambda m_L} C(\lambda, L, K; m_\lambda, m_L, \mu)\; Y_{\lambda m_\lambda}(\hat{\mathbf{r}})\; Y_{L m_L}(\hat{\mathbf{R}}) \tag{2}$$

with C and Y being Clebsch-Gordan coefficients and spherical harmonics, respectively. $\mathbf{R}$ is the vector pointing from the center of mass of the H_2-molecule to the rare-gas atom and $\hat{\mathbf{r}}$ is the unit vector specifying the orientation of the H_2-molecule. The r-dependence does not appear in Eq. (1) since it is averaged over. In other words, we do not consider vibrational excitations of H_2. The polarized (isotropic) light scattering [1] is due to the trace components $\Delta\alpha^{(0)}_{\lambda\lambda}(R)$ which are shown with $\lambda = 0, 2, 4$ for H_2–He and H_2–Ar in Figs. 1 and 2, respectively. The dipole-induced-dipole (DID) approximation for the 022 component is also displayed for both systems, showing that it works better for the Ar-system than for the He-system. In addition the pair correlation function (PCF) and collision diameter (σ) are presented.

In our calculations of light scattering intensity we have used the interaction potentials from [4] and [5] for H_2–He and H_2–Ar, respectively. First, we assume that those interaction potentials are isotropic, i.e. $V(\hat{\mathbf{r}},\mathbf{R}) = V(R)$. This is straightforward to do if they have been expanded in angular functions in the standard way [6]. Then the spectra can be computed by evaluation and proper thermal weighting of the squared matrix elements [7]

$$\left| \int \psi_i(R)\; \Delta\alpha^{(0)}_{\lambda\lambda}(R)\; \psi_f(R)\; dR \right|^2$$

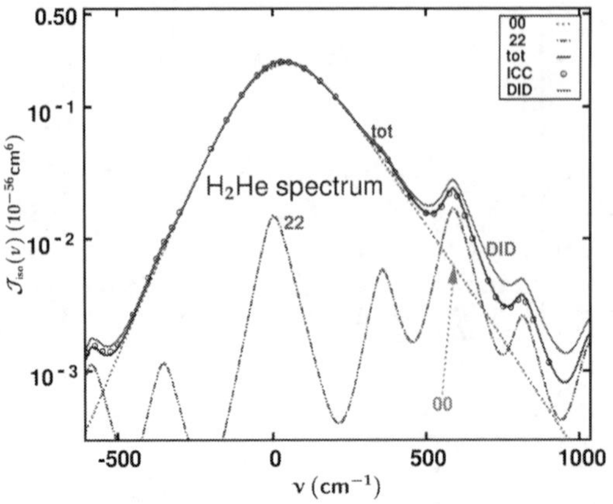

Figure 3. H_2–He polarized light scattering spectrum. Reprinted with permission from [3]. Copyright (2014), American Institute of Physics.

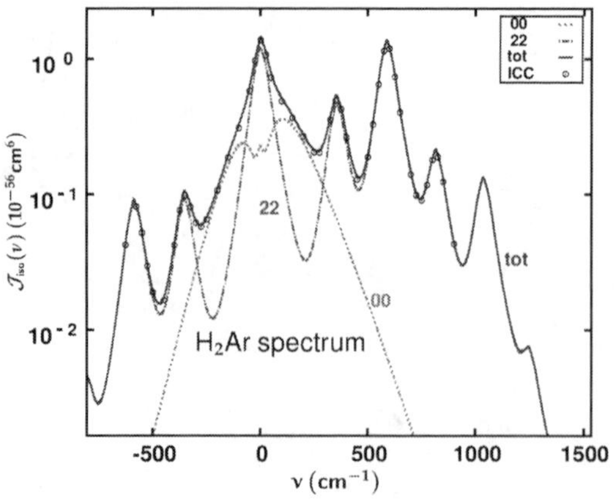

Figure 4. H_2–Ar polarized light scattering spectrum. Reprinted with permission from [3]. Copyright (2014), American Institute of Physics.

where ψ_i and ψ_f are the initial and final continuum wave functions, respectively. In this case it is thus sufficient to solve a one-dimensional Schrödinger equation in the R-coordinate to obtain the continuum functions. In Figs. 3 and 4 the calculated polarized light scattering spectra are shown. The 000 component of the trace polarizability accounts for the vast majority of the intensity in the H_2–He case, while for H_2–Ar it is the 022 component that dominates. For H_2–He the result based on the DID 022 component is also displayed, which causes an overestimation of the intensity for frequency shifts larger than about 500 cm^{-1}.

Second, we account for the full (anisotropic) potential, $V(\hat{\mathbf{r}}, \mathbf{R})$, by using the close-coupling scheme to obtain S-matrix elements corresponding to light scattering [8, 9]. The electromagnetic field and its coupling with the molecular system (through the interaction-induced polarizability) are included explicitly in the Hamiltonian. This results in a coupled system of Schrödinger equations in the R-coordinate. The anisotropy of the potential may be excluded in this scheme and we have labeled such calculations ICC (isotropic close-coupling) and included the resulting spectra in Figs. 3 and 4. Those calculations are clearly consistent with the conventional quantum calculations, as they should be, and we can trust that both methods are correctly implemented. In Fig. 5 comparisons of spectra obtained with and without the anisotropy of the interaction potential are displayed for both systems. It turns out that the anisotropy only affects the spectra by a few percent for both the H_2–He and H_2–Ar systems, with a slightly larger effect on the latter.

3. Conclusions

The polarized Raman intensities for the H_2–rare-gas systems studied in this work are on the same order of magnitude as the intensities due to the corresponding similar pairs H_2–H_2, He–He, and Ar–Ar, which have been studied before, e.g. in Refs. [7, 10, 11]. It should thus be possible to extract the H_2–rare-gas isotropic spectra from light scattering experiments with hydrogen gas mixed with helium or argon. The conventional quantum calculations, which are based on the distorted wave approximation and an isotropic interaction potential, agree with close-coupling calculations when the anisotropy of the potential is excluded in the latter. Full close-coupling calculations show that the anisotropy of the interaction potential only plays a role on the few percent level for the isotropic Raman spectra of the H_2–He and H_2–Ar systems.

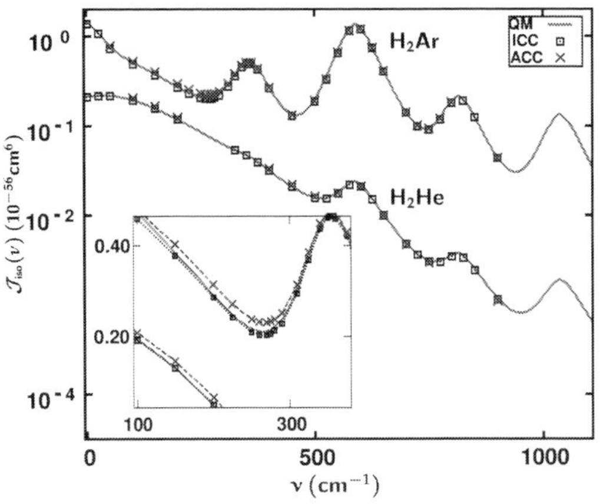

Figure 5. Isotropic light scattering intensity for H_2–He and H_2–Ar, displaying the influence on the spectra from the anisotropy of the potential. QM indicates conventional quantum calculations. ICC and ACC indicate close-coupling calculations with isotropic and anisotropic interaction potential, respectively. Reprinted with permission from [3]. Copyright (2014), American Institute of Physics.

References

[1] Berne B J and Pecora R 1976 *Dynamic Light Scattering* (New York: John Wiley & Sons)
[2] Frommhold L 1981 *Adv. Chem. Phys.* **46** ed Prigogine I and Rice S (New York: John Wiley and Sons, Inc.) pp 1 – 72 see also an update in Canad. J. Phys. **59** (1981) 1459
[3] Głaz W, Bancewicz T, Godet J L, Gustafsson M, Maroulis G and Haskopoulos A 2014 *J. Chem. Phys.* **141** 074315
[4] Schaefer J and Koehler W E 1985 *Physica A* **129** 469
[5] Williams H L, Szalewicz K, Jeziorski B, Moszynski R and Rybak S 1993 *J. Chem. Phys.* **98** 1279
[6] Arthurs A M and Dalgarno A 1960 *Proc. Roy. Soc. (London)* **Ser. A, 256** 540
[7] Brown M S, Wang S K and Frommhold L 1989 *Phys. Rev. A* **40** 2276
[8] Gustafsson M, Frommhold L and Meyer W 2000 *J. Chem. Phys.* **113** 3641
[9] Gustafsson M, Frommhold L, Li X and Hunt K 2009 *J. Chem. Phys.* **130** 164314
[10] Moszynski R, Heijmen T, Wormer P and vd Avoird A 1996 *J. Chem. Phys.* **104** 6997
[11] Gaye O, Chrysos M, Teboul V and Duff Y L 1997 *Phys. Rev. A* **55** 3484

Spectral line-shapes of oxygen B-band transitions measured with cavity ring-down spectroscopy

S Wójtewicz, A Cygan, P Masłowski, J Domysławska, P Wcisło, M Zaborowski, D Lisak*, R S Trawiński and R Ciuryło

Institute of Physics, Faculty of Physics, Astronomy, and Informatics, Nicolaus Copernicus University, Grudziadzka 5, 87-100 Torun, Poland

E-mail: `dlisak@fizyka.umk.pl`

Abstract. Results of line-shape measurements of self- and N_2-broadened P9 P9 transition of the oxygen B band are presented. Spectra were acquired using the optical frequency comb-assisted Pound-Drever-Hall-locked frequency-stabilized cavity ring-down spectrometer (PDH-locked FS-CRDS). In the line-shape analysis the line narrowing described by Dicke narrowing or and the speed dependence of collisional broadening were taken into account. The multi-spectrum fitting technique was used to minimize numerical correlations between line-shape parameters. Collisional broadening and shifting coe cients are reported with sub-percent uncertainties. Influence of the spectral line-shape model used in data analysis on determined line intensities and collisional broadening is discussed.

1. Introduction

Systematic progress in experimental techniques of molecular spectroscopy and spectral line shape investigation is stimulated by increasing demands in high precision remote sensing of atmosphere composition, metrological applications and testing fundamental physics. Spectroscopic molecular databases, *e.g.* [1,2], containing line-by-line parameters, are in most cases based on the Voigt profile (VP). This simplification generally does not allow to reproduce experimental data with sub-percent accuracy. It is well established that such physical effects as velocity-changing collisions, which lead to Dicke narrowing of line, and dependence of the collisional broadening and shifting on molecules speed, called speed-dependent effects, must be taken into account [3,4] to model the line shape with sub-percent accuracy. As recently proposed [5,6] and recommended in Ref. [7] the new generation spectroscopic databases may easily incorporate these two effects and additionally effect of correlation between VC and phase state-changing collisions [8] by using the partially correlated speed-dependent Nelkin-Ghatak profile PCSDNGP [9] with quadratic speed dependence [10] instead of the Voigt profile. This profile, also called Hartmann-Tran profile (HTP), can be calculated almost as fast as Voigt profile [5, 11]. On the other hand it is known that simple models of quadratic speed-dependence, as well as the hard-collision approximation of velocity-changing collisions used in PCSDNGP are not optimal for description of many molecular systems.

In this paper we report progress in systematic study of spectral line shapes of O_2 transitions from the B-band ($b\,^1\Sigma_g^+(v=1) \leftarrow X\,^3\Sigma_g^-(v=0)$). In our earlier papers [12–16] self-broadened lines were studied. High signal-to-noise ratio experimental spectra enabled us to study advanced line-shape effects, such as velocity-changing collisions, speed-dependent effects, and correlations between velocity- and phase-changing collisions. This analysis is now extended to the case of perturbation by a mixture $N_2 - O_2$, in which both self and foreign broadening has to be considered. This is nontrivial for the speed-dependent profiles, however when quadratic model of speed dependence is used a gas mixture can be treated as single perturber with effective speed dependence parameters [17].

2. Experimental setup

Experimental line shapes of O_2 were measured with PDH-locked FS-CRDS described in Refs. [16, 18]. The length of the high-finesse ring-down cavity is stabilized to the HeNe laser with frequency stability of 1.5 MHz. The probe laser (ECDL) is locked to the consecutive TEM_{00} modes of the cavity with the PDH method [19]. This enables to maximize power and repetition rate of the ring-down signals [20, 21] and obtain typical SNR of spectra as high as 10^4, or even 2×10^5 with long-term averaging [14]. The absolute frequency of the probe laser is measured at each point of the spectrum by the optical frequency comb [13]. This scheme leads to a combined standard uncertainty of the spectrum frequency axis lower than 200 kHz.

To prepare a sample of $N_2 - O_2$ gas mixture individual gases with purity of 99.999% were mixed in a vacuum chamber equipped with a fan to assure uniform mixing. For the investigated P9 P9 transition the optimal gas composition was 3% of O_2 in N_2. The pressures were measured with Baratron gauges having relative uncertainties of 0.05% and temperature of the ring-down cavity was stable to 0.1 during a single spectrum measurement.

3. Results and discussion

Systematic study of subtle line-shape effects requires to use a multi-spectrum fitting technique to avoid problem of numerical correlations between fitted parameters [22]. Also for this reason spectra were measured in a wide pressure range (10 - 400) Torr. In Fig. 1 experimental profiles

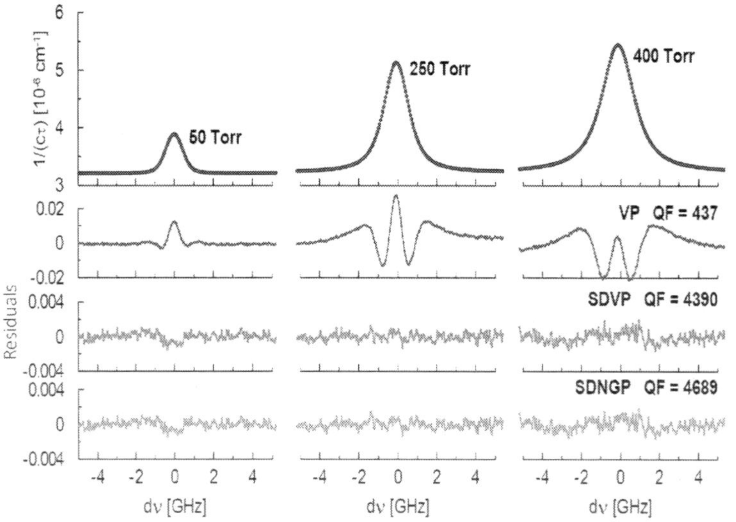

Figure 1. Experimental profiles of O_2 P9 P9 line broadened by $N_2 - O_2$ mixture, and residuals from multi-spectrum fits of VP, SDVP and SDNGP.

of O_2 P9 P9 line broadened by mixture of $N_2 - O_2$, are shown with residuals from fits of Voigt profile (VP), speed-dependent Voigt (SDVP) and the speed-dependent Nelkin-Ghatak profile (SDNGP).In the latter two models the quadratic speed dependence was used. For each residuals the quality of the fit (QF) factor, defined as a peak absorption divided by the standard deviation from residuals [14], was calculated. As seen in Fig. 1 incorporation of the speed-dependent effects to the model improves the QF by an order of magnitude. Taking both the speed dependence and velocity-changing collisions into account in the SDNGP further improves the QF, but only by 7% comparing to the SDVP.

In Fig. 2 relative differences of line intensity S and collisional width γ_L obtained from multi-spectrum fits of different models to N_2-broadened P9 P9 oxygen line are presented. The

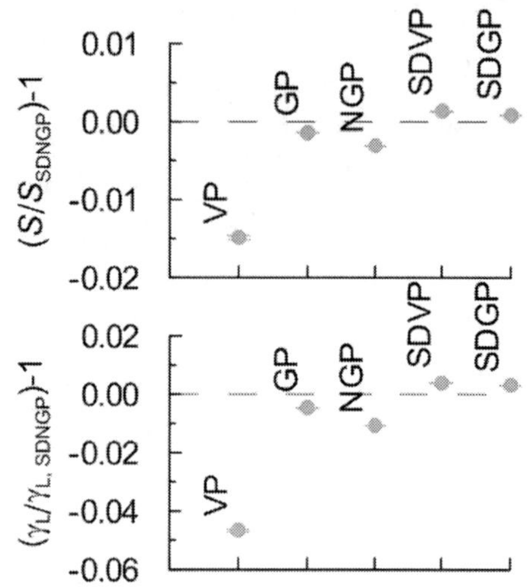

Figure 2. Relative differences of line intensities S and collisional widths γ_L obtained from multi-spectrum fits of different models to N_2-broadened P9 P9 oxygen line. The reference profile is SDNGP.

reference profile corresponding to zero in both plots is the SDNGP. Here additionally to the previously described models we considered also the Galatry (GP), Nelkin-Ghatak (NGP) and the speed-dependent Galatry (SDGP) [23, 24] profiles. Clearly all the speed-dependent profiles give consistent S and γ_L to within 0.2%, while the profiles that consider velocity-changing collisions, but not speed-dependent effects differ by up to 1%, depending on velocity-changing model. The Voigt profile leads to systematic errors of 1.5% for S and 5% for γ_L.

Comparison of the line-shape parameters for self- and N_2-broadened P9 P9 oxygen line obtained from the SDVP fits is given in Table 1. Parameters for N_2 broadening were calculated from results for $N_2 - O_2$ mixture with use of self-broadening results. Because of low gas pressures used for self-broadening experiment the a_S could not be determined. Comparing a_W parameters it seems that the speed-dependent effects are significantly stronger in case of perturbation by N_2 than O_2.

Table 1. Line-shape parameters for self- and N_2-broadened P9 P9 oxygen line with their standard deviations, obtained from the SDVP fits. γ_L/N and δ/N in units (10^{-19}GHz (molec. cm^3)). Line intensity S in units (10^{-25}cm^{-1}/(molec. cm^{-2})).

perturber	S at T=296	γ_L/N	δ/N	a_W	a_S
O_2	5.411(27)	1.118(5)	-0.0956(14)	0.0497(10)	–
N_2	–	1.231(9)	-0.1183(9)	0.0935(10)	-0.0267(5)

4. Conclusions

We presented systematic line-shape study of the oxygen B-band transition P9 P9. It revealed that satisfactory quality of the fits of the model profiles to the experimental data for line perturbed by N_2 can be achieved with the speed-dependent Voigt profile. The in uence of Dicke narrowing is not evident, as the speed-dependent Nelkin-Ghatak profile leads to nearly equal quality of the fit as SDVP. This can be explained in terms of correlations between velocity- and phase-changing collisions, which effectively reduce Dicke narrowing.

High-precision line-shape parameters of the investigated speed-dependent Voigt profile, given in the paper allows one to reproduce experimental spectra with relative precision better than 1 4000, which is ten times better than the commonly used Voigt profile.

Acknowledgments

The research is part of the program of the National Laboratory FAMO in Toruń, Poland, and is supported by the Polish National Science Centre Project nos. DEC-2011 01 B ST2 00491 and MO-2012 05 N ST2 02717. The research is also supported by the Foundation for Polish Science TEAM and HOMING PL S Projects co-financed by the E European Regional Development Fund. A. Cygan is supported by the Foundation for Polish Science START Project.

References

[1] Rothman L S *et al.* 2013 *J. Quant. Spectrosc. Radiat. Transf.* **130** 4
[2] Jacquinet-Husson N *et al.* 2011 *J. Quant. Spectrosc. Radiat. Transf.* **112** 2395
[3] Duggan P, *et al.* 1997 *J. Molec. Spectrosc.* **186** 90
[4] Ciuryło R, Pine A S, Szudy J 2001 *J. Quant. Spectrosc. Radiat. Transf.* **68** 257
[5] Ngo N H, Lisak D, Tran H, Hartmann J-M 2013 *J. Quant. Spectrosc. Radiat. Transf.* **129**, 89
[6] Ngo N H, Lisak D, Tran H, Hartmann J-M 2014 *J. Quant. Spectrosc. Radiat. Transf.* **134** 105
[7] Tennyson J, *et al.* 2014 *Pure. Appl. Chem.* in press.
[8] Rautian S G and Sobelman I I 1967 *Sov. Phys. Usp.* **9** 701
[9] Pine A S 1999 *J. Quant. Spectrosc. Radiat. Transf.* **62** 397
[10] Rohart F, Mader H, Nicolaisen H-W 1994 *J. Chem. Phys.* **101** 6475
[11] Tran H, Ngo N H, Hartmann J-M 2013 *J. Quant. Spectrosc. Radiat. Transf.* **129** 199
[12] Wójtewicz S, Lisak D, Cygan A, Domysławska J, Trawiński R S and Ciuryło R 2011 *Phys. Rev. A* **84** 032511
[13] Domysławska J, *et al.* 2012 *J. Chem. Phys.* **136** 024201
[14] Cygan A, *et al.* 2012 *Phys. Rev. A* **85** 022508
[15] Domysławska J, *et al.* 2013 *J. Chem. Phys.* **139** 194312
[16] Wójtewicz *et al.* 2014 *J. Quant. Spectrosc. Radiat. Transf.* **144** 36
[17] Lisak D, Cygan A, Wcisło P and Ciuryło R 2015 *J. Quant. Spectrosc. Radiat. Transf.* **151** 43
[18] Cygan A, *et al.* 2013 *Eur. Phys. J.-Spec. Top.* **222** 2119
[19] Drever R W P, *et al.* 1983 *Appl. Phys. B* **31** 97
[20] Cygan A, *et al.* 2011 *Rev. Sci. Instrum.* **82** 063107
[21] Cygan A, *et al.* 2011 *Meas. Sci. Technol.* **22** 115303
[22] Benner C, *et al.* 1995 *J. Quant. Spectrosc. Radiat. Transf.* **53** 705
[23] Ciuryło R and Szudy J 1997 *J. Quant. Spectrosc. Radiat. Transf.* **57** 411
[24] Priem D, *et al.* 2000 *J. Molec. Struct.* **517** 435

Absorption of Ultrashort Electromagnetic Pulses on Broadened Dipole Transitions

S Yu Svita, V A Astapenko
Moscow Institute of Physics and Technology, 141700, Institutskii per.9, Dolgoprudnyi, Russia

E-mail: sergey.svita@gmail.com

Abstract. The study is devoted to the theoretical analysis of ultrashort electromagnetic pulses (USP) absorption on broadened dipole transitions. Calculations are made in the frame of perturbation theory with the use of the basic formula for energy absorbed during all time of the action of USP on dipole transition. Dependences of absorbed energy upon pulse duration and carrier frequency are obtained and analyzed for different types of spectral line shape and USP parameters.

1. Introduction

The development of technology of generation of ultrashort electromagnetic pulses (USP) with controlled shape and duration down to several tens of attoseconds [1] make urgent the consideration of peculiarities of their interaction with a substance: atoms, clusters, nanoparticles, etc. The present study is devoted to the theoretical analysis of ultrashort electromagnetic pulses absorption on broadened dipole transitions.

2. General formulas for radiation absorption probability

Derivation of basic formula for total probability of photoabsorption for all time of pulse action is presented in paper [2]. Here we outline the main points of this derivation. Using standard perturbation approach, one can obtain the following expression for the photoexcitation probability by an electric field with amplitude E(t) in the dipole approximation:

$$W_{tot} = \sum_n W_{n0} = \frac{1}{\hbar^2} \int_{-\infty}^{\infty} \int_{-\infty}^{\infty} \langle 0 | \hat{d}_i(t') \hat{d}_k(t) | 0 \rangle E_i(t') E_k(t) \, dt \, dt' . \quad (1)$$

In terms of the the correlator of the dipole moments, $K_{ik}(\tau) = \langle 0 | \hat{d}_i(t') \hat{d}_k(t) | 0 \rangle$, Equation (1) can be rewritten as

$$W_{tot} = \frac{1}{\hbar^2} \int_{-\infty}^{\infty} \int_{-\infty}^{\infty} K(\tau) E_i(t+\tau) E_i(t) \, dt \, d\tau . \quad (2)$$

After Fourier transformation of the functions under the integral in Eq. (2) we have

$$W_{tot} = \frac{1}{2\pi\hbar^2}\int_{-\infty}^{\infty} K(\omega')|\mathbf{E}(\omega')|^2 \, d\omega' . \tag{3}$$

Here

$$K(\omega) = \frac{1}{3}\int_{-\infty}^{\infty}\langle 0|\hat{d}_i(t)\hat{d}_i(t+\tau)|0\rangle \exp(i\,\omega\,\tau)\,d\tau \tag{4}$$

is the Fourier transform of the track of the dipole moment correlator.

It can be shown that function $K(\omega)$ is connected with the photoionization cross-section $\sigma(\omega)$ via the relation (see for details [2]):

$$K(\omega) = \frac{\hbar c}{2\pi\,\omega}\sigma(\omega). \tag{5}$$

Here c denotes the speed of light in free space. Substituting this expression for $K(\omega)$ into Eq. (3) we obtain the basic formula

$$W = \frac{c}{(2\pi)^2}\int_0^{\infty}\sigma(\omega')\frac{|\mathbf{E}(\omega')|^2}{\hbar\,\omega'}\,d\omega' , \tag{6}$$

where $\mathbf{E}(\omega')$ is Fourier transform of electric field strength in USP. It is obvious that the expression (6) is meaningful only for $W < 1$, within the framework of applicability of the perturbation theory.

3. Numerical calculations

We consider three types of transition broadening: homogeneous (Lorentzian), Gaussian and Holtsmark. The spectral profiles of corresponding broadening are given by well-known expressions:

$$G_{\hom}(\omega,\Omega,\Delta) = \frac{\Delta/\pi}{(\omega-\Omega)^2+\Delta^2} , \tag{7}$$

$$G_{inh}(\omega,\Omega,\Delta) = \frac{1}{\sqrt{2\pi}\cdot\Delta}\cdot\exp\left[-\frac{(\omega-\Omega)^2}{2\cdot\Delta^2}\right], \tag{8}$$

$$G_{Holt}(\omega,\Omega,\Delta) = \frac{1}{2\cdot\Delta}\cdot H_i\left(\left|\frac{\omega-\Omega}{\Delta}\right|\right), \tag{9}$$

$$H_i(\beta) = \frac{2}{\pi}\cdot\beta\cdot\int_0^{\infty} x\cdot\sin(\beta\cdot x)\cdot\exp\left(-x^{\frac{3}{2}}\right)dx , \tag{10}$$

where Ω is the central frequency, and Δ is the spectral width of transition.

In this study we use a corrected Gaussian pulse [3], because it has a zero constant component (the Fourier transform equals to zero at zero frequency). The Fourier transform of the corrected Gaussian pulse equals

$$E_{cor}(\omega',\omega,\tau,\varphi) = i\,E_0\,\tau\sqrt{\frac{\pi}{2}}\cdot\frac{\omega'^2\,\tau^2}{1+\omega^2\,\tau^2}\cdot\left[e^{-i\varphi-(\omega-\omega')^2\tau^2/2} - e^{i\varphi-(\omega+\omega')^2\tau^2/2}\right], \tag{11}$$

where ω is the carrier frequency, ω' is the current frequency, τ is the pulse duration, and E_0 is the amplitude of electric field strength.

The absorption cross-section of the broadened transition $\sigma_{21}(\omega')$ is equal to

$$\sigma_{21}(\omega') = \frac{2\pi^2 e^2}{mc} f_{21}\, G_{21}(\omega'), \tag{12}$$

where f_{21} is the oscillator strength and $G_{21}(\omega')$ is the spectral profile of the transition.

Let us express the pulse duration in terms of a number of periods at the carrier frequency ω that fit into the effective pulse duration [2] ($\Delta t_p = \sqrt{\pi} \cdot \tau$) to obtain

$$n = \frac{\Delta t_p}{T} = \frac{\omega \cdot \tau}{2\sqrt{\pi}}. \tag{13}$$

Thus the dimensionless parameter n (the number of cycles in a pulse), together with the time τ, can be used as measure of the pulse duration.

The expression for the absorbed energy during the USP action on broadening transitions follows from Equations (6) and (12), and one finds

$$\Delta \mathrm{E}(n) = \frac{e^2 f_{21}}{2m} \cdot \int_0^\infty G_{21}(\omega') \cdot \left(\left|E_{cor}(\omega', \omega, n)\right|\right)^2 d\omega'. \tag{14}$$

The spectral function $G_{21}(\omega')$ is given by Eqs. (7) to (10) for the different types of line broadening.

4. Results and conclusions

The results of the calculation of the absorbed energy dependence as a function of pulse duration for different line shapes (Lorentzian, Gaussian and Holtsmark) are shown in Figure 1 for nonresonant and Figure 2 for resonant carrier frequency of USP. One can see from these figures that the considered dependence is non-linear in contrast with the long pulse limit.

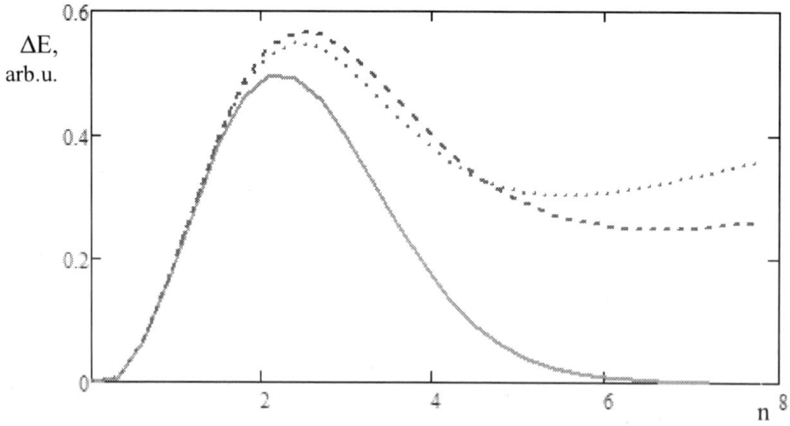

Figure 1. Absorbed energy dependence upon USP duration for Gaussian (solid line), homogeneous (Lorentzian) (dotted line) and Holtsmark broadening (dashed line). Where n is number of carrier frequency cycles, spectral width of the transition is 0.01 a. u., relative detuning is 0.15 and amplitude of USP electric field strength is assumed 0.1 a.u.

Figure 1 shows that for sufficiently large detuning of the carrier frequency from the central frequency of the transition the dependence is a response curve with a maximum. It is established that for the same parameters the maximum is reached when the number of cycles of carrier frequency n is equal to 2…2.5, and for sufficiently large value of n that are considered, the dependence reduces to a linear function. Figure1 further illustrates that for the case of Gaussian broadening the considered dependence goes to zero in the limit of a long pulse.

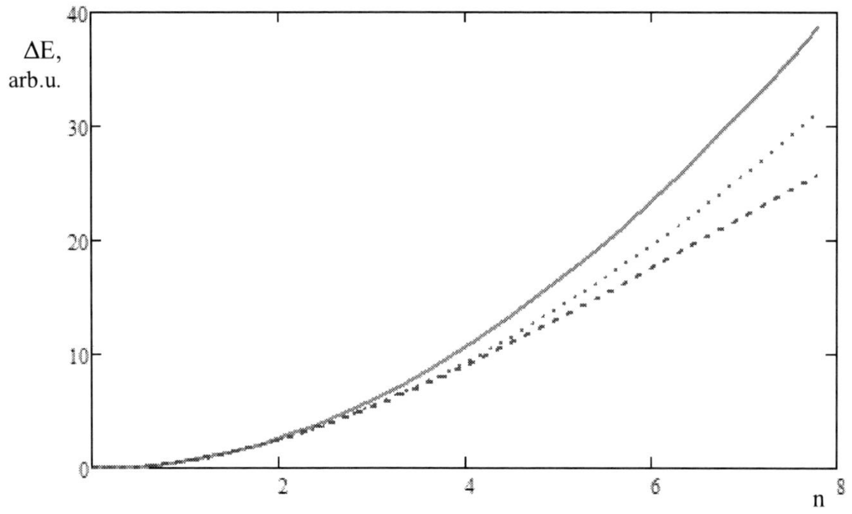

Figure 2. Absorbed energy dependence upon USP duration for Gaussian (solid line), homogeneous (dotted line) and Holtsmark broadening (dashed line). Parameters are the same as in Fig.1, but the amplitude of incident pulse is 0.01 in atomic units, and the relative detuning equals zero.

In summary, we computed the absorption of USP energy by the dipole transition with different spectral shapes. Three types of broadening are considered: Lorentz type, Gaussian type and Holtsmark type. The dependence of the absorbed energy upon pulse duration is analyzed for nonresonant and resonant cases. We conclude that for sufficiently large detuning of the USP carrier frequency from resonance, this dependence is described by a response curve with a maximum, and the form of the response is determined by the type of broadening.

References

[1] Zhao K, Zhang Q, Chini M, Wu Y, Wang X, Chang Z 2012 *Optics Letters* **37** 3891
[2] Astapenko V A 2010 *Phys. Lett. A* **374** 1585
[3] Lin Q, Zheng J, Becker W *Phys. Rev. Lett.* **97** 253902

XXII International Conference on Spectral Line Shapes 2014 IOP Publishing
Journal of Physics: Conference Series **548** (2014) 012030 doi:10.1088/1742-6596/548/1/012030

Review of the phenomenon of dips in spectral lines emitted from plasmas and their applications

E Oks[1,7], **E Dalimier**[2,3], **A Faenov**[4,5] **and O Renner**[6]

[1]Physics Department, Auburn University, AL 36849, USA
[2]Sorbonne Universités Pierre et Marie Curie, Paris
[3]LULI Ecole Polytechnique, Palaiseau, France
[4]Joint Institute for High Temperatures, Russian Academy of Sciences, Moscow, Russia
[5]Institute for Academic Initiatives, Osaka University, Osaka, Japan
[6]Institute of Physics, Academy of Science, Prague, Czech Republic

E-mail: goks@physics.auburn.edu

Abstract. The review covers theoretical and experimental studies of two kinds of dips (local depressions) in spectral line profiles emitted by plasmas: Langmuir-wave-caused dips (L-dips) and charge-exchange-caused dips (X-dips). Positions of L-dips (relative to the unperturbed wavelength of a spectral line) scale with the electron density N_e roughly as $N_e^{1/2}$, while positions of X-dips are almost independent of N_e. L-dips and X-dips phenomena are interesting and important both fundamentally and practically. The fundamental interest is due to a rich physics behind each of these phenomena. As for important practical applications, they are as follows. Observation of L-dips constitutes a very accurate method to measure the electron density in plasmas – the method that does not require the knowledge of the electron temperature. L-dips also allow measuring the amplitude of the electric field of Langmuir waves – the only one spectroscopic method available for this purpose. In the most recent laser plasma experiments, L-dips were found to be a spectroscopic signature of the two-plasmon decay instability. This instability causes hot-electron generation and is a critical part in laser-driven inertial confinement fusion program. As for observations of X-dips, they serve to determine rates of charge exchange between multicharged ions. This is an important reference data virtually inaccessible by other experimental methods. The rates of charge exchange are essential for magnetic fusion in tokamaks, for population inversion in the soft x-ray and VUV ranges, for ion storage devices, as well as for astrophysics (e.g., for the solar plasma and for determining the physical state of planetary nebulae).

1. Langmuir-Wave-Caused Dips

1.1. Theory of L-dips

Langmuir-wave-caused dips (L-dips) in profiles of spectral lines in plasmas were discovered experimentally and explained theoretically for dense plasmas, where one of the electric fields **F** experienced by hydrogenic radiators is quasi-static. This field can be the ion micro-field and (or) a low frequency electrostatic turbulence.

There is a rich physics behind each of these phenomena. L-dips result from a resonance between the Stark splitting $\omega_F = 3n\hbar F/(2Z_r m_e e)$ of hydrogenic energy levels, caused by a quasistatic field **F** in a plasma, and the frequency ω_L of the Langmuir waves, which practically coincides with the plasma electron frequency $\omega_p(N_e) = (4\pi e^2 N_e/m_e)^{1/2} \approx 5.641\times10^4\,[N_e(cm^{-3})]^{1/2}$: $\omega_F = s\omega_p(N_e)$, s = 1, 2, … .

[7] To whom any correspondence should be addressed

Even for the most common case of s=1, it is actually a multi-frequency resonance phenomenon despite the electric field of the Langmuir wave is considered to be single-frequency (monochromatic): $\mathbf{E}_0 \cos\omega_p t$. Its multi-quantum nature has been revealed in paper [1]: it is a resonance between many quasienergy harmonics of the combined system "radiator + oscillatory field" caused simultaneously by all harmonics of the total electric field $\mathbf{E}(t) = \mathbf{F} + \mathbf{E}_0 \cos\omega_p t$, where vectors $\mathbf{F}$ and $\mathbf{E}_0$ are not collinear.

The history of dips covers a long period from 1977 to 2013, during which they have been studied experimentally at different plasma sources, such as gas-liner pinch, laser-produced plasmas, Z-pinch plasmas. These experiments, performed by various groups, required specific configurations and improved high-resolution X-ray spectrometers. The theory of L-dips from [2-5] provided a diagnostic tool for measuring the electric field amplitude E_0 of the Langmuir waves and an independent method for measuring the electron density N_e. The resonance $\omega_F = s\omega_p(N_e)$ translates into specific locations of L-dips in spectral line profiles depending on N_e and provides a simple, but accurate N_e-diagnostic.

Each L-dip represents a structure consisting of the dip itself (the primary minimum of intensity) and two surrounding bumps. The bumps are due to a partial transfer of the intensity from the wavelength of the dip to adjacent wavelengths. The total structure can lead to a secondary dip (or a small shoulder).

The half-width of the L-dip, controlled by the amplitude E_0 of the electric field of the Langmuir wave, is given by $\delta\lambda_{1/2} \approx (3/2)^{1/2}\lambda_0^2 n^2 E_0/(8\pi m_e ecZ_r)$, where λ_0 is the unperturbed wavelength of the spectral line. Thus, by measuring the experimental half-width of L-dips, one can determine the amplitude E_0 of the Langmuir wave.

1.2. Experimental Observations of L-dips

1.2.1. Gas-Liner Pinch Experiments in Germany. The first observations of L-dips were made in a gas-liner pinch [6]. It is characterized by a density (1-3)×10^{18} cm^{-3} and a relatively low temperature 10-13 eV.

The electron densities were obtained by an independent diagnostic: coherent Thomson scattering. The first-order dips (due to the one-quantum resonance $\omega_F = \omega_p$) and the second-order dips (due to the two-quantum resonance $\omega_F = 2\omega_p$ were observed in the profile of the Ly α line of hydrogen in a helium plasma and their positions were in agreement with the theory. The detailed bump-dip-bump structure in the profile of each Stark component was revealed for the first time (figure 1).

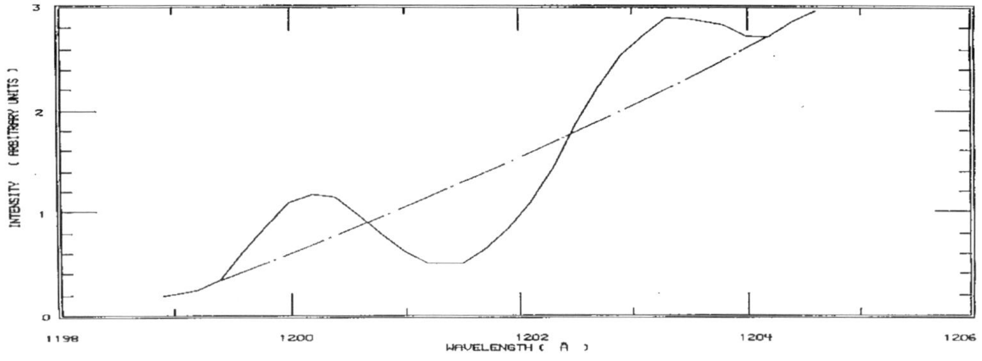

Figure 1. Bump-dip-bump structure in the H I Ly α line observed at a gas-liner pinch [6].

The gas-liner pinch experiments allowed two important diagnostics: the electron density N_e from the location of the dips and the Langmuir wave amplitude from the half-width of the L-dips. It turned out that measuring N_e from the L-dips yielded the same high accuracy as the much more complicated diagnostic using the coherent Thomson scattering.

1.2.2. Laser Produced Plasma Experiments in France. The experiment was performed at the nanosecond Nd : glass laser facility at LULI. A single laser beam with the intensity 2×10^{14} W/cm^2 was focused onto a structured target: an Al strip sandwiched between magnesium substrate [7, 8]. The Al plasma was well confined and the transverse emission was optimized. The plasma parameters $N_e=5\times10^{22}$ cm^{-3} and T_e=300 eV were estimated by hydrodynamic simulations. They confirm a resonant coupling between the ion micro-field F and the Langmuir field E. A Vertical-geometry Johann Spectrometer VJS with high spectral (4200) and spatial resolutions was specially designed for revealing fine structures in the red wing of Al XIII Ly γ line.

Theoretically expected L-dip positions were confirmed by the experiment (figure 2). The simultaneous production of a pair of symmetric spectra with this spectroscopic diagnostic provided a reference point for the computational reconstruction of raw data, thus increasing the confidence in the identification of the dips.

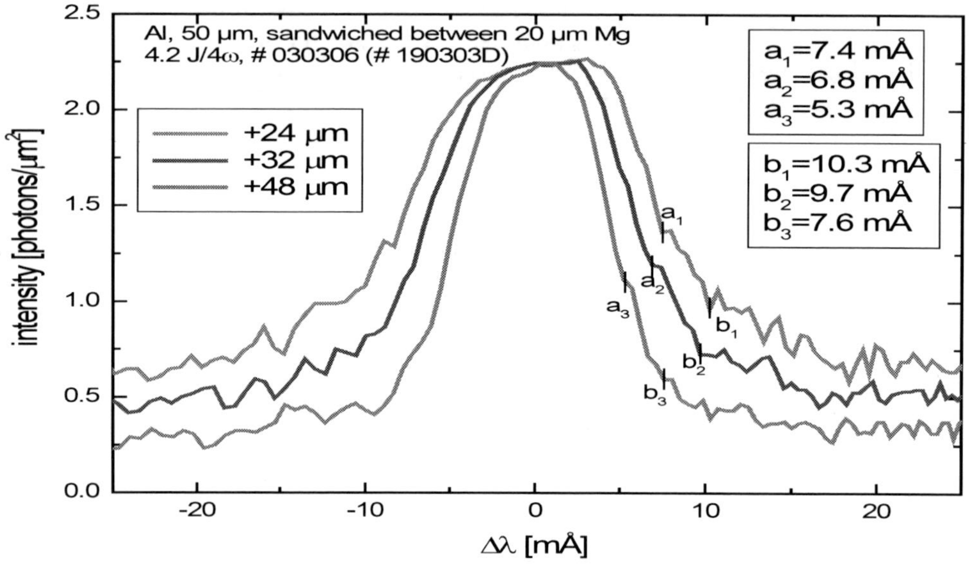

Figure 2. L-dips in the experimental profiles of Al XIII Ly γ line from a plasma produced by the LULI laser [7, 8]. Only the red dips are visible, the blue dips are merged with the noise. The different spectra correspond to the emission from different distances from the target (i.e., different densities).

This experiment in a laser produced plasma confirmed the dependence of the dips positions on the density, thus allowing a density diagnostic. The density values were similar to the ones obtained from line broadening simulations (IDEFIX code at LULI and PIM PAM POUM code at PIIM-Marseille).

1.2.3. The latest experimental application of the theory of L-dips in Z-pinch plasma in China at Chongqing University. At the Yang accelerator Z-pinch aluminum plasmas, characterized by a density $N_e=5\times10^{21}$ cm^{-3}, the electron temperature T_e=500 eV and the ion temperature T_i= 10 keV, were obtained by imploding aluminum wire-array [9]. The experimental Stark broadened profiles of Al XIII Ly α and Al XIII Ly γ lines were recorded with a high spectral resolution (2500) of a uniform dispersion mica crystal spectrograph. These lines, emitted from different regions, exhibited bump-dip-bump structures. They corresponded to slightly different densities and to different radial temperature gradients. The relationship between the positions from the line center of red Langmuir dips and electron density was studied. The electron densities deduced from the red dips were compared to those derived from measurements of the Plasma Polarization Shift (PPS) of the entire spectral line. There was a 5% difference in the electron densities obtained the two methods, which was due to the uncertainty of determining the average T_e required for the PPS method (but not required by the method based on the L-dips).

2. Charge-Exchange-Caused Dips

2.1 Theory of X-dips

The charge-exchange-caused-dips (X-dips) in profiles of spectral lines are due to the Charge Exchange (CE) in a plasma containing ions of at least two different nuclear charges Z and Z'≠ Z. The history of X-dips covers a long period from 1995 to 2012. Until recently the theoretical studies focused at the X-dips are caused by CE at quasi-crossings of the energy terms E(R) of the quasi-molecule Z-e-Z' made up of a H-like radiating ion Z and a perturbing fully stripped ion Z' [10–12].

Two independent complementary mechanisms explain the formation of X-dips in spectral line profiles [12]: one through the behavior of the transition energies ΔE(R) and another through the dynamical broadening $\gamma(R) = \gamma_{CE}(R) + \gamma_{nonCE}(R)$. The bump-to-dip ratio of intensities is a function of the ratio of the two parts of the dynamical broadening $r = \gamma_{CE}(R_c)/\gamma_{nonCE}(R_c)$.This function was calculated analytically in [12]. As r increases, the bumps move away from the center of the X-dip and their intensities decrease. The bump-to-dip ratio measurement r_{exp} allows deducing the rate of CE [12] as follows $< v\, \sigma_{CE}(v)> = \gamma_{nonCE}(R_c)\, r_{exp}/N_i$, where N_i is the ion density. The quantity $\gamma_{nonCE}(R_{cr})$, representing the frequency of inelastic collisions with electrons and ions leading to virtual transitions from the upper state of the radiator to other states, can be calculated for given plasma parameters N_e, N_i, T_e, and T_i by using one of few contemporary theories (presented, e.g., in book [13]). The experimental determination of the rates of CE for multicharged ions from using X-dips is an important reference data virtually inaccessible by other experimental methods.

2.2 Experimental Observations of X-dips

2.2.1. The first observation at the gas-liner pinch, Germany. The X-dip was observed in the blue side of the H α line emitted by hydrogen atoms (Z=1) perturbed by fully stripped helium (Z'=2) – for a relatively small range of electron densities around 10^{18} cm^{-3} [10]. For lower densities the quasi-crossing distance R_c was much lower than the mean inter-ionic distance R_i and for higher densities the dynamical broadening $\gamma_{nonCE}(R_c)$ was smoothing the dip structure, both reasons being unfavorable for experimental observations. In these experiments the electron density N_e was measured by Thomson scattering and it was verified that the positions of the X-dips did not depend on N_e.

2.2.2. Laser Produced Plasma Experiments in France. The X-dips were observed in laser-produced plasmas characterized by $N_e=10^{22} - 3\times10^{22}$ cm^{-3} [12,14]. The setup was implemented at LULI using the same nanosecond laser at 10^{14}W/cm^2 and the same high-resolution Vertical-geometry Johann Spectrometer (R=8000) as in the experiments devoted to the observation of L-dips [7,8]. The targets used for the observation of X-dips were aluminum carbide Al_4C_3 strips inserted in carbon substrate. The emission from the heterogeneous plasma made up of Al and C ions exhibited spectroscopic signatures of CE. X-dips were observed for the first time in the experimental profile of the Ly γ line of Al XIII (Z=13) perturbed by fully stripped carbon CVI (Z'=6), as shown in figure 3.

From the experimental bump-to-dip ratio, the rate of charge exchange between a hydrogenic aluminum ion in the state n=4 and a fully stripped carbon was found to be [12]:

$< v\, \sigma_{CE}(v)> = (5.2 \pm 1.1)\times10^{-6}$ cm^3/s.

Journal of Physics: Conference Series **548** (2014) 012030 doi:10.1088/1742-6596/548/1/012030

2.2.3. Laser-produced plasma experiments in Prague. The X-dips were studied in Plasma Wall Interaction (PWI) experiments performed at PALS [15]. A plasma jet of aluminum ions, produced by the nanosecond iodine laser (intensity 3×10^{14} Wcm^{-2}), incident on a foil, interacted with a massive carbon target. The high spectral and spatial resolution Vertical-geometry Johann Spectrometer was adjusted for the observation of X-dips in the experimental profile of the Ly γ line of Al XIII (Z=13). The X-dips were clearly visible in the red wing as the signatures of CE phenomena accompanying the PWI. The electron densities simulated by codes PALE [16] and MULTIF [17] could reach 5×10^{22} cm^{-3} which is higher than those achieved at LULI. A weak dependence of the positions of the dips on N_e was detected and explained by simulations [12]. Also the L-dip was visible in the far-red wing, thus providing a spectroscopic diagnostic of the electron density.

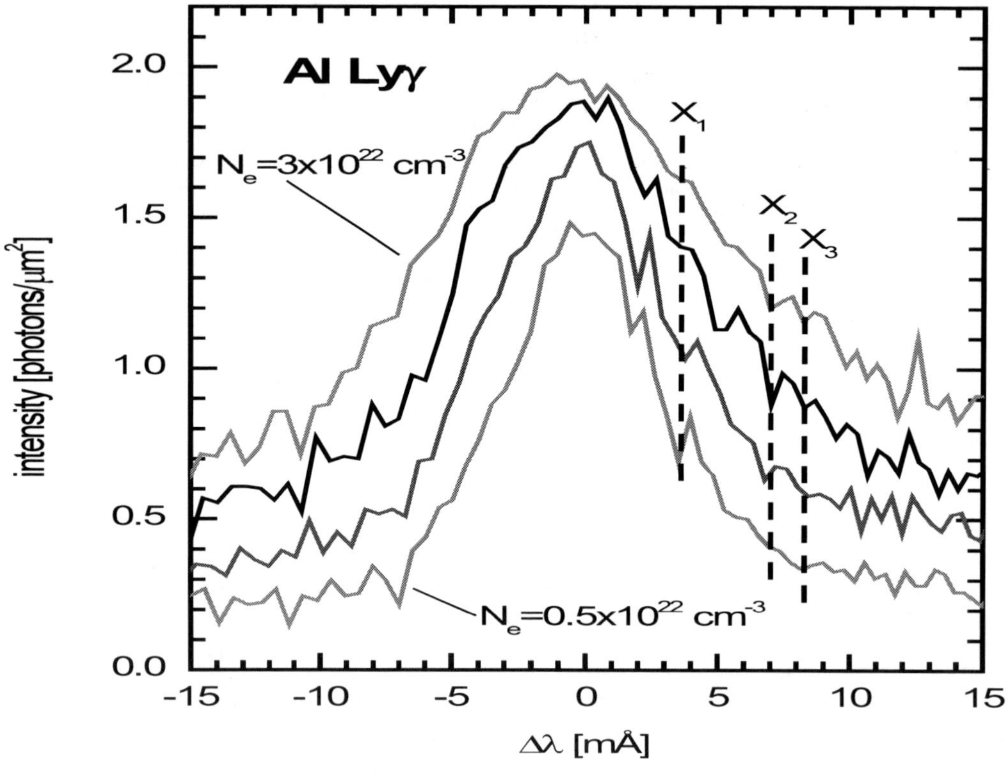

Figure 3. Observation of the X-dip structures X_1, X_2, X_3 in the red wing of the Al XIII Ly γ line perturbed by fully-stripped carbon in laser-produced plasma experiments at LULI. The L-dips are also visible but less pronounced than the X-dips.

More details on all theoretical and experimental studies noted above can be found in the review [18].

3. The latest experiment: two-plasmon decay instability's signature in spectral lines and spectroscopic measurements of CE rate in a femtosecond laser-driven cluster-based plasma

This is the first study of both L-dips and X-dips in spectral lines from *femtosecond* laser-driven *cluster*-based plasma [19]. The experiments were performed at Kansai Photon Science Institute, Japan Atomic Energy Agency. Two Ti:sapphire laser facilities were used: wavelength approximately 800 nm, pulse duration 40 fs. The clusters were created when a gas of a high initial pressure was expanded into vacuum through a specially designed supersonic nozzle, which consisted of three coaxial conical surfaces. CO_2-clusters with a diameter of about 0.5 mm for pure CO_2 (gas pressure 20 bar) and 0.22 mm for the mixed gas of 90% He + 10% CO2 (gas pressure 60 bar) were produced in both experiments. The spectral resolution of a focusing spectrometer was ~ 4000.

From the theoretical point of view, we found a *unique opportunity*. For the line Ly ε of O VIII perturbed by He^{+2}, the three possible X-dips should occur practically at the same location in the profile – merging into one *"X-superdip"* located at (30 ± 3.5) mA in the red wing. The experimental results far exceeded the expectations: we observed not only the X-superdip, but also L-dips (figure 4). The latter allowed an accurate experimental determination of the electron density (as well as of the amplitude of the electric field of the Langmuir wave) leading to the conclusion that we observed *the spectroscopic signature of the two-plasmon decay instability (TPD).*

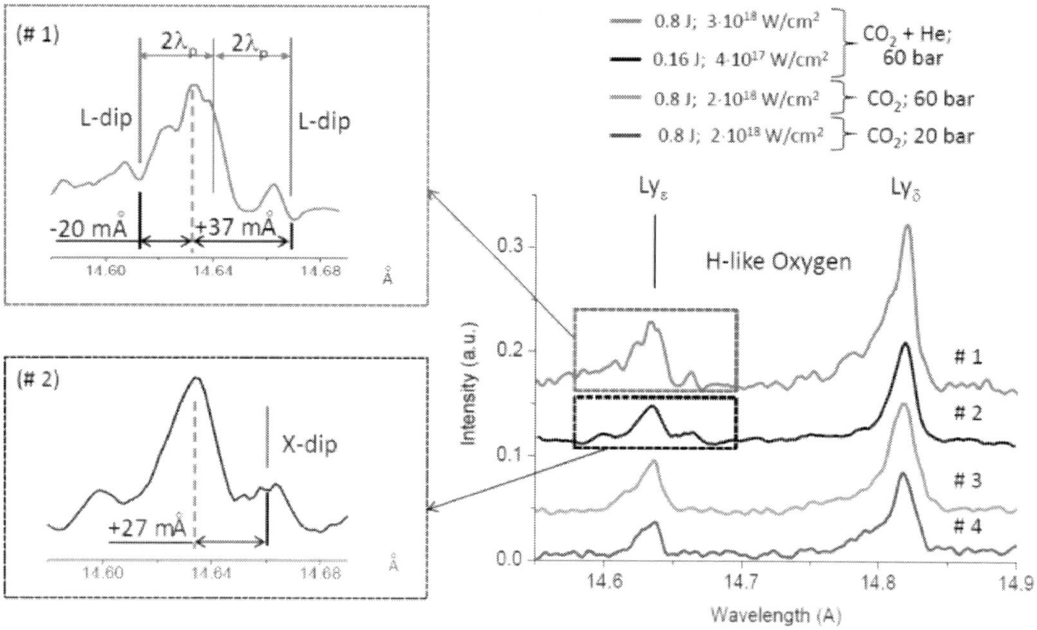

Figure 4. Typical spectra of the O VIII Ly δ and Ly ε lines obtained in femtosecond laser-driven cluster-based experiments. Enlarged spectra of the O VIII Ly ε line in the inset show the positions of the L- and X- dips in more detail.

The value of the plasma electron frequency ω_p deduced from the separation of the experimental L-dips is $\omega_p = 1.26\times10^{15}$ s^{-1}, while the laser frequency is $\omega_{laser} = 2.4\times10^{15}$ s^{-1}. *It is seen that* $\omega_{laser} = 2\omega_p$ (within the accuracy of 5%). This means that we observed in the spectral line profile a *signature of the TPD instability*, occurring at the quarter-critical density: the decay of one quantum of the laser radiation into two Langmuir quanta (plasmons).

The TPD instability is a significant concern in the inertial confinement fusion (ICF). Hot electrons generated by TPD can preheat the target shell and degrade the implosion. Understanding and controlling TPD hot-electron generation is critical for ICF. Besides, TPD can be used for the temperature diagnostic in the coronal region of the ICF plasmas.

The ability to detect TPD via L-dips opens up a *new avenue for diagnosing TPD.* The most important is that this is the only one non-perturbing method to measure the amplitude of the Langmuir waves resulting from TPD and thus to *test theories of TPD.*

In shot #1 the interaction of laser radiation of the very high laser intensity 3×10^{18} W/cm^2 with CO_2-He mixture caused Langmuir waves strong enough to produce pronounced L-dips. In distinction, in shot #2 the laser intensity was by an order of magnitude smaller (4×10^{17} W/cm^2). Therefore, Langmuir waves (if any) were significantly weaker and did not manifest as identifiable L-dips in the profile. The electron density deduced from the separation of the two L-dips is $N_e=5.0\times10^{20}$ cm^{-3}. It is consistent with N_e deduced from the shift of the center of gravity of the two L-dips. From the halfwidth of the L-dips in shot #1 we found the amplitude of the electric field of the Langmuir wave was E_0 = 40 MV/cm.

The theory predicts also a possible *X-superdip* at (30±3.5) mA in the red wing of the O VIII Ly-epsilon line. This X-dip is clearly visible in shot #2: it is marked by the solid vertical line in the inset in the figure below. The central minimum of the experimental x-dip is at 27 mA (in a good agreement with the theory) and is surrounded by two bumps, as predicted by the theory. In the profile from shot #1, the possible X-dip at 27 mA was practically “swallowed” by the near bump of the L-dip. More accurately, in the profile from shot #1, the bump between the central minimum of the L-dip at 37 mA and the central minimum of the X-dip at 27 mA is a superposition of the near bump of the L-dip and of the far bump of the X-dip. This is why this bump is so intense.

As for shots #3 and #4 without the admixture of He, there is no X-dip at 27 mA. This is expected for the plasma without He and thus further confirms the identification of the X-dip in CO_2–He mixture. Whether or not there were Langmuir waves in the plasma from shots #3 and #4 is hard to determine reliably because the modulations of the profile at the possible locations of the L-dips are the level of the noise. We note that the shot #1, where the L-dips were reliably identified, had the laser intensity by 50% greater than shots # 3 and #4.

From the experimental bump-to-dip-contrast a_{exp} of the X-dip in profile #2, we obtained the rate of charge exchange between the ion O^{+8} in the state of N=6 and the ion He^{+2} in the state of n’=2:
$\langle v\, \sigma_{CE}(v)\rangle = (1.5 \pm 0.3)\times10^{-6}$ cm^3/s. This is a new fundamental data virtually inaccessible by other experimental methods.

4. The latest theoretical development on x-dips

Up to now the X-dips were considered to be possible only in spectral lines of H-like ions (perturbed by fully stripped ions). However, recently we showed that the X-dips are also possible in spectral lines of He-like ions from laser-produced plasmas [20]. We presented a table containing 15 prospective He-like spectral lines and 10 corresponding solid targets for observing x-dips in laser-produced plasmas. For completeness we presented also a similar table for H-like lines: 16 prospective H-like spectral lines and 11 corresponding solid targets. This should very significantly extend the range of fundamental data on charge exchange between multicharged ions that can be obtained only via the X-dip phenomenon.

Acknowledgements

A.F. was partly supported by the RFBR research projects14-22-02089 and 14-02-91171-GFEN_a.

References

[1] Gavrilenko V P and Oks E 1981 *Sov. Phys. JETP* **53** 1122
[2] Zhuzhunashvili A I and Oks E 1977 *Sov. Phys. JETP* **46** 2142
[3] Oks E and Rantsev-Kartinov V A 1980 *Sov. Phys. JETP* **52** 50

[4] Gavrilenko V P and Oks E 1987 Sov. *Phys. Plasma Phys.* **13**, 22

[5] Oks E 1995 *Plasma Spectroscopy: The Influence of Microwave and Laser Fields,* Springer Series on Atoms and Plasmas, vol. 9 (New York: Springer)

[6] Oks E, Böddeker St and Kunze H-J 1991 *Phys. Rev. A* **44** 8338 (1991)

[7] Renner O, Dalimier E, Oks E, Krasniqi F, Dufour E, Schott R, and Förster E 2006 *J. Quant. Spectrosc. Rad. Transfer* **99** 439

[8] Krasniqi F, Renner O, Dalimier E, Dufour E, Schott R and Förster E 2006 *Eur. Phys. J. D* **39** 439

[9] Jian L, Shali X, Qingguo Y, Lifeng L and Yufen W 2013 *J. Quant. Spectrosc. Rad. Transfer* **116** 41

[10] Boeddeker St, Kunze H-J and Oks E 1995 *Phys. Rev. Lett.* **75** 4740

[11] Oks E and Leboucher-Dalimier E 2000 *J. Phys. B: Atom. Mol. Opt. Phys.* **33** 3795

[12] Dalimier E, Oks E, Renner O and Schott R 2007 *J. Phys. B: Atom. Mol. Opt. Phys.* **40** 909

[13] Oks E 2006 *Stark Broadening of Hydrogen and Hydrogenlike Spectral Lines in Plasmas: The Physical Insight* (Oxford, UK: Alpha Science International)

[14] Leboucher-Dalimier E, Oks E, Dufour E, Sauvan P, Angelo P, Schott R, and Poquerusse A 2001 *Phys. Rev. E, Rapid Communications* **64** 065401

[15] Renner O, Dalimier E, Liska R, Oks E, and Šmíd M 2012 *J. Phys. Conf. Ser.* **397** 012017

[16] Liska R, Limpouch J, Kucharik M, and Renner O 2008 *J. Phys. Conf. Ser.* **112** 022009

[17] Chenais-Popovics C et al 1997 *Phys. Plasmas* **4** 190

[18] Dalimier E, Oks E, and Renner O 2014 *Atoms* **2** 178

[19] Oks E, Dalimier E, Faenov A et al, *J. Phys. B: Atom. Mol. Opt. Phys.*, Fast Track Communications, 2014.

[20] Dalimier E and Oks E 2014 *J. Phys. B: Atom. Mol. Opt. Phys.* **47** 105001

Journal of Physics: Conference Series **548** (2014) 012031 doi:10.1088/1742-6596/548/1/012031

Nanomaterials induced plasma spectroscopy

A M EL Sherbini, A Abdel Galil, S H Allam and Th M EL Sherbini

Faculty of Science, Cairo University, Department of Physics,
Laboratory of Lasers and New Materials (LLNM), Giza, Egypt

E-mail: elsherbiniaa@gmail.com; elsherbinia@sci.cu.edu.eg

Abstract. LIBS of nanosecond pulsed Nd: YAG laser (1064 nm) produced plasma from a set of nanomaterial and bulk targets (ZnO, Fe_3O_4, Ag_2O, TiO_2, SiO_2 and Al_2O_3) is investigated at laser fluencies in the range 86 J/cm^2 to 2.5 J/cm^2. The optical emission spectra is recorded at the gate and time delay of 1 µs after the onset of the plasma in air having a constant spot size of 0.9 mm. Nanoparticle targets revealed salient enhanced spectral emission compared to their bulky counterparts. Atomic spectral lines average and integral radiance tend to decrease exponentially with laser fluence. Yet, plasma parameters measurements indicated unnoticeable variation of relative electron density and temperature. Therefore, self-absorption corrected enhanced spectral emission was plausibly attributed solely to variation in the inherent nanoparticle relative concentrations. Viable explanations were elaborated based on changes in the intrinsic physical properties of the nanomaterial under high power laser irradiation.

1. Introduction

LIBS and LIPS acronyms stand for respectively "laser induced breakdown spectroscopy" and "laser-induced plasma spectroscopy" [1, 2], and indicate a technique that has been successfully applied to a wide range of applications, mainly in the field of spectrochemical analysis [3-10]. Although this technique has several advantages, e.g., fast outcomes, freedom from sample preparation and very wide dynamic range, it still suffers from several physical problems that should be carefully addressed, for instance, effects of self-absorption of the strongly emitted spectral lines [11] and the limited sensitivity of the technique [12, 13]. Several schemes were proposed to promote the limit of detection LOD values including the use of double pulse technique [14], femtosecond lasers [15] or even a mixed hyper technique [16].

In contrast, nanomaterials exhibited unique physical properties [17] which proved to be promising to improve signal to noise ratio pertinent to optical signals from trace elements with minute concentrations [18-21].

The first Nanoparticle-Enhanced LIBS (NELIBS) of plant leaf surface coated with thin layer of silver nanomaterials was reported by Ohta *et al.* [18]. Localized surface plasmon resonance (LSPR) was invoked to justify the experimental findings. However, a systematic study of the temporal behavior of the spectral lines emission from ZnO nanomaterial plasma [19] indicated the production of large concentrations (removed mass) to be responsible for the enhanced signals. Nonetheless, the relative plasma parameters (electron density and temperature) tend to bear little impact on observed signal enhancement [18-21].

Significant NELIBS of plasma spectral line emissions from noble-metal colloidal solution [20] were rather correlated with the increase in the mass ejected from nanomaterial suspensions as well as the decrease in the threshold of plasma ignition.

Very recently, De Giacomo *et al.* [21] argued that the basic mechanism of Nanoparticles-Enhanced LIBS (NELIBS) is associated with the laser ablation evolution process.

[1] To whom any correspondence should be addressed: A. M. EL Sherbini, Faculty of Science, Cairo University, Department of Physics, Laboratory of Lasers and New Materials (LLNM), elsherbiniaa@gmail.com; elsherbinia@sci.cu.edu.eg

In general, it was shown that the physical properties (e.g., the thermal, electrical and optical properties) of the nanomaterials are changed upon the decrease of the nanoparticle size [17]. This decrease was attributed to nanoparticle fragmentation into smaller ones via different mechanisms during irradiation of nanomaterials by short laser pulses [22-28].

In this work, we present the results of systematic spectroscopic study of spectral line emission enhancement from different nanomaterials-based plasmas in comparison with bulky ones at different laser fluencies and the possible phenomenological explanation in terms of concomitant changes in the physical properties of the nanomaterials during laser irradiation.

2. Experimental arrangement

The experimental setup is shown in Figure 1 and described in details elsewhere [18]. It comprises A Nd:YAG laser working at the fundamental wavelength of 1064 nm that delivers an energy of ~ 650 mJ with pulse duration of 5 ns. The laser beam was focused with a 10 cm quartz convex lens to a circular focal spot of diameter 0.9 ±0.1 mm, hence a maximum fluence of ~ 86J/cm^2. The laser fluence at the target surface was tuned by changing the incident laser pulse energy (at fixed laser spot size) using a set of glass absorber sheets. The emitted light from plasma was collected at the entrance hole of the SE200 Echelle-type spectrograph via a quartz optical fiber of diameter of 25 μm, positioned at 5 mm from the laser-plasma axis. The gate and delay times of the camera were kept fixed at constant value of 1 μs.

The nanomaterials (ZnO, Fe_3O_4, TiO_2, SiO_2 and Al_2O_3) are purchased from MK_{nano}® in a powder form as obtained without heat or chemical treatments. The silver nanomaterial (Ag_2O) was purchased from Sigma Aldrish® on the form of a suspension in a buffer liquid and put on the form of drops on a substrate and left until the puffer solution is evaporated.

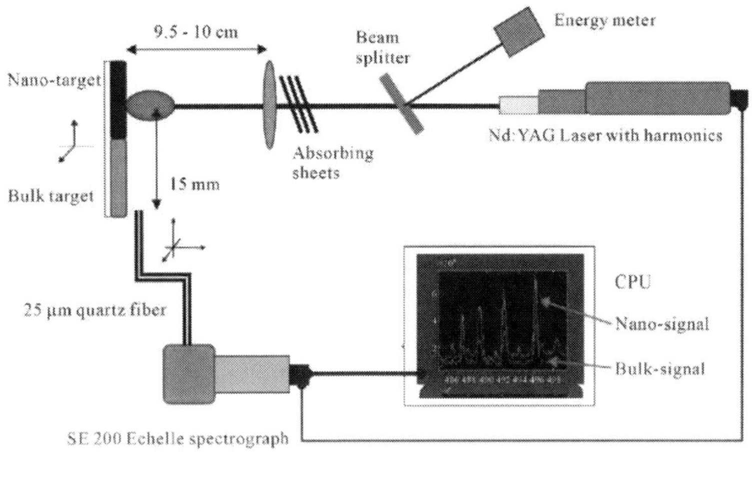

Figure1. Experimental setup.

Both targets (nano & bulk targets) were fixed on a xyφ holder at the same level. Each recorded spectrum was taken under a fresh target condition for three times, which enable us to estimate the error margins. The ratio of nanomaterial to bulk spectral line signal radiance peaks is averaged over different pertinent spectral lines. This procedure is repeated over three shots and the standard deviation about the mean value was taken as the standard error.

3. Optical opacity of plasma to spectral lines

In order to eliminate the effect of plasma opacity to the emitted lines, emitted spectral lines self-absorption should be corrected [11, 18]. This turns out to be crucial [11, 18, 32] to any reliable measurements of the plasma electron temperature as well as relative population density of the ground states (relative concentration). This is usually implemented by using an SA factor [11],

$SA_{line} = \left(n_{e-line}/n_{e-H_\alpha}\right)^{-1.7858}$ with values between zero (complete opacity) to one (perfectly optically thin spectral line devoid of any self-absorption). The corrected spectral intensity can be evaluated using the simple relation $I_o(\lambda_o) = \left(I_{exp}(\lambda_o)/SA_{line}\right)$ where $I_{exp}(\lambda_o)$ is the experimentally measured peak of spectral radiance at the line center of wavelength (λ_o) [32]. The plasma densities n_{e-H_α} and n_{e-line} are measured using the Stark broadening of the usually observed H_α-line while for the spectral lines other than hydrogen one can use the relation $n_{e-line} \approx (\Delta\lambda_{s-line}/2\omega_{s-line})N_{ref}$ [30], where, $\Delta\lambda_{s-line}$ is the experimentally measured FWHM of Lorentzian line width and ω_{s-line} is the Stark broadening parameter tabulated at certain reference electron densities N_{ref} [cf.31].

4. Results and discussion

Optical emission spectra recorded for different nanomaterial (blue colored) and bulk targets (red colored) at a variety of experimental conditions are displayed in Figures 2a to 2f. The maximum enhancement factor about 37 can be clearly related to nanomaterial silver oxide. The nanomaterial plasma emission atomic spectral lines are favorably enhanced in a descending order as Si I, Zn I, Ti I, Fe I, Ag I and Al I respectively.

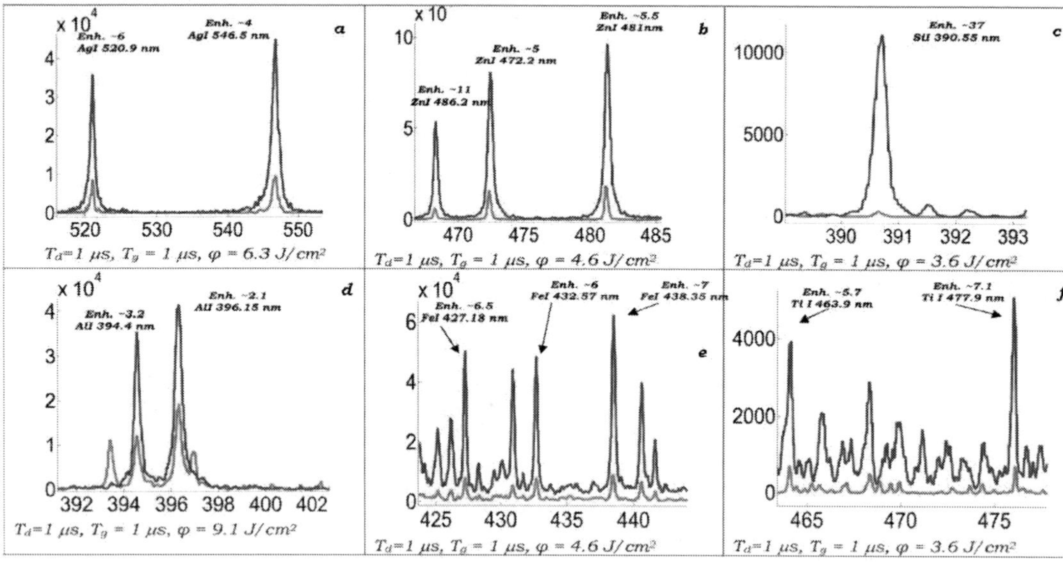

Figure 2. Demonstration of the enhancement factors (as shown) exhibited by the nanomaterial (blue colored) and bulk (red colored) plasma emission spectral lines at different experimental conditions. (a) Ag_2O, (b) ZnO, (c) SiO, (d) Al_2O_3, (e) Fe_3O_4 and (f) TiO_2.

To investigate the effect of laser fluence on the enhancement factor, ZnO nanomaterial and bulk targets plasma emission at 468.2, 472.2 and 481 nm wavelengths were recorded at different laser fluencies as shown in Figure 3. Average peak and integral spectral radiance enhancement factors apparently tend to exponentially decrease as the laser fluence increases Figures 4a, b. The observed signals were not actually corrected against self-absorption effects. It is worth noting that the spectral lines are not equally enhanced which may be related to their different inherent oscillator strengths or to selectively favorable surface plasmon resonance [21].

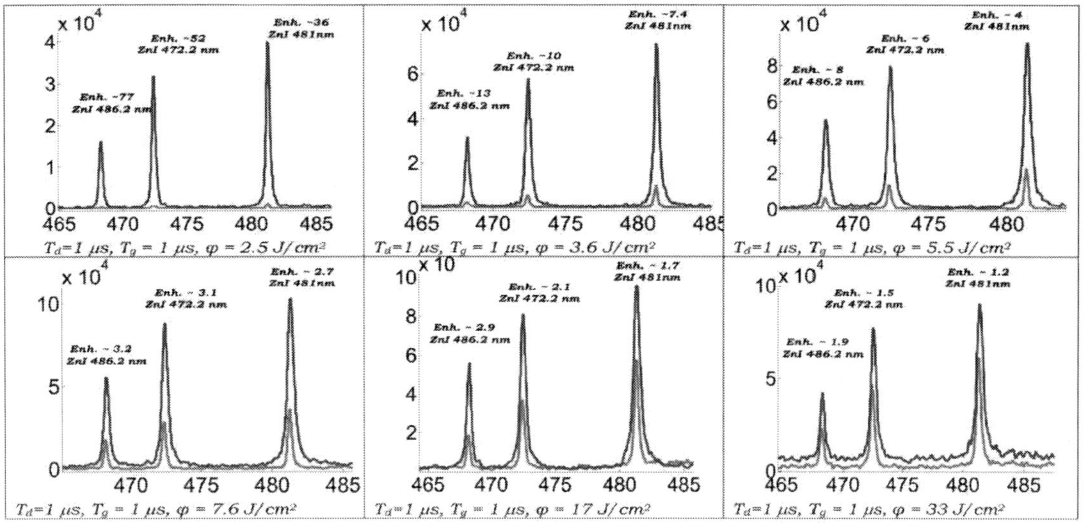

Figure 3. Enhancement factors at different laser fluencies associated with Zn I-lines.

Observed enhancement in spectral line radiance was pursued through the variation of the peak spectral radiance itself as a function of laser fluence. Plasma evolution seems to be influenced slightly differently by nanomaterial and bulk targets. It can be seen from Figure 4c that plasma emerged from the nanomaterial targets tend to decay relatively more slowly than that of the bulk-based plasma.

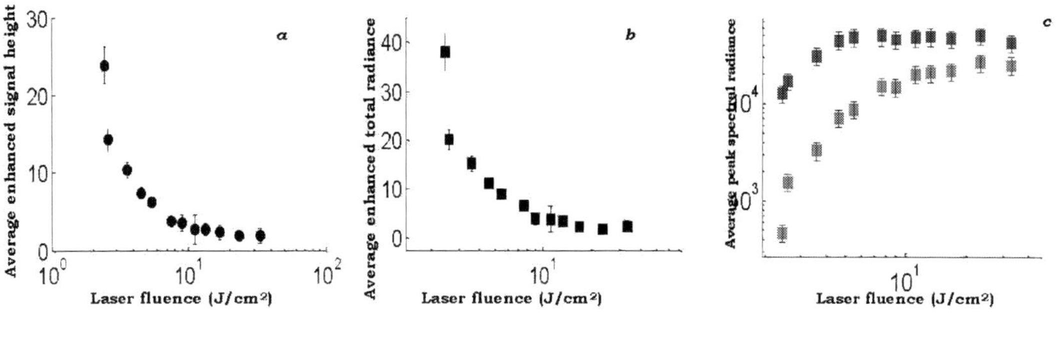

Figure 4. (a) The variation of the average enhancement in peak spectral radiance (signal height) with laser fluence. (b) The variation of the integrated radiance (area under curve). (c) The variation of the peak spectral radiance from nano-based plasma ((blue upper squares) and bulk-based plasma (lower red squares) with laser fluence.

The role played by plasma characteristics, plasma electron density, temperature and stoichiometric concentrations was assessed via optical emission spectroscopy (OES) [18]. Boltzmann plots were performed on peak spectral radiance of the Zn I-lines at 328.23, 330.3, 334.5, 468, 472.2, 480 and 636.23 nm after correction against the effect of self-absorption using available atomic data [33], while the Stark broadening parameters of the different Zn I-lines were taken from Ref. [31].

Figure 5b indicates the quality of the previously suggested method of correction process of spectral radiance against the effect of self-absorption which results in a straight line Boltzmann plot [29].

Journal of Physics: Conference Series **548** (2014) 012031 doi:10.1088/1742-6596/548/1/012031

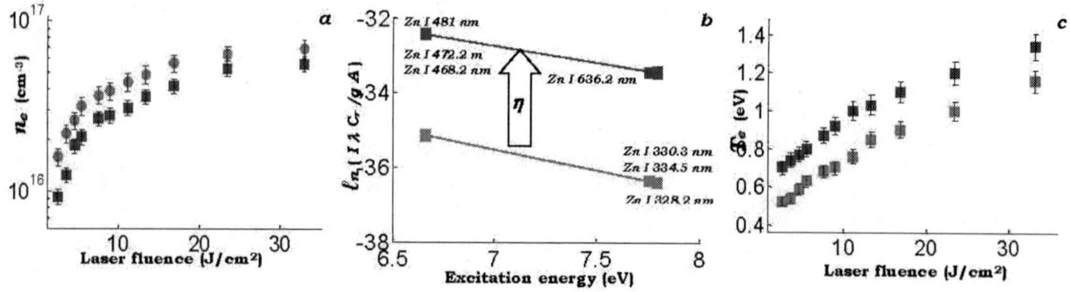

Figure 5. (a) The variation of the electron density with laser fluence for nano-based plasma (lower blue squares) and bulk-based plasma (upper red solid circles). (b) Multi-elemental Boltzmann plot of Zn I lines associated with nano-based (upper blue line) and bulk-based (lower red line) targets at laser fluence of 17 J/cm^2. (c) The variation of the plasma electron temperature for nano-based (upper blue squares) and bulk-based (lower red squares) plasmas.

Notwithstanding the similar trend exhibited by both hotter nanomaterial and denser bulk targets plasma temperature and density in Figures 5a, c [19-21], Boltzmann elemental analysis reveals higher levels of ground state atomic population in favor of nanomaterial targets, see Figure 5b.

The experimental findings indicate that plasma characteristics in terms of plasma density and temperature have little impact on the observed enhancement phenomena in fair agreement with previous reports [18-21]. The relative nanomaterial to bulk plasma stoichiometric concentrations is evaluated in terms of the relative concentration factor $R_c = \exp(\eta) = (N_{oN}/N_{oB})$ [19].

A comprehensive relative nanomaterial to bulk data was elaborated as shown in Figure 6 and Figure 7 for ZnO and other nanomaterials respectively. Close examination of Figure 6 reveals a close correlation between the estimated relative concentration and enhancement factor whereby suggesting that the relative concentration enhancement might be the main reason behind the plasma emission spectral line enhancement [19-21].

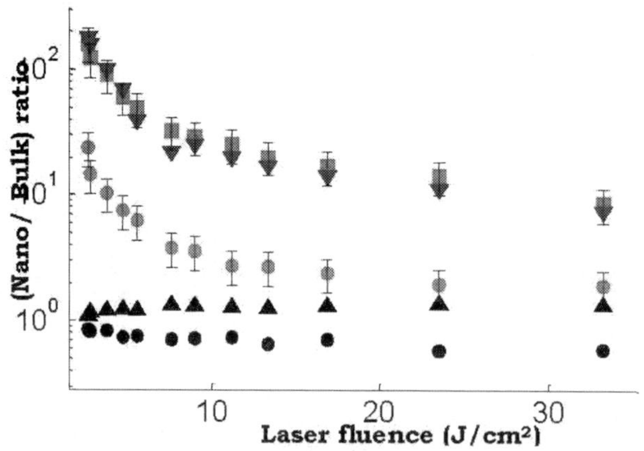

Figure 6. The variation of ZnO relative electron density (black solid circles), the relative electron temperature (black triangles), enhancement factors with self-absorption (red solid circles), self-absorption free (red squares) and relative concentration (blue inverted triangles).

As clearly seen from Figure 7, a noticeable relative spectral radiance enhancement is set on at threshold laser fluence about ~ 10 J/cm^2 followed by significant increase reminiscent of internal transformation in connection with laser pulse irradiation.

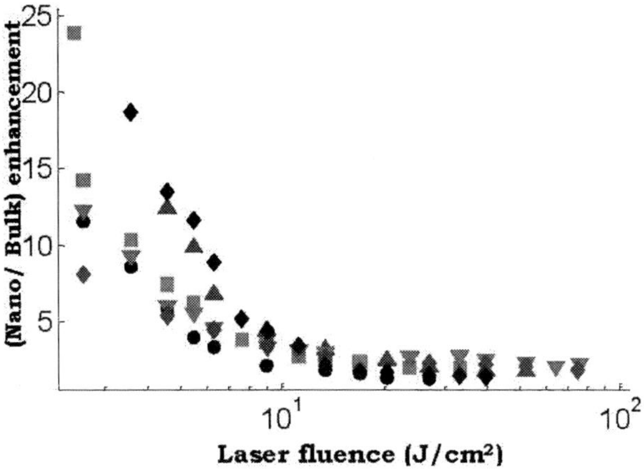

Figure 7. Shown is the variation of the enhanced peak spectral radiance with laser fluence; ZnO (red squares), Fe_3O_4 (black disks), TiO_2 (blue triangles), Al_2O_3 (Inverted green triangles), Ag_2O (blue hex) and SiO_2 (black hex).

Recent experimental findings using transient absorption spectroscopy [22-28] have shown that nanoparticle size is diminished on par with the incident laser fluence due to fragmentation into smaller particles during laser irradiation. Moreover, the time required for this transient size modulation was measured for the gold nanoparticles and amounts to ~ 0.1 ns [26-27] and in the range of femtosecond regime for TiO_2 nanoparticles [28].

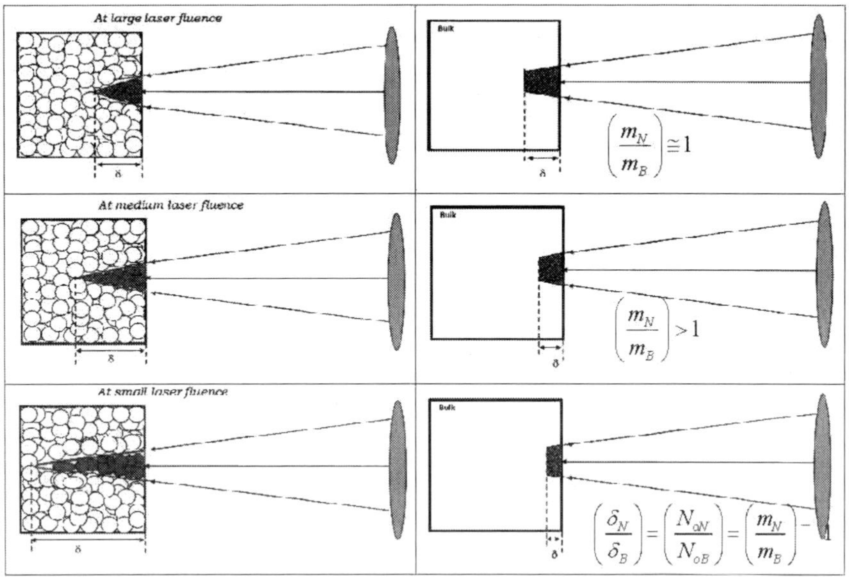

Figure 8. An illustration to possible mechanisms of the interaction of the laser light pulses with nano-based target in comparison with bulk-based target at different laser fluencies.

In accordance with these experimental findings, nanoparticles fragmentation at high laser fluencies may account for the decline in the observed spectral line radiances. Apparent is the decrease in laser fluence due to the larger surface area of fragmented nanoparticles [22-28] accompanied by a sudden increase in optical conductivity of the nanoparticles [17, 34-35] with less laser light penetration. Figure 8 would justify this decrease. Moreover, inefficient mass ejection is influenced by spatial confinement of laser beam as the refractive index increases [17, 22] assisted by even higher specific heat values [17] at high laser fluencies.

A more theoretical and experimental investigation is needed in particular the use of electron microscopy to monitor relative mass ejected from nanomaterial and bulk targets.

5. Conclusion

The enhanced emission from the ZnO and various nanomaterials as Fe_3O_4, Ag_2O, TiO_2, SiO_2 and Al_2O_3 was examined experimentally with the help of optical emission spectroscopy (OES). High laser fluencies are paradoxically quenching the attainment of high spectral radiance enhancement. Besides, the observed enhanced emission was found to be independent of variations in plasma parameters (electron density and temperature). Higher ground state population density in nanomaterial targets proved to be paramount in spectral line radiance enhancement. The interaction of the high peak power lasers with nanomaterials strongly influence the intrinsic physical properties of irradiated targets which require further investigation.

Acknowledgement

This paper was presented as an Invited Talk at the 22nd International Conference on Spectral Line Shapes ICSLS22 at the University of Tennessee Space Institute, Tullahoma, TN 37388, USA. The authors thank the local and international committee for their support.

References

[1] Radziemski L and Cremers D, 2012 *Spectrochim. Acta Part B* **87** 3
[2] Rusak D A, Castle BC, Smith B W, Winefordner J D 1997 *Crit. Rev. Anal. Chem.* **27** 257
[3] Song K, Lee Y I, Sneddon J 1997 *Appl. Spectrosc. Rev.* **32** 183
[4] Berman L M and Wolf P J 1998 *Appl. Spectrosc.* **52** 438
[5] Hilbk-Kortenbruck F, Noll R, Wintjens P, Falkb H, Becker C 2001 *Spectrochim. Acta Part B* **56** 933
[6] Jankowsk M, Liwin´ski G S 2003 *Journal of Cultural Heritage* **4** 65
[7] Chappé M, Hildenhagen J, Dickmann K, Bredol M 2003 *Journal of Cultural Heritage* **4** 264s
[8] Miziolek A, Palleschi AV and Schechter I 2006 *Laser-Induced Breakdown Spectroscopy (LIBS) Fundamentals and Applications* (Cambridge: University Press)
[9] Noll R 2012 *Laser-Induced Breakdown Spectroscopy Fundamentals and Applications* (Berlin: Springer-Verlag)
[10] Cremers D A and Radziemski L J 2006 *Handbook of laser-induced breakdown spectroscopy* (New York: John Wiley & Sons Ltd).
[11] EL Sherbini A M, EL Sherbini Th M, Hegazy H, Cristoforetti G, Legnaioli S, Palleschi V, Pardini L, Salvetti A, Tognoni E 2005 *Spectrochim. Acta Part B* **60** 1573
[12] Sobral H, Sanginés R 2014 *Spectrochim. Acta Part B* **94-95** 1
[13] Gautier C, Fichet P, Menut D, Lacour J-L, L'Hermite D, Dubessy J 2004 *Spectrochim. Acta Part B* **59** 975
[14] Babushok VI, DeLucia Jr FC, Gottfried JL, Munson CA, Miziolek AW 2006 *Spectrochim. Acta Part B* **61** 999
[15] Gurevich E L, Hergenröder R 2007 *Appl. Spectrosc.* **61** 233A
[16] Scaffidi J, Pender J, Pearman W, Goode S R, Colston Jr. BW and Varter JC 2003 *Appl. Opt.* **42** 6099
[17] Pokropivny V, Lohmus R, Hussainova I, Pokropivny A,Vlassov S 2007 *Introduction to nanomaterials and nanotechnology - Special lecture course for bachelors, M.Sc, post-graduates and specialists in nanotechnology* (Tartu: University Press).
[18] Ohta T, Ito M, Kotani T, Hattoti T 2009 *Appl. Spectrosc.* **63** 555
[19] EL Sherbini A M, Aboulfotouh A M, Rashid F F, Allam S H, EL Dakrouri A, EL Sherbini Th M 2012 *WJNSE* **2** 181
[20] De Giacomo A, Gaudiuso R, Koral C, Dell'Aglio M, De Pascale O 2013 *Anal. Chem.* **85** 10180

[21] De Giacomo A, Gaudiuso R, Koral C, Dell'Aglio M, De Pascale O 2014 *Spectrochim. Acta Part B* **98** 19–27
[22] Link S, and EL SayedA M 2000 *Int . Rev. Phys. Chem.* **3** 409
[23] Amendola V, Meneghetti M 2009 *Phys. Chem. Chem. Phys.* **11** 3805
[24] Sendova M, Sendova-Vassilev M, Pivin J. C. Hofmeister H, Coffey K, Warren A 2006 *Journal of Nanoscience and Nanotechnology* **6** 748
[25] Giammanco F, Giorgetti E, Marsili P and Giusti A, 2010 *J. Phys. Chem. C* **114** 3354
[26] Werner D, Hashimoto S 2011 J. *Phys. Chem. C* **115** 5063
[27] Hashimoto S, Werner D, Uwada T 2012 *J. Photochemistry Photobiology C: Photochemistry Rev.* **13** 28
[28] Alnassar S I, Akman E, Oztoprak B G, Kacar E, Gundogdud O, Khaleel A, Demir A 2013 *Optics & Laser Technology* **51** 17
[29] El Sherbini A M, Hegazy H, El Sherbini Th M, 2006 *Spectrochim. Acta Part B* **61** 532
[30] Konjević N 1999 *Physics Reports* **316** 339
[31] Konjević N, Dimitrijević M S, Wiese W L 1984 *J. Phys. Chem. Ref. Data* **13** 619
[32] El Sherbini A M, Al Aamer A A, Hassan A T, El Sherbini Th M 2012 *Optics and Photonics J.* **2** 278
[33] http://www.nist.gov/pml/data/asd.cfm
[34] Kosacki I, Anderson H U 2000 *Ionics* **6** 294
[35] Kosacki I, Anderson H U 1996 *Appl. Phys. Lett.* **69** 4171

Journal of Physics: Conference Series **548** (2014) 012032 doi:10.1088/1742-6596/548/1/012032

Laser-induced breakdown spectroscopy for Stark broadening and shift experiments: Measurement of Fe II and Ni II Stark shifts

C Aragón[1], J A Aguilera[1], J Manrique[2]

[1]Departamento de Física, Universidad Pública de Navarra, Campus de Arrosadía, E-31006 Pamplona, Spain
[2]Facultad de Farmacia, Universidad CEU San Pablo, Urbanización Montepríncipe, Boadilla del Monte, E-28668 Madrid, Spain

E-mail: carlos.aragon@unavarra.es

Abstract. We present experimental studies on characterization of laser-induced plasmas relevant for the measurement of Stark widths and shifts by laser-induced breakdown spectroscopy. The selection of samples for plasma generation is investigated. It is shown that fused glass samples provide spectra with much higher line-to-background ratio for many lines of interest, compared to alloys. The influence of self-absorption on the line width is studied as a function of the concentrations of the emitting element in the sample and time of the plasma evolution. New Stark shifts for several Fe II and Ni II are measured.

1. Introduction

Since the advent of laser-induced breakdown spectroscopy (LIBS), an interest has existed in performing Stark broadening and shift experiments for characterization of the electron density of laser-induced plasmas used in different applications [1]. Precise measurement of the electron density relies on the knowledge of accurate data for Stark widths and shifts. However, these parameters are still not available for many transitions of several atoms and ions, so there is a need of new experimental data, which are also useful for the diagnostics of other types of plasmas and to verify the calculations. In the case of Fe II and Ni II, our group has reported recently measurements of Stark widths for a high number of transitions [2-5], which have been compared to the relatively few data found in the literature. Regarding Stark shifts, we have not found previous results for Fe II or Ni II.

Laser-induced plasmas are spectroscopic sources of high electron density which may be obtained in a relatively easy way from gases, liquids and electrically conducting or non-conducting solid samples. Because of these properties, laser-induced plasmas are particularly suitable for the measurement of Stark widths and shifts [6]. However, a detailed characterization of the laser-induced plasma is necessary to obtain precise measurements. The previous step of plasma characterization allows the control of self-absorption, the selection of suitable samples for plasma generation and the estimation of the effect of plasma inhomogeneity.

In this work, we describe experimental work performed by our group in which LIBS is used for characterization of laser-induced plasmas with the purpose of measuring Stark parameters. Also, new data for Fe II and Ni II Stark shifts measured by LIBS are presented.

2. Experiment

The experimental set-up is similar to that used previously [5], so it is only described briefly. The laser-induced plasmas are generated by focusing a Nd:YAG laser (wavelength 1064 nm, pulse width 4.5 ns, repetition rate 20 Hz) onto a sample placed in air at atmospheric pressure. The laser pulse energy has varied depending on the type of sample used: For alloys, the pulse energy was set at 100 mJ by means of an optical attenuator whereas for fused glass samples, a lower pulse energy of 60 mJ has been employed. The focal length of the focusing lens was 128 mm and the lens-to-sample distance was 118 mm and 124 mm for pulse energies of 100 mJ and 60 mJ, respectively. The emission of the plasma is collected and focused by a mirror system onto the entrance slit of a Czerny-Turner spectrometer equipped with a time-resolved intensified CCD (1200 × 256 effective pixels). The grating (3600 and 1200 lines mm^{-1}) and the slit width (20 and 50 μm) are selected according to the spectral resolution needed. In each spectrum measured the emission from 100 laser shots is accumulated while the sample rotates at 100 rev min^{-1}.

3. Results and discussion

3.1. Samples used

Two different types of samples have been used to generate the laser-induced plasmas. Alloy samples have been prepared from the pure metals in powder form by melting in an induction furnace [2-4]. Alloys have the advantage that they can be

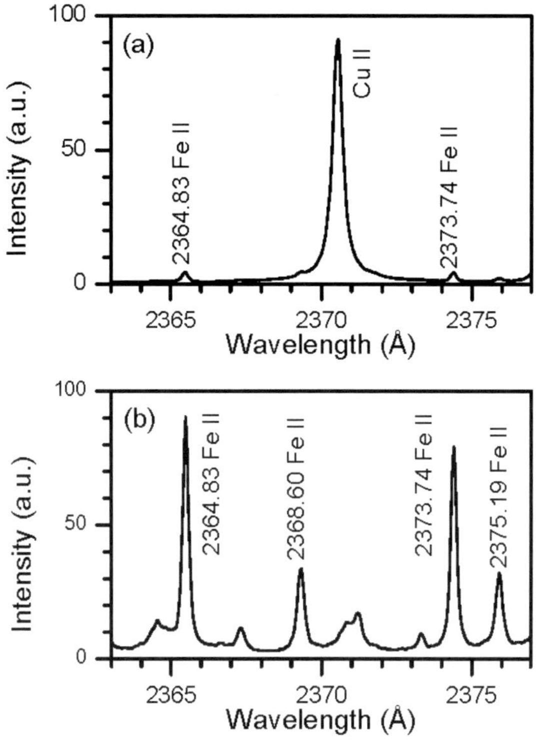

Figure 1. Spectra from laser-induced plasmas generated from a Fe-Cu alloy sample (a) and a Fe_2O_3 fused glass sample (b). The iron concentration in both cases is 1.0 at. %.

reused in different measurements by sanding the surface to eliminate the craters formed by the laser ablation. On the other hand, we have prepared fused glass samples from pure oxides in powder form by borate fusion in a fluxer [5]. These samples may be prepared from several types of compounds, including oxides and salts, which allows including non-metallic elements. In figure 1, spectra containing Fe II lines of interest obtained using a Fe-Cu alloy and a Fe_2O_3 fused glass are compared. The iron concentration in the sample is 1.0 at.% in both cases. As can be seen, for the alloy sample, the Cu II lines from the copper matrix overlap with the Fe II lines. Conversely, the absence of lines from the borate matrix in the case of the fused glass, with the exception of a few intense B I lines and weak Li I lines appearing at other spectral regions, allows the measurement of Fe II lines as 2368.60 Å, which may hardly be observed in the alloy spectrum. Furthermore, even lines well resolved from the Cu II line as 2364.83 Å show a higher line-to-background ratio in the case of the fused glass spectrum. This is shown in figure 2, where the line-to-background ratio for this line, obtained as the height of the line divided by the intensity of the background, is plotted as a function of the time of plasma evolution for the two types of samples.

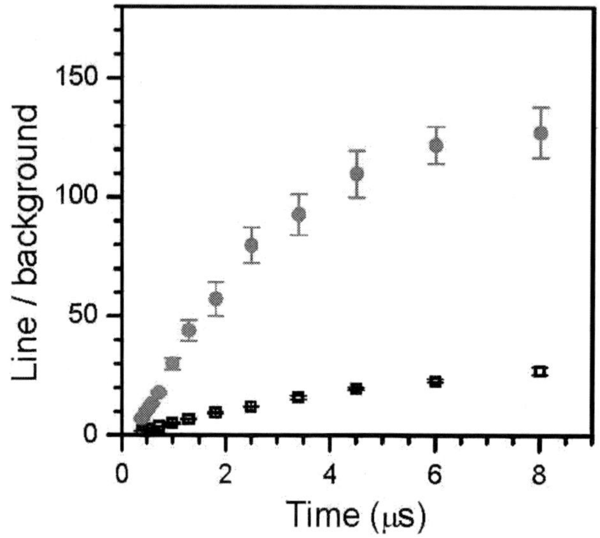

Figure 2. Line-to-background ratio for the Fe II line at 2364.83 Å as a function of the time of plasma evolution for alloy samples (open squares) and fused glass samples.

As can be seen, the line-to-background ratio increases monotonously with time, a behavior expected from the decrease of the continuum emitted by the plasma and the reduction of the line width due to the decrease of the electron density, which makes that the line intensity, in spite of suffering also a decrease, is distributed among a narrower wavelength range. For the purpose of measuring Stark widths and shifts, however, only spectra at initial times may be used, as the electron density must be high enough so that the Stark width is significantly higher than the instrumental width. In the case of lines with relatively low Stark widths and shifts, such as the Fe II lines, for our instrumental width of 0.135 Å, the useful time interval goes from around 0.5 μs to 3.5 μs. We may observe in figure 2 that, at this time interval, the line-to-background ratio is significantly higher for fused glass samples than for alloy samples.

3.2. *Investigation of self-absorption*

Self-absorption is one of the main causes of systematic error in the measurement of Stark widths by LIBS, as it distorts the line profiles, leading to an increase of the line width [7]. However, the possibility of LIBS for varying the concentration of the emitting element in the sample allows controlling this effect. We have studied the influence of self-absorption on the line width using fused glass samples prepared from iron oxide having Fe concentrations in the range 0.1-0.8 at.%. As self-absorption depends also strongly on the time of plasma evolution, the investigation has also included the dependence on this parameter. Figure 3 shows the ratio of the line width (FWHM) w to the electron density N_e for the intense Fe II line at 2599.40 Å plotted as a function of the Fe concentration in the sample, for different instants of the plasma evolution. The electron density of the plasma at each instant has been measured as described in previous works [4,8].

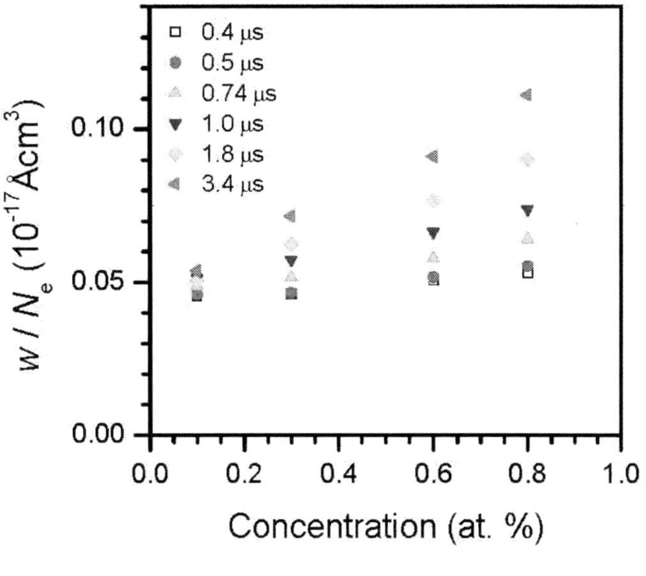

Figure 3. Ratio between the line width w (FWHM) and the electron density N_e for the Fe II line at 2599.40 Å as a function of the Fe concentration in the sample for different instants of the plasma lifetime.

As can be seen in figure 3, the ratio w/N_e, which should be approximately constant in the absence of self-absorption, increases with the concentration in the sample, the increase being steeper for higher delay from the laser pulse. This figure shows the importance of controlling self-absorption to obtain an accurate measurement of the Stark width. For the concentration 0.1 at.% finally used in the measurement, the values of w/N_e at different instants converge to a single value within the experimental error, so the effect of self-absorption on the Stark width is small, leading to an error which is estimated as lower than 10% for the ensemble of Fe II lines investigated.

3.3. *Measurement of Stark shifts*

Figure 4 shows a typical measurement of the profile of a Fe II line at different instants of the plasma evolution. The spectra have been measured at six time windows having different delays from the laser pulse, centred at instants of the plasma lifetime ranging from 0.6 to 3.4 μs. The width of the time windows increases with the delay from 0.06 to 0.6 μs. In these plots, the intensity is normalized to the

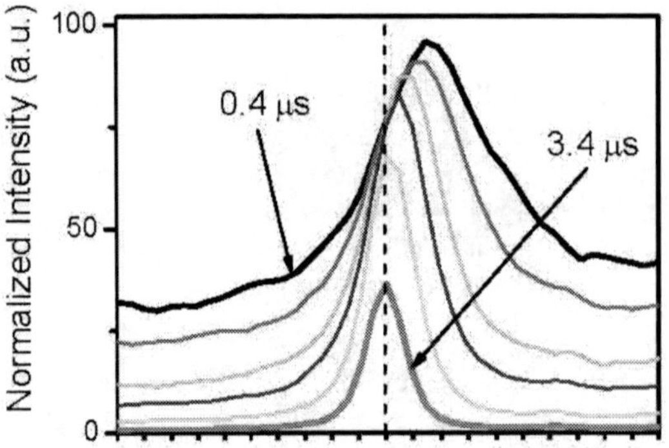

Figure 4. Spectra of the Fe II line at 2368.60 Å obtained at different time windows from 0.4 μs to 3.4 μs. The dashed line indicates the wavelength of the unshifted profile.

width of the time window used in each case. As the delay from the laser pulse increases, a decrease of the line width and a decreasing red shift are observed, both related to the temporal decay of the electron density. The wavelength of the unshifted profile, indicated by the dashed line, is obtained from the spectrum at 10-μs delay, when the Stark shift is negligible. To obtain the shifts, the position of the maximum is determined by fitting the line profiles to Voigt profiles. The electron density of the plasma at each time window is determined from the Stark broadening of the H_α line [4,8].

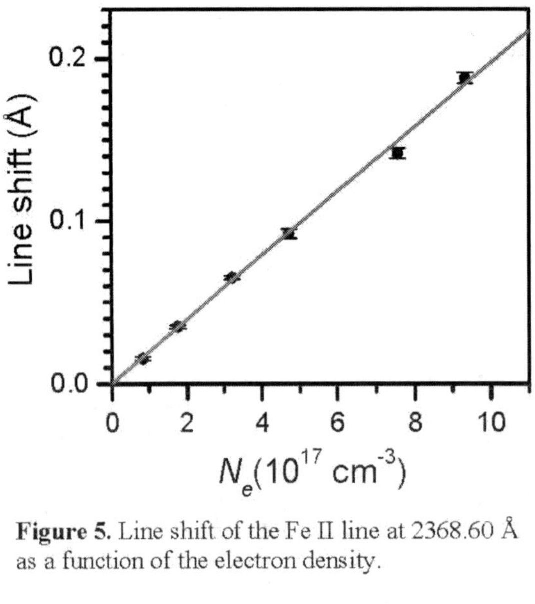

Figure 5. Line shift of the Fe II line at 2368.60 Å as a function of the electron density.

A typical plot of the line shift vs. the electron density is shown in figure 5. As can be seen, a proportionality relation is obtained, which means that the weak dependence of the Stark shift on temperature has not been observed. Therefore, the final results are obtained as the slopes of the linear fittings of these plots with zero intercept, this value corresponding to the Stark shift at an electron density of 10^{17} cm^{-3}.

The results for the Stark shifts of Fe II and Ni II lines are presented in tables 1 and 2, respectively. The experimental relative error of the Stark shifts is mainly due to the uncertainty of the electron density, which has been estimated as 11%. However, the minimum absolute error due to resolution is 0.1 pm. To our knowledge, these are the first experimental results for the Stark shifts of these lines and no previous theoretical results have been reported.

Table 1. Stark shift at electron density $N_e = 10^{17}$ cm^{-3} of Fe II spectral lines.

Transition	Multiplet	λ (Å)	d^a (pm)
$3p^63d^7$ - $3d^6(^5D)4p$	$a\,^4F$ - $z\,^4F^\circ$	2331.307	2.06
	$a\,^4F$ - $z\,^4D^\circ$	2368.596	1.96
		2375.193	1.87
$3d^6(^5D)4s$ - $3d^6(^5D)4p$	$a\,^4D$ - $z\,^4P^\circ$	2591.543	0.48
$3p^63d^7$ - $3d^6(^3H)4p$	$a\,^2H$ - $z\,^2H^\circ$	2233.917	1.73

[a]Temperature range: 12000-17600 K. The relative error of d is 11%, with a minimum absolute error of 0.1 pm.

Table 2. Stark shift at electron density $N_e = 10^{17}$ cm^{-3} of Ni II spectral lines.

Transition	Multiplet	λ (Å)	d^a (pm)
$3d^8(^3F)4s$ - $3d^8(^3F)4p$	^4F - $^4G^\circ$	2216.477	0.51
	^4F - $^4F^\circ$	2138.582	0.48
		2210.379	0.48
	^2F - $^2D^\circ$	2356.403	0.48
$3d^8(^3P)4s$ - $3d^8(^1D)4p$	^4P - $^2F^\circ$	2220.399	0.47
$3d^8(^1D)4s$ - $3d^8(^1D)4p$	^2D - $^2P^\circ$	2213.195	0.40
$3d^8(^1G)4s$ - $3d^8(^1G)4p$	^2G - $^2G^\circ$	2107.953	0.74
		2113.518	0.70

[a]Temperature range: 12000-17600 K. The relative error of d is 11%, with a minimum absolute error of 0.1 pm.

4. Conclusions

Laser-induced plasmas are spectroscopic sources suitable for measurement of Stark widths and shifts. Generation of the plasmas from fused glass samples allows obtaining spectra from many atoms and ions showing higher line-to-background ratios compared to spectra obtained from alloy plasmas. The selection of the concentration of the emitting element in the sample allows controlling the effect of self-absorption on the line width.

Acknowledgment
This work has been supported by the project FIS2011-29521 of the Spanish Ministerio de Economía y Competitividad.

References

[1] Radziemski L J, Loree T R, Cremers D A, Hoffman N M 1983 *Anal. Chem.* **55** 1246
[2] Aragón C, Vega P, Aguilera J A 2011 *J. Phys B:At Mol Opt Phys.* **44** 055002
[3] Aguilera J A, Manrique J, Aragón C, 2011 *J. Phys B:At Mol Opt Phys.* **44** 245701
[4] Aguilera J A, Aragón C, Manrique J. 2013 *JQSRT* **114** 151
[5] Aragón C, Aguilera J A, Manrique J. 2014 *JQSRT* **134** 39
[6] Colón C, Hatem G, Verdugo E, Ruiz P, Campos J 1993 *J. Appl. Phys.* **73** 4752
[7] Konjević N, Ivković M, Jovićević S 2010 *Spectrochim. Acta Part B* **65** 593
[8] Aragón C, Aguilera J A 2010 *Spectrochim. Acta Part B* **65** 395

Journal of Physics: Conference Series **548** (2014) 012033 doi:10.1088/1742-6596/548/1/012033

An Algorithm For Computation Of Diatomic Spectra

J O Hornkohl[1], A C Woods[2] and C G Parigger[2]

[1] 344 Turkey Creek Road, Tullahoma, TN 37388, USA

[2] Center for Laser Applications, University of Tennessee Space Institute, 411 B.H. Goethert Parkway, Tullahoma, TN 37388, USA

E-mail: cparigge@tennessee.edu

Abstract. In principle, an atomic or molecular spectrum would be computed as follows: Upper and lower Hamiltonians would be enumerated in a complete basis, and numerically diagonalized to give the upper and lower energy eigenvalues and eigenvectors. The transition moments for the appropriate operator, *e.g.*, the electric dipole transition moments, would be evaluated from the eigenvectors. The vacuum wavenumbers $\tilde{\nu}$, *i.e.*, energy eigenvalue differences, would be found for all non-vanishing transition moments. And the line strengths for each spectral line of wavenumber $\tilde{\nu}$ would be determined as sum of the squares of the transition moments over all transitions producing the same $\tilde{\nu}$. A line list that includes line strengths would be generated by repeating the above computations over the required range of upper and lower total angular momentum quantum numbers. The spectrum from $\tilde{\nu}_{\mathrm{min}}$ to $\tilde{\nu}_{\mathrm{max}}$ would be separated into a number of pixels, subsequently, the contribution of each line to each pixel is calculated using the line list. We show how this algorithm can be implemented for a diatomic spectrum if the required molecular parameters are available.

1. Introduction

This work reports our algorithm used to compute the synthetic spectra of diatomic molecules. Figure 1 shows a measured spectrum of interest in for example Laser-Induced Breakdown Spectroscopy (LIBS) [1, 2, 3, 4, 5], and it also shows two computed C_2 Swan spectra fitted to the experimental spectrum. With one caveat, the algorithm is the textbook method for computation of a spectrum. The caveat is that a comprehensive set of accurate diatomic parameters is required for its implementation. We describe our algorithm and give an explanation of why it is based on the Wigner-Witmer diatomic eigenfunction and not the Born-Oppenheimer approximation.

2. Computation of a diatomic spectrum

Except for the baseline corrections, the synthetic spontaneous emission spectra in Fig. (1) were computed using the following algorithm.

- For given upper total angular momentum quantum number J' and lower quantum number J, the upper and lower Hamiltonian matrices H' and H are computed in the Hund's case a basis and numerically diagonalized to obtain the eigenvalues F' and F and eigenvectors U' and U,

$$F' = U'^{\dagger}\, H'\, U', \tag{1}$$

$$F = U^{\dagger}\, H\, U. \tag{2}$$

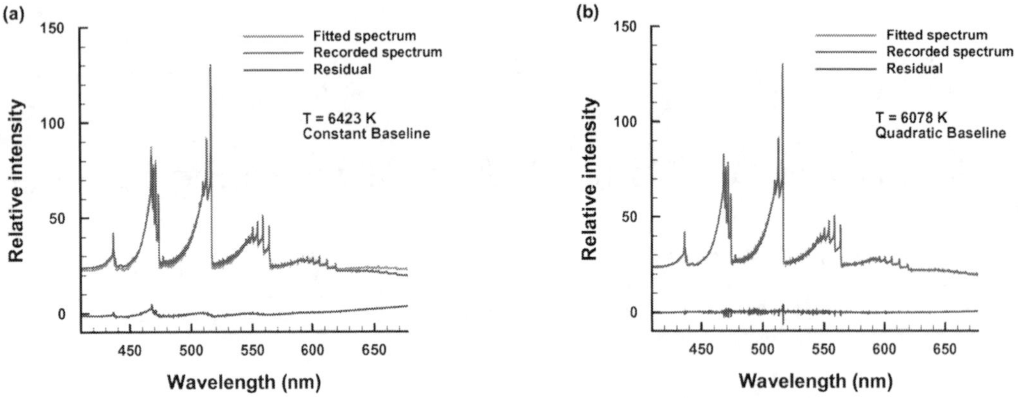

Figure 1. A recorded C_2 Swan spectrum [6] compared with two computed Swan spectra containing (a) a small constant and (b) a quadratic baseline. The temperature and baseline values were obtained by trial and error implemented with the Nelder-Mead extremum algorithm by varying the temperature for here a thermal distribution of excited states.

- The Hönl-London factors $S_{J'J}$ are computed [7],

$$S_{J'J} = (2J+1)\left|\sum_{a'}\sum_{a}\langle J\,|a\rangle\,\langle a|C^{J'\Omega'}_{J\Omega q\kappa}|a'\rangle\,\langle a'\,|J'\rangle\right|^2 \tag{3}$$

  and the term difference, *i.e.*, line position, is computed for each non-vanishing Hönl-London factor,

$$\tilde{\nu} = F' - F \qquad \text{if } S_{J'J} \neq 0. \tag{4}$$

- A line list which includes line strengths is created by repeating the above for all required J' and J.
- A thermal distribution of excited states and a Gaussian spectrometer slit function having a specified full-width at half-maximum (FWHM) are assumed.
- The spectrum is separated into a number of pixels, and the the contribution of each line in the computed line list to each pixel is computed.

Use of the above algorithm requires the Wigner-Witmer diatomic eigenfunction,

$$\langle \mathbf{r}_1, \mathbf{r}_2, \ldots, \mathbf{r}_N, \mathbf{r}\,|nvJM\rangle = \sum_{\Omega=-J}^{J} \langle \rho, \zeta, \mathbf{r}'_2, \ldots, \mathbf{r}'_N, r\,|nv\rangle\, D^{J*}_{M\Omega}(\phi, \theta, \chi). \tag{5}$$

in which the total angular momentum states are *exactly* separated instead of a Born-Oppenheimer eigenfunction in which segregation of electronic and nuclear coordinates is enforced thereby preventing separation of the total angular momentum. The quantum numbers in Eq. (5) are the vibrational quantum number v, the total angular momentum quantum number J, the magnetic quantum number M for the z-component of $\mathbf{J}$, the magnetic quantum number Ω for the z' component of $\mathbf{J}$, and n which is a symbol representing all other required quantum numbers and labels for continuum indices. In Eq. (5), primed and unprimed coordinates are related by coordinate rotation,

$$\begin{bmatrix} x' \\ y' \\ z' \end{bmatrix} = \mathcal{D}(\phi,\theta,\chi) \begin{bmatrix} x \\ y \\ z \end{bmatrix}, \tag{6}$$

The Euler angles ϕ, θ, and χ and the coordinate rotation matrix $\mathcal{D}(\phi,\theta,\chi)$ used here,

$$\mathcal{D}(\phi,\theta,\chi) = \begin{bmatrix} \cos\phi\cos\theta\cos\chi - \sin\phi\sin\chi & \sin\phi\cos\theta\cos\chi + \cos\phi\sin\chi & -\sin\theta\cos\chi \\ -\cos\phi\cos\theta\sin\chi - \sin\phi\cos\chi & -\sin\phi\cos\theta\sin\chi + \cos\phi\cos\chi & \sin\theta\sin\chi \\ \cos\phi\sin\theta & \sin\phi\sin\theta & \cos\theta \end{bmatrix}. \tag{7}$$

are those used almost universally in the quantum theory of angular momentum[8, 9, 10, 11, 12, 13, 14]. Kovács[15] uses a different set of Euler angles. Primes do not appear on ρ (the distance of one of the electrons from the internuclear axis), ζ (this distance of this electron above or below the plane perpendicular to the internuclear axis and passing through the origin at the center of mass of the nuclei), and r (the internuclear distance) because they are scalars whose value is the same in all coordinate systems. The polar coordinates r, θ, and ϕ are those of the fictitious particle of reduced mass of the two nuclei who motion has replaced their motion. There are N electrons and their coordinates are labeled 1, 2, ..., N. The cylindrical coordinates ρ, χ, and ζ are those of one of the electrons. In accord with standard practice in angular momentum theory, the quantum numbers J, M, and Ω refer to the *total* angular momentum. The spectroscopic quantum numbers F, M_F, and Ω_F replace J, M, and Ω in Eq. (5) when it is written in spectroscopic notation.

In the Wigner-Witmer diatomic eigenfunction the Euler angles ϕ, θ, and χ are both the parameters of coordinate rotation and angles of physical rotation. The diatomic molecule is perhaps the most complicated actual system in which a single set of Euler angles can serve both purposes. For example, in polyatomic theory the Euler angles ϕ, θ, and χ are physical rotations describing a frame to which the nuclei are attached, and are not the parameters of coordinate rotation α, β, and γ one would use to demonstrate rotational invariance and conservation of the total angular momentum in a polyatomic model which included vibrational angular momentum.

The Wigner-Eckart theorem breaks the transition moment into two parts, the so-called reduced matrix element whose value is controlled by the initial and final total angular quantum numbers and the degree of transition tensor operator, and Clebsch-Gordan coefficient whose value is controlled by the magnetic quantum numbers and the indices of the components of the tensor operator. The exact separation of the total angular momentum in the Wigner-Witmer eigenfunction simplifies and improves the accuracy of computation of the Hönl-London factors. Line positions $\tilde{\nu}$ and Hönl-London factors are normally independently evaluated. In the present algorithm, they are simultaneously computed. Of the myriad upper and lower term differences, only those for which the Hönl-London factor is non-vanishing become spectral lines.

The Born-Oppenheimer approximation is not totally eliminated from computation of a diatomic spectrum. It is needed to break the electronic-vibrational eigenfunction into the product of electronic and vibrational eigenfunction. The Wigner-Witmer diatomic eigenfunction exactly separates the diatomic line strength into the product of the electronic-vibrational strength $S_{n'v',nv}$ and the Hönl-London factor $S_{J',J}$,

$$S_{n'v'J',nvJ} = S_{n'v',nv}\, S_{J',J}, \tag{8}$$

$$S_{n'v',nv} = \left| a_0 + a_1 \bar{r}_{v'v} + a_2 \bar{r}^2_{v'v} + \ldots \right|^2 q_{v'v}. \tag{9}$$

The Born-Oppenheimer approximation separates the electronic-vibrational strength into the product of the square of the electronic transition moment,

$$R_e(r) = a_0 + a_1 r + a_2 r^2 + \ldots, \tag{10}$$

evaluated at the r-centroids $\bar{r}^n_{v'v}$,

$$\bar{r}^n_{v'v} = \frac{\langle v' | r^n | v \rangle}{\langle v' | v \rangle}, \tag{11}$$

and with the Franck-Condon factors $q_{v'v}$,

$$q_{v'v} = \langle v' | v \rangle^2. \tag{12}$$

Whereas experimental and calculated line positions are often known with an accuracy of $1 : 10^6$ or better, spectral line intensity is rarely recorded with an accuracy of better than 1%. Thus, use of the Born-Oppenheimer approximation for calculation of the electronic-vibrational strength, but not for calculation of line positions, does not introduce significant error.

3. Determination of the molecular parameters

Implementation of the algorithm described above requires a comprehensive set of accurate molecular parameters. This is not a minor proviso, and the situation is further complicated by the fact the use of the Wigner-Witmer eigenfunction modifies the manner in which diatomic parameters are determined from a accurate line position measurements such as those provided by Fourier transform spectroscopy. The basic idea in both current practice [16, 17] and our determination of diatomic parameters is the calculation of matrix elements of the form

$$\langle a_i | H_k | a_j \rangle = \langle n_i v_i J_i M_i \Omega_i \Lambda_i S_i \Sigma_i | H_k | n_j v_j J_j M_j \Omega_j \Lambda_j S_j \Sigma_j \rangle \tag{13}$$

in which $|a\rangle$ is a Hund's case a basis function,

$$|a\rangle = |nvJM\Omega\Lambda S\Sigma\rangle = \sqrt{\frac{2J+1}{8\pi^2}} \langle \rho, \zeta, \mathbf{r}'_2, \ldots, \mathbf{r}'_N, r | nv \rangle \, D^{J*}_{M\Omega}(\phi, \theta, \chi) \, |S\Sigma\rangle \tag{14}$$

and H_k is a term from the diatomic Hamiltonian

$$H = \sum_k H_k. \tag{15}$$

Use of the Wigner-Witmer eigenfunction breaks required matrix elements of the Hamiltonian into two parts, one the angular momentum part which can be calculated exactly and the electronic-vibrational part which, except for the very simplest molecules, cannot be calculated with spectroscopic accuracy. Use of a Born-Oppenheimer eigenfunction breaks calculation into three parts, an electronic part, a part consisting of a sum over an infinite number of Born-Oppenheimer vibrational states, and a rotational part. In general, none of the three can be done exactly. Van Vleck transformations, parity partitioning, and the concept of an "effective Hamiltonian" reduce the dimensions of the Hamiltonian matrix. We simply compute a Hamiltonian matrix composed of case a matrix elements, and let the dimension of the matrix be determined by the range of Ω in Eq. (5) be that required make computed line positions $\tilde{nu}_{\text{cal}}$ equal the experimental positions $\tilde{\nu}_{\text{exp}}$ to within the estimated accuracy of the $\tilde{\nu}_{\text{exp}}$. With the exceptions (1) our Hamiltonian of unmodified case a matrix elements replaces the effective Hamiltonian used by others and (2) instead of using coded selection rules we use non-vanishing of the Hönl-London factor as the only selection rule, our determination of molecular parameters is identical to that described by

[18]. Trial values of upper and lower parameters are assumed, the line positions are computed, corrections to the parameters are computed from the errors in the computed line positions,

$$\Delta\tilde{\nu} = \tilde{\nu}_{\mathrm{cal}} - \tilde{\nu}_{\mathrm{exp}}, \tag{16}$$

and the process is repeated until the corrections to the parameters become negligibly small. The algorithm for finding molecular parameters by fitting computed line positions to measured line positions is described in more detail in [19].

4. Discussion

The Wigner-Witmer paper[20] appeared about a year after the Born-Oppenheimer paper[21]. The Born and Oppenheimer work treats all molecules, but the Wigner-Witmer paper is strictly limited to the diatomic molecule. Although Born's formulation[21, 22] in terms of

$$\kappa = \left(\frac{m}{M_0}\right)^{1/4} \tag{17}$$

in which m is the electronic mass and "where M_0 can taken as any one of the nuclear masses or their mean"[22] is rarely used, the Born-Oppenheimer approximation became the foundation of molecular theory. Although the Born-Oppenheimer approximation is applicable to all molecules, in their final section Born and Oppenheimer show that the polar angle and azimuthal angle of the polar coordinates r, θ, and ϕ of the internuclear vector in a diatomic molecule are exactly separable in the spherical harmonic $Y_{\ell m}(\theta, \phi)$. The Wigner complex conjugate of the D-symbol, $D^{j*}_{m\omega}(\phi, \theta, \chi)$ is the mathematical extension of the spherical harmonics,

$$Y_{\ell m}(\theta, \phi) = \sqrt{\frac{2\ell+1}{4\pi}}\, D^{\ell *}_{m\,0}(\phi, \theta, 0) \tag{18}$$

which allows one to deal with the total angular momentum $|jm\rangle$ rather that just the orbital angular momentum $|\ell m\rangle$. The first half of the Wigner-Witmer paper is the logical extension of the Born-Oppenheimer $Y_{\ell m}(\theta, \phi)$ result to include all three Euler angles in diatomic eigenfunction. The two-part Wigner-Witmer paper became famous for its second part which gives correlation rules relating the electronic states of a diatomic molecule to the $L-S$ coupled states of the separated atoms. Oddly, the exact diatomic eigenfunction with which Wigner and Witmer determined their correlation rules has been ignored. About 40 years passed between publication the Wigner-Witmer paper and entry of the Wigner D-function into the literature of diatomic theory. By then, the Wigner-Witmer diatomic eigenfunction had apparently been forgotten.

References

[1] Musazzi S and Perini U, eds., 2014 *Laser-Induced Breakdown Spectroscopy* (Heidelberg: Springer)
[2] Cremers D A and Radziemski L J 2006 *Handbook of Laser-Induced Breakdown Spectoscopy* (West Sussex: Wiley)
[3] Noll R 2011 *Laser-Induced Breakdown Spectroscopy* (Heidelberg: Springer)
[4] Singh J P and S. N. Thakur, eds. 2007 *Laser-Induced Breakdown Spectroscopy* (Amsterdam: Elsevier)
[5] Miziolek A W, Palleschi V, and Schechter I, eds., 2006 *Laser-Induced Breakdown Spectroscopy (LIBS)* (Cambridge: Cambridge University Press)
[6] Nemes L, Keszler A M, Hornkohl J O, Parigger C G 2005 *Appl. Opt.* **44** 3661
[7] Hornkohl J O, Parigger C G, Nemes L 2005 *Appl. Opt.* **44** 3686
[8] Rose M E 1995 *Elementary Theory of Angular Momentum* (Mineola, NY: Dover)
[9] Edmonds A R 1974 *Angular Momentum in Quantum Mechanics, 2nd ed.* (Princeton, NJ: Princeton University Press)

[10] Brink D M and Satchler G R 1993 *Angular Momentum, 3rd ed.* (Oxford: Oxford University Press)
[11] Biedenharn L C and Louck J D 2009 *Angular Momentum in Quantum Physics* (Cambridge: Cambridge University Press)
[12] Zare R N 1988 *Angular Momentum* (New York: Wiley)
[13] Thompson W J 1994 *Angular Momentum* (New York: Wiley)
[14] Varshalovich D A, Moskalev A N, and Khersonskii V K 1988 *Quantum Theory of Angular Momentum* (Singapore: World Scientific)
[15] Kovács I 1969 *Rotational Structure in the Spectra of Diatomic Molecules* (New York: Elsevier)
[16] Lefebvre-Brion H and Field R W 2004 *The Spectra and Dynamics of Diatomic Molecules* (New York: Elsevier/Academic)
[17] Brown J M and Carrington A 2003 *Rotational Spectroscopy of Diatomic Molecules* (Cambridge: Cambridge University Press)
[18] Zare R, Schmeltekopf A, Harrop W, and Albritton D 1973 *J. Mol. Spectrosc.* **46** 37
[19] Nemes L and Irle S, eds., 2011 *Spectroscopy, Dynamics and Molecular Theory of Carbon Plasmas and Vapors* (New Jersey: World Scientific)
[20] Wigner E and Witmer E E 1928 *Z. Phys.* **51** 859; Hettema H 2000 *Quantum Chemistry: Classic Scientfic Papers* (Singapore: World Scientific) 287
[21] Born M and Oppenheimer R 1927 *Ann. Phys.* **84** 457; Hettema H 2000 *Quantum Chemistry: Classic Scientfic Papers* (Singapore: World Scientific) 1
[22] Born M and Huang K 1954 *Dynamical Theory of Crystal Lattices* (Oxford: Clarendon)

Journal of Physics: Conference Series **548** (2014) 012034 doi:10.1088/1742-6596/548/1/012034

Line Shape Measurement and Modelling for Plasma Diagnostics

G Revalde[1], N Zorina[2] and A Skudra[2]

[1]Institute of Technical Physics, Riga Technical University, Āzenes str. 14/24, Riga, Latvia

[1]Ventspils University College, Inženieru str.101, Ventspils, Latvia

[2]Institute of Atomic Physics and Spectroscopy, University of Latvia, Skunu str. 4, Rīga, Latvia, LV 1050

gitar@latnet.lv

Abstract. In this paper we discuss different methods of narrow spectral line shape measurements for a wide spectral range by means of high-resolution spectrometers such as the Fabry-Perot spectrometer, Zeeman spectrometer and Fourier transform spectrometer as well as a theoretical model for spectral line shape modelling and solving of the inverse task based on Tikhonov's regularization method. Special attention is devoted to the line shape measurements for the optically thin light sources filled with Hg, Ar, Xe, Kr for their use in high precision analysers for detection of heavy metals and benzene.

1. Introduction

Line shape diagnostics are essential when light sources with well-defined shapes of the emission lines are required. In this paper we present the spectral line shape measurement and calculation methods created for optimisation needs for special types of low pressure lamps for their application in atomic absorption spectrometers (AAS). Novel type AA analysers are developed for detection of heavy metals, benzene, toluene and other pollutants in air, water and food in real time [1].High selectivity and very low limits of detection depend on the quality of the spectral line shape.

Spectral line shapes are known as important tools for emission plasma diagnostics in different type of plasma since the form of the line is determined by all plasma processes. However many processes act at the same time, and it is not easy to resolve partial effects on the total line shape. In addition, the processes often correlate with each other. We have to take into account also the influence of spectral apparatus. Reconstructing the real line shape from the measured one is a so called inverse ill-posed task since small uncertainties in the measurement give large deviations in solution. Since it is a complex task, sometimes consideration of the instrument function is omitted. It can be done if the width of the instrument function is much smaller than the real spectral line shape which is true in general for high temperatures and dense plasmas. However in the case of low–pressure plasma the instrument function is on the same order as experimental profile and it has to be taken into account.

In this work we will focus on the deconvolution of narrow real spectral line shapes from experimental profiles registered by different spectrometers with high precision. The influence of the

Journal of Physics: Conference Series **548** (2014) 012034 doi:10.1088/1742-6596/548/1/012034

instrument functions on Hg emission lines from low-temperature discharges will be discussed for Fabry-Perot, Zeeman and Fourier transform spectrometers. The influence of the instrument function on the FWHM of spectral line shapes and gas temperature estimation will be elaborated.

2. Experimental details

The developed light sources contain reservoir with working element, capillary and excitation electrodes located outside the capillary and exciting the discharge at 100 MHz frequency. The discharge can be excited also in the bulb of the lamp. Set-up of capillary light source is shown in Fig.1. In this work, experiments were conducted with mercury light sources with different buffer gases Ar, Kr and Xe. To avoid the splitting and consequently the broadening of the spectral line, the isotopes of the metal are used as filling elements.

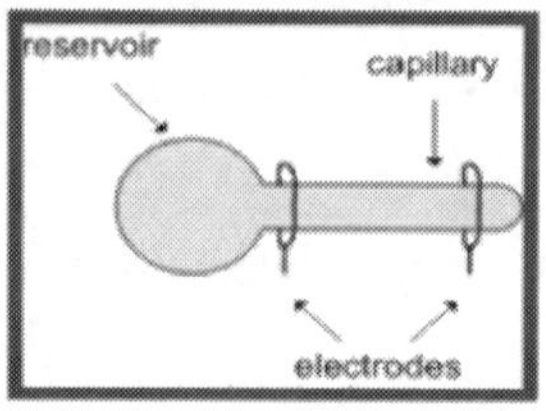

Figure 1. Set-up of a capillary light source

2.1. Fabry-Perot Interferometer

Since the working lines, selected for the needs of AAS, are narrow and often in the UV, the registration instrument should have high sensitivity in broad spectral region. Direct spectral line shape's measurements were done by three different methods in this work. In Fig. 2, the experimental arrangement of a high resolution scanning Fabry-Perot spectrometer is illustrated. The heart of this experimental arrangement is the Fabry-Perot interferometer, consisting of two parallel mirrors, placed into a vacuum chamber which is connected with the pumping system and evacuated. When the air is gradually let to refill the vacuum chamber the refraction index of the mirrors is changing and the spectrum lines are registered by a PMT. The instrument function of the Fabry-Perot interferometer is so called Airy function which can be approximated also by Voigt profile [2].

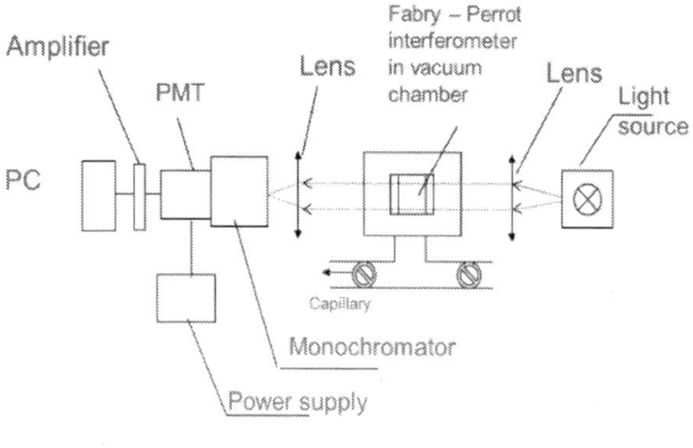

Figure 2. Set-up of a Fabry Perot high resolution spectrometer.

The finesse of the interferometer is restricted by the effective refraction coefficient of the mirrors which includes both, the refraction coefficient and inaccuracy of the adjustment. For example at the effective refraction coefficient R = 80%, the full width at half of maximum (FWHM) of the instrument function $\Delta\nu_{instu}$ amounts to 0.071 cm^{-1}.

2.2. Zeeman Spectrometer

Since The principle set-up of the scanning Zeeman spectrometer is shown in Fig. 3(a). This type of spectrometer is very useful to measure the resonance line shapes in UV region [3, 4]. Examples of the measured Hg VUV resonance line of 185 nm wavelength are shown in Fig.3 (b). By our knowledge this is the only available spectrometer which makes possible direct registration of the Hg 185 nm spectral line shape. The light from the plasma source traverses the same metal vapour cell. The frequency scanning is ensured by changing the magnetic field strength in the absorption cell. The signal is registered by the PMT.

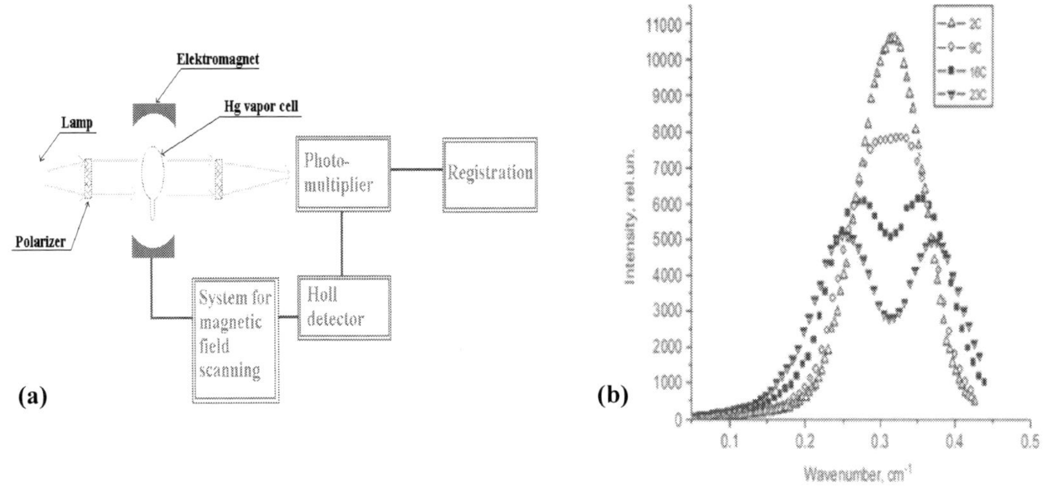

Figure 3. (a) Set-up of Zeeman high resolution spectrometer for the line shape measurements based on absorption of light in the cell placed in magnetic field; **(b)** experimentally measured Hg resonance line shape of 185 nm emitted from the Hg isotope 198 lamp in dependence of the Hg cold spot temperature (metal vapour pressure) [4].

For the Zeeman spectrometer, the instrument function is absorption profile of the metal vapor in the absorption cell. If the cell is hold at the room temperature of 20° C, the Doppler broadened absorption profile has FWHM of 0.036 cm^{-1} which gives better resolution than for the Fabry-Perot interferometer.

2.3. Fourier Transform Spectrometer

The Fourier Transform Spectrometer model Bruker IFS 125 HR can be used to measure spectral line shapes in wide spectral range from 320 - 910 nm, about 2000 pixels/nm. The theoretical instrument function of the Fourier spectrometer is function sinc(x) = sin(x)/x shown in Fig.4. However the measured function using the He laser line of wavelength 632.8 nm, shown in Fig.5 looks differently.

Since the reconstruction takes into account the theoretical or measured prompt function and since we encountered issues with finding stable solutions, we first applied a fitting procedure. Subsequently, we found that the best approximation of the instrument function is a Lorentz profile with FWHM of 0.02 cm^{-1} [5]. Thus all further calculations were performed with this function. Experimentally measured spectral line shapes of 435.8 nm for two Hg isotope lamps, 202 and 198 isotopes, respectively, depend on the discharge power. Fig. 6 displays these results.

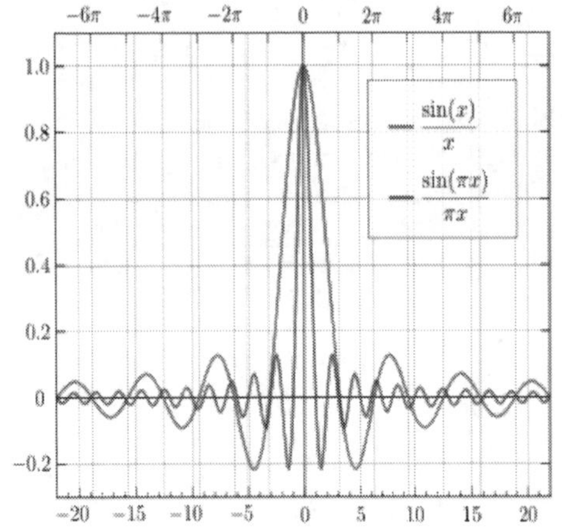

Figure 4. Theoretical instrument function of Fourier transforms spectrometer.

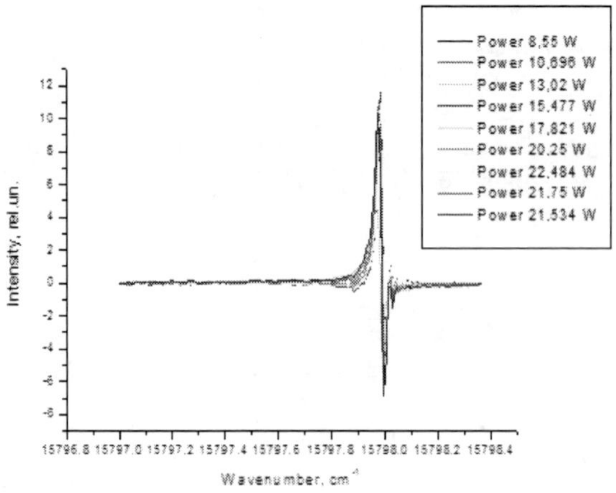

Figure 5. Experimentally measured Fourier transform spectrometer instrument function.

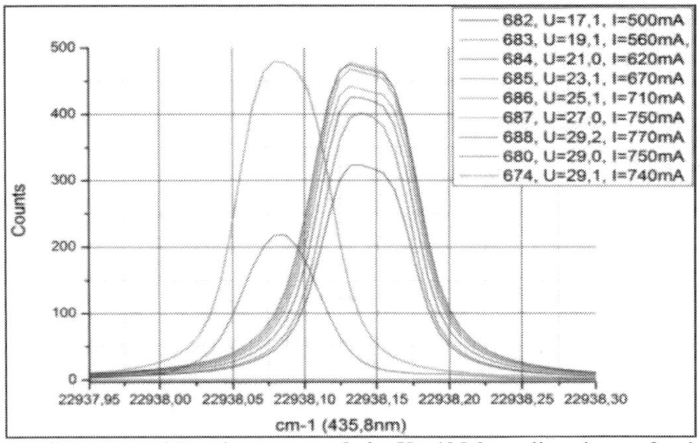

Figure 6. Experimentally measured Fourier spectra of the Hg 435.8 nm line shapes for the two Hg isotopes 198 and 202.

3. Theoretical Approach

As is well known, the experimentally registered spectral line shape, influenced by the instrument function, can be expressed in form of a Fredholm integral equation of first kind:

$$\int_a^b A(x,s)y(s)ds = f(x), \quad c \le x \le d, \quad (1)$$

where *f(x)*- measured spectral line profile; *y(s)*- real spectral line profile; *A(x,s)*-instrument function and a, b and c, d -the limits of the real and experimental profiles, accordingly.

To calculate the real spectral line profile from experimental line, it is necessary to solve this inverse ill-posed task where small experimental uncertainties can cause large deviations in the solution. Our experience showed that the Tikhonov's regularisation algorithm [6] is one of the most useful tools for solving the task.

Assuming, that the values on the right side and integral equation (1) kernel are known with accuracy:

$$\left.\begin{aligned} \left\|\tilde{f}-f\right\|_F &\le \delta; \\ \left\|\tilde{A}-A\right\| &\le \xi \end{aligned}\right\} \qquad (2)$$

where δ is the error of right part of (1) or the error of experimental $f(x)$ profile, ξ is the error of kernel of (1) or the error of instrument function $A(x,s)$, Tikhonov proved, that the initial, ill-posed task can be transformed into a task of searching for the minimum of a smoothing functional:

$$M_\alpha\left[y,\tilde{f}\right]=\inf_{y\in Y} M_\alpha\left[y,\tilde{f}\right], \qquad (3)$$

where the smoothing functional $M_\alpha\left[y,\tilde{f}\right]$, so called Tikhonov's functional, is given in the form:

$$M_\alpha\left[y,\tilde{f}\right]=\left\|\tilde{A}y-\tilde{f}\right\|_F^2+\alpha\Omega[y]. \qquad (4)$$

The stabilising functional Ω is described by the following expression:

$$\Omega[y]=\|y\|_Y^2 \qquad (5)$$

$\alpha>0$ is the regularisation's parameter; the number discrepancy is denoted by $\left\|\tilde{A}y-\tilde{f}\right\|_F^2$.

According to Tikhonov's method, instead of the initial ill posed problem (1) we get well posed task, which is described by the Fredholm integral equation of second kind. After changing the integrals to finite sums, one can write the system of linear equations are

$$\sum_{i=1}^{N} b_i\left(\sum_{m=1}^{N} b_m \tilde{A}_{mk}\tilde{A}_{mi}\right)y_i+\alpha y_k=\sum_{m=1}^{N} b_m\tilde{A}_{mk}\tilde{f}_m, \quad k=\overline{1,N} \qquad (6)$$

where A_{ij} – elements of N xN size matrix A, which approximates kernel; f_i – vectors-column with initial dates; y_i – vector-column of solution; b_i – the coefficients.

The system of Equations (6) can be solved with classical methods. All calculations in this work were done for zero order of regularization (q=0) using programs written in MathCad. Regularisation parameter α was obtained in supposition, that the error of kernel is equal to zero (ξ=0) in cases when the instrumental function is known analytically. Typically we use two independent methods to obtain the regularisation parameter α [7,8].

However the deconvoluted real spectral line shape still needs to be modelled because we want to obtain plasma parameters. In the low temperature plasma the real spectral line shape can be approximated Voigt profile which is a convolution of the Doppler and Lorentz profiles. Normally, the Doppler width (FWHM), $\delta\nu_D$, and Lorentz width, $\delta\nu_L$, are used as the parameters characterizing the Voigt profile. Voigt profile includes the basic factors causing the spectral line broadening in a low-pressure discharge: natural, temperature, collisional. The main problem is to take into account the self-absorption. Spectral line modelling is further described in [2-4]. Along the line of sight the intensity distribution in the line shape $I(\nu)$ can be written as:

$$I(\nu) = I_0 \int_{-\infty}^{+\infty} \bar{n}_e(r) P_e(\nu, r) \exp\left[-s \int_r^{\infty} P_a(\nu, x) \bar{n}_a(x) dx \right] dr$$
$$\bar{n}_a(r) = \frac{n_a(r)}{N_a}; \quad \bar{n}_e = \frac{n_e(r)}{N_e} \qquad (7),$$

where

$$N_a = \frac{1}{2} \int_{-\infty}^{+\infty} n_a(r) dr; \qquad N_e = \int_{-\infty}^{+\infty} n_e(r) dr \,.$$

It is important to know the distributions of the absorbing and emitting particles $n_a(r)$ and $n_e(r)$. For our calculations we use several approximations: uniform distribution of emitting and absorbing particles (homogenous light source), light source with spatially resolved emitting and absorbing particles (inhomogeneous light source), experimentally determined distributions as well as model supposed by Cowan and Dicke which gives some kind of intermediate distributions of the excitation function of the light source [9].

4. Results and discussion

Mercury 202 isotope 253.7 nm line shapes, recorded with a Zeeman spectrometer and the Fabry-Perot spectrometer are shown in Fig. 7.

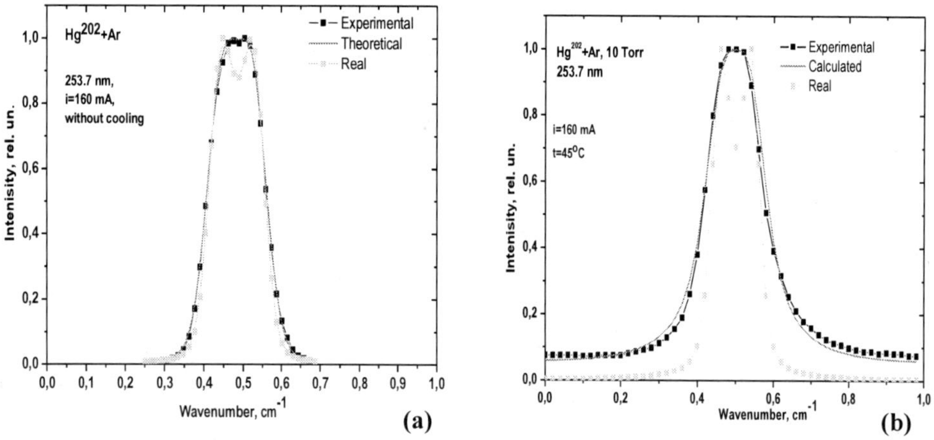

Figure 7. Comparison of mercury 202 isotope 253,7 nm line shapes, registered with Zeeman **(a)** and Fabry-Perot spectrometers **(b)** and the calculated real spectral line (green squares).

As we can observe, in both cases the dip due to the self-absorption is covered by the instrument functions of the spectrometers despite the fact that the instrument functions in both cases are narrow (the broadening is on the same order as the line shape itself). The Fabry-Perot spectrometer gives more influence on the wings of the shape. For the calculation of the real profile, the model of the inhomogeneous light source excitation function was applied. In Fig. 8, the results of deconvolution procedure are shown for Hg 546.1 nm visible triplet line.

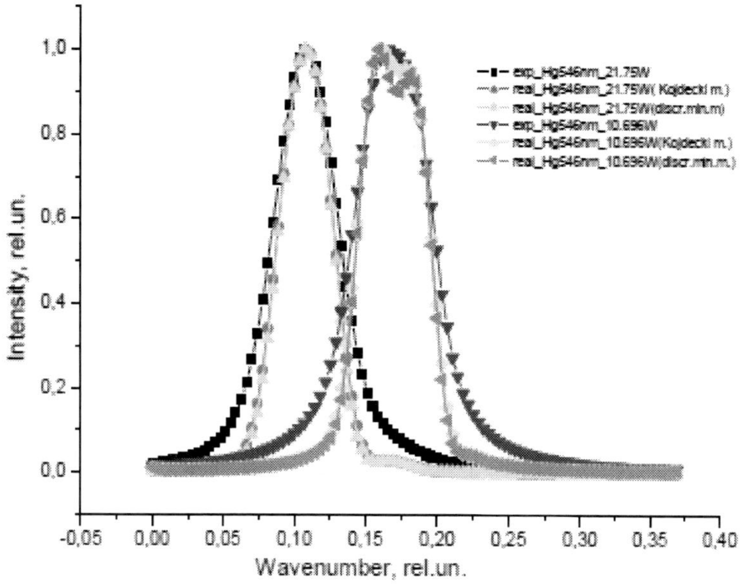

Figure 8. Experimental Fourier spectra of the Hg 546.1 nm line and deconvoluted shapes for the two Hg isotope lamps, 202 and 198.

The results are given for two methods of finding the regularization parameter α, the Kojdecki method [7] and minimum of discrepancy method [8]. Both methods give similar results. It is interesting to observe that the line, emitted from the 198 isotope lamp, is self-absorbed because the lamp contains higher concentration of mercury.

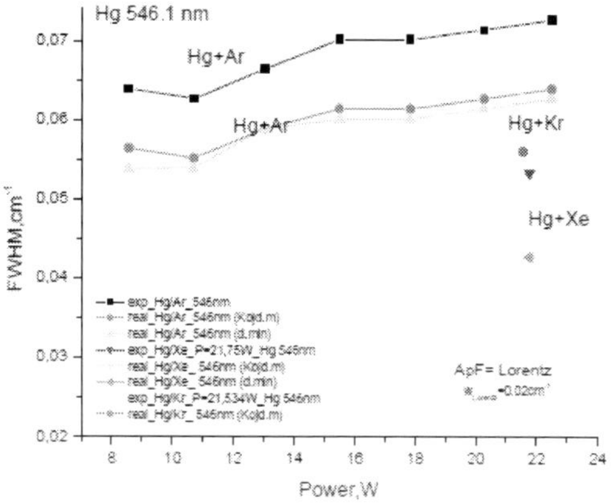

Figure 9. FWHM in dependence from the discharge power, calculated from the deconvoluted line shapes of Hg line 546.1 nm emitted from the Hg/Ar, Hg/Kr, Hg/Xe discharge plasmas.

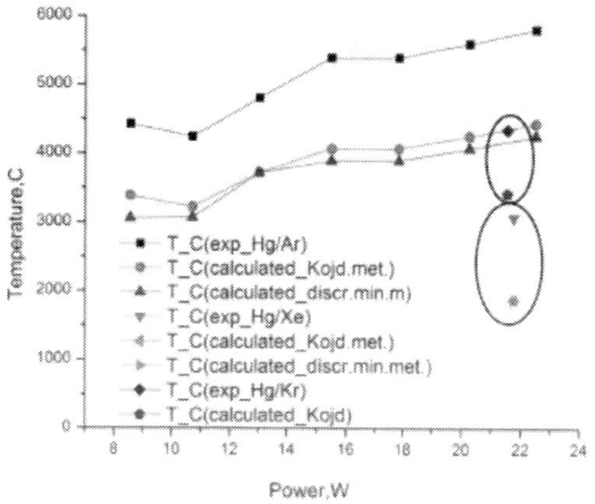

Figure 10. FWHM in dependence from the discharge power, calculated from the deconvoluted line shapes of 435.8 nm Hg line emitted from the Hg/Ar, Hg/Kr, Hg/Xe discharge plasmas.

We can obtain the real FWHM for all the registered lines that allows us to determine temperature of the emitting particles. Examples of the results of the calculated FWHM of the 546.1 nm line and temperatures (T) are shown in the Figs. 9 and 9. It is clearly demonstrated that the neglecting instrument function can lead to large error of about 1500° C in the temperature calculations. Clear dependence of the temperature of the type of buffer gas can be observed – the heavier the buffer gas, the colder the plasma. Adding Kr instead of Ar decreases temperature of about 900° C, adding Xe instead of Ar decreases temperature about 2000° C.

5. Conclusion

Narrow spectral line shapes, emitted from the light sources, developed for the atomic absorption spectrometry for detection of heavy metals and benzene, are analysed by three different measurement methods in UV and visible spectral regions. The regularisation method and mathematical modelling is used to extract the real spectral line shape. Neglecting the instruments function gives an error of about 1500° C for the discharge temperature. Clear influence of buffer gas type was found on the mercury discharge temperature.

Acknowledgement

This research has been partly supported by the European Social Fund within the project "Elaboration of Innovative Functional Materials and Nanomaterials for Application in Environment Control Technologies" 1DP/1.1.1.2.0/13/APIA/VIAA/30.

References

[1] Ganeev A, Gavare Z, Khutorshikov V I, Khutorchikov S V,Revalde G, Skudra A, Smirnova GM, Stankov N R 2003 *Spectrochim. Acta Part B* **58** 879
[2] Skudra A, Revalde G 1999 *J. Quant. Spectrosc. Rad. Transfer* **61** 717
[3] Revalde G, Denisova N, Gavare Z, Skudra A 2005 *J. Quant. Spectrosc. Rad. Transfer* **94** 311
[4] Revalde G, Skudra A, Zorina N, Sholupov S 2007 *J. Quant. Spectrosc. Rad. Transfer* **107** 164
[5] Revalde G, Zorina N, Skudra A, Gavare Z 2014 *Romanian Reports in Physics* **66** (in press: *N4*)
[6] Tikhonov A, Arsenin V 1979 *The solution's methods of the ill-posed problem* (Moscow: Nauka)
[7] Verlan F, Sizikov V S 1986 The integral equations: the methods, algorithms, program (Kiev: Naukova Dumka)
[8] Kojdecki M F 2000 New criterion of regularization parameter choice in Tikhonov's method *Biuletyn WAT* (Biul. Mil. Univ. Technol.) **49** № 1 (569) 47
[9] Cowan R D, Dieke G H 1948 *Rev. Mod. Phys.* **2** 418

Application of Saturated Absorption Spectroscopy to Plasma Diagnostics

S Nishiyama[1], M Goto[2], H Wang[1], and K. Sasaki[1]

[1]Graduate school of Engineering, Hokkaido University, Sapporo 060-8628, Japan

[2]National Institute for Fusion Science, Toki 509-5292, Japan

shu@eng.hokudai.ac.jp

Abstract. In this work, we utilized saturation spectroscopy to the Balmer-alpha line of atomic hydrogen in a linear magnetized plasma source with the intention of applying it to plasma diagnostics. The fine-structure of the Balmer-alpha line was observed clearly in weak magnetic field strength. Most of the peaks are assigned as transitions of the fine-structure of the Balmer-alpha line and its cross-over peaks. However, irregular cross-over peaks also observed and the population exchange among the 2s and the 2p states was suggested. Saturation spectra with Zeeman-splitting were also observed, however, the spectra became complicated with many small peaks.

1. Introduction

Saturation spectroscopy is a kind of laser absorption spectroscopy, and it achieves a Doppler-free spectral resolution. This technique is widely used in the field of fundamental spectroscopy for the precise determination of the wavelengths of transitions [1,2]. However, there are a few works of application of saturation spectroscopy to plasma diagnostics [3-5].

In this work, we developed a saturation spectroscopy system to investigate the basic properties of saturation spectrum at the Balmer-alpha line of atomic hydrogen in magnetized plasmas. This study intends to be applied to an analysis of spatial distribution of ionization in the Large Helical Device at the National Institute for Fusion Science. The Doppler-free spectral resolution is required for Zeeman-splitting analysis of high temperature hydrogen plasmas because the wider Doppler broadening masks the complex fine structure of the Zeeman-splitting spectrum [6], which carries information of the spatial distribution of ionization in the confinement magnetic field of helical devices [7]. In the following, we report the basic properties of the saturation spectrum of Balmer-alpha line with and without Zeeman-splitting in various conditions.

2. Experimental setup

The experimental apparatus is schematically shown in Figure 1. The plasma source was linear machine with a uniform magnetic field along the axis. The helical antenna was wound around a 1.6 cm inner diameter and 30 cm length quartz glass tube. The antenna was connected to an RF power supply at 13.56MHz via a matching circuit. One end of the quartz glass tube was connected to a diffusion chamber and the overall length of the plasma was 60 cm. The diffusion chamber was

Journal of Physics: Conference Series **548** (2014) 012035 doi:10.1088/1742-6596/548/1/012035

evacuated by a turbo-molecular pump and was fed hydrogen gas via a mass-flow controller. The hydrogen gas pressure in the chamber was settled to 6.7 to 53 Pa (50 to 400 mTorr).

The light source for saturation spectroscopy was a tunable CW diode laser (New Focus Vortex II). The frequency of the laser was scanned over the range of 120 GHz in 25 ms. This frequency range was wide enough to cover the Doppler broadened Balmer-alpha spectrum. The main part of the laser beam was amplified by a diode laser amplifier, TOPTICA BoosTA, as a pump beam. The power of the pump beam was 110 mW at the optical window of the diffusion chamber. A half-wavelength plate was used to adjust the direction of polarization perpendicular to one of the probe beams to reduce influences of scattered pump laser light. The other part of the laser beam was divided into two laser beams, a probe beam and a reference beam. The power of the probe beam was 220 μW at the end of the quartz discharge tube. These beams passed axially through the plasma column. The probe beam and the pump beam were injected into the plasma from opposite directions and were overlapped with precision to obtain clear saturation signal. The reference beam also passed through the plasma column parallel to the probe beam, separated by the distance of 8 mm. It was used for compensation of temporal fluctuation of plasma density. The transmitted intensity of the probe beam and the reference beam were detected using avalanche photo diodes. A part of the master oscillator beam was picked up to measure the frequency scan using a Fabry-Pérot spectrum analyser. The signals from the avalanche photo diodes and spectrum analyser were recorded using a digital oscilloscope.

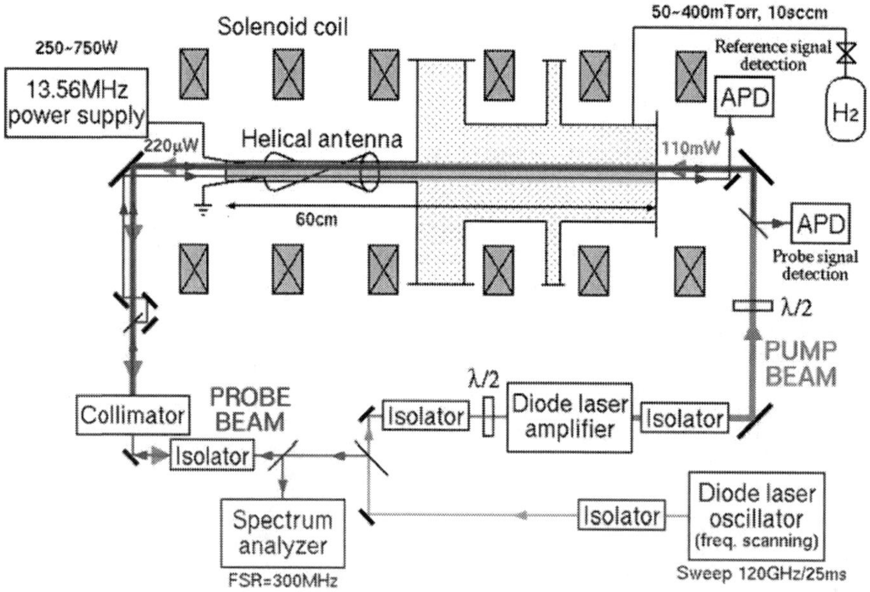

Figure 1. Experimental apparatus.

3. Results and discussion

3.1. Saturation spectrum with Zeeman-splitting

A typical result of absorption spectra with and without the pump laser beam is shown in Figure 2. The hydrogen gas pressure and applied RF power were 13 Pa (100 mTorr) and 750 W, respectively. The horizontal axis shows the frequency deviation from 656.2819 nm. The vertical axis shows the absorbance αl which is obtained by the Lambert-Beer law αl = -ln(I_t / I_0) where I_t and I_0 are the transmitted and the incident probe beam intensities, respectively. In Figure 2, the outer smooth curve with two broad peaks shows the absorbance without the pump laser beam. It shows good agreement with a spectrum calculated using theoretical transition coefficients of each fine-structure component of the Balmer-alpha line and assumed Doppler broadening at 500 K temperature. With the pump laser beam, the absorbance spectrum shown in Figure 2 has many sharp dips caused by saturation of absorption with the pump beam. The absorbance difference $\Delta\alpha l = \alpha_s l - \alpha_0 l$, where $\alpha_s l$ and $\alpha_0 l$ are the absorbance with and without pump beam, respectively, shows fine resolution enough to separate the fine-structure of the Balmer-alpha line. In Figure 2, Zeeman-splitting is not found in this weak magnetic field, 6 mT (60 Gauss), since the expected separation of Zeeman-splitting is less than 200 MHz which is close to homogeneous line-width.

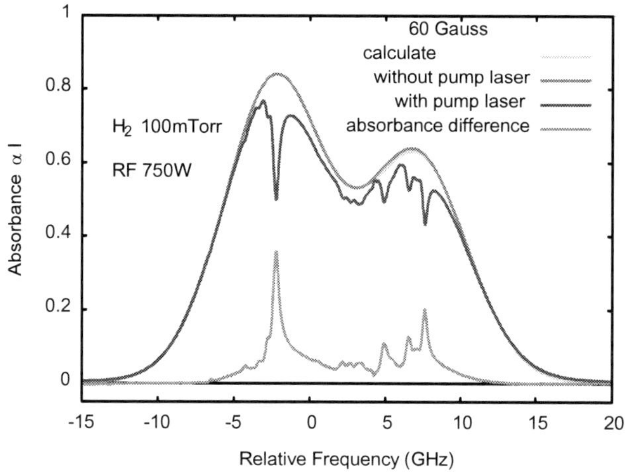

Figure 2. Absorption spectra of Balmer-alpha line with and without pump beam. Theoretical spectrum is almost overlapped by the experimental spectrum without pump laser.

The experimental result of saturation spectra with various magnetic fields are shown in Figure 3. The vertical axis shows relative absorbance difference $\Delta\alpha$ / α_0 which is approximately equal to S / 2, where S is called as saturation parameter. The hydrogen gas pressure and applied RF power were 6.7 Pa (50 mTorr) and 750 W, respectively. Figure 3 shows Zeeman-splitting fine-structure spectra of the Balmer-alpha line. Because the direction of magnetic field is parallel to the direction of propagation of the pump and the probe laser beams, only the σ components of Zeeman-splitting are considered. The major peak at -2 GHz, the $2P_{3/2}$-$3D_{5/2}$ transition, shows clear split separation dependency to the magnetic field strength. However, in the case of relatively intense magnetic field, 130 mT (1300 Gauss), the saturation spectrum become complicated and shows many small peaks.

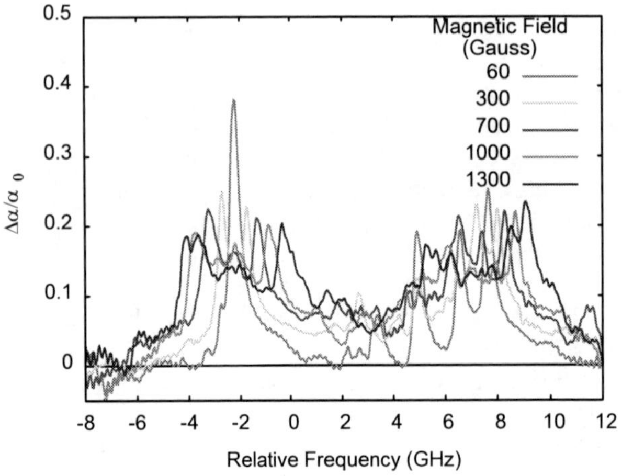

Figure 3. Saturation spectra with various magnetic field strength. The peaks around -2GHz are the $2P_{3/2}$ – $3D_{5/2}$ transition and its Zeeman-splitting.

3.2. Assignment of saturation peaks

To understand the structure of these saturation spectra, the assignment of peaks of saturation spectrum without Zeeman-splitting was carried out. Figure 4 shows the saturation spectrum with weak magnetic field strength 6 mT (60 Gauss) and various hydrogen gas pressures. Thick vertical bars indicate the each transition frequency and relative magnitude of the B coefficient of the fine-structure component. The fine-structure of the Balmer-alpha line consists of seven transitions shown in Figure 4 with labels of the upper and the lower energy levels. It is confirmed that five peaks can be assigned as these transitions ($2P_{3/2}$-$3D_{3/2}$, $2P_{3/2}$-$3D_{5/2}$, $2S_{1/2}$-$3P_{1/2}$, $2S_{1/2}$-$3P_{3/2}$, and $2P_{1/2}$-$3D_{3/2}$). The saturation peak of $2P_{3/2}$-$3S_{1/2}$ and $2P_{1/2}$-$3S_{1/2}$ cannot be recognized.

According to the theory of saturation spectroscopy [1], fictitious saturation peak, called as cross-over peak, should be considered. A cross-over peak arises at the midpoint frequency between two transition lines with common lower energy level and overlapping wings of Doppler-broadening. In Figure 4, the thin vertical bars indicate the expected frequency of cross-over peaks and some peaks are understood as the cross-over peaks. However, there are many irregular peaks which are pointed out by the down arrows in Figure 4. These irregular peaks are observed at the midpoint frequency between two transition lines, similar to the ordinary cross-over peak, but the lower energy levels of the two transition lines are different.

The lower energy levels of the Balmer-alpha line of atomic hydrogen are composed of three levels: $2S_{1/2}$, $2P_{1/2}$, and $2P_{3/2}$. All pairs of the lower energy levels show the irregular cross-over peaks. The ordinary cross-over peak is caused by sharing of the hole burning, which is population dip in the velocity space by the pump laser, on the velocity distribution function with two transitions. The irregular cross-over peaks suggest that the velocity distribution function of different two energy levels have common hole burnings. This means that the frequently population exchange or transfer without velocity change is occurred among the $2S_{1/2}$, $2P_{1/2}$ and $2P_{2/3}$ states. A possibly mechanism of this population transfer is electron impact transfer between 2s and 2p states. It is known that the electron impact transfer from the 2s state to the 2p state has significant rate coefficient [8]. In our analysis of the saturation parameter for each fine-structure of the Balmer-alpha line, the $2S_{1/2}$ state has significant relaxation frequency although the $2S_{1/2}$ state is metastable. This result suggests that the population of the $2S_{1/2}$ state is transfer to the other states. The details of the analysis of saturation parameter will be reported elsewhere.

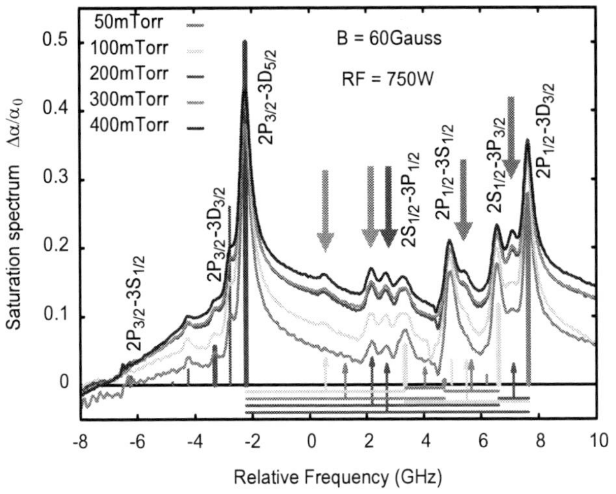

Figure 4. Saturation Spectra without Zeeman-splitting. The thick vertical bars indicate the frequency and intensity of the fine-structure components. The thin vertical bars indicate the expected ordinary cross-over peaks. The down arrows point the irregular cross-over peaks. The horizontal bars show the pairs of transitions corresponding to the irregular cross-over peaks.

4. Conclusion

We applied saturation spectroscopy to the Balmer-alpha line of atomic hydrogen in a linear magnetized plasma source with the intention of applying it to plasma diagnostics. We observed a clear saturation spectrum with high frequency resolution sufficient to resolve the fine-structure of the Balmer-alpha line with weak magnetic field strength. Saturation spectra with Zeeman-splitting also observed with the intense magnetic field up to 130 mT (1300 Gauss), however, the spectra became complex with many small peaks. Most of the peaks of the saturation spectrum without Zeeman-splitting are assigned as transitions of the fine-structure and ordinary cross-over peaks. However, irregular cross-over peaks also found. This suggests that a frequently population exchange or transfer without velocity change occurrs among the 2s and 2p states.

References

[1] Demtröder W 2008 *Laser Spectroscopy* (Berlin: Springer-Verlag)
[2] Levenson M D 1982 *Introduction to Nonlinear Laser Spectroscopy* (New York: Academic Press)
[3] Weber E W, Frankenberger R and Schillng M 1983 *Appl. Phys. B* **32** 63
[4] Weber E W, Humpert E J 1981 *Phys. Lett.* **83** 386
[5] Weber E W 1979 *Phys. Rev. A* **20** 2278
[6] Iwamae A, Hayakawa M, Atake M and Fujimoto T 2005 *Phys. Plasmas* **12** 042501
[7] Goto M, Sawada K, Fujimoto M 2002 *Phys. Plasmas* **19** 4316
[8] Janev R K, Langer W D, Evans K Jr. and Post D E Jr 1987 *Elementary Processes in Hydrogen Helium Plasmas* (Berlin: Springer-Verlag)

The Effect of Wedge Angle on the Evolution of a Stagnation Layer in a Colliding Plasma Experiment

C Fallon, P Hayden, N Walsh, ET Kennedy and JT Costello
School of Physical Sciences and NCPST, Dublin City University, Dublin 9, Ireland
E-mail: fallonc5@mail.dcu.ie

Abstract. Colliding plasmas are steadily gaining significance in hohlraum studies, pulsed laser deposition and laser-induced breakdown spectroscopy for a number of reasons, not least the levels of control they offer over the properties of the slab of plasma that accumulates at the collision front, i.e. the stagnation layer. We present here some results of a time and space resolved optical-spectroscopic study of colliding plasmas formed at the front surfaces of flat and inclined Cu slab targets as a function of the wedge angle between them for angles ranging from 100^0 to 180^0 (i.e., laterally colliding plasmas). Presented here are the kinetics of atomic/ionic spatial distributions throughout the stagnation layers, both of which have been found to vary significantly with wedge angle.

1. Introduction

Although colliding laser produced plasmas are not a new area of study[1], there is renewed interest in a number of international centres [2-15] due to their potential application in inertial confinement fusion [16, 17], pulsed laser deposition (PLD) [18, 19], laser induced breakdown spectroscopy (LIBS) [20,21], and other applications. When two laser-produced plumes collide the outcome usually sits between two extremes. If the so-called seed plasmas have a large relative velocity and low density at the collision front the plasma plumes will tend to interpenetrate. Alternatively, when the relative velocity between the plasmas is small and the density at the collision front is large, plasma constituents will tend to decelerate rapidly leading to the formation of a so-called stagnation layer. This is known as 'hard' stagnation, whereas the intermediate regime between interpenetration and hard stagnation is known as soft stagnation. To help determine which regime to expect, one can make use of the collisionality parameter [22,23] defined as

$$\zeta = D/\lambda_{ii} \quad (1)$$

where D is the separation between the two seed plasmas and λ_{ii} is the ion-ion mean free path given by [22]

$$\lambda_{ii}(1 \rightarrow 2) = \frac{m_i^2 v_{12}^4}{4\pi q^4 z^4 n_i ln\Lambda_{1\rightarrow 2}} \quad (2)$$

where the indices 1 and 2 refer to the individual plasma plumes, m_i is the ion mass, v_{12} is the relative collision velocity, q is the fundamental electronic charge, z is the average ionisation state of the plasma, n_i is the average plasma ion density, and $ln\Lambda_{1\to2}$ is the so-called Coulomb logarithm.

It is clear from equations 1 and 2 that control over seed plasma properties such as v_{12} - and, to a lesser extent, the distance D between them - can yield a significant degree of control over the collisionality parameter ζ, which determines not only the degree of stagnation but also plays a significant role in determining many of the properties of the stagnation layer such as temperature, density, composition, degree of ionisation, geometry, etc. A change of the distance D is a simple experimental task, and by varying the angle of inclination of the seeds we can gain a degree of control over v_{12}. Here we present some results of a systematic study of the effects on the stagnation layer when the angular attitude of the target slabs is varied.

2. Experimental Setup

Figure 1, which is split into four parts, shows the setup used for the experiments described in this paper. Figure 1 (a) shows a Surelite III Nd:YAG laser which operates at a wavelength of 1064 nm and generates up to 800 mJ per pulse (reduced to 210 mJ using a Brewster's window and half-wave plate) in 6 ns FWHM pulses. By passing part of the beam through a wedge prism while the remainder passes over it [24], the beam is split into two parts. It then passes through a lens which focuses each beam to a spot of diameter $\approx 100\mu m$, as shown in figure 1 (b). The distance between the two foci is related to the lens and wedge prism by:

$$d = f\gamma(n-1) \tag{3}$$

where n is the refractive index of the wedge prism (here glass with $n = 1.5$), γ is the acute angle of the wedge in radians, and f is the focal length of the lens.

A rotating target is used to refresh the target material and is placed inside a vacuum chamber maintained by a turbomolecular pump at a base pressure of 1×10^{-7} mBar. Two seed plasmas are created, which collide to form the stagnation layer (under suitable conditions). Both broadband optical imaging (figure 1 (c)) and spatially and temporally resolved spectroscopy (figure 1 (d)) are carried out simultaneously. Broadband imaging uses a variable zoom lens to focus the image onto an Andor DH534 camera which is mainly used to align and monitor the highly sensitive setup while spectroscopic measurements are undertaken. The setup is described in detail in [25].

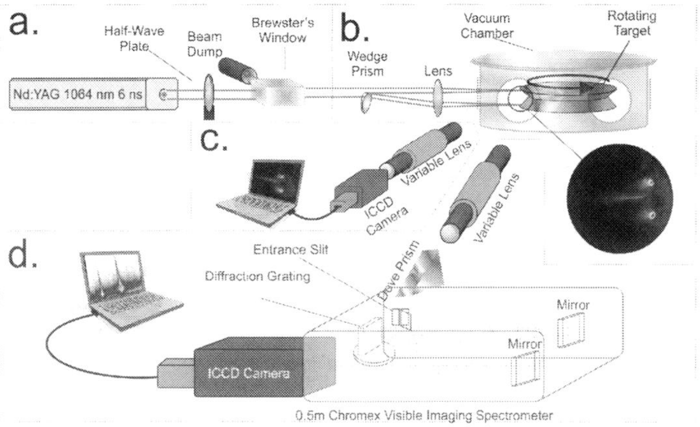

Figure 1: Experimental setup for the optical analysis of the interaction region between two laser produced plasmas.

3. Results & Discussion

Information about the evolution of a plasma over time can be obtained by tracking the plume front position as a function of time delay [20,24]. If, instead, we follow the maximum point of the intensity of the emitting atoms and ions - taken to be the brightest part of each emission line (assuming no attenuation due to opacity) - it is possible to trace the movement of each of the species throughout the plume. This is achieved by recording and plotting the intensity at each point along the forward expansion direction of each space and time-resolved spectral line to yield a curve showing the peak of emission intensity for atoms and ions along the plume. Figures 2 and 3 show examples of these plots, obtained by analysis of the Cu I line at 521.82nm and the Cu II line at 508.83nm respectively, revealing the differing movements of each atomic/ionic fluid.

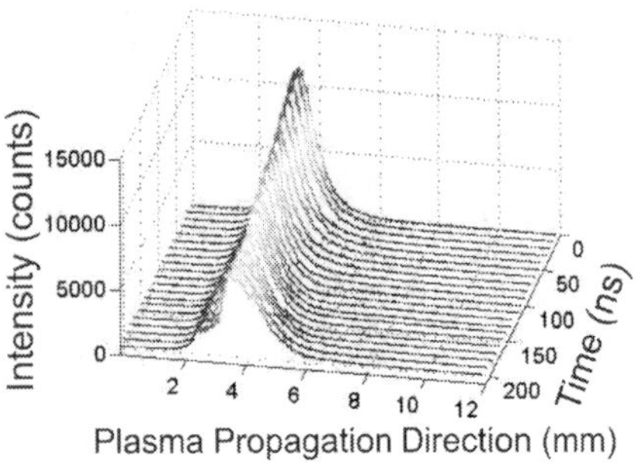

Figure 2: Motion of the peak intensity point of atomic copper (λ = 522 nm) in the stagnation layer.

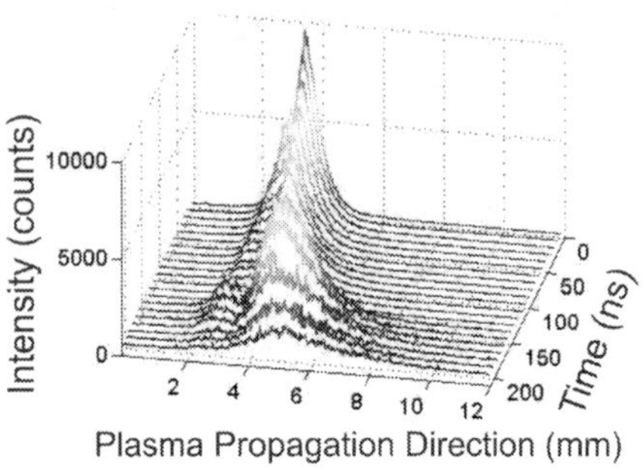

Figure 3: Motion of the peak intensity point of singly charged copper (λ = 509 nm) in the stagnation layer.

Figure 4 (a) shows the emission image obtained by colliding seed plasmas created on laterally facing (parallel) targets corresponding to a 180^0 target wedge angle, or the flat configuration. As the two seed plasmas expand, they collide at relatively low velocities along the lateral expansion plane (parallel to the target surface).

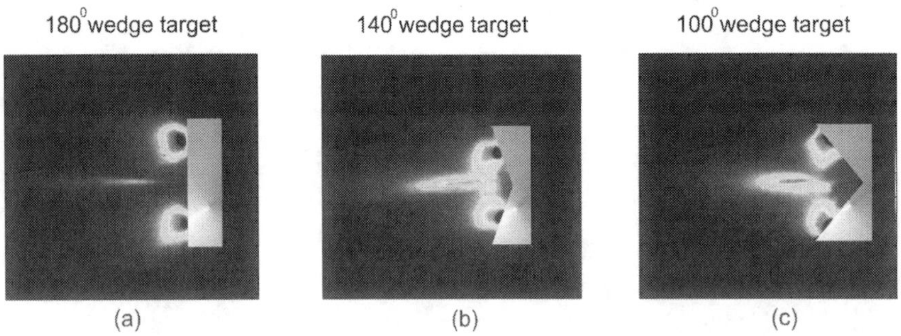

Figure 4: Time integrated images of the stagnation layer formed using 3 separate wedge angles: (a) 180^0 (b) 140^0 & (c) 100^0.

By employing a 140^0 wedge-shaped target (figure 4 (b)) the two seed plasmas now collide with a larger component of their forward expansion velocity directed orthogonally (towards each other). The relative collision velocity is thus higher, and the ion-ion mean free path (MFP) becomes larger, resulting in a higher degree of plume interpenetration. Figure 4 (c) shows the case where the wedge angle is decreased to 100^0, thus further increasing the degree of plume interpenetration here. The shape of the stagnation layer illustrates the degree of interpenetration in each case, where hard stagnation creates tight and well defined stagnation layers. It is clear from the figures that the most tightly defined stagnation layer is formed in the case of laterally colliding plasma plumes.

As described above, the point of maximum intensity for selected emission lines from Cu atoms and singly charged ions is traced as a function of time. The results are shown in figure 5 for three separate wedge angles (100^0, 140^0 & 180^0) at a seed distance of D = 1.3 mm for both neutral atoms and singly charged ions. Shown also is a linear fit to each data set, where appropriate.

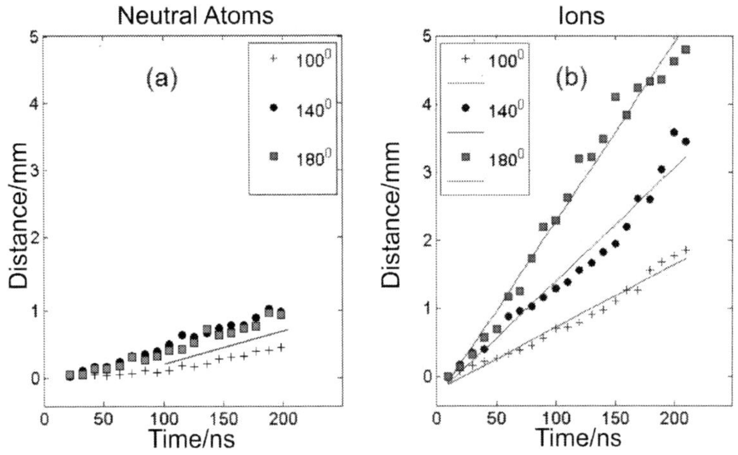

Figure 5: Comparison of the movement of the peak of intensity of neutral emitting atoms and singly charged ions in a stagnation layer generated by focussing two 105 mJ 6 ns laser pulses onto three different wedge systems.

A quick comparison of both graphs reveals that the singly charged ions move at a significantly higher velocity relative to the atoms for all three wedge angles. It is well established that the initial expansion velocity of electrons from the seed plumes far exceeds that of the heavier ions [26]. This leads to the generation of a strong electric field caused by the steep gradient in the spatial distribution of charged particles as the electrons attempt to separate from the ions (ambipolar diffusion). The resulting net field slows the electrons while accelerating the ions, and has little effect on the neutral atoms.

Two mechanisms contribute to growth in the forward expansion direction of the ions in the stagnation layer. Firstly, the time of flight for the material from each seed plasma to reach more and more distant points on the collision plane, and secondly the fact that material that arrived earlier has a component of forward momentum and will thus move outwards along the stagnation layer. A comparison of the velocity of the peak of intensity of singly charged ions is shown in table 1. It can be seen that in the case of the 100^0 wedge angle in particular, the orthogonal component does not appear until times $\geq$ 60 ns, i.e. a linear fit applies only after this time. In this case, interpenetration at the lower layer region is increased; hence growth in the forward expansion direction is not evident until later times. Applying the linear fit to each case, the Cu^+ ion growth rates range from 9.2 $\times 10^5$ cm/s for the case of the 1.3 mm, 210 mJ, 100^0 wedge case to 26.1 $\times 10^5$ cm/s for the 1.3 mm, 210 mJ flat target case. In the flat target case, the stagnation layer grows predominantly in the forward expansion direction as time progresses. This is supported by figure 4 (a), where the stagnation layer formed is narrow and elongated. There is little or no growth along its lateral dimension, giving a strong indication that the degree of interpenetration is quite low. By comparison, for neutral atomic species, as stated earlier, the average velocity is almost independent of wedge angle at a value of $\sim 3 \times 10^5$ cm/s.

Table 1: Velocity* of peak of intensity of singly charged ions.

Target Geometry	ions/(cm/s)
1.3 mm, 210 mJ, 180^0 Wedge	26.1×10^5
1.3 mm, 210 mJ, 140^0 Wedge	16.6×10^5
1.3 mm, 210 mJ, 100^0 Wedge	9.2×10^5

** Velocity here does not mean expansion rate in the normal sense of plasma plume expansion. Rather it refers to the propagation of the peak intensity value along the stagnation layer as material from the seed plasma plumes reaches the collision plane at a delay determined by the distance to be travelled and the concomitant times-of-flight.*

In summary, the stagnation layer growth is different for separate ionic species, and the growth of each specie throughout the layer is dependent on two mechanisms: accumulation at the collision plane at a rate determined by the time of flight of seed plume species to the collision plane, and the movement of stagnation layer species along the collision plane due to their natural component of forward directed momentum. Thus, for any one wedge angle the atom and ion growth rates are different and hence fluid growth rates in the stagnation layer will be different, as will the initial spatial distribution for each as the layer grows.

3.1. Comparison of a Stagnation Layer with a Single Seed Plasma

We have demonstrated above that the properties of the stagnation layer can be controlled in a manner not available to single laser plasma plumes. Both the lifetimes and intensities of spectral lines in laser produced plasmas can be increased by creating relatively moderate density high-energy plasmas which can overcome the problem of flux loss due to opacity, which leads to the attenuation of discrete emission lines with a concomitant reduction in line contrast, signal-to-noise ratio (SNR), and signal-to-background ratio (SBR). The latter is a key parameter in determining the limit-of-detection (LOD) of the LIBS technique. Here we compare with a single plasma plume the stagnation layer of most interest for LIBS, a layer created by the 1.3 mm seed separation, 210 mJ total pulse energy, 100^0 wedge case, where relatively low species growth rates occur.

Figure 6 shows a comparison of the growth rates of species in a stagnation layer created using the laser and target geometries described above, along with a single plume using the same total pulse energy. The velocity of both the peak of the neutral atom distribution and of the singly charged ion distribution is higher in the case of the single pulse plume, and the ions move at almost twice the velocity in the forward expansion direction. The stagnation layer is confined spatially for a longer period of time which slows its evolution through time and should

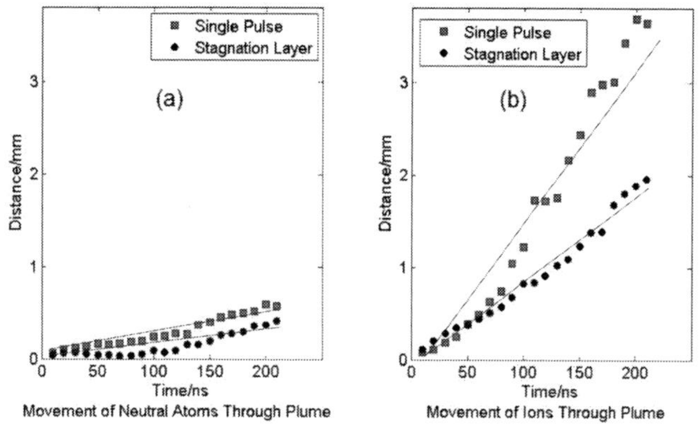

Figure 6: Comparison of the movement of the peak of intensity of neutral emitting atoms and singly charged ions for a single pulse plasma created using a 210 mJ total pulse energy with a stagnation layer created using a 100^0 wedge, 210 mJ total pulse energy and 1.3 mm seed separation.

cause the density gradient to be smaller than the single pulse case. Hence one should expect the temperature and density profiles to be more uniform throughout the stagnation layer. This topic will be the subject of a follow-up paper.

4. Conclusions

Spatially resolved and temporally integrated as well as spatially and temporally resolved imaging spectra have been used to study the interaction of counter-propagating laser-produced plasmas. The results presented show a range of different behaviours in the interaction dynamics between each of the plasma plumes for different initial target wedge angles. A study of the progression of the peak of the intensity of singly charged ions along the stagnation layer as a function of time for a range of initial conditions shows that the velocity of ions varies significantly as a function of the wedge angle between the seed plasma plumes. Figure 4 (b) and (c) illustrate the progression from 'hard stagnation' to 'soft stagnation' of the layer, where the increase in the lateral dimension of the stagnation layer, i.e. the width, and the decrease in length is apparent. The decreasing growth rates of the peak position of singly charged ions (Table 1) as the wedge angle closes confirm this. It was demonstrated that the geometry of the stagnation layer can be controlled by variation of the wedge angle, a property which has implications for double-pulse (DP-) LIBS where the size and shape of the main plasma are important for efficient coupling of laser energy to the plasma.

Acknowledgments

Work supported by Science Foundation Ireland (Grant No. 12/ IA/ 1742), the FP7 Erasmus Mundus Joint Doctorate EXTATIC under framework partnership agreement FPA-2012-0033 and support from the HEA PRTLI 5 INSPIRE Graduate Programme. Work is partly associated with the EU COST Actions MP1203 and MP1208. The authors wish to thank Des Lavelle for help in the design and construction of the experimental system.

References

[1] Rumsby P T and Paul J W M 1974 *Plasma Physics* **16** 969
[2] Al-Shboul K F, Harilal S S, Hassan S M, Hassanein A, Costello J T, Yabuuchi T, Tanaka K A, Hirooka Y 2014 *Plasma Physics* **21** 013502
[3] Li C K *et al* 2013 *Phys. Rev. Lett.* **111** 235003
[4] Merritt E C, Moser A L, Hsu S C, Loverich J, Gilmore M 2013*Phys. Rev. Lett. 111* 085003
[5] Ross, J S, Park H-S, Berger R, Divol L, Kugland N L, Rozmus W, Ryutov D and Glenzer S H 2013 *Phys. Rev. Lett.* **110** 145003
[6] Thoma C, Welch D R and Hsu S C 2013 *Physics of Plasmas* **20** 8
[7] Singh R P, Gupta S L and Thareja R K 2013 *Spectrochimica Acta Part B-Atomic Spectroscopy* **88** 54
[8] Yeates P, Fallon C, Kennedy E T and Costello J T 2013 *Physics of Plasmas* **20** 9
[9] Gambino N, Hayden P, Mascali D, Costello J, Fallon C, Hough P, Yeates P, Anzalone A, Musumeci F and Tudisco S 2013 *Applied Surface Science* **272** 69
[10] Kuramitsu Y,Sakawa Y, Morita T, Gregory C D, Waugh J N, Koenig M, Woolsey N and Takabe H 2013 *High Energy Density Physics* **9** 222
[11] Rosenberg M J, Ross J S, Li C K, Town R P J, Séguin F H, Frenje J A, Froula D H and Petrasso R D 2012 *Phys. Rev. E* **86** 056407
[12] Park H S *et al* 2012 *High Energy Density Physics* **8** 38
[13] Harilal S S, Polek, M P and Hassanein A 2011*IEEE Transactions On Plasma Science* **39** 2780
[14] Hirooka Y, Omoto N, Oishi T and Tanaka K A 2012 *Fusion Engineering And Design* **87** 1760
[15] Mascali D, Tudisco S, Bonanno A, Gambino N, Ivanovski S, Anzalone A, Gammino S, Miracoli R and Musumeci, F 2010 *Radiation Effects And Defects In Solids* **165** 730
[16] Brueckne K A and Jorna S 1974 *Reviews Of Modern Physics* **46** 325
[17] Hurricane *et al* 2014 *Nature* **506** 343
[18] Irissou E, Vidal F, Johnston T, Chaker M, Guay D and Ryabinin A N 2006 *Journal Of Applied Physics* **99** 3
[19] Camps E, Escobar-Alarcon L, Castrejon-Sanchez V H, Camacho-Lopez M A and Muhl S 2007 *Applied Surface Science* **254** 185
[20] Khater M A, van Kampen P, Costello J T, Mosnier J P and Kennedy E T 2000 *J Phys. D: Appl Phys.* **33** 2252
[21] Hahn D W and Omenetto N 2010 *Applied Spectroscopy* **64** 335
[22] ChenaisPopovics C *et al* 1997 *Physics Of Plasmas* **4** 190
[23] Rambo PW and Denavit J 1992 *Journal Of Computational Physics* **98** 317
[24] Harilal S S 2007 *Journal of Applied Physics* **102** 123306
[25] Fallon C 2013 *Optical diagnostics of colliding laser produced plasmas: towards next generation plasma light sources* PhD Thesis, DCU
[26] Okano Y, Hironaka Y, Nakamura K G and Kondo K 2003 *Applied Physics Letters* **83** 1536

Diatomic Molecular Emission Spectroscopy of Laser-induced Titanium Plasma

A C Woods and C G Parigger

University of Tennessee Space Institute, 411 B. H. Goethert Pkwy, Tullahoma, TN, 37388, USA

E-mail: cparigge@utsi.edu

Abstract. Previous research regarding laser-induced breakdown spectroscopy (LIBS) of titanium normally focuses on the atomic and ionic Ti spectral transition lines. However, after a characteristic time subsequent to laser ablation, these lines are no longer discernable. During this temporal regime, the diatomic molecular transition lines of titanium monoxide (TiO) are prominent in the laser-induced plasma (LIP) emissions. TiO has long been studied in the contexts of stellar emissions, allowing for some of the molecular transition bands to be accurately computed from theory. In this research, optical emission spectroscopy (OES) of laser-induced plasma (LIP) generated by laser ablation of titanium is performed in order to infer temperature as a function of time subsequent to plasma formation. The emission spectra of the resulting ablation plume is imaged as a function of height above the sample surface. Temperatures are inferred over time delays following plasma formation ranging from 20 μs–200 μs. Computed TiO $A^3\Phi - X^3\Delta$, $\Delta v = 0$ transition lines are fit to spectral measurements in order to infer temperature. At $t_{delay} = 20$ μs–80 μs, the observed plume contains two luminescent regions each with a distinctly different temperature. As the plume evolves in time, the two regions combine and an overall temperature increase is observed.

1. Introduction

Laser-induced breakdown spectroscopy (LIBS), as an experimental technique, has grown in popularity in large part due to the simplicity of the required equipment. The traditional LIBS apparatus consists of a single laser, serving as an excitation source, and a spectrometer equipped with a detector for gathering spectral emissions [1]. The success and growth of future LIBS applications require a deeper understanding of the fundamental processes occurring as a plasma is observed at various time delays, given various experimental conditions [2]. To this extent, much of the ongoing fundamental research utilizing LIBS, seeks to observe and analyze specific atomic and molecular species present in the laser-induced plasma (LIP) [3, 4, 5, 6, 7]. By closely examining the properties of spectra from specific atoms and molecules at various regions of the plasma, as it evolves in time, a description of the characteristic processes going on throughout the plasma can be obtained. Of particular interest is the time domain for which molecular recombination overcomes electron collisions as the dominant mechanism concerning plasma emissions.

As LIP evolves in time, the characteristic emitting species dominating the spectra will evolve as well. An initial LIBS spectrum, typically within the first tens of nanoseconds, is dominated by a spectral continuum contribution. Mechanisms, such as inverse Bremsstrahlung and radiative recombination involving free electrons, are responsible for this spectral continuum. As the plasma expands and cools, ionic spectral line transitions appear in the plasma emissions, superposed to the continuum contribution. The more highly ionized species will become prominent first, followed by the lesser ionized species and ultimately the neutral atomic transition lines [8, 9, 10]. Three-body recombination occurring with plasma expansion causes a rapid decrease of the electron density in the plasma [11]. As a consequence of the decreasing electron density in the plasma, each emission line will become progressively more narrow during the plasma expansion [12].

After a characteristic time, depending on electron density, the elements in the plasma and the plasma expansion, emissions from radiative molecular transitions will become discernible as the prominence of atomic spectral lines observed in the plasma decreases [13]. Of particular interest for this research is the diatomic molecular radiative transitions of the plasma. Typically, this phenomenon is attributed to the recombination of atoms to form molecules. However, the detection of diatomic transitions of certain species at far earlier stages of plasma evolution when compared to other diatomic transitions in similar conditions leads to the conclusion that not all diatomic transitions are a product of recombination. In these scenarios, it is hypothesized that the molecule was present prior to plasma formation. However, many diatomic molecules observed in laser-induced plasma are unstable in typical laboratory environments. For such molecules, LIBS offers a suitable method for observation. The spectral transitions of such molecules can be collected typically for tens of microseconds. Often times, these emissions are quite luminous, from the ultra-violet to the infrared.

For the purposes of analyzing the laser ablation plasma above the Ti surface at time delays on the order of tens to hundreds of microseconds, TiO spectral transitions are utilized to infer temperature. This required the computation of new, accurate line-strength tables. Diatomic quantum theory is utilized in computing line-strengths of spectral transitions [4, 14]. The specifics concerning the computation of the line-strengths used in this research are presented elsewhere [16, 17]. Given a few parameters, such as resolution and temperature, the spectral line-strengths can be used generate a synthetic spectrum representative of the radiant transitions of a diatomic molecule. In order to infer micro-plasma parameters, the experimentally collected spectra believed to contain the diatomic spectral signatures are fit with the computed spectra for the molecule [15, 16, 17].

Previous studies observed the TiO $\mathrm{B}^3\Pi \rightarrow X^3\Delta$ and $\mathrm{A}^3\Phi \rightarrow X^3\Delta$, $\Delta v = 0$ transition bands in the ablation plasma [17, 18]. The collected spectra were then fit with computed spectra in order to infer temperature as a function of delay time. Temperature inferences, resulting from the use of an intensified linear diode array arrangement to gather time-resolved spectra, typically contained a local minimum occurring between a 40 μs and 60 μs delay time. At 60 μs, when the expanding plasma is thought to be cooling, this unexpected result indicated an increase in temperature. In order to further investigate the situation, the current research observes the TiO $A^3\Phi - X^3\Delta$, $\Delta v = 0$ transition band at time delays ranging from 20 μs–200 μs. Utilizing a gated two-dimensional ICCD detector, the gathered emissions are time-resolved and imaged as a function of height above an ablation surface. The gathered spectra are then fit with computed spectra representing the TiO $A^3\Phi - X^3\Delta$ band to infer temperature.

2. Experiment

A Ti 6-4 sample is ablated by radiation from a Nd:YAG laser with 160 mJ per pulse and a nominal pulse width of 13 ns. The laser radiation is supplied at 10 Hz and focused to approximately a 1 mm spot size. The beam propagates vertically downward onto the flat Ti surface such that the beam path is parallel to the slit aperture of a Jobin-Yvon spectrometer. The resulting ablation plume emissions are imaged onto the slit in a 1:1 manner. The spectrometer is equipped with an Andor iStar ICCD camera, providing both temporal and spatial resolution. In order to enhance the signal to noise ratio, the 1024 × 1024 pixels provided by the detector are consolidated into 32 spectra (1024 pixels across) gathered along the height of the plume for each image. Additionally, 200 accumulations are used to produce each image. At t_{delay} = 20–200 μs, spectra in the range of 705 nm–715 nm are collected every 20 μs utilizing a 2 μs gate width. In this spectral region, at these time delays, TiO transition lines are collected. The gathered spectra are then fit with computed spectra from the TiO $A^3\Phi - X^3\Delta$, $\Delta v = 0$ transition band in order to infer temperature.

3. Results

Figure 1 provides an example of an experimentally obtained spectrum fit with a computed spectrum for the TiO γ transition. The collected spectrum of Fig. 1 represents a single horizontal slice of a larger image. Fig. 2 is an example of such an image. The plot just below the image in Fig. 2 represents the vertical sum of the pixels in the image. Such a summation is likened to using a linear diode array detector. While spectra computed from theory are used to infer the TiO transition temperature along the height of the image, the spectrum obtained by integrating the height of the detector can be similarly analyzed. The result is typically a temperature corresponding to the inferred temperature at the hottest region of the image.

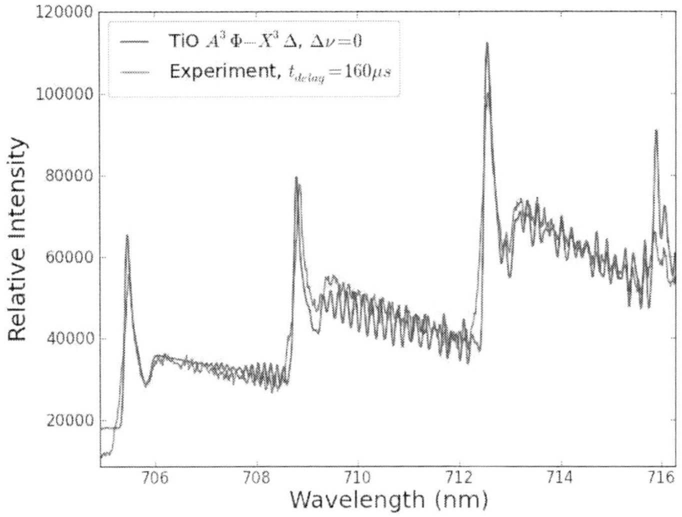

Figure 1. The above plot represents a horizontal slice of one the gathered images fit with computed spectra corresponding to T = 3400 K. The measured spectrum was collected at t_{delay} = 160 μs.

By studying Figures 2–6, some very interesting phenomena becomes apparent. The temperature inferred from the TiO molecular transitions increases with height along the ablation plume. This phenomena is consistent throughout t_{delay} =20–200 μs. More interestingly, it is clear that between t_{delay} = 20–60 μs there are two distinct regions of the plasma. One region rests several millimeters above the sample surface, while the other appears to rise from the surface before the two ultimately combine. After the two regions combine (~100μs), there is a noticeable bulge in uppermost part of the intensity profile.

When the temperature inferences made from emissions of the two regions are compared, it is apparent that the inferred temperature of the TiO emissions from the region closer to the

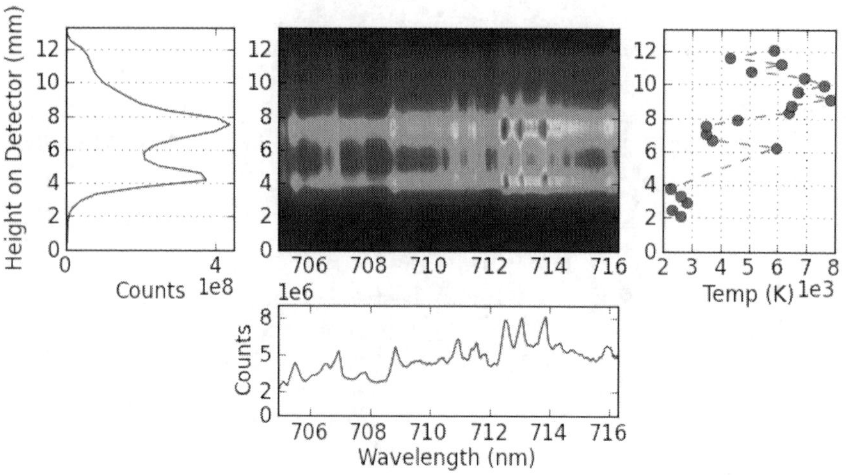

Figure 2. The above image represents spectra of laser ablation plasma above a titanium surface recorded at $t_{delay} = 40$ μs with a 2 μs gate width.

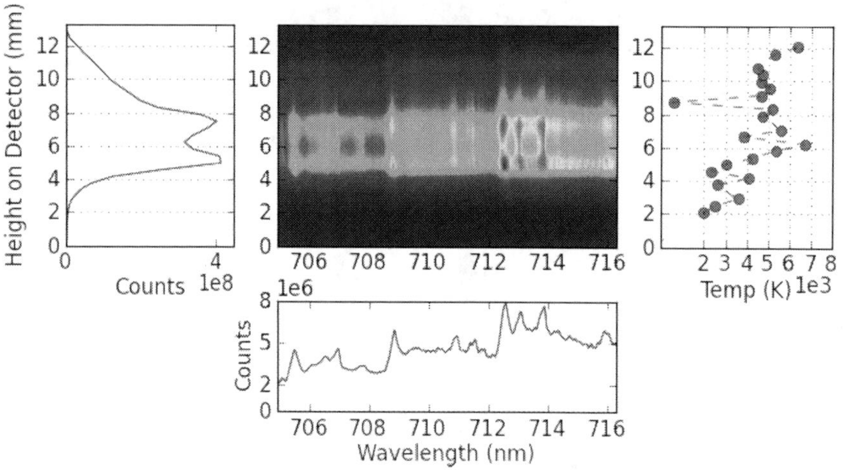

Figure 3. The above image represents spectra of laser ablation plasma above a titanium surface recorded at $t_{delay} = 60$ μs with a 2 μs gate width.

sample surface is distinctly lower than that of the region further above the surface. This is most clearly evident in Figures 2 and 3. Particularly when examining the inferred temperatures of Fig. 2, it is important to keep in mind that we are struggling to make temperature inferences resulting from fitting computed TiO spectra to the collected spectra on account of signal to noise issues. Notice at the earlier time delays, the characteristic structure of the TiO $A^3\Phi - X^3\Delta$ band is hardly discernable. Much of the spectral signatures observed in this time domain are atomic line shapes. When inspecting the inferred temperature more confidence is given to those values inferred from the brighter regions of the plume. These regions are made brighter by the luminescent TiO transitions. This conveniently allows for succinct comparison between the two brightest regions of Fig. 2, as the TiO spectral signatures in the dimmer regions are less discernable from the other spectral contributions.

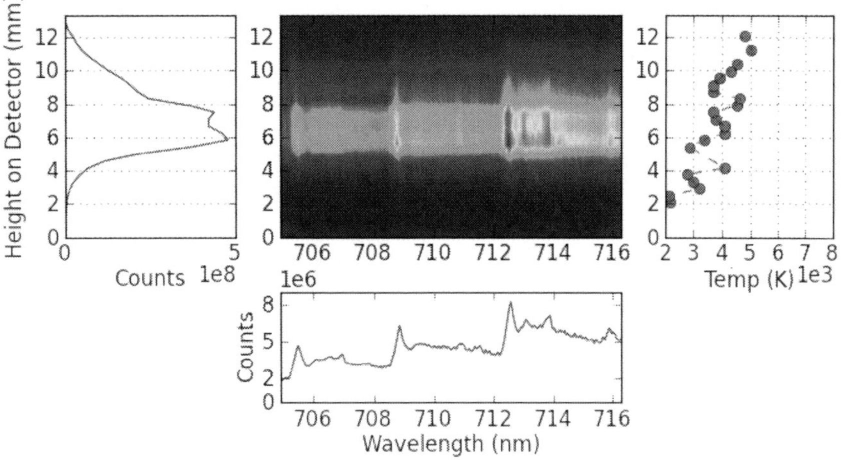

Figure 4. The above image represents spectra of laser ablation plasma above a titanium surface recorded at $t_{delay} = 80$ μs with a 2 μs gate width.

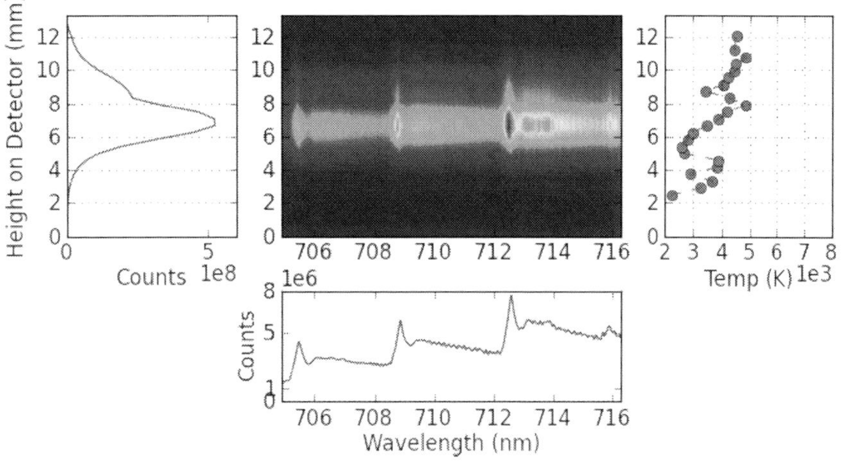

Figure 5. The above image represents spectra of laser ablation plasma above a titanium surface recorded at $t_{delay} = 100$ μs with a 2 μs gate width.

Examining the plots to the left of the images in Figures 2 - 6 provides insight into the luminosity along the height of the plasma. The motion of the lower peak towards the upper peak can be used as a measure of the kinetics of the luminous lower region of the plasma. These plots for Fig. 2 - 4 also illustrate an increase in luminosity of the lower region as it approaches the upper. At later time delays, $t_{delay} = 100$–200 μs, a hump is observed at the top luminous region of the ablation plume. The height of the plasma corresponding to the slight dip below the hump is determined to be hotter than the regions above or below the dip. While hump appears to be a consequence of a snowplow effect at the top of the plasma, two-dimensional spatial imaging of the plume would possibly provide an explanation. The current apparatus observes primarily along the axial distance perpendicular to the surface; however, imaging in the radial direction would likely provide a crown-shaped profile, as in Fig. 8 of Aguilera *et al.*

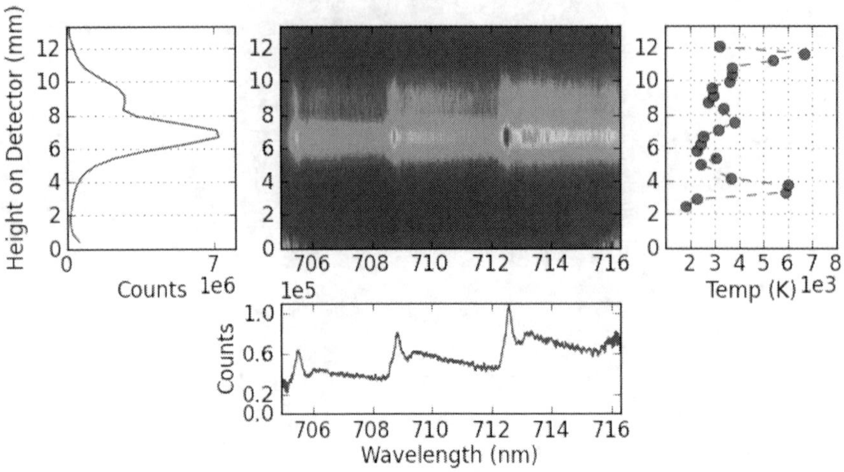

Figure 6. The above image represents spectra of laser ablation plasma above a titanium surface recorded at $t_{delay} = 200$ μs with a 2 μs gate width.

[19].

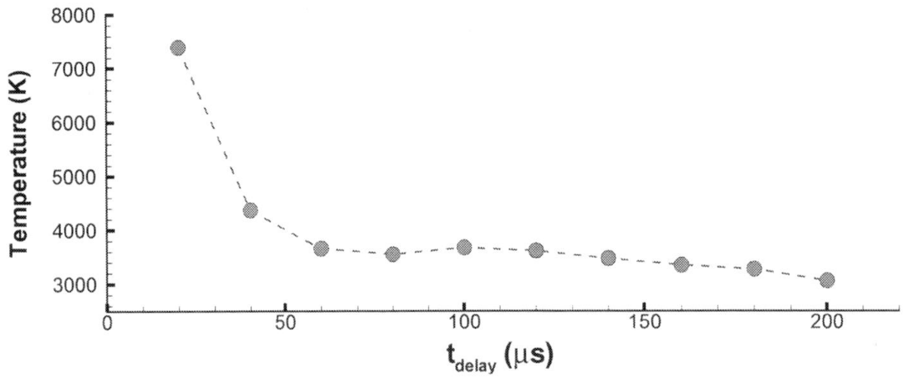

Figure 7. The plot above illustrates the temperature as function of time delay inferred from the analysis of the spectra resulting from integrating along the height of the detector.

4. Discussion

By comparing the inferred temperature from the upper region of Figure 2, we find remarkable agreement with the results obtained previously at this same time delay ($t_{delay} = 40$ μs) using this same window (2 μs) [18]. If we make the same comparison with the upper region of Fig. 3 and Woods *et al.* [18] at $t_{delay} = 60$ μs, there is some agreement for the inferred temperatures here as well. When the spectra obtained by integrating over the height of the images are analyzed, the inferred temperatures typically correspond to a temperature slightly elevated from that of the upper luminescent region (see Figure 7).

Examining Figure 7, a temperature increase of ≈ 100 K occurs at $t_{delay} = 100$ μs. Previous investigations concerning both the TiO $B^3\Pi \rightarrow X^3\Delta$ and $A^3\Phi - X^3\Delta$ systems have also found an unexpected temperature increase at later time delays [17, 18]. Inspection of the inferred temperatures along the height of the image over t_{delay} = 40–100 μs indicate that the lower region undergoes drastic heating, as the cooler TiO molecules of the lower luminescent region interact with the warmer TiO molecules of the upper region. This interaction results in an overall increase in TiO temperature in the plasma.

While more investigations are required to fully characterize this bright region rising from the surface, some hypotheses can be formed. By the ability to fit computed TiO spectra to this emitting region, it is determined that the TiO molecule is the dominant emitter. In the upper region, the TiO transitions are likely the result of recombination. This is consistent with the delay time in which the transitions become discernible given the excitation source and the evolution of the temperature and electron density, as determined by the measurements made within the first hundreds of nanoseconds. The drastic temperature difference at t_{delay} = 40 μs between the upper and lower luminescent regions remains largely unexplained. As the cooler plasma temperatures are typically found to be in the lower regions, perhaps a sudden flurry of recombination occurs near the sample surface, overtaking electron impacts as the driving mechanism for emissions. This trend then rises along the plasma until it reaches the upper region. However, because the temperatures inferred from the TiO transitions in the lower region are starkly less than those of the upper region and gradually increase as the regions interact, an alternate theory would maintain that the mechanism causing the radiative transitions of the diatomic molecule is different between the two regions. During laser ablation at this intensity, the sample surface will become molten for a characteristic duration [20]. Considering the lower region is much cooler and appears to heat as it leaves the surface, possibly the TiO emissions in this region are emanating from bulk titanium being ejected from the sample surface.

5. Acknowledgements

The authors would like to thank the University of Tennessee Space Institute and the Center for Laser Applications for support of this research.

References

[1] Parigger CG, Woods AC, Witte MJ, Swafford LD, and Surmick DM 2014 *Journal of Visualized Experiments: JoVE* **84**
[2] Amoruso S, Bruzzese R, Spinelli N, and Velotta R 1999 *J. Phys. B* **32** R131
[3] Parigger CG and Oks, E 2010 *Int. Rev. Atom. Mol. Phys.* **1** 13
[4] Hornkohl JO, Nemes L, and Parigger CG 2011 ed Nemes L and Irle S (World Scientific) chapter 4 pp. 113–165
[5] Witte MJ, Parigger CG, Bullock NA, Merten JA, Allen SD 2014 *Appl. Spec.* **68** 367
[6] Parigger CG, Swafford LD, Woods AC, Surmick DM, and Witte MJ 2014 *Spectrochim. Acta Part B* **99** 28
[7] Parigger CG, Woods AC, Surmick DM, Swafford LD, and Witte MJ 2014 *Spectrochim. Acta Part B* **99** 15
[8] Albert O, Roger S, Glinec Y, Loulergue JC, Etchepare J, Boulmer-Leborgne C, Perrière J, and Millon E 2003 *Appl. Phys. A* **76** 319
[9] De Giacomo A 2003 *Spectrochim. Acta Part B* **58** 71
[10] Capitelli M 2004 *Spectrochim. Acta Part B* **59** 271
[11] Casavola A, Colonna G, and Capitelli M 2003 *Appl. Surf. Sci.* **208** 85
[12] Griem H 1974 *Spectral Line Broadening by Plasma* (New York: Academic Press)
[13] De Giacomo A, Dell'Aglio M, Santagata A, and Teghil R 2005 *Spectrochim. Acta Part B* **60** 935
[14] Parigger CG and Hornkohl JO 2011 *Spectrochim. Acta Part A* **81** 404
[15] Hermann J, Perrone A, and Dutouquet C 2001 *J. Phys. B* **34** 153
[16] Parigger CG, Woods AC, Keszler A, Nemes L, and Hornkohl JO 2012 *Int. Symp. On High Power Laser Ablation* (AIP Publishing) **1464** p 628
[17] Woods AC, Parigger CG, and Hornkohl JO 2012 *Opt. Lett.* **37** 5139
[18] Woods AC, Parigger CG, and Hornkohl JO 2012 *Int. Rev. Atom. Mol. Phys.* **3** 103
[19] Aguilera JA and Aragón C 2004 *Spectrochim. Acta Part B* **59** 1861
[20] Bogaerts A and Chen Z 2005 *Spectrochim. Acta Part B* **60** 1280

Zeeman splitting of laser-driven soft x-ray laser line by the enhancement of magnetic field in plasma

T Kawachi[1], N Hasegawa[1] A Iwamae[2] and H Yoneda[3]
[1]Quantum Beam Science Center, Japan Atomic Energy Agency (JAEA), 8-1-7, Umemidai, Kizugawa, Kyoto, 619-0215, Japan
[2]Faculty of Engineering, Fukui University, 3-9-1Bunkyo, Fukui, 910-8517, Japan
[3]Institute of Laser Science, University of Electro-Communications, 1-5-1, Chofu, Tokyo, 182, Japan

E-mail: kawachi.tetsuya@jaea.go.jp

Abstract. We observed a laser-driven plasma soft x-ray laser (SXRL) line at the wavelength of 18.895 nm from the nickel-like molybdenum ions under the existence of external magnetic field. In this experiment, 4 mm-diameter magnetic coil driven by an electrical pulsed power supply provided the external magnetic field of 15 T along the direction of the plasma column, and we could obtain the left-handed circular and right-handed circular polarization components separately. The experimental Zeeman splitting indicated that the strength of the magnetic field was enhanced by a factor of more than 10 than that of the applied field, which implied that the compression of magnetic field occurred in the dynamics of intense laser-plasma interaction.

1. Introduction

Polarized light sources have been widely used as powerful probe in many research fields such as material science, plasma science, biological and medical science and so on. In particular, the circular dichroism measurement has been established in the visible and hard x-ray region, which enables us to obtain the information of the 3-D structure of the materials such as classification of the optical isomer (chirality) [1]. In these wavelength regions transmission optics as phase control devises of the circularly polarized light are available, whereas in the soft x-ray region, effective method to generate and control the circular polarization is not established yet. Recently, the circularly polarized soft x-ray has been generated in synchrotron undulators [3]. However this requires still large facility, and it is difficult to obtain the right-handed and left-handed polarization at the same time.

In this paper we propose an alternative method to obtain circularly polarized intense soft x-ray source using Zeeman splitting of laser-driven plasma soft x-ray lasers (SXRL) [4-6]. We took the transition of the $3d^9 4p$ ($J = 1$) - $3d^9 4d$ ($J = 0$) of the nickel-like molybdenum ions at a wavelength of 18.895 nm as an example, which was well-known as a strong lasing line in SXRL in collisional excitation scheme. We applied the external magnetic field to the gain-medium plasma and investigated the polarization components of the lasing line. This study is valuable as the demonstration of compact intense polarization x-ray source, and at the same time, it is quite interesting in terms of plasma physics, *i.e.*, the influence of the laser-plasma interaction on the magnetic fields in plasma. In the following, the details of the experiment and the analysis are described.

2. Experimental arrangement

Figure 1 shows the simplified experimental arrangement. A thin rod molybdenum target (the cross section was 1 mm × 1mm square and length was 30 mm) was set at the center of the small magnetic coil. Nd:glass laser at a wavelength of 1053 nm was weakly focused on the target surface under the grazing incident pumping (GRIP) configuration [7,8]. The laser light consisted of the pre-pulse and main pulse. The role of the pre-pulse is to generate a pre-formed plasma, by which the following main pulse can be effectively absorbed in plasma to generate population inversion. The total energy on the target was 12 J, and the energy ratio of these pulses was 1 : 4, respectively. The pulse separation was 2.0 ns, and the duration was 400 ps and 7 ps, respectively. The direction of the polarization of Nd:glass laser was perpendicular to the target surface. The grazing incidence angle was $\theta = 14$ degree, and the focal width and length were 60 μm and 5 mm, respectively. The electron density at which the laser light was mainly absorbed was given by $n_{cr} \sin^2\theta = 6 \times 10^{25}$ m^{-3}, where n_{cr} is the critical electron density at the wavelength of 1053 nm (= 10^{27} m^{-3}). The magnetic coil, the diameter of 4 mm and 5 mm-length, consisted of winding cupper sheet covered by polyimide foil. It is driven by a pulse power supplies with 20kV to obtain the magnetic field of 15 T. The raise up time of the magnetic field is around 800 ns, and the maximum field strength was sustained for several tens ns, which is sufficiently long compared with duration of the XRL pulse. (< 10 ps [9]).

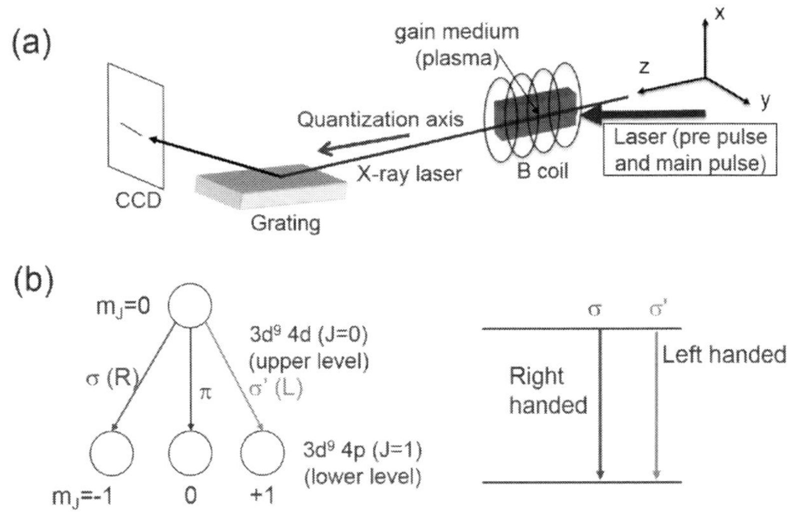

Figure1. Simplified experimental set-up; (a), and Kastler diagram of the lasing line; (b).

In the following context, we define the direction of the magnetic field, *i.e.*, z-axis in Fig. 1, as the quantization axis. Figure 1(b) shows the Kastler diagram of the lasing transition, $3d^9 4d$ ($J = 0$) levels → $3d^9 4p$ ($J = 1$). The lower lasing level ($3d^9 4p$ level) has three magnetic sublevels with $m_J = -1, 0, 1$, and the SXRL has three polarization components, *i.e.*, left-handed circular (σ'), linear (π) and right-handed circular polarization (σ). The SXRL beam is generated in the z-direction, therefore only the σ and σ' components could be observed. Under the sufficient magnetic field, the degeneracy of the lower magnetic sublevels can be resolved by the Zeeman effect, and we can extract the two polarization components separately. The advantage of this method is that the right- and left-handed circularly polarized SXRL are obtained at the same time.

The Zeeman splitting of the SXRL line was observed by a high-resolution spectrograph, HIREFS (Hettrick scientific) [10]. Figure 2 shows the optical arrangement of this spectrograph. The 3 μm-width entrance slit of the spectrograph was put at a distance of 70 mm from the edge of the

SXRL gain medium plasma, and the image of the SXRL at the slit was relayed to the detector position through a concave grating and two grazing imaging mirrors. The magnification factor in the direction of *x*-axis in Fig. 1 (a) (= the direction of the plasma expansion) and *y*-axis (= the direction of the wavelength dispersion) were 2.4 and 3.2, respectively. The inverse linear dispersion of HIREFS was measured to be 0.0785 nm / mm on the focal plane by using the Carbon Balmer α line (18.220 nm) and Ni-like Mo XRL line (18.895 nm). The back-illuminated CCD camera (Princeton, PI-SX:1K, pixel size 13 μm) was used for the detector, which corresponded to the spectral resolution of 0.00126 nm ($d\lambda/\lambda = 6.7 \times 10^{-5}$) [11]. The spatial resolution of HIREFS in the direction of the wavelength dispersion was 9.6 μm, therefore the resolution in the wavelength was mainly determined by the CCD pixel size.

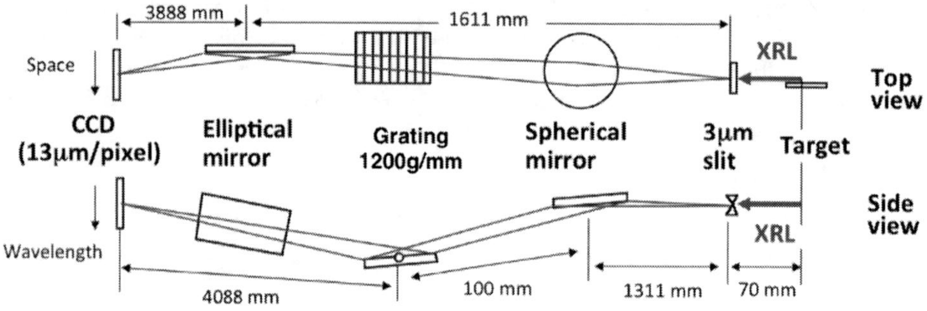

Figure 2. Schematic diagram of the optical arrangement of HIREFS spectrograph.

3. Results and discussion

At first we observed the gain narrowing of the nickel-like molybdenum SXRL line. We changed the plasma length from 3 mm to 5 mm and obtained apparent gain narrowing effect as shown in Fig. 3. The experimental uncertainty for the 3 mm case was larger than that of 5 mm case. This was because the intensity for the 3 mm case was weak due to small amplification effect. The linewidth of the lasing line for the 3 mm plasma length was 36mA. Approximately this linewidth could be treated as the intrinsic linewidth of the lasing line which was the convolution of homogeneous and inhomogeneous broadening. Our theoretical population kinetics model including plasma parameters, *i.e.*, electron temperature and density, implied that the homogeneous and inhomogeneous broadening was 28 mA and 15 mA, respectively. These values were almost consistent with the total linewidth obtained in the experiment.

Figure 4 (a) shows the spectrum of the lasing line without (upper) and with (lower) the external magnetic field. The vertical and horizontal axes show the wavelength and the refraction angle of the SXRL, respectively. Apparent split was obtained under the presence of the magnetic field, and these two components correspond to the left-handed circular (σ') and right-handed circular polarization (σ). The separation of the two components becomes larger as the refraction angle increases. The SXRL beam is typically refracted due to the density gradient in the gain medium plasma. Since the density gradient becomes steep as the electron density increases, the large refraction angle of the SXRL indicated that the source of the SXRL is in higher density region. That is, the magnetic field strength is higher as the electron density increases. Figure 4 (b) shows the spectral profiles of SXRL line at the position of (A) ~ (E) with the magnetic field in Fig. 4 (a). Under the assumption of linear Zeeman effect, the amount of the split is described as $2\Delta\varepsilon = \{J(J+1)\}^{1/2}\mu_B B$ ($= 1.6 \times 10^{-4} B$ [eV]), where μ_B is the Bohr magnetron and B is the magnetic field strength. This leads to the magnetic field strength of (A) 40 T, (B) 120 T, (C) 180 T, (D) 220 T, (E) 270 T, respectively; these values are quite high compared with the applied field, and this implies an alternative mechanism to enhance the magnetic field in plasma.

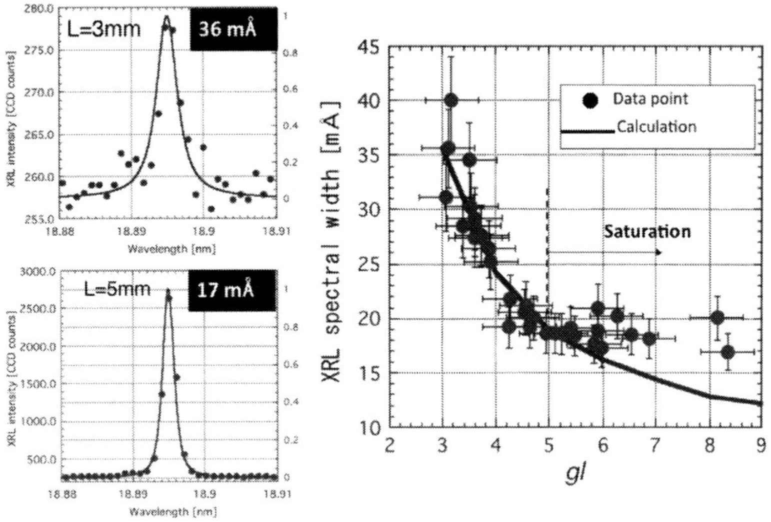

Figure 3. The linewidth of the lasing line with the plasma length of L=3 mm and L=5 mm. Apparent gain narrowing effect can be seen; (left-hand side). The linewidth of the SXRL line as the function of the gain-length product. For the plasma length longer than 5 mm, we can see the saturation behaviour; (right hand side).

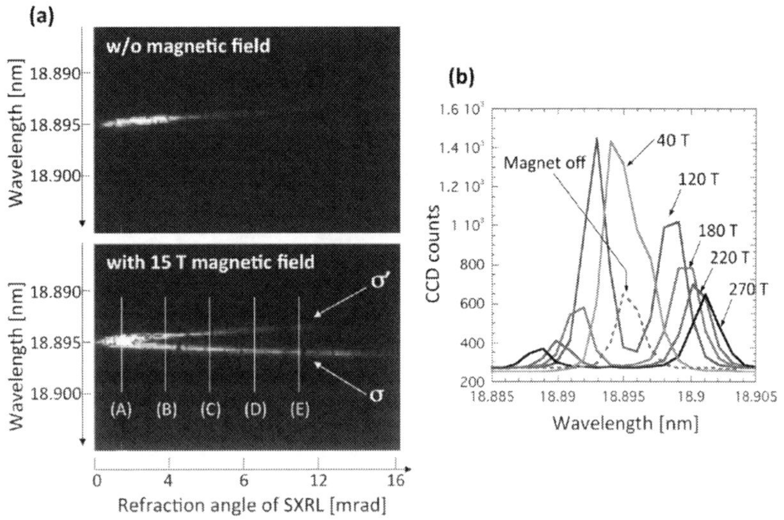

Figure 4. The obtained spectrum of the lasing line without (upper) and with (lower) the external magnetic field. The vertical and horizontal axes show the wavelength and the refraction angle of the SXRL, respectively; (left-hand side). The spectral profiles of SXRL line at (A) ~ (E) in (a). The Zeeman split of these profiles correspond to the magnetic field of (A) 40 T, (B) 120 T, (C) 180 T, (D) 220 T, (E) 270 T. These values are larger by a factor of 8~20 than that of the applied magnetic field.

Enhancement of compression of the magnetic field in laser-produced plasmas has been reported in several previous works, and the compression mechanisms are explained by the shock wave [12] and ponderomotive force [13, 14]. The spatial compression of the magnetic field requires 'frozen in' condition in the magnet hydrodynamics (MHD) equation. This means that the change of the electron density distributions due to the compression has to be faster than the relaxation time, t_c, of the magnetic field which is given by $t_c = \mu R^2/\eta$ [s], where η, μ and R are an resistivity, a magnetic permeability and the size of the gain region, respectively. η is in proportion to the collision frequency and is in proportion to $T_e^{-3/2}$, therefore large t_c can be obtained in high temperature plasmas such as the SXRL gain medium.

Indeed our theoretical investigation using 1D hydrodynamics simulation, HYADES [15] shows that sufficient plasma condition, $T_e = 0.8$ keV, $n_e = 6 \times 10^{25}$ m^{-3} and $R \sim 30$ μm, can be obtained at the position of the generation of SXRL gain at the time of 5-10 ps after the main laser pulse. From this estimation, the relaxation time of the magnetic field leads to be $t_c \sim 16$ ns ($\eta \sim 7 \times 10^{-8}$ Vm/A, $R \sim 30$ μm). This relaxation time is sufficiently long compared with the gain duration and the time for hydrodynamics motion near the critical density region, that is, and the spatial compression of the magnetic field can occur under sufficiently fast change of the electron density profile in plasma. It is noted that the hydrodynamics simulation indicated that the shock wave is not effective in the present plasma condition, rather, the possibility of the drift of the plasma by pondermotive force of laser pulse is realitic.

Under the present experimental condition (GRIP scheme), the ponderomotive force (F_p) originated from the gradient of the laser electric field pushes the plasma electrons in the direction of the density gradient (= x-direction), and the plasma is drifted to the higher density region with the lines of the magnetic force. We estimate the amount of the drift as follows: F_p is described as follows:

$$F_p = -\frac{1}{4} n_e \cdot m_e \cdot c_0^2 \nabla \left(\frac{qE}{m_e \omega_{Laser} c_0} \right)^2 \left[N/m^3 \right] \quad , \tag{1}$$

where, n_e, m_e, c_0, ω_{Laser}, q and E are electron density, electron mass, light velocity, angular velocity, electric charge and the strength of the electric field, respectively. We assume the electric field of the pump laser is Gaussian profiles, *i.e.*, $E(r, t) = E_0 \exp[-4\ln 2\,(r+r_0)^2 / \phi^2] \exp[-4\ln 2\,(t-\tau)^2 / \tau^2)$, then F_p can be described as

$$F_p = 5.86 \times 10^{-5} \frac{n_e I_0}{\omega_{Laser}^2} \frac{r + r_0}{\Phi^2} exp\left(\frac{-4ln2(r + r_0)^2}{\Phi^2} \right) exp\left(\frac{-4ln2(r + \tau)^2}{\tau^2} \right) \quad . \tag{2}$$

Here, ϕ, τ, t, r, and r_0 are the focal spot size of the pump laser, the duration, the time, the distance between the electron position and the laser axis at time t and that of the initial condition, respectively. Under the present laser parameters, $\lambda_{Laser} = 1053$ nm, $\phi = 60$ μm, $\tau = 7$ ps, $I_0 = 4.4 \times 10^{19}$ W/m^2, the peak of F_p is obtained to be 1.2×10^{14} N/m^3 at $r = 20$ μm and $n_e = 6 \times 10^{25}$ m^{-3}.

The pressure of the ponderomotive force was estimated from F_p and the thickness of the plasma (~ 100 μm), *i.e.*, 1.2×10^{10} N/m^3. It was larger than the internal pressure of plasma (= 8.2×10^9 N/m^2, at $T_e = 0.8$ keV) estimated from the result of HYADES code. The drift length of the electron depends on the initial position of the electrons (r_0) as shown in Fig. 5. Since this calculation does not include the repulsive force due to the compression, the validity of the result is limited to the qualitative estimation. In Fig. 5, for the initial positions of $r_0 = 30.0$ and 40.0 μm (initial distance = 10.0 μm), the drifted positions are estimated to be 60 and 61 μm (distance = 1 μm) at the time of 7 x 10^{-12} sec. Therefore more than 10 times spatial compression of the magnetic field can be expected. Present estimation indicates that the combination of the external magnetic field and the following spatial compression of magnetic field by the intense laser pulse is quite effective to generate higher transient magnetic field and is applicable for the extraction of the right- and left-handed circularly polarized XRL. For the detailed understanding of this dynamics, further analysis is needed.

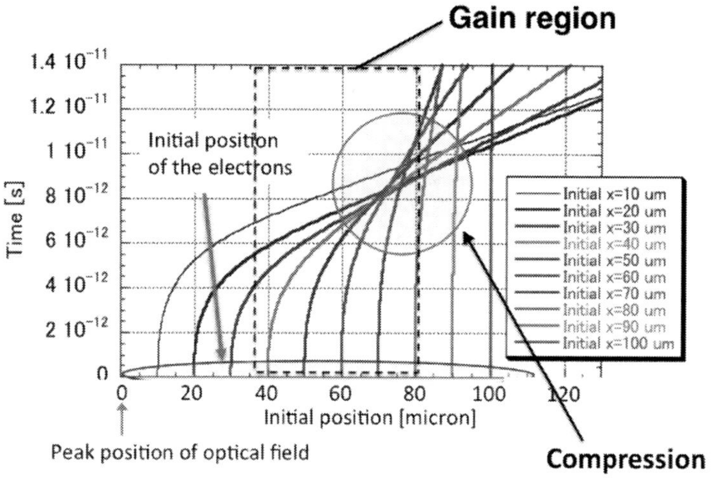

Figure 5. Rough estimation of the drift of the electrons for various initial positions. Due to the pondermotive force, electrons at each position are pushed into the higher density region and plasma compression occurs. For the time $t > 8 \times 10^{-12}$ s, several lines begin to cross each other, and this time region is out of the validity of the calculation. This is because the repulsive force due to the compression is not included in the model calculation.

4. Summary

In summary, we have demonstrated the extraction of right- and left-handed circularly polarized x-ray lasers using the Zeeman splitting. Under the external magnetic field of 15 T together with the spatial compression of the magnetic field by the ponderomotive force of the laser pulse, the effective magnetic field is enhanced by around 10 times larger, and we could observe the resolvent of the degeneracy of the magnetic sublevels of the nickel-like molybdenum XRL line at the wavelength of 18.895 nm. Present analysis is limited to the quantitative analysis, and detailed analysis including the coupling of ponderomotive force with hydrodynamics motion of the plasma is requires for the deep understanding of this dynamics.

Acknowledgement

This work was partly supported by Grand-in-Aid of Scientific Research (B) (No.25289244) and Grant-in-Aid for Young Scientists (B) (No. 19740350, 22740267) from the Japan Society for the Promotion of Science (JSPS). We thank Professor H. Nishimura, Institute of Laser Engineering (ILE) in Osaka University for his help to use HIREFS spectrometer and thank the x-ray laser crew in JAEA for their support to conduct the experiment.

References

[1] Fasman G D (Eds.) 1996 *Circular Dichroism and the Conformational Analysis of Biomolecules* (New York: Plenum Press)

[2] Hirano K, Okitsu K, Momose A and Amemiya Y 2001 *Advances in X-Ray Chemical Analysis, Japan* **33** 25

[3] Tanaka M, Nakagawa K, Agui A, Fujii K, Yokoya A 2005 *Phys. Scr. T* **115** 873

[4] Kalachnikov M P, Nickles P V, Schnürer M, Sandner W, Shlyaptsev V N, Danson C, Neely D, Wolfrum E, Zhang J, Behjat A, Demir A, Tallents G J, Warwick P J and Lewis C L S 1998 *Phys. Rev. A* **57** 4778

[5] Zeitoun Ph, Faivre G, Sebban S, Mocek T, Hallou A, Fajardo M, Aubert D, Balcou Ph, Burgy F, Douillet D, Kazamias S, de Lacheze-Murel G, Lefrou T, le. Pape S, Mercere P, Merdji H, Morlens A S, Rousseau J P and Valentin C 2004 *Nature* **431** 23

[6] Luther B M, Wang Y, Larotonda M A, Alessi D, Berrill M, Marconi M C, Rocca J J and Shlyaptsev V N 2005 *Opt. Lett.* **30** 165

[7] Keenan R, Dunn J, Shlyaptsev V N, Smith R F, Patel P K and Price D F 2003 *Proc. SPIE* **5197** 213

[8] Luther B M, Wang Y, Larotonda M A, Alessi D, Berrill M, Marconi M C, Rocca J J and Shlyaptsev V N 2005 *Opt. Lett.* **30** 165

[9] Ochi Y, Kawachi T, Hasegawa N, Sasaki A, Nagashima K, Sukegawa K, Kishimoto M, Tanaka M, Nishikino M and Kado M 2004 *Appl. Phys. B* **78** 961

[10] Hettrick M C, Underwood J H, Batson P J, and Eckart M J 1988 *Appl. Opt.* **27** 200

[11] Li Y, Nilsen J, Dunn K and Osterheld A L 1998 *Phys. Rev. A* **58** R2667

[12] Tomasel F G, Shlyaptsev V N, and Rocca J J 1996 *Phys. Rev A* 54 2474

[13] Sudan R N 1993 *Phys. Lett.* **70** 3075

[14] Sandhu A S., Dharmadhikari A K, Rajeev P P, Kumar G R, Sengupta S, Das A, and Kaw P K 2002 *Phys. Lett.* **89** 225002

[15] Pert G J 1983 *J. Fluid Mech.* **131** 401

Study of spectra for lanthanides atoms with relativistic many-body perturbation theory: Rydberg resonances

A A Svinarenko

Odessa State Environmental University, L'vovskaya str. 15, Odessa-16, 65016, Ukraine

E-mail: quantsvi@mail.ru

Abstract. Theoretical studies of Rydberg autoionization resonances in spectra of lanthanides atoms (ytterbium) are carried out within the relativistic many-body perturbation theory in the generalized relativistic energy approach (Gell-Mann and Low S-matrix formalism). The accurate theoretical results on the autoionization $4f^{13}[^2F_{7/2}]6s^2np[5/2]_2$, $4f^{13}$ $[^2F_{7/2}]$ $6s^2nf[5/2]_2$ resonances energies and widths are presented and compared with experimental data, obtained by using laser polarization spectroscopy method.

1. Introduction

In this work, we present the results of our theoretical studies of spectra and spectroscopic parameters for heavy atoms, namely, lanthanides atoms (see, for example [1]). Investigations of spectra, optical and spectral, radiative and autoionization characteristics for heavy elements atoms and multicharged ions are traditionally of a great interest for further developments of quantum atomic optics and atomic spectroscopy and different applications in the plasma chemistry, astrophysics, laser physics etc. (see Refs. [1–23]). Different atomic spectroscopy methods have been used in describing radiative and autoionization characteristics of atomic systems. The well known classical multi-configuration Hartree-Fock method (relativistic effects are taken into account in the Pauli approximation or Breit Hamiltonian) allowed us to obtain a great number of useful spectral information about light and/or not-heavy atomic systems, yet it results in only qualitative description of spectra of the heavy atoms and ions. The multi-configuration Dirac-Fock method is the most reliable for multielectron systems with a large nuclear charge. In this calculation approach, the one- and two-particle relativistic effects are taken into account practically precisely. In essence, special attention should be given to two very general and important computer systems for relativistic and Quantum Electro-Dynamics (QED) calculations of atomic and molecular properties developed at Oxford, UK, known as GRASP ("GRASP", "Dirac"; "BERTHA") (see Refs. [1-4] and references therein). Relativistic many-body perturbation theory is effectively applied to the computation of spectra of low-lying states for lanthanides atoms [5-10].

Interpretation of spectra, radiative and autoionization characteristics for heavy atoms is quite challenging because it is required to correctly account for the exchange-correlation and relativistic corrections. Here we present the results of the Rydberg Yb $4f^{13}[^2F_{7/2}]6s^2np[5/2]_2$, $4f^{13}$ $[^2F_{7/2}]$ $6s^2nf[5/2]_2$ states energies, widths and compare with experimental laser polarization spectroscopy data [23].

2. Advanced energy approach and relativistic many-body perturbation theory

As the method of computing has been presented in detail previously, here we summarize only the key elements. A model relativistic energy approach in a case of the multielectron atom has been proposed by Ivanov-Ivanova *et al.* [6-8] and its generalized gauge-invariant version has been developed in Refs. [9-12]. The approach is based on the Gell-Mann and Low S-matrix formalism, and the relativistic many-body perturbation theory that uses the optimized one-quasiparticle representation and an accurate account of the relativistic and exchange-correlation effects. In the relativistic case, the Gell-Mann and Low formula expressed an energy shift ΔE through the QED scattering matrix including the interaction of the photon vacuum field and of the laser field. The wave function zeroth basis is found from the Dirac equation with a potential, which includes *ab initio* optimized model potentials (Ivanov-Ivanova type [6]) or DF potentials, the electric potential of a nucleus – we usually use the Gaussian form of the charge distribution in a nucleus [4]. The correlation corrections of the perturbation theory of second and higher orders are taken into account with the polarization and screening potentials (see Refs. [10-16]).

Generally speaking, the majority of complex atomic systems possess a dense energy spectrum of interacting states with essentially relativistic properties. In the theory of the non-relativistic atom a convenient field procedure is applied for calculating the energy shifts ΔE of degenerate states. This procedure is connected to secular matrix M diagonalization [7, 8]. In constructing M, the Gell-Mann and Low adiabatic formula for ΔE is used. In contrast to the non-relativistic case, the secular matrix elements are already complex in the second order of the electrodynamical perturbation theory (first order of the interelectron interaction). Their imaginary part of ΔE is connected to radiative decay. In this approach, the whole calculation of the energies and decay probabilities of a non-degenerate excited state is reduced to the calculation and diagonalization of the complex matrix M. In published work by different authors, the Re{ΔE} calculation procedure has been generalized for the case of nearly degenerate states, whose levels form a more or less compact group. One of these variants has been introduced previously [4, 13, 14]: For a system with a dense energy spectrum, a group of nearly degenerate states is extracted and their matrix M is calculated and diagonalized. If the states are well separated in energy, the matrix M reduces to one term, equal to ΔE. The non-relativistic secular matrix elements are expanded in a Perturbation Theory (PT) series for the interelectron interaction. The complex secular matrix M is represented in the form [7,8]:

$$M = M^{(0)} + M^{(1)} + M^{(2)} + M^{(3)}. \tag{1}$$

Here, $M^{(0)}$ is the contribution of the vacuum diagrams of all order of PT, and $M^{(1)}$, $M^{(2)}$, $M^{(3)}$ those of the one-, two- and three- quasiparticle diagrams respectively. $M^{(0)}$ is real matrix, proportional to the unit matrix, but it determines only the general level shift. We assumed $M^{(0)} = 0$. The diagonal matrix $M^{(1)}$ can be represented as a sum of the independent one-quasiparticle contributions. For simple systems (such as alkali atoms and ions) the one-quasiparticle energies can be taken from the experiment. Substituting these quantities into (1) one could have summed all the contributions of the one -quasiparticle diagrams of all orders of the formally exact QED perturbation theory. However, the necessary experimental quantities are not often available. The first two order corrections to Re{$M^{(2)}$} have been analyzed [4] using Feynman diagrams. The contributions of the first-order diagrams have been completely calculated. In second order, there are two kinds of diagrams: Polarization and ladder ones. The polarization diagrams take into account the quasiparticle interaction through the polarizable core, and the ladder diagrams account for the immediate quasiparticle interaction. Some of the ladder diagram contributions as well as some of the three-quasiparticle diagram contributions in all PT orders have the same angular symmetry as the two-quasiparticle diagram contributions of the first order. These contributions have been incorporated by a modification of the central potential, which must now include the screening (anti-screening) of the core potential of each particle by the two others. The additional potential modifies the one-quasiparticle orbitals and energies. Therefore, the secular matrix can be written as follows:

$$M \to \tilde{M}^{(1)} + \tilde{M}^{(2)}, \tag{2}$$

where $\tilde{M}^{(1)}$ is the modified one-quasiparticle matrix (diagonal), and $\tilde{M}^{(2)}$ the modified two-quasiparticle one. $\tilde{M}^{(1)}$ is calculated by substituting the modified one-quasiparticle energies, and $\tilde{M}^{(2)}$ by means of the first PT order formulae for $M^{(2)}$, putting the modified radial functions of the one-quasiparticle states in the radial integrals.

In QED theory, the photon propagator D(12) plays the role of the interaction. Naturally the analytical form of D(12) depends on the gauge, in which the electrodynamical potentials are written. The interelectron interaction operator that accounts for the Breit interaction has been taken as follows:

$$V(r_i r_j) = exp(i\omega r_{ij}) \cdot \frac{(1 - \alpha_i \alpha_j)}{r_{ij}} + V_{ee}^{w}, \tag{3}$$

where, as usual, α_i are the Dirac matrices. In general, the results of all approximate calculations depend on the gauge. Yet the correct result must be gauge-invariant. The gauge dependence of the amplitudes of the photo processes in the approximate calculations is well known. This is investigated in detail by Grant, Armstrong, Aymar and Luc-Koenig, Glushkov-Ivanov *et al.*, see Refs. [1-4,9-15]. Grant has investigated the gauge connection with the limiting non-relativistic form of the transition operator and has formulated the conditions for approximate functions of the states, in which the amplitudes of the photo processes are gauge invariant. These results remain true in the energy approach because the final formulae for the probabilities coincide in both approaches. Glushkov-Ivanov developed a new relativistic gauge-conserved version of the energy approach [11]. Here we apply this approach for generating the optimized relativistic orbitals basis in the zeroth approximation of the many-body perturbation theory [1]. Below we will be interested in studying the spectra of the autoionization resonances in the ytterbium atom and calculate their energies and widths. The excited states of the ytterbium atom can be treated as the states with two-quasiparticles above the electron core [Xe]$4f^{14}$.

Within the standard energy approach [8-11], the autoionization width is determined by the square of an electron interaction matrix element having the form:

$$V_{1234}^{\omega} = [(j_1)(j_2)(j_3)(j_4)]^{1/2} \sum_{\lambda\mu} (-1)^{\mu} \begin{pmatrix} j_1 j_3 & \lambda \\ m_1 - m_3 & \mu \end{pmatrix} \times \mathrm{Re}\, Q_\lambda(1234) \tag{4}$$

The real part of the electron interaction matrix element is determined using expansion in terms of Bessel functions [17-19, 26]:

$$\frac{\cos|\omega| r_{12}}{r_{12}} = \frac{\pi}{2\sqrt{r_1 r_2}} \sum_{\lambda=0} (\lambda) J_{\lambda+1/2}(|\omega| r_<) J_{-\lambda-1/2}(|\omega| r_>) P_\lambda(\cos \widehat{\mathbf{r}_1 \mathbf{r}_2}). \tag{5}$$

The Coulomb part Q_λ^{Qul} is expressed in terms of radial integrals R_λ, angular coefficients S_λ:

$$\mathrm{Re}\, Q_\lambda^{\mathrm{Qul}} = \frac{1}{Z} \mathrm{Re}\{R_l(1243) S_\lambda(1243) + R_\lambda(\tilde{1}24\tilde{3}) S_\lambda(\tilde{1}24\tilde{3}) + R_\lambda(1\tilde{2}4\tilde{3}) S_\lambda(1\tilde{2}4\tilde{3}) + R_\lambda(\tilde{1}\tilde{2}\tilde{4}\tilde{3}) S_\lambda(\tilde{1}\tilde{2}\tilde{4}\tilde{3})\} \tag{6}$$

As a result, the autoionization decay probability is expressed in terms of $\mathrm{Re}Q_\lambda(1243)$ matrix elements. We present an example below:

$$\mathbf{Re}\, R_\lambda(1243) = \iint dr_1 r_1^2 r_2^2 f_1(r_1) f_3(r_1) f_2(r_2) f_4(r_2) Z_\lambda^{(1)}(r_<) Z_\lambda^{(1)}(r_>). \tag{7}$$

Here, f is the large component of radial part of single electron state Dirac function and the function Z is connected with the Bessel functions. The angular coefficient is defined in the standard way as above [26]. The Breit part of Q is defined in similar way as above, but our interest focuses on the real part. The Breit interaction is known to change considerably the Auger decay dynamics in some cases (see, for example, Refs. [8]). Determination of the radiation decay probabilities (oscillator strengths) simplifies to calculating the imaginary matrix elements of the interaction. The calculation of radial integrals $\mathbf{Re}R_\lambda(1243)$ is reduced to the solution of a system of differential equations according to the Ivanova-Ivanov method [26]. The system of differential equations includes also equations for functions $f/r^{|æ|-1}$, $g/r^{|æ|-1}$, $Z_\lambda^{(1)}, Z_\lambda^{(2)}$. The formulas for the autoionization (or "Auger") decay probability include the radial integrals $R_\alpha(\alpha k\gamma\beta)$, where one of the functions describes electron in the continuum state. When calculating this integral, the correct normalization of the function Ψ_k needs attention. The correctly normalized function should have the following asymptotic behaviour for $r\to 0$:

$$\left.\begin{matrix} f \\ g \end{matrix}\right\} \rightarrow (\lambda\omega)^{-1/2} \begin{cases} \left[\omega + (\alpha Z)^{-2}\right]^{-1/2} sin(kr+\delta), \\ \left[\omega - (\alpha Z)^{-2}\right]^{-1/2} cos(kr+\delta). \end{cases} \quad (10)$$

It is important to note that the calculation is carried out in the jj-coupling scheme representation. The transition to the intermediate coupling scheme has been accomplished by diagonalizing the secular matrix. Indeed, only Re{M} should be diagonalized. The imaginary part is converted by means of the matrix of eigenvectors $\{C_{mk}\}$, obtained by diagonalization of Re{M}:

$$\mathrm{Im} M_{mk} = \sum_{ij} C^*_{mi} M_{ij} C_{jk} \quad (12)$$

M_{ij} are the matrix elements in the jj-coupling scheme, and M_{mk} in the intermediate coupling scheme representation. Other details can be found in Refs. [6-11,13-19].

3. Results

In Table 1 we present the experimental data (Jong-Hoon et al [22,23]) and theoretical results for energies and widths of the excited (autoionization) states of the $4f^{13}$ $[^2F_{7/2}]6s^2np[5/2]_2$ and $4f^{13}$ $[^2F_{7/2}]$ $6s^2$ nf $[5/2]_2$ states (because of excitation of the 4f shell) .

Table 1. Energies E (cm^{-1}) and widths Γ(cm^{-1}) of the $4f^{13}$ $[^2F_{7/2}]6s^2np[5/2]_2$ and $4f^{13}$ $[^2F_{7/2}]$ $6s^2$ nf $[5/2]_2$ states.

	$4f^{13}$ $[^2F_{7/2}]$ $6s^2$ $np[5/2]_2$				$4f^{13}$ $[^2F_{7/2}]$ $6s^2$ $np[5/2]_2$			
N	Exp. E_{exp}	Exp. Γ_{exp}	Theory E	Theory Γ	Exp. E_{exp}	Exp. Γ_{exp}	Theory E	Theory Γ
12	70120.5	1.5	70121	1.7	70963.6	0.5	70965	0.7
13	70482.0	0.4	70483	0.5	71105.0	0.4	71107	0.5
15	70914.8	1.2	70916	1.4	71312.2	1.4	71313	1.6
20	71428.1	0.6	71429	0.7	71559.1	0.8	71561	0.9
25	71612.5	1.3	71611	1.5	71672.5	0.5	71673	0.8
26	71633.3	0.6	71631	0.8	71687.5	0.5	71689	0.7
30	71698.8	0.5	71697	0.7	71732.4	0.4	71734	0.5
46	-	-	71797	0.4	-	-	71813	0.3

The smallness of the resonance widths requires attention. In our interpretation, narrow resonances are related to the complex energy structure of the 4f-shell atoms. Consequently, unusual behavior occurs for autoionization resonances and their decay mechanisms, especially in comparison with the standard spectroscopy (for inert gases, alkali atoms). We emphasize that that our theory provides a physically reasonable agreement with experimental results. Finally, it should be noted that the studied autoionization resonances (Rydberg states are more preferable) can be very useful, for example, in new laser photo-ionization schemes of separating the Yb isotopes and nuclear isomers [24].

Acknowledgements.
The author would like to thank Dr. Christian G. Parigger for invitation to make contributions to ICSLS-22 (Tullahoma, Tennessee, USA, 2014), help in preparing the manuscript and invaluable advice. The useful comments of the anonymous referees are acknowledged too.

References

[1] Glushkov A V, Khetselius O Yu and Svinarenko A A 2013 *Phys. Scr.* **T153** 014029
[2] Quiney H 2002 *New Trends in Quantum Systems in Chemistry and Physics, Series: Progress in Theoretical Chemistry and Physics* vol 6 (Berlin: Springer) p 135
[3] Bell K L, Berrington K, Crothers D, Hibbert A and Taylor K T 2002 *Supercomputing, Collision Processes and Application, Series: Physics of Atoms and Molecules* (Berlin: Springer) p 213
[4] Glushkov A V, Khetselius O Yu, Loboda A V, Svinarenko A A, Gurnitskaya E P, Florko T A, Sukharev D E and Lovett L 2008 *Frontiers in Quantum Systems in Chemistry and Physics, Series: Progress in Theoretical Chemistry and Physics* vol 18, ed S Wilson, P J Grout, J Maruani, G Delgado-Barrio and P Piecuch (Berlin: Springer) p 505
[5] Safronova U I, Johnson W R , Safronova M S and Albritton J R 2002 *Phys. Rev.* A **66** 022507
[6] Ivanova E P and Ivanov L N 1979 *Atom. Data. Nucl. Data. Tabl.* **24** 95
[7] Ivanov L N and Letokhov V S *Com. Mod. Phys.* D **4** 169
[8] Bekov G, Vidolova-Angelova E, Ivanov L N, Letokhov V S and Mishin V I 1981 *JETP.* **80** 866
[9] Ivanova E P and Glushkov A V 1986 *J. Quant. Spectrosc. Rad. Transfer* **36** 127
[10] Glushkov A V, Malinovskaya S, Prepelitsa G and Ignatenko V 2005 *J. of Phys.: Conf. Ser.* **11** 199
[11] Glushkov A V and Ivanov L N 1992 *Phys. Lett.* A **170** 33
[12] Glushkov A V 2012 *Advances in the Theory of Quantum Systems in Chemistry and Physics, Series: Progress in Theoretical Chemistry and Physics* vol 28, ed K Nishikawa, J Maruani, E Brandas, G Delgado-Barrio and P Piecuch (Berlin: Springer) p 131
[13] Khetselius O Yu 2012 *J. of Phys.: Conf. Ser.* **397** 012012
[14] Malinovskaya S V, Glushkov A V, Khetselius O Yu, Loboda A V, Lopatkin Yu M, Svinarenko A A, Nikola L V and Perelygina T B 2011 *Int. J. Quant. Chem.* **111** 288
[15] Glushkov A, Malinovskaya S, Prepelitsa G and Ignatenko V 2005 *J. of Phys.: Conf. Ser.* **11** 199
[16] Glushkov A V, Ambrosov S V, Loboda A V, Gurnitskaya E P and Prepelitsa G P 2005 *Int. J. Quant.Chem.* **104** 562
[17] Glushkov A V, Ambrosov S, Ignatenko A and Korchevsky D 2004 *Int. J. Quant. Chem.* **99** 936
[18] Glushkov A V, Khetselius O Yu and Malinovskaya S V 2008 *Europ. Phys. J.* ST **160** 195
[19] Glushkov A V 2012 *J. of Phys.: Conf. Ser.* **397** 012011
[20] Khetselius O Yu 2009 *Phys.Scr.* **T.135** 014023
[21] Khetselius O Yu 2009 *Int. J. Quant. Chem.* **109** 3330
[22] Glushkov A V, Malinovskaya S V, Sukharev D E, Khetselius O Yu, Loboda A V and Lovett L 2009 *Int. J. Quant. Chem.* **109** 1717
[23] Jong-hoon Yi, Park H and Lee J 2001 *J. Korean Phys.Soc.* **39** 916
[24] Letokhov V S and Ivanov L N 1975 *Quantum Electr.* **2** 585

Emission spectroscopy of nitric oxide in laser-induced plasma

J O Hornkohl[1,2], J P Fleischmann, D M Surmick[1], M J Witte[1], L D Swafford[1], A C Woods[1], and C G Parigger[1]

[1]University of Tennessee Space Institute, 411 B.H. Goethert Parkway, Tullahoma, TN 37388, USA

[2]Hornkohl Consulting, Tullahoma, TN 37388, USA

E-mail: dsurmick@utsi.edu, cparigge@utsi.edu

Abstract. In this paper, we investigate the laser-induced breakdown spectra of nitric oxide. Nitric oxide spectra are studied from laser-induced plasma emissions from plasma initiated both in laboratory air. Temperatures are inferred from the spectroscopic emissions using two methods. Spectra are fit to theoretical calculations of the diatomic spectra using the method of diatomic line strengths. For a time delay of 25 μs the temperature is found to be 6800 Kelvin. Comparisons are also provided to previously determined temperatures using a non-equilibrium air radiation fitting (NEQAIR) program.

1. Introduction

Analysis of spectra from laser-induced plasma is of interest in many fields of study and is commonly employed with the use of laser-induced breakdown spectroscopy (LIBS) which has become a popular experimental technique given its ease, low costs, and minimal sample preparation. One of the many applications of LIBS is use in diagnostic studies, such as combustion studies where quantitative analysis is applied to both atomic and molecular spectra [1]. In such works diatomic species of interest were CN, N_2, and NH [2, 3]. Of particular interest to combustion are quantitative calculations and analysis of nitric oxide (NO) spectra for studies of nitrogen containing flames and exhaust plumes of rockets. Considered is the near UV emission of the NO γ-band in the 140-340 nm range [4].

In order to quantitatively study the emission of diatomic molecules, a proper comparison to theory is required. We chose to employ the diatomic line strength method for calculations of theoretical NO spectra. The line strength is a popular method for calculating diatomic, molecular spectra due to its symmetric nature [5]. The formal definition of the line strength is the total sum of degenerate states that produce the same spectral line

$$S_{ul} = \sum_u \sum_l |< u \mid T_k^{(q)} \mid l >|^2 \qquad (1)$$

where u and l represent the upper and lower states and $T_k^{(q)}$ is the kth component of the irreducible tensor operator of rank q that gives rise to the spectral transition. This operator is the electric dipole operator and the line strength may be expressed as

$$S_{ul} = S(n'v', nv)S(J', J) \qquad (2)$$

where $S(n'v', nv)$ is the electronic-vibration transition strength and $S(J', J)$ is the Hönl-London factor. Expressing the line strength as in Equ. 2 allows for a direct analytical calculation of the line strength, and in turn, the intensity of a diatomic emission spectra which may be expressed as

$$I_{ul} = \frac{16\pi^3 c(a_0 e)^2 C_{abs} N_0}{3\epsilon_0 Q} C_\nu \tilde{v_{ul}}^4 S_{ul} e^{-hF_u/kT} \tag{3}$$

where a_0 is the Bohr radius, e is the electron charge, c is the speed of light, Q is the partition function, N_0 is the total population of the species, C_{abs} and C_ν are the absolute and relative calibration factors, h is Planck's constant, F_u is the upper term value, k is Boltzmann's constant, and T is the temperature. A simple, yet rigorous selection rule results from the expression in Equ. 2. Given the relative ease in analytically calculating the Hönl-London factor and the ability for the Hönl-London factor to be analytically zero, allowed transitions are those with a non-vanishing line strength. In this study we apply this method to infer the emission temperature of the NO emission spectra from measurements made in the nominal 230-240 nm range.

2. Experimental Details

To study the laser-induced breakdown spectra of nitric oxide, laser-induced plasma were initiated in laboratory air. Optical breakdown is initiated by tightly focussing 1064 nm, 12 nanosecond pulsed laser radiation from a Q-switch, Nd:YAG laser operating in its fundamental mode with an average energy per pulse of 190 mJ. Light from the plasma emission is dispersed by a 0.64 meter Jobin-Yvon czerny turner spectrometer installed with 3600 grooves/mm grating. Spectra are recorded with an intensified, linear diode array coupled to an optical multichannel analyzer. Time synchronization for the laser pulse firing rate and the measurement rate of the linear diode array is achieved with the use of waveform, pulse, and delay generators. Exact details on the timing apparatus used in this experiment and in other, similar laser-induced plasma studies involving atomic and diatomic spectra is provided in reference [6]. Prior to analysis, all spectra were properly calibrated for the detector background and sensitivity response.

3. Results

The NO spectra are fit to theoretical calculations tabulated from the line strength method. Comparisons between experiment and theory are made using a Nelder-Mead algorithm, which is used for its ability to use fit multiple parameters simultaneously, including a variable baseline offset. The Nelder-Mead minimizes parameters by iteratively reducing the size of a geometric simplex where the number of vertices in the simplex is one more than the number of parameters to be fit [7, 8]. The parameters considered while fitting were the temperature, spectral resolution, and offset. The spectral resolution parameter is held constant for each fit and is determined from the physical limitations of the experimental apparatus. Given the wavelength region and the high groove density of the grating used, the spectral resolution of the experimental spectra was 0.056 nm which also accounts for slit width effects. The constant offset parameter was allowed to vary. NO spectra were collected and analyzed for 25, 50, and 75 μs time delays for air breakdown experiments and the inferred temperatures were found to be 6800, 6100, and 5800 Kelvin for 25, 50, and 75 μs time delays, respectively. The fitting results are depicted in Figure 1.

The temperature results from using the diatomic line strength are in accordance with previously completed NO LIBS experiments in which breakdown was initiated in a 1:1 mixture of N_2 and O_2 gas near atmospheric pressure. The spectra from that work were analyzed with the use of the non-equilibrium air radiation NASA Gordon-McBride program NEQAIR and the temperatures were found to be 6700 and 6000 Kelvin for 25 and 50 μs time delays, respectively [2, 9, 10]. It is important to note, however, that the approximate 1 nm spectral resolution of the previous study was significantly lower than that of the presented results. As can be seen in

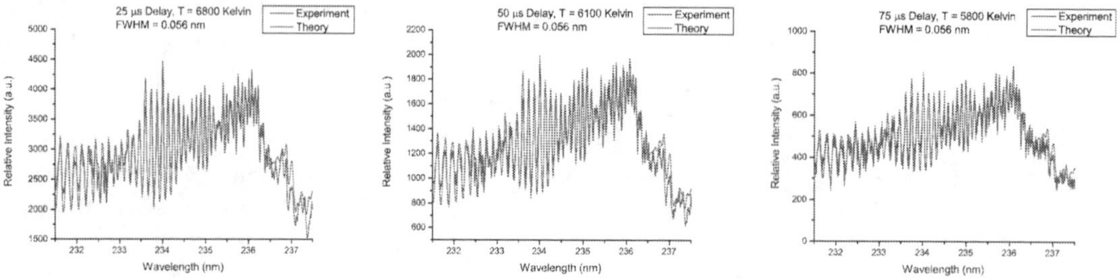

Figure 1. Fits to NO laser-induced plasma spectra at 25, 50, and 75 μs time delays with inferred temperatures of 6800, 6100, and 5800 Kelvin, respectively.

the fits shown in Figure 1, the diatomic line strength method has the capability to calculate and fit highly resolved spectra which gives the method the ability to include many rotational lines. Taken in this context the line strength method for calculating diatomic spectra provides a viable option for quantifying diatomic spectraand should be viewed as useful for combustion studies which often employ laser-induced florescence as a diagnostic method with spectra that often have comb-like structures. In such spectra the highly resolved nature of a spectra calculated from the line strength is applicable [11]. This is found to be particularly useful in studies of combustion with high temperature flames [12].

4. Conclusions

In this work, we have inferred the temperature of laser-induced plasma following optical breakdown in laboratory air from NO emission spectra. The plasma temperature was inferred by fitting experimental to theoretically calculated spectra with the use of a Nelder-Mead algorithm. The theory spectra were calculated using the method of diatomic line strengths and the temperatures are found to be 6800, 6100, and 5800 Kelvin for 25, 50, and 75 μs time delays. In comparison to previous experiments in which NO spectra were fit using the NEQAIR program, the determined temperatures are found to closely agree for similar time delays. Given the ability for the diatomic line strength method to calculate highly resolved spectra, the ability to observe highly resolved spectra, and close agreement with other methods, the diatomic line strength method is a viable method to quantify diatomic, molecular spectra.

Acknowledgments

This work is supported by the Center for Laser Applications at the University of Tennessee Space Institute.

References

[1] Drakes JA, Pruitt DW, Howard RP, Hornkohl JO 1997 *J. Quant. Spectrosc. Radiat. Transf.* **57** 23
[2] Lewis JWL, Parigger CG, Hornkohl JO, and Guan G 1999 *AIAA-99-0723*
[3] Plemmons DH, Parigger C, Lewis JWL, and Hornkohl JO 1998 *Appl. Opt.* **37** 2493
[4] Ochkin VN 2009 (Weinheim:Wiley)
[5] Hornkohl JO, Parigger CG and Nemes L 2005 *Appl. Opt.* **44** 3686
[6] Parigger CG, Woods AC, Witte MJ, Swafford LD, and Surmick DM 2014 *J. Vis. Exp.* **84**
[7] Nelder JA and Mead R 1965 *The Comp. J.* **7** 308
[8] Lagarias JC, Reeds JA, Wright MH and Wright PE 1998 *SIAM J. Optim.* **9** 112
[9] Parigger CG, Lewis JWL, Plemmons DH, and Hornkohl JO 1996 *Laser Appl. Chem. Bio. Env. Anal.* **3** 85
[10] Park C 1885 NASA TM 86707, Ames Reasearch Center, Moffet Field, CA 94035
[11] Hammack SD, Carter CD, Gord JR, and Lee T 2012 *Appl. Opt.* **51** 8817
[12] Schultz C, Sick V, Heinze J, and Stricker W 1997 *Appl. Opt.* **36** 3227

Journal of Physics: Conference Series **548** (2014) 012041 doi:10.1088/1742-6596/548/1/012041

Experimental Stark parameters of Cr II spectral lines

J. Manrique[a], J.A. Aguilera[b], C. Aragón[b],

[a]Facultad de Farmacia, Universidad CEU San Pablo, Urbanización Montepríncipe, Boadilla del Monte, E-28668 Madrid, Spain
[b]Departamento de Física, Universidad Pública de Navarra, Campus de Arrosadía, E-31006 Pamplona, Spain

E-mail: jamanros@ceu.es

Abstract. Stark broadening and shift parameters of Cr II lines with wavelengths in the range 2000-3500 Å have been measured by laser-induced breakdown spectroscopy. The plasmas were generated from fused glass samples selected with different chromium concentrations to avoid self-absorption which would lead to distorted line profiles. The spectra have been measured at different instants of the plasma evolution from 0.6 to 3.4 μs, at which the temperature and electron density are in the ranges 12000-16300 K and (0.89-8.2) × 10^{17} cm^{-3}, respectively. The electron density has been obtained from the Stark broadening of the H_α line.

1. Introduction

Stark broadening and shift parameters are essential for the analysis of the astrophysical and laboratory plasmas. In the particular case of chromium, a considerable number of lines have been identified in the spectra of A-type stars [1-2] where the Stark broadening is the most important pressure broadening mechanism. Despite their relevance, there is a need of experimental data for Cr II Stark widths and shifts since only measurements for three lines, performed with a T-tube in 1984 [3], are available. Theoretical data have been calculated for seven Cr II multiplets from the 4s-4p transition [4] using the semiclassical perturbation approach whereas recently nine resonant $3d^5$-$3d^4$4p multiplets have been obtained within the same approach [5]. In the latter work, after having applied the available data for stellar atmosphere modelling, the authors conclude the need of new Stark parameters for the Cr II lines. The aim of this paper is to provide reliable experimental Stark widths and shifts for several multiplets not previously measured and verify the available values in the literature. Our group has performed laser-induced breakdown spectroscopy to measure Stark parameters, see for example [6], with a procedure based on the curve of growth methodology which allows to control the self-absorption of the lines [7]. Recently the use of fused glass samples has been shown to improve the signal-background ratio in these type of measurements [8]. Hence in the present work we utilize this method and report data for Stark widths and shifts of ionized chromium lines with wavelengths in the range 2000-3500 Å.

2. Experiment

The experimental system consists in a Nd:YAG laser (wavelength 1064 nm, pulse energy 60 mJ, pulse width 4.5 ns, repetition rate 20 Hz) focused onto a sample placed in air at atmospheric pressure to

Journal of Physics: Conference Series **548** (2014) 012041 doi:10.1088/1742-6596/548/1/012041

generate the plasma. The focusing lens has 128 mm focal length whereas the lens-to-sample distance is 124 mm. The light emitted from the plasma is collected in a direction forming a small angle with the laser beam by a system of flat and concave mirrors forming an image on the entrance slit of a 0.75 m focal length Czerny-Turner spectrometer. The slit width (20 and 50 μm) and the grating (3600 and 1200 lines mm^{-1}) are selected depending on the spectral resolution needed. Spectra were recorded by a time-resolved intensified CCD (1200 × 256 effective pixels) at different delays from the laser pulse in synchronism with the electronic trigger of the laser Q-switch. For every period of data acquisition the spectrum of 100 laser shots is accumulated while the sample rotates at 100 rev min^{-1}.
In these measurements three fused glass samples have been used with chromium atomic concentrations of 0.05, 0.1 and 0.2 per cent, previously prepared from borate fusion of pure chromium oxide. One extra sample was prepared with low Fe II concentration to measure plasma temperature.

3. Results and discussion

The spectra have been recorded with six time windows centred at instants from 0.6 to 3.4 μs, having widths from 0.06 to 0.6 μs, and with an extra window centred at 10 μs to obtain a negligible line shift. Plasma temperature was determined by the Boltzmann plot of seven well known Fe II spectral lines, whereas the electron density was obtained by using the Stark broadening of the H_α line [9], whose emission arises from the small water content of air surrounding the sample. In order to perform the latter measurements it was necessary to use the grating of 1200 lines mm^{-1} and a spectrometer slit of 50 μm. During the plasma evolution the temperature and electron density decrease from 16300 to 12000 K and from 8.2 to 0.89 × 10^{17} cm^{-3} respectively, as shown in figures 1 and 2.

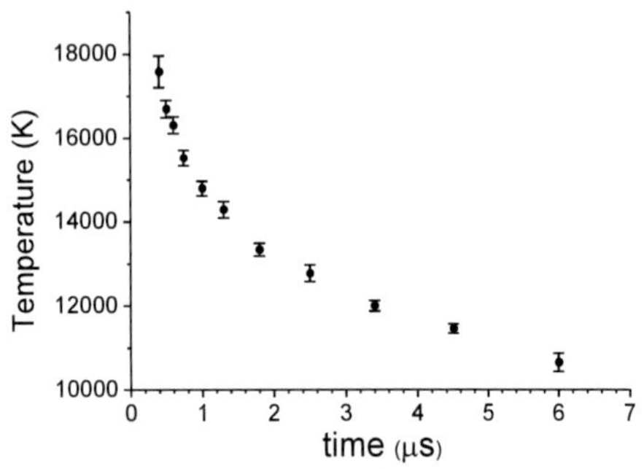

Figure 1. Plasma temperature obtained by Boltzmann plot from seven Fe II spectral lines.

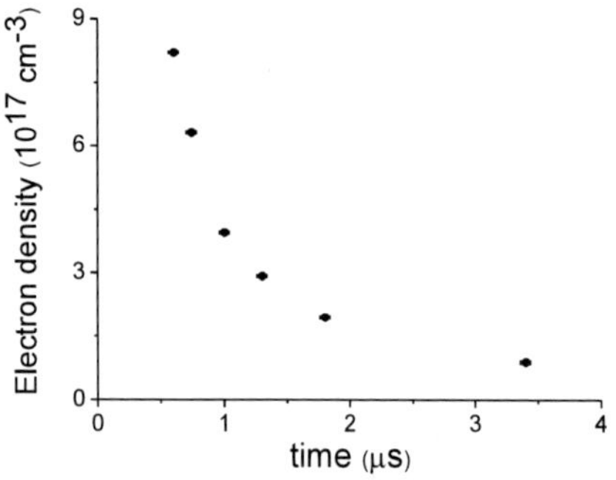

Figure 2. Electron density measured using the H_α broadening.

Self-absorption has been studied with the help of the k_t coefficient as defined in [7], thus atomic data related to the spectral line, plasma temperature and emitter ion density are required together with geometrical factors. For the lines studied, a chromium atomic concentration of 0.05 per cent was needed to ensure self-absorption lower than ten per cent for the more intense lines. Other two samples with 0.1 and 0.2 per cent concentration were used with less intense lines in order to improve the signal-background ratio.

Line widths and shifts have been obtained by fitting the recorded spectral lines to Voigt profiles using home-made software. As is well known, the Lorentzian component corresponds to the Stark width. The Gaussian component has been determined as the combination of the instrumental width and the Doppler width, the first value being measured as 13.5 pm using a low-pressure Hg lamp, and the second one being calculated as 3 pm for a typical temperature of 14000 K. Other broadening mechanisms are negligible in comparison with the electron impact contribution. In figure 3 the fitting of the 2061.575 Å line at the time window centred at 1 μs is shown as an example.

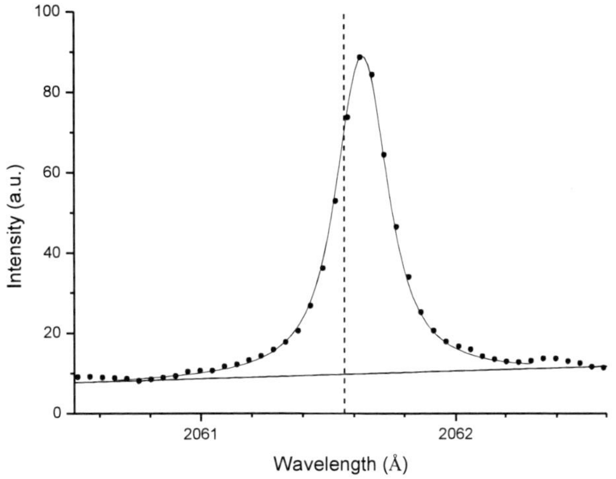

Figure 3. Experimental spectrum of the Cr II line at 2061.575 and fitting to a Voigt profile. The dashed line shows the position of the unshifted line.

The obtained values were proportional to the electron density during the evolution of the plasma, which denotes that the weak dependence of Stark width and shift on temperature is not observed in our range. Therefore, the final value of Stark width (FWHM) and shift at an electron density of 10^{17} cm^{-3} corresponds to the slope of the linear fitting of these magnitudes as shown in figures 4 and 5.

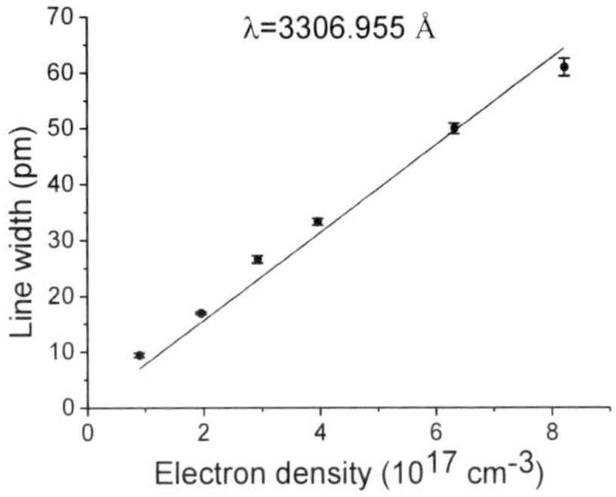

Figure 4. Line width of Cr II line at 3306.955 Å as a function of the electron density.

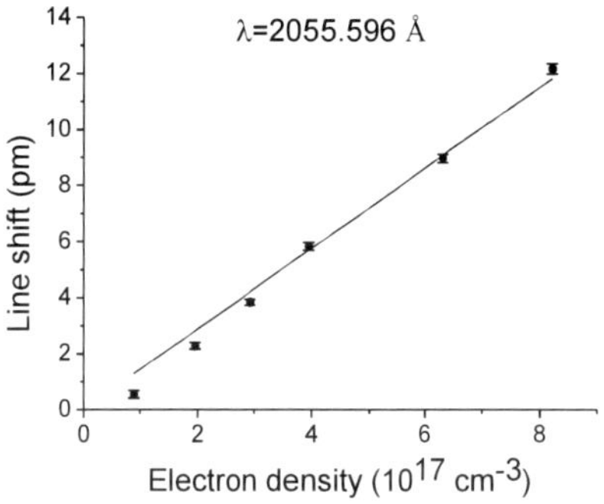

Figure 5. Line shift of Cr II line at 2055.596 Å as a function of the electron density.

The results for transitions belonging to several multiplets, including resonance lines, are presented in table 1 together with previous data from the literature. The complete results, including 83 widths and 49 shifts have been reported in [10].

Table 1. Stark widths *w* (pm) and shifts *d* (pm) at electron density 10^{17} cm^{-3} of Cr II lines.

				Experimental				Theoretical			
Transition	Multiplet	λ(Å)	k_t (10^{-30} m^3)	w	d	w^a	d^a	w^b	w^c	d^b	d^c
$3d^5$ - $3d^4(^5D)4p$	$a\,^6S$ - $z\,^6P^*$	2055.596	6.32	3.7	1.4				3.42		-0.0402
		2061.575	4.76	3.4	1.4				3.42		-0.0402
$3d^4(^5D)4s$ - $3d^4(^5D)4p$	$a\,^4D$ - $z\,^4F^*$	3132.053	7.60	7.2	-0.1	34	-10	27.8		-9.06	
		3124.973	5.47	7.4	-0.1	26	-12	27.8		-9.06	
		3120.359	3.67	7.5	-0.2	22.6	-11	27.8		-9.06	
		3118.646	2.30	7.6	-0.2			27.8		-9.06	
		3147.220	0.832	7.5	-0.2			27.8		-9.06	
		3128.692	0.856	7.1	-0.2			27.8		-9.06	
$3d^5$ - $3d^4(a\,^3F)4p$	$c\,^2F$ - $y\,^2G^*$	3306.955	0.189	7.8	0.0						
$3d^4(^3D)4s$ - $3d^4(^3D)4p$	$b\,^2D$ - $w\,^2F^*$	2941.957	0.418	6.6	0.7						

[a] Temperature 13700 K, experimental [3].
[b] Data interpolated to a temperature of 14000 K, semiclassical perturbation approach [4].
[c] Data interpolated to a temperature of 14000 K, semiclassical perturbation approach [5].

4. Conclusions

We present measurements of Stark widths and shifts of Cr II lines carried out by LIBS, some of them firstly measured. The curve of growth methodology and the use of fused glass samples have allowed to control the self-absorption of the lines.

Acknowledgment
This work has been supported by the project FIS2011-29521 of the Spanish Ministerio de Economía y Competitividad.

References

[1] Adelman J, Philip A G 1996 *MNRAS* **282** 1181
[2] Caliskan H, Adelman J 1997 *MNRAS* **288** 501
[3] Rathore B A, Lakićević I S, Ćuk M, Purić J 1984 *Phys. Lett. A* **100** 31
[4] Dimitrijević M S, Ryabchikova T, Simić Z, Popović L Č, Dačić M 2007 *A&A* **469** 681
[5] Simić Z, Dimitrijević M S, Sahal-Bréchot S 2013 *MNRAS* **432** 2247
[6] Aguilera J A, Aragón C, Manrique J 2013 *JQSRT* **114** 151
[7] Aragón C, Peñalba F, Aguilera J A 2005 *Spectrochim. Acta Part B* **60** 879
[8] Aragón C, Aguilera J A, Manrique J 2014 *JQSRT* **134** 39
[9] Gigosos M A, Cardeñoso V 1996 *J. Phys. B: At. Mol. Opt. Phys* **29** 4795
[10] Aguilera J A, Aragón C, Manrique J 2014 *MNRAS* **438** 841

Journal of Physics: Conference Series **548** (2014) 012042 doi:10.1088/1742-6596/548/1/012042

Differential Spectral Imaging of the CN Violet Band in Laser-Induced Plasmas on TNT Simulant Molecules

J Merten[1], M Jones, S Hoke[1] and S Allen[3]

[1]Department of Chemistry and Physics, Arkansas State University, P.O. Box 419, State University, AR, 72467, USA

[2]Anton Paar, 15421 Vantage Parkway W., Houston, TX, 77032, USA

[3]Department of Mechanical Engineering, Embry Riddle Aeronautical University, 600 S. Clyde Morris Blvd., Daytona Beach, FL, 32114, USA

jmerten@astate.edu

Abstract. Dual channel emission imaging of m-nitrobenzoic acid and benzoic acid was performed in order to visualize the morphology of the CN violet band emission of a TNT analogue. The CN channel was corrected for continuum emission using a simultaneously imaged background channel. Simultaneous dual channel imaging alleviated problems with shot to shot variation in the plasma morphology due to the friable substrates and showed differences between plasmas formed on the two targets.

1. Introduction

Although LIBS is primarily an elemental technique, there have been attempts to use the plasma's molecular emission to identify complex molecules in the ablated sample. As such, the origin of diatomic emission from the LIBS plasma is relevant to any potential selectivity. Previous studies have suggested that C_2 and CN emission may arise from native CN bonds cleaved from molecular substrates, especially after 266nm or ultra-short pulse ablation [1]. Some researchers have suggested that the temporal development of the CN emission indicates the origin of the emitting molecules, with single exponential decays arising from native-cleaved CN bonds and more complex development indicating that the CN arises from reactions in the plasma. More recently, investigations of the morphology of the molecular and atomic emission from the plasma have suggested that the location of the emission in the plasma may carry information about the ablated target [2]. In order to better understand the origin of the CN emission following single laser shots on pressed pellets of a TNT analogue, we present differential spectral imaging of CN emission with simultaneous spectra after ablation of m-nitrobenzoic acid (NBA) and benzoic acid (BA). The former is used as a convenient analogue of TNT with the latter used as a control as it contains no native C-N bonds.

2. Instrument and data processing

The optical arrangement is summarized in Figure 1. The pressed pellet samples were ablated with 20 mJ of 355 nm emission provided by a Continuum Surelite. The spot diameter was approximately 500 μm at the target. LIBS plasma emission was collimated with an f=12.5 cm NUV achromat. A microscope slide picked off a portion of the collimated light for imaging onto the spectrometer slit. The remaining light was split into CN and background correction arms with a dichroic beamsplitter.

Journal of Physics: Conference Series **548** (2014) 012042 doi:10.1088/1742-6596/548/1/012042

Spectral channels were isolated with interference filters. The slightly magnified (1.2x) images were collected simultaneously on the same ICCD face (Andor USB Istar, 2048x512). Simultaneous collection of signal and background images was necessary to minimize irreproducibility of plasma morphology caused by the friable pressed powder samples. Data were processed in Labview, correcting relative efficiencies of the two channels using the image of a fiber optic coupled to a standard deuterium lamp, while image positions (and rotation) were registered using the dual images of a geometric alignment grid. Figure 2 shows the relative spectral positions of the two channels. Because a narrow-band filter overlapping the CN band was not available, the CN channel integrates some atomic emission, particularly at early times.

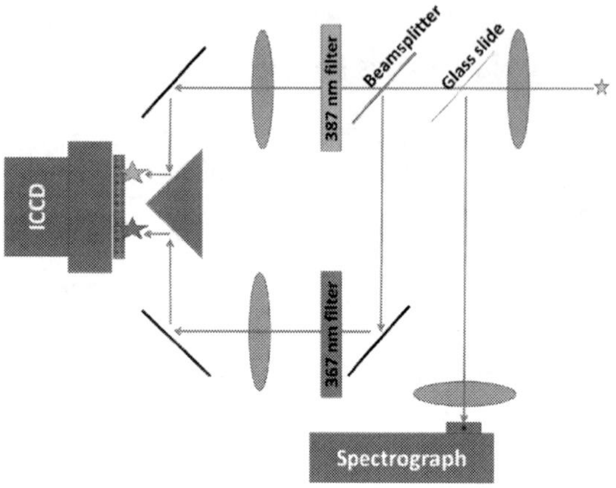

Figure 1. Optical arrangement used to acquire single-shot, dual-channel images with simultaneous spectra. The image at the spectrograph is rotated such that the plasma image propagates along the slit. The spectrum is integrated over the height of the plasma which is centered laterally on the slit.

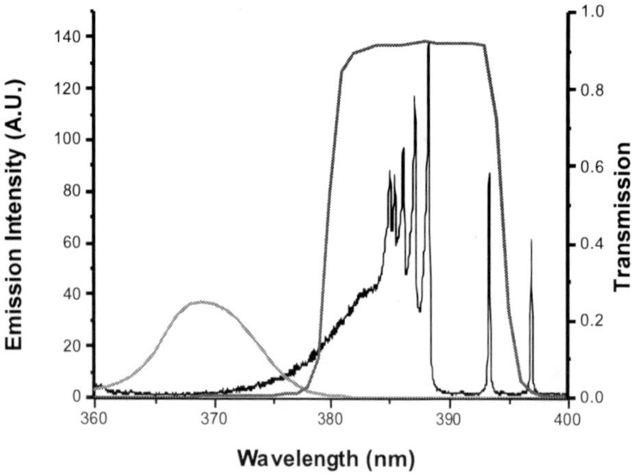

Figure 2. Position of CN (λ_{center}=388 nm) and background channel (λ_{center}=370 nm) relative to the CN violet band $\Delta v=0$ emission.

3. Imaging and conclusions

In both the BA and NBA plasmas, the imaging at the earliest time is prone to artifacts as the CN emission sits atop a substantial continuum, with additional structure at the long wavelength side of the CN band, see Figures 3 to 6. The net NBA image at 50 ns is of poor quality, though comparison with the 100 ns image suggests that the indicated net CN emission may be real. In the earliest BA image, there is clear evidence of a symmetrical CN emission front near the presumed atmosphere/plasma interface, in contrast to the more irregular emission seen in the early NBA image.

The net CN imaging at later times is less prone to artifacts, with a decreased and less structured background. The net CN images on BA continue to show more of a dome shape, while the NBA emission remains more chevron-shaped and dense. Though the plasmas produced on the two substrates produce different image morphologies, it is not clear if this is a result of different chemistry in the plasma or simply physical/mechanical differences in the laser-target interaction and subsequent fluid flow.

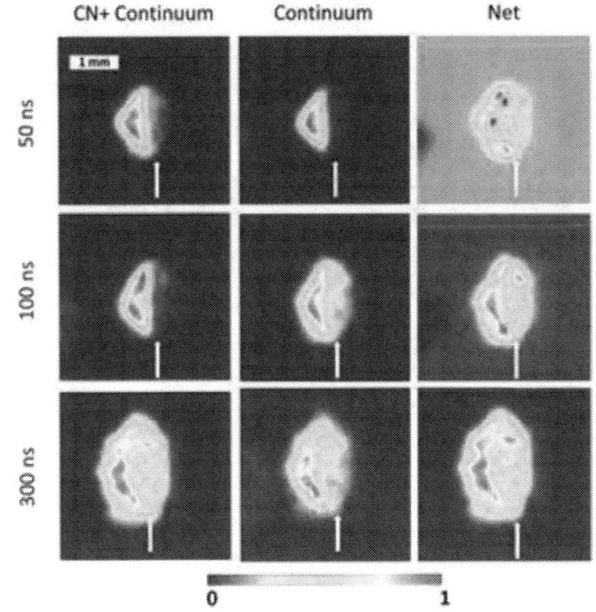

Figure 3. Nitrobenzoic acid imaging and spectra at indicated delays. The plasma propagates from right to left with time. The laser is incident from the image left. Each image is scaled separately such that absolute comparisons are not possible. The arrow indicates the approximate location of the sample surface. Gate *widths* are 50, 100, and 200 ns, respectively, from early to late.

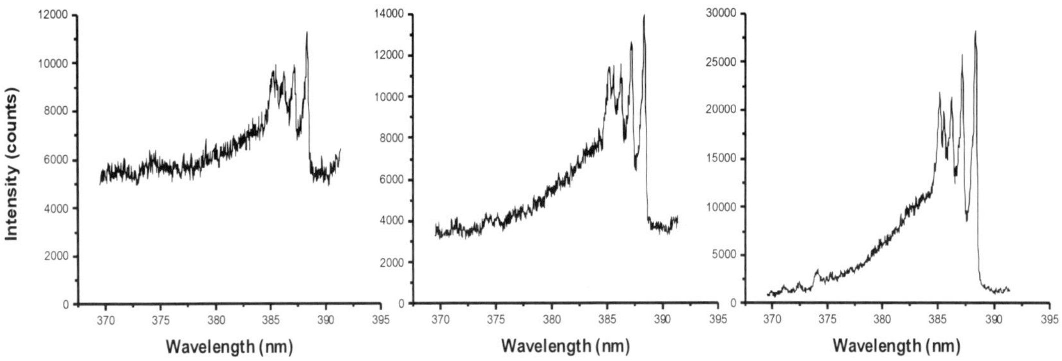

Figure 4. Nitrobenzoic acid CN emission spectra collected simultaneously with images in Figure 3. Gate positions and widths are as in Fig. 3.

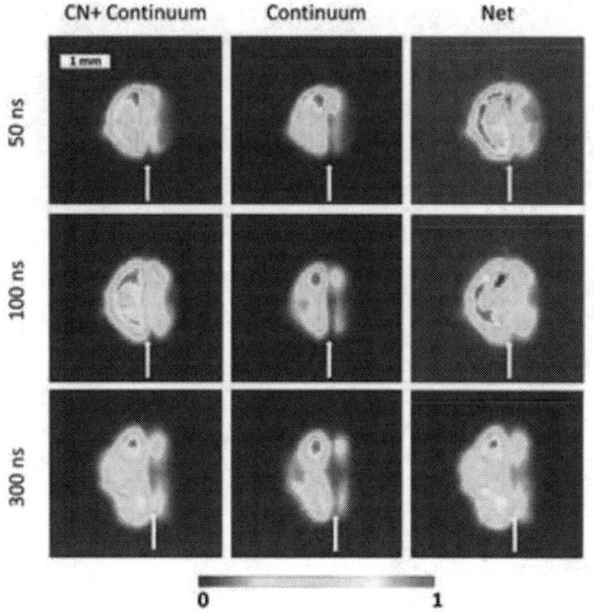

Figure 5. Benzoic acid imaging and spectra at indicated delays. The plasma propagates from right to left with time. The laser is incident from the image left. Each image is scaled separately such that absolute comparisons are not possible. The arrow indicates the approximate location of the sample surface. Gate *widths* are 50, 100, and 200 ns from early to late.

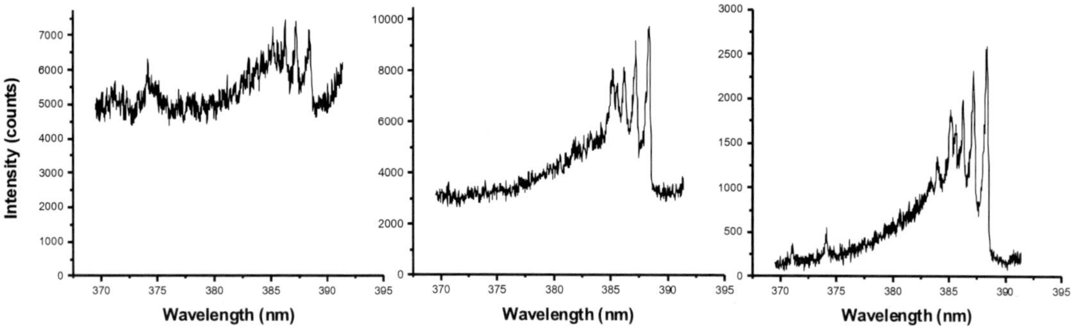

Figure 6. Benzoic acid spectra taken simultaneously with images in Figure 5. Gate delays and widths are as in Fig. 5.

References

[1] Baudelet M, Guyon L, Yu J, Wolf J, Amodeo T, Frejafon E, and Laloi P 2006 *Applied Physics Letters* **88** 063901; Boueri M, Baudelet M, Yu J, Mao X, Mao S, and Russo R 2009 *Applied Surface Science* **255** 9566; St-Onge L, Sing R, Bechard S, and Sabsabi M 1999 *Applied Physics A-Materials Science & Processing* **69** S913

[2] Grégoire S, Motto-Ros V, Ma Q, Lei W, Wang X, Pelascini F, Surma F, Detalle V, and Yu J 2012 *Spectrochimica Acta Part B: Atomic Spectroscopy* **74–75** 31

Hydrogen Alpha Self-Absorption Effects in Laser-Induced Air Plasma

C G Parigger, L D Swafford, D M Surmick, M J Witte, A C Woods, and G Gautam

University of Tennessee Space Institute, 411 B.H. Goethert Parkway, Tullahoma, TN 37388, USA

E-mail: cparigge@tennessee.edu

Abstract. Time-resolved spectroscopy measurements of the hydrogen alpha Balmer series line following laser-induced optical breakdown in laboratory air are designed to investigate in detail the determination of electron density from Stark-broadened spectral line shapes. Comparisons of results obtained from H_β and H_γ lines indicate higher electron density inferred from H_α early in the plasma decay, suggesting self-absorption occurs. However, detailed comparisons for time delays of 300 and 400 ns after optical breakdown reveal the minute extent of self-absorption in air breakdown experiments from (i) differences of electron density determined from the N^+ lines and the H_α line, and/or from (ii) differences in recorded data sets with/without the mirror for the various time delays in the experiments.

1. Introduction

In this work, we systematically vary the time delay from 100 - 1000 ns, and employ a gate width of 10 ns. Optical breakdown is generated by focusing pulsed Nd:YAG laser radiation parallel to the spectrometer slit using f# = 4 optics. The emission from the decaying plasma is spatially and temporally resolved with a Czerny-Turner spectrometer and an ICCD camera [1, 2]. Rather than using a spherical mirror [3], we utilize a flat mirror and a plano-convex lens in the self-absorption studies. Typical results for the full-width half-maxima for H_α and the two N^+ lines on either side of the H_α line lead to an electron density of $\simeq 10\times10^{23}$ m^{-3}, indicating hardly any H_α self-absorption.

The self-absorption analyses include comparisons of recorded data with and without the duplicating mirror for both the continuum radiation and the measured plasma emissions of atomic lines above the continuum [3]. This approach works well for line-of-sight measurements, for homogeneous distributions, and in optically thin conditions [3, 4, 5]. For optically-thick conditions, the observed line will no longer be linearly related to the absorption coefficients. Determination of the electron density from ionized nitrogen lines [6] and comparison with the inferred electron density from hydrogen alpha emissions will allow us to investigate the level of self-absorption as well. In previous studies, H_α self-absorption was indicated by comparisons with H_β and H_γ measurements [1].

2. Experiment

Measurements of spatially and temporally resolved spectra with/without mirror are reported here. The laser-induced plasma is generated by focusing the radiation from the top. The emission spectra are captured using a 10 ns gate width in steps of 100 ns from the optical breakdown events. A set of 276 spectra was recorded along the 7.15 mm slit height that was utilized for this experiment. The recorded data from 81 consecutive breakdown events in air are portrayed as images of wavelength versus slit height.

3. Results

In the recorded spectra, hydrogen alpha emissions begin to emerge for time delays of 300 ns, and are clearly recognizable for the explored range of 400 to 1000 ns. For time delays of 100 ns to 700 ns, ionized nitrogen lines are clearly recognizable at 648.2 nm and 661.1 nm, i.e., at the low and high wavelength side of the 656.28 nm H_α line.

The addition of a mirror allows us to investigate self-absorption effects of the line-of-sight emission spectroscopy diagnostic. Optical radiation from the plasma is retro-reflected and imaged as well onto the spectrometer slit. Detailed comparisons are aimed to reveal the level of self-absorption due to (i) differences of electron density determined from the N^+ lines and the H_α line, and/or due to (ii) differences in recorded data sets with/without the mirror for the various time delays in the experiments.

Addition of a mirror and subsequent comparisons of continuum radiation and of recorded emission line is a well-accepted method for determination of the level of self-absorption [3]. The experimental factor, $K_{\lambda,\mathrm{corr}}$, to correct for self absorption [3, 4, 5] in the optically-thin case,

$$K_{\lambda,\mathrm{corr}} = \frac{\ln\{y_\lambda\}}{y_\lambda - 1}\,, \quad \text{with} \quad y_\lambda = \frac{R_\lambda - 1}{R_C - 1}\,, \tag{1}$$

can be evaluated from the data independent of the detector's sensitivity. The ratio of the continuum radiation, R_C, is determined from data with and without the mirror. The wavelength dependent ratio, R_λ, is obtained as ratio of signal above the continuum for data w/ and w/o the mirror. The self-absorption correction factors at line centers of the two N^+ lines amount to $\simeq 1.0$ for the 300 ns data sets. At a time delay of 300 ns, the hydrogen alpha line is too broad for application of the correction factor method; therefore, comparisons of electron densities determinend from N^+ and H_α widths is discussed in the next sections. For a time delay of 800 ns, Equation (1) is used to evaluate $K_{\lambda,\mathrm{corr}}$ for H_α. Figure 1 shows the recorded, filtered data w/ and w/o the mirror. The illustrated data have been detector-background subtracted, sensitivity corrected and wavelength calibrated. The line center correction factors amount to $\simeq$ 1. Self-absorption of H_α is insignificant for 800 ns time delay.

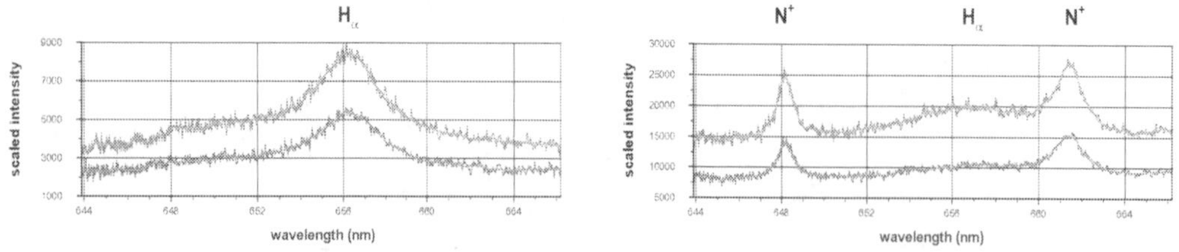

Figure 1. Spectra w/ and w/o mirror (Left) time delay 800 ns, slit height: 5.15 mm; (Right) time delay 300 ns, slit height: 6.15 mm. Smoothed line shapes are Savitzky-Golay filtered.

Measurements of electron density from ionized nitrogen lines utilize published tables for Stark widths and shifts [6]. In previous analysis we utilized the N^+ 399.5 line [7], here we use the two N^+ lines at 648.2 nm and 661.1 nm. Figure 1 shows spectra at slit heights of 6.15 mm. For the 648.2 nm line, the width is 0.86 nm, leading to an electron density of 12.7×10^{23} m^{-3}. For the 661.1 nm line, the width is 0.98 nm, leading to an electron density of 12.1×10^{23} m^{-3}. The accuracies of the electron densities from the N^+ lines amount to 30% [6].

Electron density from Stark widths and shifts of the hydrogen alpha Balmer series line can be inferred using Stark width tables [8], results from convergent theory [9, 10] and empirical formulae [11]. Comparisons of the accuracy of H_α and of H_β fitting shows it is generally better to use H_β lines [12]; yet for measurements of electron density larger than $\simeq 7 \times 10^{23}$ m^{-3} the H_β line can not be used [10]. For optical breakdown in air, this implies that H_β diagnostic will be useful for time delays larger than $\simeq 1.4$ μs. Figure 1 also shows broad H_α emission spectra. The inferred widths from the broad profiles are 8 to 10 nm, the shifts are 0.8 to 1.0 nm. H_α widths of 9 nm and H_α shifts of 1 nm imply electron densities of $14 - 20 \times 10^{23}$ m^{-3} and $10 - 20 \times 10^{23}$ m^{-3} [10]. In turn, the N^+ line widths show electron densities of $12 - 13 \times 10^{23}$ m^{-3}, suggesting significant self-absorption of H_α occurs for time delays on the order of 300 ns.

4. Conclusions

Detailed investigation and comparisons of time-resolved spectroscopy measurements of the N^+ and H_α lines allow one to determine the hydrogen alpha self-absorption effects in Laser-induced air plasma. The N^+ lines yield smaller electron densities than the H_α line for early time delays, yet the differences diminish for data that is averaged along the spectrometer's slit dimension. Results show differences in electron density along the plasma with noticeable H_α red-shifts for early time delays. There are indications that significant self-absorption of H_α indeed occurs for a time delay of 300 ns from optical breakdown, for 400 ns hardly any self-absorption is evident, and for 800 ns time delay self-absorption effects are insignificant for optical breakdown in laboratory air. From the recorded data sets, one can infer plasma expansion speed on the order of 300 to 100 μm in 100 ns or 3 to 1 km/s, in the range of 300 to 1000 ns after optical breakdown.

Acknowledgments

The authors thank for support of this work by the Center for Laser Applications at the University of Tennessee Space Institute.

References

[1] Parigger C G 2010 *Int. Rev. At. Mol. Phys.* **1** 129
[2] Parigger C G 2013 *Spectrochim. Acta Part B* **79-80** 4
[3] Moon H, Herrera K K, Omenetto N, Smith B W, Winefordner J D 2009 *Spectrochim. Acta Part B* **64** 702
[4] Konjević N 1999 *Phys. Rep.* **316** 339
[5] El Sherbini A M, El Sherbini Th M, Hegazy H, Cristoforetti G, Legnaoili S, Palleschi V, Pardini L, Salvetti A, Tognoni E 2005 *Spectrochim. Acta Part B* **60** 1573
[6] Konjević N, Lesage A, Fuhr J R, Wiese W L 2002 *J. Phys. Chem. Ref. Data* **31** 819
[7] Parigger C G, Woods A C, Surmick D M, Swafford L D, Witte M J 2014 *Spectrochim. Acta Part B* **99** 15
[8] Griem HR. 1964 *Plasma Spectroscopy*, McGraw-Hill, New York 1964
[9] Oks E. 2006 18th ICSLS: AIP Conf. Proceedings **874** 19
[10] Parigger CG, Plemmons DH, Oks E. 2003 *Appl. Opt.* **42** 5992
[11] Konjević N, Ivković M, Sakan N. 2012 *Spectrochim. Acta Part B* **76** 593
[12] Parigger C G, Swafford L D, Woods A C, Surmick D M, Witte M J 2014 *Spectrochim. Acta Part B* **99** 28

Theoretical Investigations of the Effect of N-Doping on Growth and Field Emission Properties of Carbon Nanotubes (CNTs)

I Santolia, A Tewari, S C Sharma and R Sharma

Department of Applied Physics, Delhi Technological University,
Shahbad Daulatpur, Bawana Road, Delhi-110 042, India

isha.santolia@gmail.com

Abstract. In this work, we investigate effects of doping of hetero-atoms such as Nitrogen (N) on the growth and field emission properties of Carbon Nanotubes (CNTs) in low temperature plasma. A theoretical model is developed to incorporate kinetics of electron, ions and neutral atoms with N as dop ing elements in complex and low temperature plasma. Results of numerical calculations of the radius of N-doped CNT are presented for typical glow discharge plasma parameters. It is found that the radius of CNT is reduced with N-doping. The field enhancement factor of CNT is also estimated from the obtained results.

1. Introduction

The effect of the experimental parameters on the structure of nitrogen containing Multi-Wall Nanotubes (MWNTs) produced by an aerosol assisted Chemical Vapor Deposition (CVD) method has been studied by Koos *et al.* [1] previously. Ayala *et al.* [2] have demonstrated the effect of doping of carbon nanotubes with nitrogen. Kim *et al.* [3] have shown in their investigations that enhanced CNT growth in an N_2 or NH_3 environment is due to nitrogen incorporation into the CNT wall. Sharma and Tewari [4] have investigated the effect of plasma parameters on the growth and field emission properties of Carbon Nanotubes. The variation of field enhancement factor has been experimentally verified by Srivastava et al. [5] who also studied the enhanced field emission characteristics of nitrogen-doped carbon nanotube films showing field emissions at a level of 2.65–3.55 V/μm.

2. Model

Plasma containing electrons, positively charged ions of type *A*, B, and *C*, and neutral atoms of type *A*, *B*, and *C*, is considered. Here *A* corresponds to carbon, *B* to neon, and *C* to nitrogen as doping material. The initial radius of spherical CNT tip, r_0 , can be estimated by evaluating the accumulation of electrons and positively charged ions on the CNT,

$$n_e\left(\frac{T_e}{m_e}\right)^{\frac{1}{2}}\exp\left(-\frac{e^2}{r_0k_BT_e}\right)=\left(1+\frac{e^2}{r_0k_BT_i}\right)\left[n_{iA}\left(\frac{T_i}{m_{iA}}\right)^{\frac{1}{2}}+n_{iB}\left(\frac{T_i}{m_{iB}}\right)^{\frac{1}{2}}+n_{iC}\left(\frac{T_i}{m_{iC}}\right)^{\frac{1}{2}}\right], \quad (1)$$

with n_e and T_e the electron density and temperature, respectively. The ion temperature is denoted by T_i, the Boltzmann constant by k_B , and the electron charge by e. The densities of the ions are indicated by n_{iA} , n_{iB} and n_{iC} with masses m_{iA} , m_{iB} and m_{iC}, respectively.

2.1 Growth rate equation for the mass of CNT

The growth rate is evaluated using

$$\frac{d}{d\tau}m_{ct} = (m_A\gamma_A n_{Act} + m_{iA}\gamma_{iA}n_{iAct} + m_C\gamma_C n_{Cct} + m_{iC}\gamma_{iC}n_{iCct})\,, \quad (2)$$

where m_{ct} is the mass of the CNT of radius a. With the density of the CNT, ρ_{ct}, the mass of the CNT can be expressed in terms of the density $m_{ct}=4/3\ \pi\ r^3\ \rho_{ct}$. The sticking coefficients for the atomic and ionic species are γ_A , γ_{iA} , γ_C and γ_{iC}, respectively. The two terms in Equation (2) correspond to gains in mass due to collection of atomic and ionic species A and C.

3. Results and discussion

For the developed model, calculations are carried out by taking the same parameters for the undoped spherical CNT tip as taken by Sharma and Tewari [4]. Calculations are for doped-nitrogen CNTs using boundary conditions and parameters at τ=0, n_{ct} $=10^6$ cm^{-3}, $n_{iA0}=n_{iB0}=n_{iC0}=10^9$ cm^{-3}, $n_{A0}=n_{B0}=5\times10^8$ cm^{-3}, $n_{C0}=5\times10^8$ cm^{-3}, $n_{e0}=10^9$ cm^{-3}, T_{e0}=0.5 eV, T_{i0}= 2500K, $T_{n0}=T_{ct}$=2000K. Further, $m_{iA}\approx m_A$=12 amu (carbon), $m_{iB}\approx m_B$=20 amu (neon) and $m_{iC}\approx m_C$=14 amu (nitrogen).

Figure 1 illustrates the variation of normalized radius r/r_0 of spherical CNT tip with time for undoped CNTs and nitrogen-doped CNTs. The value of normalized radius is found to be more for undoped CNTs, and it reduces with N-doped CNTs. This confirms better field emission for N-doped CNTs. Fig. 2 illustrates the variation of field enhancement factor for undoped and N-doped CNTs. The field enhancement factor, β, for N-doped CNTs has been calculated. The results show higher value of β for N-doped CNTs.

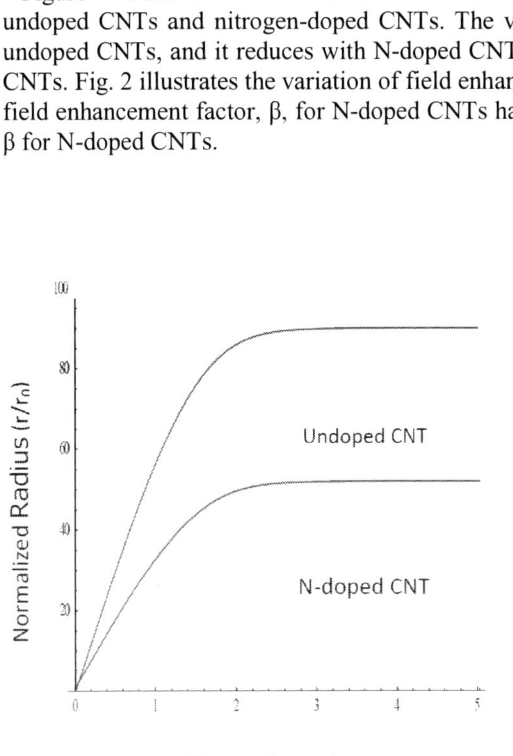

Figure 1. Variation of the normalized radius r/r_0 for undoped and N-doped CNTs.

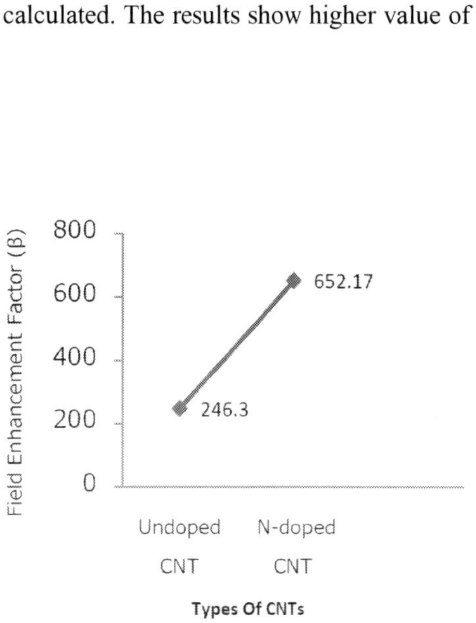

Figure 2. Variation of the field enhancement factor for undoped and N-doped CNT's.

Journal of Physics: Conference Series **548** (2014) 012044 doi:10.1088/1742-6596/548/1/012044

4. Conclusions

The model developed for calculating the growth of spherical CNT tip with nitrogen as doping material predicts a diminished radius with nitrogen-doped as compared to undoped CNTs. Moreover, the computed field enhancement factors β for both doped and undoped CNTs show a maximum value on the order of 650 for nitrogen–doped CNTs. The obtained results have also been experimentally verified by Srivastava *et al.* [5].

References

[1] Ayala P, Arenal R, Rummeli M, Rubio A 2010 *Carbon* **48** 575
[2] Koos AA, Dowling M, Jurkschat K, Crossley A, Grobert N 2009 *Carbon* **47** 30
[3] Tae–Young K, Kwang-Ryeol L, Kwang Yong E, Kyu-Hwan O 2003 *Chem. Phys. Lett.* **372** 1851
[4] Sharma S C, Tewari A 2011 *Phys. Plasm.* **18** 06503
Srivastava S K, Shukla A K, Vankar V D and Kumar V 2005 *Thin Solid Films* **492** 124

Journal of Physics: Conference Series **548** (2014) 012045 doi:10.1088/1742-6596/548/1/012045

Effects of Strong Langmuir Turbulence on Hydrogen Lines

I Hannachi[a], M Meireni[b], H Capes[b], F Guzman[b], M Koubiti[b], Y Marandet[b], M T Meftah[c], L Mouret[b], J Rosato[b], R Stamm[b]
[a]University of Batna, Department of Physics, Batna, Algeria
[b]Aix-Marseille University and CNRS, PIIM, 13397 Marseille, France
[c]University of Ouargla, Department of Physics, 30000 Ouargla, Algeria

E-mail: roland.stamm@univ-amu.fr

Abstract We consider conditions of strong Langmuir turbulence, for which a wave packet cycle is observed. The dynamics of such intense Langmuir electric fields may affect the line broadening of simple atoms. We use a simple stochastic model to evaluate the effect of such turbulent fields on the Lyman α line in low density plasmas.

1. Introduction

Following the pioneering work of Baranger and Mozer [1], the effect of plasma oscillations on spectral line shapes in plasmas have long been studied [2]. A problem of interest is the case of strongly non linear wave turbulence, observed in conditions where the energy of the Langmuir wave is not small compared to the plasma thermal energy. Such conditions may be reached during the interaction of the solar wind with the planet magnetospheres, or in laser plasmas. They may occur also each time an energetic electron beam interacts with plasma. In the strong Langmuir turbulence regime, Langmuir wave packets will form, collapse, dissipate and reform, thus creating a wave packet cycle [3]. With the help of a stochastic model for the Langmuir wave packet cycle [2], the effect on a line shape of non linear wave collapse may be investigated. Using the Lyman α line of hydrogen emitted in the center of mass as a benchmark, we have analyzed the changes in the line shape for different parameters found in the strong Langmuir turbulence regime.

2. Renewal process for the wave packet cycle in the strong Langmuir turbulence regime

We consider the case of electron-plasma or Langmuir waves, which are electrostatic waves in unmagnetized plasma. In the linear regime, the Langmuir waves oscillate at a high-frequency close to the plasma frequency $\omega_p = \sqrt{4\pi Ne^2/m}$ (with e and m the electron charge and mass, and N the density), and coexist with transverse electromagnetic waves and low-frequency ion sound waves. The dispersion relations of high-frequency waves depend on the density through the plasma frequency, enabling the density fluctuations of the ion sound waves to affect the high-frequency waves [3]. This provides a nonlinear coupling mechanism between low and high frequency waves. Equations taking account of such a coupling where first derived by Zakharov [4]. The first Zakharov equation explains how the density fluctuations affect Langmuir waves. Weaker densities correspond to a higher refractive index, causing the high-frequency waves to refract into low density regions. A Langmuir wave packet can produce a density depression via the ponderomotive force, which is a nonlinear force proportional to the gradient of the square of the wave electric field.

In a region where the wave packet electric field is strong, this force can be shown to expel the plasma [3], thus creating the density depression and trapping the wave packet. The non-linear dynamics of such regions has been studied with the Zakharov equations [4], and particle-in cell simulations. It has been found that the three-dimensional wave packets spatially narrow and become more intense. Such simulations also reveal that for a steadily driven system, many collapsing Langmuir wave packets are present. Since they are unstable, these packets will form, collapse, dissipate, and reform, thus creating a wave packet cycle. The characteristic times for such cycles have been studied [5], and take values much larger than one in units of the inverse electron plasma frequency. If we consider an emitting atom immersed in such turbulent plasma, its energy levels will be strongly affected by the electric field of the coherent wave packets located in the vicinity.

A stochastic model already proposed for the Langmuir fields in the linear regime [2] may also be used for strong Langmuir turbulence, for which the ratio of the Langmuir field energy and plasma energy densities $W = E_L^2/8\pi NkT$ reaches values of the order of one or larger. Our stochastic model uses probability density functions (PDFs) for the lifetime of a wave packet cycle, and for the values of the electric field in the wave packet. We assume that the Langmuir field applied on an emitter is a sequence of oscillating field with random amplitude and phase, $\vec{E}(t) = \sum_j \vec{E}_j \cos(\omega_p t + \varphi_j)$, where the field $\vec{E}_j$ and phase φ_j change at times prescribed by a waiting time distribution (WTD). A Markovian choice for the WTD is a simple Poisson law $\nu \exp(-\nu t)$, with ν the jumping frequency, but memory effects may be retained with other choices for this distribution. In this work we will take a Poisson law with the jumping frequency equal to the inverse of the lifetime of a wave packet cycle [3]. Whereas the phase is uniformly distributed in the interval $(0,2\pi)$, the Langmuir field values are chosen to be distributed with a log-normal PDF. Such PDFs of fluctuating plasma parameters have been found in solar wind plasmas [6] or in tokamak plasmas [7]. We have used a log-normal PDF with a standard deviation $\sigma = 1/4$ for the underlying normal distribution. For a ratio W=1, the average Langmuir field E_L is generally much larger than the Holtsmark field $E_0 = e/r_0^2$, with r_0 the mean interparticle distance. At a density $N=10^{12}$ cm^{-3} and a temperature $T=10^4$ K, E_L is almost 50 times larger than E_0.

3. Results: Lyman α line shape in the presence of strong Langmuir turbulence

We have calculated Lyman α with a numerical integration of the Schrödinger equation, and an average over about 10^4 histories of the field. It is possible to compute pure Langmuir profiles, or profiles resulting from a convolution with a thermal Stark effect. On Fig.1(a) pure Langmuir profiles for $\nu = \omega_P/50$ are calculated for four values of the average Langmuir field expressed in units of E_0. A first doubling of E_L results in a increase of the width by a factor 2.6, whereas a second doubling increases the width only by 20%. For E_L=100 E_0, one can see on Fig.1(b) a linear behaviour of the width as a function of ν, with a doubling of the width following a doubling of ν. Compared to pure Stark profiles obtained with an impact approximation, or a simulation, the action of Langmuir wave packets may result in a strong broadening, as can be seen on Fig.1(c) where the full profile is almost three times larger than the pure Stark for a density $N=10^{12}$ cm^{-3}, E_L=100 E_0 and $\nu = \omega_P/50$. Keeping W values of the order of 1, this effect is much reduced for a higher density case $N=10^{15}$ cm^{-3}. Even for an unrealistic value $\nu = \omega_P/5$, the broadening effect of strong turbulence Langmuir waves is limited to 30%.

4. Conclusion

We have described the main features of Strong Langmuir turbulence, a regime which is expected as the energy of the Langmuir wave becomes comparable to the plasma thermal energy. The effect of wave packet cycles may be modelled by a stochastic process using PDFs for the lifetime of the cycle, and the values of the turbulent electric field. This field can reach values up to two orders of magnitude larger than the Holtsmark field in such conditions. For plasmas with a density $N=10^{12}$ cm^{-3} and a temperature $T=10^4$ K, an additional broadening by a factor of three has been calculated. The broadening effect increases linearly with the jumping frequency ν, but has saturation behaviour as the average Langmuir field is increased. For the higher density $N = 10^{15}$ cm^{-3}, the broadening

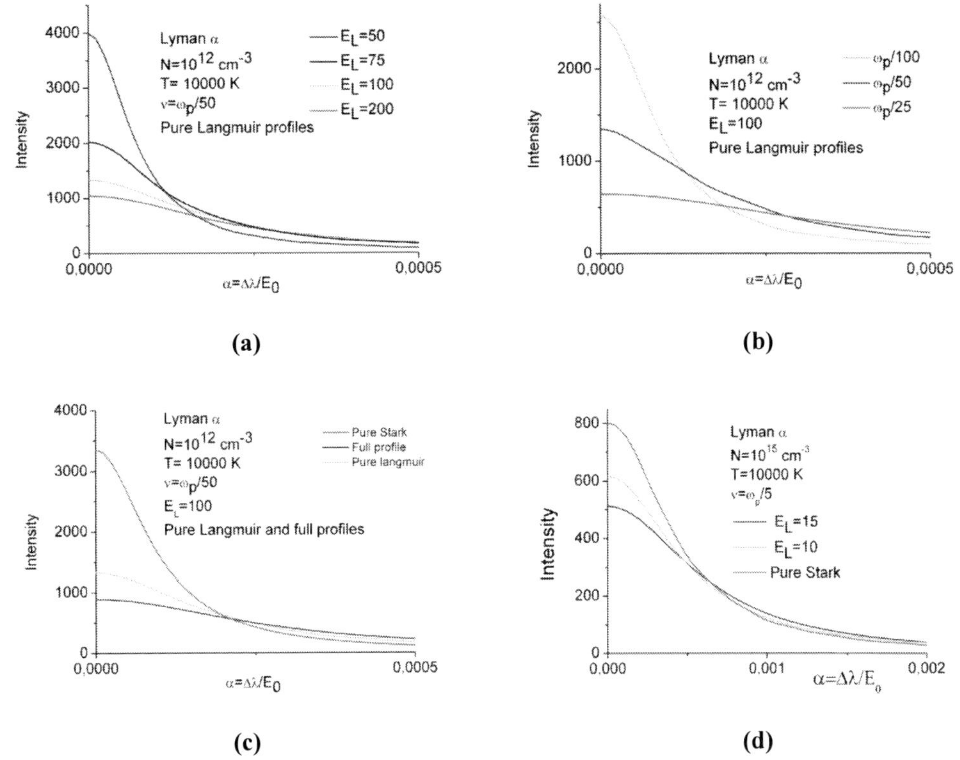

Figure 1. Lyman α profiles calculated with our simulation. Pure Langmuir profiles for different E_L (a) and ν (b) values. Pure Stark, pure Langmuir and full profiles (c). Pure Stark and full profiles (d).

effect is weaker if one uses realistic ν values. Further work will concern other lines than Lyman α, other PDFs for the electric field, the role of memory in the stochastic process, and applications to various plasmas affected by strong Langmuir turbulence.

Acknowledgments
This work was carried out within the framework of the European Fusion Development Agreement and the French Research Federation for Fusion Studies. We also acknowledge the support of the French National Research Agency (contract ANR-11- BS09-023, SEDIBA).

References
[1] Baranger M and Mozer B 1961 *Phys. Rev.* **123** 25
[2] Oks E 1995 *Plasma Spectroscopy, The influence of Microwave and Laser Fields* (Berlin: Springer Ser. on Atoms and Plasmas)
[3] Robinson P 1997 *Rev. Mod. Phys.* **69** 507
[4] Zakharov V E 1972 *Sov. Phys. JETP* **35** 908
[5] DuBois D F, Rose H and Russel D 1988 *Phys. Rev. Lett.* **61** 2209
[6] Bakshi P and Kalman G 1976 *J. Geophys Res* **47** 307
[7] Sattin F et al 2004 *Phys. Plasmas* **11** 5032

Time-resolved aluminium laser-induced plasma temperature measurements

D M Surmick and C G Parigger

University of Tennessee Space Institute, 411 B.H. Goethert Parkway, Tullahoma, TN 37388, USA

E-mail: dsurmick@utsi.edu,cparigge@utsi.edu

Abstract. We seek to characterize the temperature decay of laser-induced plasma near the surface of an aluminium target from laser-induced breakdown spectroscopy measurements of aluminium alloy sample. Laser-induced plasma are initiated by tightly focussing 1064 nm, nanosecond pulsed Nd:YAG laser radiation. Temperatures are inferred from aluminium monoxide spectra viewed at systematically varied time delays by comparing experimental spectra to theoretical calculations with a Nelder Mead algorithm. The temperatures are found to decay from 5173 ± 270 to 3862 ± 46 Kelvin from 10 to 100 μs time delays following optical breakdown. The temperature profile along the plasma height is also inferred from spatially resolved spectral measurements and the electron number density is inferred from Stark broadened H_β spectra.

1. Introduction

Time-resolved laser-induced breakdown spectroscopy (LIBS) is a valuable experimental tool for determining key plasma parameters, such as the temperature and electron number density. The laser-induced plasma created following breakdown has multiple applications, including use in forensics, geophysics, and, of particular interest, combustion diagnostics. For the purposes of studying combustion, aluminium LIBS studies are of significant as aluminium is commonly found in solid rocket propellants and high explosives, where the goal is commonly to infer temperature [1, 2]. Determinations of the electron number density of aluminium laser-induced plasma also proven to be useful in diagnostic applications, such as with aluminum alloys [3, 4]. In previous time-resolved, aluminium laser ablation studies, the temperature of laser-induced plasma were determined from Boltzmann plots of aluminium atomic emissions and were found to be in the 4000-4250 Kelvin range [1, 5]. The plasma temperature may also be determined from the temperature dependence of diatomic molecular spectra such as aluminium monoxide (AlO). Temperature results from previous studies were found to be 4000 Kelvin at 50 μs [6] and 4880 Kelvin at 45 μs [7] from AlO emissions. Electron densities were also inferred in reference [6] and were found to be 10^{23} m^{-3}. In this work, we seek to use AlO diatomic molecular emissions to characterize the temperature decay of laser-induced plasma near the surface of an aluminium target in laboratory air over a broad range of times following optical breakdown using time-resolved LIBS measurements. Inferences are also made on the temperature distribution along the plasma height and the electron number density.

2. Experiment

Time-resolved spectral measurements are used to characterize the temperature decay of laser-induced plasma initiated near the surface of an aluminium target. Laser-induced plasma are formed by tightly focussing laser radiation from a Q-switched, Nd:YAG laser operating in its fundamental mode with a 12 nanosecond pulse width and an energy per pulse of 190 mJ. Line of sight spectral measurements are made with a Jobin-Yvon HR640 spectrometer with 0.1 nm spectral resolution coupled to gated, intensified detectors. An intensified, linear diode array coupled to an optical multichannel analyzer (Princeton Instruments 1460) was used to collect AlO spectra from 10 to 100 μs time delays in 5 μs steps with gate widths of 5, 10 and 20 μs for 10-20, 20-100, and 70-100 μs delays, respectively. A gated ICCD was used to infer the temperature dependence along the plasma height at a time delay of 60 μs with a 6 μs gate width. Time resolution was achieved by synchronizing the laser repetition rate with the measurement rate of the gated detectors with the use of waveform, pulse, and delay generators. Reference 8 provides a detailed description of the apparatus [8]. Prior to analysis, all spectra were properly calibrated for detector response and background sensitivity.

3. Results

To infer the temperature as function of time following optical breakdown, spectra collected with intensified, linear diode array were fit to theoretical calculations of the AlO $B^2\Sigma^+ \rightarrow X^2\Sigma^+$ $\Delta v = 0$ transition. Comparisons were made with a Nelder-Mead fitting algorithm which was utilized for its ability to fit multiple parameters simultaneously, including a variable baseline offset. The Nelder-Mead algorithm is a minimization method in which a geometric simplex is reduced in size until a satisfactory tolerance level is reached [9, 10]. The theoretical spectra are calculated from tables of accurately calculated line strengths, where the line strength is calculated in its factorized form from the electronic transition strength, the Hönl-London factor, and Franck-Condon factor [11, 12, 13]. The baseline offset of all fitted AlO spectra were quadratic in nature and errors in the inferred temperatures were found by considering variations of the spectral resolution and baseline offset. Analysis showed that the error in the inferred temperatures was in the 45-75 Kelvin range, with the exception of the inference made at 10 μs with an error of 236 Kelvin. This resulted from the presence of H_β superposition spectra. The results of the temperature analysis are shown in Figure 1 with an exponential decay trend. The trend line shows that the plasma temperature decay slows at later times as was expected. To infer the temperature dependence along the plasma height, spatially resolved spectra with 0.013 mm resolution were fit using the Nelder-Mead algorithm. The temperature at the center of the plasma was found to be 5780 ± 236 Kelvin at 5.2 mm slit height and decreased to 5396 ± 315 Kelvin at 6.5 mm. The temperature was found to increase above the plasma edge to a value of 5820 ± 230 Kelvin at a slit height of 9.1 mm. The larger temperatures and errors are likely result from non-optimal plasma and detector conditions; however, the temperature profile suggests that aluminium particles are combusting in a plume of above the plasma, given the ignition temperature of aluminium is 2750 Kelvin and the inferred temperatures from the linear diode experiment.

At time delays of 10, 15, and 20 μs following breakdown, superposition of Balmer series H_β spectra with AlO measurements were observed. The H_β spectra may be extracted and used to infer the electron number density from the Stark broadened emission. The line width of the spectra is taken to be due to Stark effects and is inferred by fitting a Lorentizian line profile to the H_β spectra. The Stark widths are used with the empirical theory of Konjević [14] and the convergent theory of Oks [15] to infer the electron number density, N_e. To determine the N_e with the theory of Oks, published data were fit to determine the N_e vs. H_β width dependence [16]. Figure 2 shows the fitted H_β spectrum from the 10 μs delay along with its corresponding Stark width and electron number density. The noticeable dip in the residual in Figure 2 near 486.5 nm

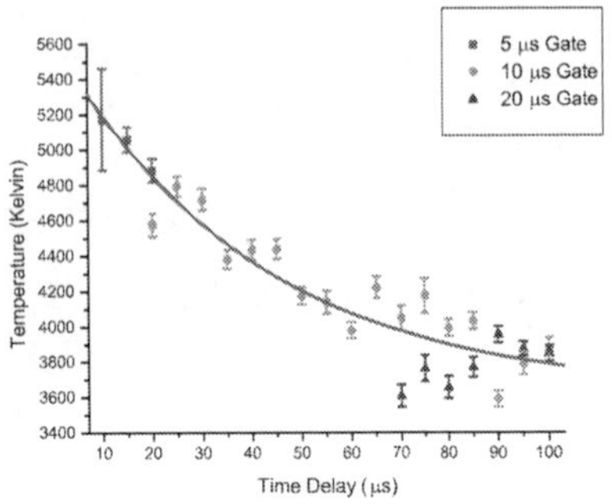

Figure 1. Shown are the inferred temperatures as a function of time with an exponential decay trend line.

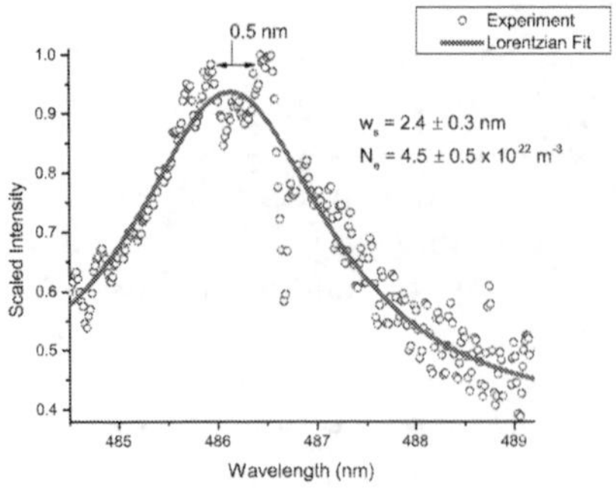

Figure 2. Shown is a fit of H_β superposition spectra at a 10 μs delay to a Lorentzian line profile.

is due to the subtraction of the the fitted AlO spectra and represents the 1-1 vibrational peak from the transition of interest. Also indicated in the figure is the 0.5 nm H_β peak separation. Though not considered in this work, this peak separation may also be used to infer the electron density [16]. The inferred N_e are found to be 4.5 ± 0.5, 3.8 ± 0.5 and $3.0 \pm 0.6 \times 10^{22}$ m^{-3}. Errors in the inferred N_e result from AlO superposition influences.

4. Conclusions

The goal of this work was to characterize the temporal temperature decay of laser-induced plasma near the surface of an aluminium target and was accomplished with time-resolved LIBS measurements. A Nelder-Mead fitting algorithm was employed to infer the temperature from the experimental spectra and errors were determined from spectral resolution and baseline offset variations. The temperatures were found to decrease as a function of time from 5173 ± 290 to 3862 ± 46 Kelvin from 10 to 100 μs time delays. The decay of the temperature was also found to slow at later time delays. Spatially resolved AlO measurements were used to infer the temperature along the height of the plasma and showed the temperature increases above the plasma edge. The electron density of the plasma was inferred from measurements of H_β spectra superimposed on AlO measurements at time delays preceding 20 μs and were found to decrease from 4.5 ± 0.5 to $3 \pm 0.5 \times 10^{22}$ m^{-3}. Observations of the slowing of the temperature decay, the temperature increase above the plasma, and H_β superposition spectra suggest that combustion occurs. Future work will seek to further quantify the presence of combustion in aluminium laser-induced plasma from spatially resolved AlO spectral measurements.

Acknowledgments

This work is supported by the Center for Laser Applications at the University of Tennessee Space Institute.

References

[1] Piehler T N, DeLucia F C, Munson C A, Homan B E, Miziolek A W and McNesby K L 2005 *Appl. Opt.* **44** 3654
[2] De Lucia F C, Harmon R S, McNesby K L, Winkel R J, and Miziolek A W 2003 *Appl. Opt.* **42** 6148
[3] Sasabi M and Cielo P 1995 *Appl. Spectrosc.* **44** 3654
[4] Rai A K, Yueh F and Signh J P 2003 *Appl. Opt.* **42** 2078
[5] Lightstone J M, Carney J R, Boswel C J and Wilikinson J 2007 *AIP Conf. Proc.* **955** 1255
[6] Parigger C G, Hornkohl J O and Nemes L 2007 *Appl. Opt.* **46** 4026

[7] Surmick D M and Parigger C G 2014 *Appl. Spectrosc.* **68**
[8] Parigger C G, Woods A C, Witte M J, Swafford L D and Surmick D M 2014 *J. Vis. Exp.* **84**
[9] Nelder J A and Mead R 1965 *The Comp. J.* **7** 308
[10] Lagarias J C, Reeds J A, Wright M H and Wright P E 1998 *SIAM J. Optim.* **9** 112
[11] Dors I G, Parigger C and Lewis J. 1998 *Opt. Lett.* **23** 1778
[12] Parigger C G and Hornkohl J O 2011 *Spectrochim. Act. A* **81** 404
[13] Hornkohl J O, Parigger C G and Nemes L. 2005 *Appl. Opt.* **44** 3686
[14] Konjević N, Ivković M and Sakan N 2012 *Spectrochom. Act. B* **76** 16
[15] Ispolatov Y and Oks E 1994 *J. Quant. Spectrosc. Rad. Transf.* **51** 129
[16] Parigger C G, Plemmons D H and Oks E 2003 *Appl. Opt.* **42** 5992

Advanced relativistic model potential approach to calculation of radiation transition parameters in spectra of multicharged ions

A A Svinarenko, A V Ignatenko, V B Ternovsky, V V Nikola, S S Seredenko and T B Tkach

Odessa State Environmental University, L'vovskaya str.15, Odessa-16, 65016, Ukraine

E-mail: quantsvi@mail.ru

Abstract. The combined relativistic energy approach and relativistic many-body perturbation theory with the zeroth order optimized one-particle approximation are used for calculation of the Li-like ions (Z=11-42,69,70) energies and oscillator strengths of radiative transitions from the ground state to the low-excited and Rydberg states, in particular, $2s_{1/2} - np_{1/2,3/2}$, $np_{1/2,3/2}$-$nd_{3/2,5/2}$ (n=2-12). A comparison of the calculated oscillator strengths with available theoretical and experimental data is performed.

1. Introduction

The research on the spectroscopic and structural properties of highly ionized atoms has a fundamental importance in many fields of atomic physics (spectroscopy, spectral lines theory), astrophysics, plasma physics, laser physics and so on. It should be mentioned that the correct data about radiative decay widths, probabilities and oscillator strengths of atomic transitions are needed in astrophysics and laboratory, thermonuclear plasma diagnostics and in fusion research. In this light, a special interest includes studying the spectral characteristics of the He-, Li- *etc.* like ions. There have been sufficiently many reports of calculations and compilation of energies and oscillator strengths for the Li-like ions and other alkali-like ions (see, for example, [1–16]). Particularly, Martin and Wiese have undertaken a critical evaluation and compilation of the spectral parameters for Li-like ions (Z=3-28) [1,2]. The results of the high-precision non-relativistic calculations of the energies and oscillator strengths of $2s$ - $2p$ for Li-like systems up to Z = 50 are presented in Refs. [9-14]. The Hylleraas-type variational method and the $1/Z$ expansion method have been used. Chen Chao and Wang Zhi-Wen [14] have listed the nonrelativistic dipole-length, -velocity and -acceleration absorption oscillator strengths for the 1s22s–1s22p transitions of the Li I isoelectronic sequence on the basis of calculation within a full core plus correlation method with using multiconfiguration interaction wave functions. Fully variational nonrelativistic Hartree-Fock wave functions have been used by Bièmont in calculation of the 1s2n2L (n <8 =s, p, d or f; 3 < Z<22) states of the Li I isoelectronic sequence [16]. In many papers, the Dirac-Fock method, model potential approach, quantum defect approximation in the different realizations have been used for calculating the energies and oscillator strengths of the Li-like ions (see Refs.[1-17]). The consistent Quantum Electrodynamics (QED) calculations of the energies, ionization potentials, hyperfine structure constants for the Li-like ions are reported in Refs. [18-21]. However, it should be stated that for Li-like ions with higher Z, particularly for their high-excited (Rydberg) states, there is not enough precise information available in the literature. In our paper the combined relativistic energy approach [22-26] and relativistic many-body perturbation theory with the zeroth order optimized one-particle representation [26] are used for calculation the Li-like ions (Z=11-42,69,70) energies and oscillator strengths of radiative transitions from ground state to low-excited and Rydberg states. The comparison of the calculated oscillator strengths with available theoretical and experimental data is performed.

2. The theoretical method

In the relativistic energy approach [4, 5, 22-25] the imaginary part of electron energy shift of an atom is connected with the radiation decay possibility (transition probability). An approach, based on the Gell-Mann and Low formula with the QED scattering matrix, is used in treating the relativistic atom. The total energy shift of the state is usually presented in the form:

$$\Delta E = \mathrm{Re}\Delta E + i\,\Gamma/2 \tag{1}$$

where Γ is interpreted as the level width, and the decay probability P = Γ. For the α-n radiation transition the imaginary part of electron energy in the lowest order of perturbation theory is determined as [4]:

$$\mathrm{Im}\,\Delta E = -\frac{1}{4\pi}\sum_{\substack{\alpha>n>f\\ [\alpha<n\le f]}} V^{|\omega_{\alpha n}|}_{\alpha n\alpha n}, \tag{2}$$

where $\omega_{\alpha n}$ is a frequency of the α-n radiation, (α>n>f) for electron and (α<n<f) for vacancy. The matrix element V is determined as follows:

$$V^{|\omega|}_{ijkl} = \iint dr_1 dr_2 \Psi^*_i(r_1)\Psi^*_j(r_2)\frac{\sin|\omega|r_{12}}{r_{12}}(1-\alpha_1\alpha_2)\Psi^*_k(r_2)\Psi^*_l(r_1) \tag{3}$$

The separated terms of the sum in (3) represent the contributions of different channels and a probability of the dipole transition is:

$$\Gamma_{\alpha_n} = \frac{1}{4\pi}\cdot V^{|\omega_{\alpha_n}|}_{\alpha_n\alpha_n} \tag{4}$$

The corresponding oscillator strength: $gf=\lambda^2\Gamma_{\alpha n}/6.67\cdot10^{15}$, where g is the degeneracy degree, λ is a wavelength in angstrems (Å). When calculating the matrix elements in Equation (3) one should use the angle symmetry and write the expansion for potentials $\sin|\omega|r_{12}/r_{12}$ in spherical functions as follows [2]:

$$\frac{\sin|\omega|r_{12}}{r_{12}} = \frac{\pi}{2\sqrt{r_1 r_2}}\sum_{\lambda=0}^{\infty}(\lambda)J_{\lambda+\frac{1}{2}}(|\omega|r_1)J_{\lambda+\frac{1}{2}}(|\omega|r_2)P_\lambda(\cos\widehat{r_1 r_2}) \tag{5}$$

where J is the Bessel function of first kind and (λ)= 2λ + 1. This expansion is corresponding to usual multipole one for probability of radiative decay. Substitution of the expansion (5) to matrix element of interaction gives as follows [20]:

$$V^{\omega}_{1234} = [(j_1)(j_2)(j_3)(j_4)]^{\frac{1}{2}}\sum_{\lambda\mu}(-1)^{\mu}\begin{pmatrix} j_1 j_3 & \lambda \\ m_1 - m_3 & \mu\end{pmatrix}\times \mathrm{Im}\,Q_\lambda(1234),$$

$$Q_\lambda = Q_\lambda^{Qul} + Q_\lambda^{Br}, \qquad (6)$$

where j_i is the total single electron momentums, m_i – the projections; Q^{Qul} is the Coulomb part of interaction, Q^{Br} - the Breit part. Their detailed definitions are presented in Refs. [4,20]. The relativistic wave functions are calculated by solution of the Dirac equation with the potential, which includes the "outer electron- ionic core" potential and polarization potential [29]. In order to describe interaction of the outer electron with the He-like core the modified model potential [4] has been used. The calibration of the single model potential parameter has been performed on the basis of the special *ab initio* procedure within relativistic energy approach [24] (see also [5]). In Ref.[18] the lowest order multielectron effects, in particular, the gauge dependent radiative contribution Im δE_{ninv} for the certain class of the photon propagator calibration is treated. This value is considered to be representative for the correlation effects, whose minimization is a reasonable criterion in the searching for the optimal one-electron basis of the many-body perturbation theory. The minimization of the density functional Im δE_{ninv} leads to the integral-differential equation that can be solved using one of the standard numerical codes. It provides the construction of the optimized one-particle representation. All calculations are performed on the basis of the numeral code Superatom-ISAN (version 93).

3. Results

We applied the above described approach to calculating the energies and oscillator strengths of transitions in spectra of the Li-like ions (Z=11-42,69,70). There are considered the radiative transitions from ground state to the low-excited and Rydberg states, particularly, $2s_{1/2} - np_{1/2,3/2}$, $np_{1/2,3/2}$-$nd_{3/2,5/2}$ (n=2-12). To test the obtained results, we compare our calculation results of the oscillator strengths values for some Li-like ions with the known theoretical and tabulated results [1,2,8]. As an example, in table 1 we present the oscillator strengths values for the 2s1/2 – 2p1/2,3/2 transitions in Li-like ions S^{13+}, Ca^{17+}, Fe^{23+}, Zn^{27+}, Zr^{37+}, Mo^{39+}, Sn^{47+}, Tm^{66+}, Yb^{67+}. The DF calculation data by Zilitis [6] and the "best" compillated (experimental) data [1,2] for the low-Z Li-like ions are listed in table 1 for comparison too. Note that the experimental data on the oscillator strengths for many (especially, high-Z) Li-like ions are missing. Overall, there is a physically reasonable agreement of the listed data. The important features of the approach used are using the optimized one-particle representation and account for polarization effects. It should be noted that an estimate of the gauge-non-invariant contributions (the difference between the oscillator strengths values calculated with using the transition operator in the form of "length" and "velocity") is about 0.3%, i.e., the results obtained with different photon propagator gauges (Coulomb, Babushkon, Landau) are practically equal.

Table 1. Oscillator strengths of the $2s_{1/2} - 2p_{1/2,3/2}$ transitions in Li-like ions.

Method	DF [6]	DF [6]	[2]	[2]	Our	Our
Ion	$2s_{1/2}$–$2p_{1/2}$	$2s_{1/2}$–$2p_{3/2}$	$2s_{1/2}$–$2p_{1/2}$	$2s_{1/2}$–$2p_{3/2}$	$2s_{1/2}$–$2p_{1/2}$	$2s_{1/2}$–$2p_{3/2}$
S^{13+}	0.0299	0.0643	0.030	0.064	0.0301	0.0641
Ca^{17+}	0.0234	0.0542	0.024	0.054	0.0236	0.0541
Fe^{23+}	0.0177	0.0482	0.018	0.048	0.0179	0.0481
Zn^{27+}	0.0153	0.0477	–	–	0.0156	0.0475
Zr^{37+}	0.0114	0.0543	–	–	0.0118	0.0540
Mo^{39+}	–	–	0.011	0.056	0.0107	0.0556
Sn^{47+}	0.0092	0.0686	–	–	0.0095	0.0684
Tm^{66+}	–	–	–	–	0.0071	01140
Yb^{67+}	0.0067	0.1170	–	–	0.0069	0.1167

In Table 2 we present the oscillator strengths values for the $2s_{1/2} - np_j$ (n=3-18, j=1/2) transitions in spectrum of the Li-like ion Zr^{37+}. The quantum defect approximation (QDA) [6,27], the DF oscillator strengths calculation results by Zilitis [6] and some tabulated (experimental) data by Martin-Weiss [1] are listed as well. It is self-understood that the QDA oscillator strengths data become more exact with the growth of the principal quantum number. At the same time the accuracy of the DF data may be decreased. The agreement between the Martin-Weiss data and our results is acceptable.

Table 2. Oscillator strengths of the $2s_{1/2} - np_{1/2}$ transitions in Zr^{37+}.

Transition	QDA [6]	DF [6]	Our data
$2s_{1/2}$–$3p_{1/2}$	13.7	13.3	13.684
–$4p_{1/2}$	-	3.22	3.232
–$6p_{1/2}$	-	-	0.682
–$8p_{1/2}$	0.258	0.257	0.260
–$10p_{1/2}$	0.126	0.124	0.125
–$16p_{1/2}$	0.0291	0.0285	0.0287
–$18p_{1/2}$	-	-	0.0216

References

[1] Martin G A, Wiese W L 1976 *Phys.Rev.A.* **13** 699; 1976 *J. Phys. Chem. Ref. Dat.* **5** 538
[2] Schweizer W, Faßbinder P, Gonzalez-Ferez R 1999 *Atom. Data. Nucl. Data. Tabl.* **72** 33
[3] Ivanova E P, Ivanov L N 1979 *Atom. Data. Nucl. Data.Tabl.* **24** 95
[4] Ivanov L N, Ivanova E P and Aglitsky E V 1988 *Phys. Rep.* **166** 315
[5] Glushkov A V, Ambrosov S V, Loboda A V, Gurnitskaya E P and Prepelitsa G P 2005 *Int. J. Quant.Chem.* **104** 562
[6] Zilitis V A 1983 *Opt. Spectr.* **55** 215
[7] Khetselius O Yu 2012 *J. of Phys.: Conf. Ser.* **397** 012012
[8] Glushkov A V 2012 *J. of Phys.: Conf. Ser.* **397** 012011
[9] Zong-Chao Yan and Drake G W F 1995 *Phys. Rev. A* **52** R4316
[10] Lianhua Qu, Zhiwen Wang and Baiwen Li 1998 *J. Phys. B* **31** 3601
[11] Yan Z C, Tambasco M and Drake G W F 1998 *Phys. Rev. A* **57** 1652
[12] Xiaoxu Guan and Baiwen Li 2001 *Phys. Rev. A* **63** 043413
[13] Khetselius O Yu 2009 *Phys.Scr.* **T.135** 014023
[14] Chen Chao and Wang Zhi-Wen 2005 *Commun. Theor. Phys.* **43** 305
[15] Glushkov A V, Malinovskaya S V, Prepelitsa G P and Ignatenko V M 2005 *J. of Phys.: Conf. Ser.* **11** 199
[16] Bièmont E 1977 *Astr. & Astrophys. Suppl.Ser.* **27** 489
[17] Malinovskaya S V, Glushkov A V, Khetselius O Yu, Loboda A V, Lopatkin Yu M, Svinarenko A A, Nikola L V and Perelygina T B 2011 *Int. J. Quant. Chem.* **111** 288
[18] Glushkov A V, Ivanov L N 1992 *Phys. Lett. A* **170** 33
[19] Yerokhin V A, Artemyev A N, Shabaev V M 2007 *Phys. Rev. A.* **75** 062501
[20] Glushkov A V, Malinovskaya S V, Sukharev D E, Khetselius O Yu, Loboda A V and Lovett L 2009 *Int. J. Quant. Chem.* **109** 1717
[21] Ivanova E P and Glushkov A V 1986 *J.Quant. Spectrosc. Rad.Transf.* **36** 127
[22] Glushkov A V, Khetselius O, Loboda A and Svinarenko A 2009 *Phys. Scr.* **T153** 014029
[23] Ivanov L N, Ivanova E P and Knight L 1993 *Phys. Rev. A* **48,** 4365
[24] Glushkov A V, Khetselius O Yu and Malinovskaya S V 2008 *Europ. Phys. J.* ST **160** 195
[25] Khetselius O Yu 2009 *Int. J. Quant. Chem.* **109** 3330
[26] Glushkov A V, Ambrosov S, Ignatenko A and Korchevsky D 2004 *Int. J. Quant. Chem.* **99** 936
[27] Seaton M J 1983 *Rep. Prog. Phys.* **46** 1983

Spectroscopy of Rydberg atoms in a Black-body radiation field: Relativistic theory of excitation and ionization

A A Svinarenko, O Yu Khetselius, V V Buyadzhi, T A Florko, P A Zaichko and E L Ponomarenko

Odessa State Environmental University, L'vovskaya str.15, Odessa-16, 65016, Ukraine

E-mail: quantsvi@mail.ru

Abstract. The combined relativistic energy approach and many-body perturbation theory with zeroth model potential approximation are used for computing Blackbody radiation ionization characteristics of the Rydberg atoms, in particular, the sodium in states with n=17,18,40-70. The calculated ionization rate values are compared with available theoretical and experimental data.

1. Introduction

A great progress in experimental laser physics and appearance of the so called tunable lasers allows one to access the highly excited Rydberg states of atoms. In fact this is a beginning of a new epoch in the atomic physics with external electromagnetic field. It has stimulated a great number of papers on the ac and dc Stark effect [1-12]. Conversely, the experiments with Rydberg atoms had very soon resulted in the discovery of an important ionization mechanism, provided by unique features of the Rydberg atoms. Relatively new topic of the modern theory is connected with consistent treating the Rydberg atoms in a field of the Blackbody radiation (BBR). It should be noted that the BBR is one of the essential factors affecting the Rydberg states in atoms [1]. The account for the ac Stark shift, fast redistribution of the levels' population and photoionization provided by the environmental BBR became of a great importance for successfully handling atoms in their Rydberg states.

The most popular theoretical approaches to computing ionization parameters of the Rydberg atom in the BBR are based on the different versions of the model potential (MP) method, quasiclassical models. We mention a simple approximation for the rate of thermal ionization of Rydberg atoms, based on the results of our systematic calculations in the Simons-Fues MP [1]. In fact, using the MP approach is very close to the quantum defect method and other semi-empirical methods, which were also widely used in the past few years for calculating atom–field interaction amplitudes in the lowest orders of the perturbation theory. The significant advantage of the Simons-Fues MP method in comparison with other models is the possibility of presenting analytically (in terms of the hypergeometric functions) the characteristics for arbitrarily high orders, related to both bound–bound and bound–free transitions. Naturally, the standard methods of the theoretical atomic physics, including the Hartree-Fock and Dirac-Fock approximations, should be used in order to determine the thermal ionization characteristics of neutral and Rydberg atoms [2]. Note that the correct treatment of the heavy Rydberg atoms parameters in an external electromagnetic field requires using strictly relativistic models. Here we apply an energy approach [11-15] and relativistic perturbation theory (PT) with the MP zeroth approximation [16-20] to computing the thermal BBR ionization characteristics of the Rydberg atoms, in particular, the sodium.

2. Ionization of the Rydberg atoms in the Blackbody radiation

The qualitative picture of the BBR Rydberg atoms ionization is in principle easily understandable. Even for temperatures on the order of T=10^4 K, the frequency of a greater part of the BBR photons ω does not exceed 0.1 a.u. One could use a single- electron approximation to calculate the ionization cross section σ_{nl} (ω). The latter appears as a product with the Planck's distribution for the thermal photon number density:

$$\rho(w,T)=\frac{\omega^2}{\pi^2c^3[\exp(\omega/kT)-1]}, \tag{1}$$

where k=1.3807×10^{-23} J·K^{-1} is the Boltzmann constant, c =2.9979·10^8 m·s^{-1} is the speed of light. Ionization rate of a bound state nl results in the integral over the Blackbody radiation frequencies:

$$P_{nl}(T)=c\int_{|E_{nl}|}^{\infty}\sigma_{nl}(\omega)\rho(\omega,T)d\omega. \tag{2}$$

The ionization cross-section from a bound state with a principal quantum number n and orbital quantum number l by photons with frequency ω is as follows:

$$\sigma_{nl}(\omega)=\frac{4\pi^2\omega}{3c(2l+1)}[lM^2_{nl\to El-1}+(l+1)M^2_{nl\to El+1}], \tag{3}$$

where the radial matrix element of the ionization transition from the bound state with the radial wavefunction $R_{nl}(r)$ to ith continuum state with wavefunction R_{El} (r) is normalized to the delta distribution of energy. The corresponding radial matrix element can be written as:

$$M_{nl\to El'}=\int_0^{\infty}R_{El'}(r)r^3R_{nl}(r)dr. \tag{4}$$

We apply a generalized energy approach [11-15] and relativistic perturbation theory with the MP zeroth approximation [16-20] to computing the Rydberg atoms ionization parameters. In relativistic theory radiation decay probability (ionization cross-section etc) is connected with the imaginary part of electron energy shift. The total energy shift of the state is usually presented in the form: ΔE = ReΔE + i Γ/2, where Γ is interpreted as the level width, and a decay probability P = Γ. The imaginary part of electron energy shift is determined as [13, 14]:

$$\mathrm{Im}\,\Delta E=-\frac{1}{4\pi}\sum_{\substack{\alpha>n>f\\ [\alpha<n\le f]}}V^{|\omega_{\alpha n}|}_{\alpha n\alpha n}, \tag{5}$$

where $\omega_{\alpha n}$ is a frequency of the α-n radiation, (α>n>f) for electron and (α<n<f) for vacancy. The matrix element V is determined as follows:

$$V^{|\omega|}_{ijkl}=\iint dr_1dr_2\Psi^*_i(r_1)\Psi^*_j(r_2)\frac{\sin|\omega|r_{12}}{r_{12}}(1-\alpha_1\alpha_2)\Psi^*_k(r_2)\Psi^*_l(r_1) \tag{6}$$

Detailed description of the matrix elements and computational procedures are presented in Refs. [12,13,15]. The relativistic wave functions are calculated by solution of the Dirac equation with the potential, which includes the "outer electron - ionic core" potential (in the Miller-Green form [21]) and exchange-polarization potential [20]. All calculations utilize the Superatom-ISAN (version 93) code.

3. Results

In Table 1 we present results of the ionization rate calculation for the Rydberg sodium atom in the states (17,18D, 18P) at temperatures of 300 K and 500 K: Th5 – our (relativistic MP theory) data, E1 – experimental data by Kleppner etal and Burkhardt etal [4], Th1- theory (nonrelativistic Simons-Fues

MP) by Glukhov-Ovsyannikova [9], Th2- theory of Lehman [8], Th3- quasiclassical model by Dyachkov-Pankratov [10] and Th4- theory by Beterov et al. [1]. Overall, there is physically reasonable agreement between the theoretical and experimental data. Obviously, the accuracy of the theoretical data is provided by a correctness of the corresponding relativistic wave functions and accounting for the exchange-correlation effects. In Table 2 we list our results of ionization rate (s^{-1}) for sodium Rydberg states (with n=40-70) induced by BBR radiation (T = 300 K).

Table 1. Theoretical and experimental values of the ionization rate (10^3 s^{-1}) of sodium Rydberg states: E1- Kleppner et al Burkhardt etal; Th1 – theory by Glukhov-Ovsyannikova; Th2- theory by Lehman, Th3- theory by Dyachkov-Pankratov; Th4- theory by Beterov et al; Th5 – this work.

T	*nL*	E1	Th1	Th2	Th3	Th4	Th5
300	17D	10^3	$1.08{\cdot}10^3$	$0.95{\cdot}10^3$	$0.9{\cdot}10^3$	$1.147{\cdot}10^3$	$1.02{\cdot}10^3$
500	18P	-	$4.18{\cdot}10^3$	-	-	-	$5.54{\cdot}10^3$
500	18D	-	$4.07{\cdot}10^3$	-	-	-	$5.46{\cdot}10^3$

Table 2. Ionization rate (s^{-1}) for the sodium Rydberg states (with n = 40-70), induced by BBR radiation (T = 300 K; our data).

Atom	40	50	60	70
Na S	142	106	61.4	29.5
Na P	804	576	311	141
Na D	707	496	268	122

References

[1] Beterov I I, Tretyakov D V, Ryabtsev I I, Entin V M, Ekers A and Bezuglov N N 2009 *New J. Phys.* **11** 013052
[2] Safronova U I and Safronova M S 2009 *Phys. Rev. A* **79** 022512
[3] Gallagher T F 1979 *Phys. Rev. Lett.* **42** 835
[4] Killian T, Kulin S, Bergeson S, Orozco L, Orzel C and Rolston S 1999 *Phys. Rev. Lett.* **83** 4776
[5] Li W, Noel M, Robinson M, Tanner P and Gallagher T 2004 *Phys. Rev. A.* **70** 042713
[6] Spencer W P, Vaidyanathan A, Kleppner D and Ducas T 1982 *Phys. Rev. A.* **26** 1490
[7] Glushkov A V, Ambrosov S, Ignatenko A and Korchevsky D 2004 *Int. J. Quant. Chem.* **99** 936
[8] Lehman G W 1983 *J. Phys. B: At. Mol. Phys.* **16** 2145
[9] Glukhov I and Ovsiannikov V 2009 *J. Phys. B: At. Mol. Phys.* **42** 075001
[10] D'yachkov L G and Pankratov P M 1994 *J. Phys. B: At. Mol. Opt. Phys.* **27** 461
[11] Ivanova E P and Ivanov L N 1979 *Atom. Data. Nucl. Data. Tabl.* **24** 95
[12] Glushkov A V 2012 *J. of Phys.: Conf. Ser.* **397** 012011
[13] Ivanov L N, Ivanova E P and Aglitsky E V 1988 *Phys. Rep.* **166** 315
[14] Glushkov A V and Ivanov L N 1992 *Phys. Lett. A* **170** 33
[15] Glushkov A V 2012 *Advances in the Theory of Quantum Systems in Chemistry and Physics. Ser. Progress in Theoretical Chemistry and Physics*, vol. 28, ed. by K Nishikawa, J Maruani, E Brandas, G Delgado-Barrio, P Piecuch (Berlin: Springer) p 131
[16] Glushkov A V, Malinovskaya S V, Prepelitsa G P and Ignatenko V M 2005 *J. of Phys.: Conf. Ser.* **11** 199
[17] Malinovskaya S V, Glushkov A V, Khetselius O Yu,, Svinarenko A A, Mischenko E V and Florko T A 2009 *Int. J. Quant. Chem.* **109** 3325
[18] Glushkov A V, Khetselius O, Loboda A and Svinarenko A 2009 *Phys. Scr.* **T153** 014029
[19] Khetselius O 2012 *J. of Phys.: Conf. Ser.* **397** 012012 ; 2009 *Int. J. Quant. Chem.* **109** 3330
[20] Malinovskaya S V, Glushkov A V, Khetselius O Yu, Loboda A V, Lopatkin Yu M, Svinarenko A A, Nikola L V and Perelygina T B 2011 *Int. J. Quant. Chem.* **111** 288
[21] Miller K J and Green A E S 1974 *J. Chem. Phys.* **60** 2617

Hydrogen Balmer Series Measurements in Laser-Induced Air Plasma

L D Swafford, D M Surmick, M J Witte, A C Woods, G Gautam, and C G Parigger

University of Tennessee Space Institute, 411 B.H. Goethert Parkway, Tullahoma, TN 37388, USA

E-mail: cparigge@tennessee.edu

Abstract. Time-resolved spectroscopy is employed to analyze micro plasma generated in laboratory air. Stark-broadened emission profiles for hydrogen alpha and beta allow us to determine plasma characteristics for specific time delays after plasma generation. Stark shift, asymmetry, and full width half maximum measurements are used to infer electron density. The measurements of hydrogen alpha and beta Balmer series line shapes are analyzed using various theory results. Our laser-induced breakdown spectroscopy arrangement uses a Q-switched Nd:YAG laser operating at the fundamental wavelength of 1064 nm that is focused for plasma generation. The hydrogen alpha and beta lines emerge from the free electron background radiation for time delays larger than 0.3 μs and 1.4 μs, respectively. Neutral and ionized nitrogen emission lines allow us to infer electron density for time delays from 0.1 to 10 μs. The electron density values are compared with results obtained from hydrogen Balmer series line shapes.

1. Introduction

Laser-Induced Breakdown Spectroscopy (LIBS) techniques are employed to measure micro-plasma emissions in laboratory air. Figure 1 illustrates the schematic experimental arrangement.

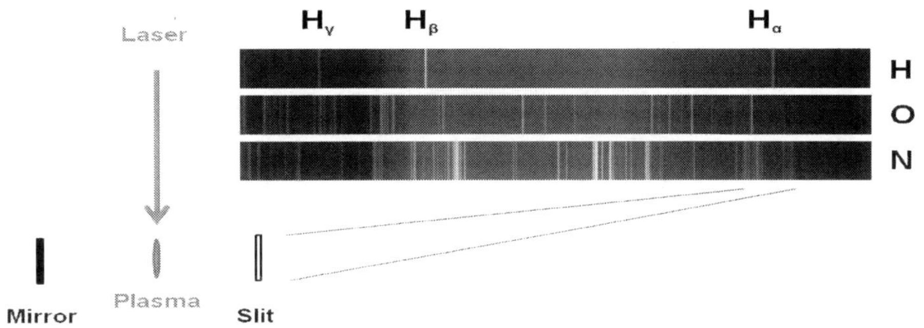

Figure 1. Schematic experimental arrangement. The laser-induced plasma is generated by focusing pulsed radiation at 1064 nm from the top. Lenses, a Czerny-Turner spectrometer and an intensified array detector (not shown) are used to collect, disperse and record spectra.

Figure 1 also shows discharge spectra for hydrogen, oxygen, and nitrogen, and the wavelength ranges of interest for measurements of H_α near 656.28 nm and H_β near 486.14 nm. Time-resolved spectral data records are obtained by use of an Intensified Linear Diode Array (Intensified LDA, model 1460 EG&G Princeton) and/or an Intensified Charge Coupled Device (ICCD, Andor model iSTAR 334T) connected to a 0.64 m Czerny-Turner Spectrometer (model Jobin Yvon 640) [1, 2, 3]. For the images reported in this work, the micro-plasma is generated parallel to the spectrometer slit by focusing Nd:YAG fundamental laser radiation at 1064 nm from the top using f# = 4 optics. Emissions from the plasma are imaged close to 1:1 onto the spectrometer slit, and in addition, we utilize a flat mirror to explore effects of self absorption. Self-absorption investigations are based on comparing results with and without the flat mirror [4, 5, 6] and by determining electron density from nitrogen lines [7].

2. Results

The emission spectra are recorded by averaging 81 consecutive plasma generation events, with the ICCD detector set to record 276 spectra spatially-resolved along the slit height. Figure 2 illustrates typical results. The images are wavelength calibrated, detector dark-count subtracted, and detector-sensitivity corrected.

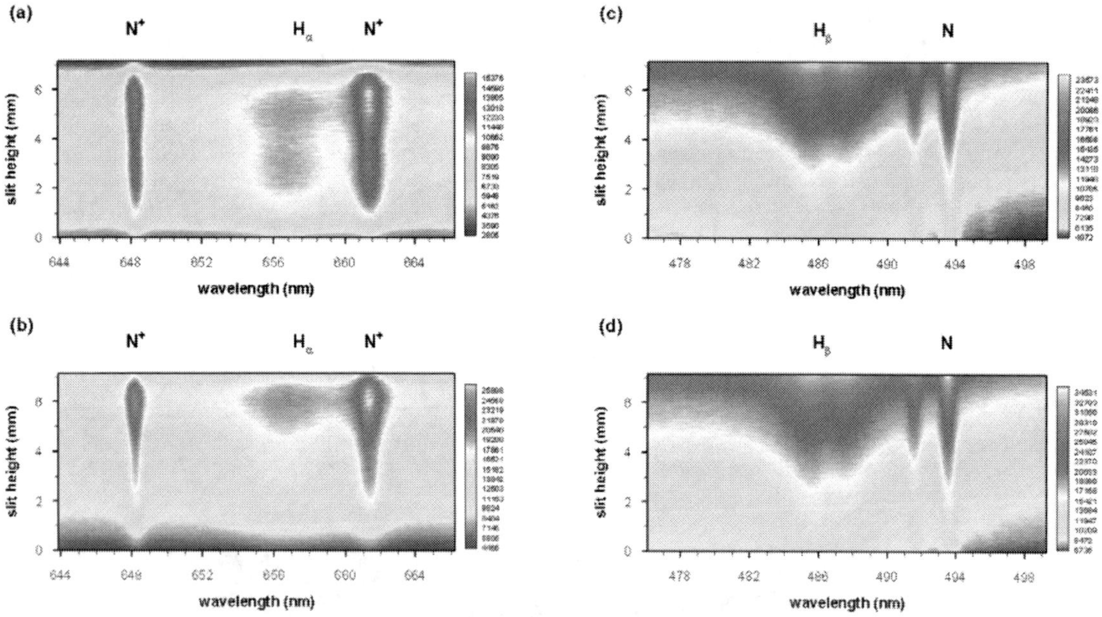

Figure 2. Recorded spectra vs. slit height. H_α (a) without mirror, (b) with mirror; $\Delta\tau = 0.3$ μs. H_β (c) without mirror, (d) with mirror; $\Delta\tau = 3.0$ μs.

A mirror and an additional lens was used to retro-reflect emissions through the laser-induced plasma. A gate-width of 10 ns was employed. For this experiment, we combined as well 2 vertical camera pixels for an effective pixel size of 26 μm $\times$ 26 μm. Early in the plasma decay, for delays of 0.3 μs, expansion speeds are typically hypersonic (or larger than Mach 5) yet for 3.0 μs time delays, speeds are supersonic. For an expansion speed of Mach 5, using a 10 ns gate width and an effective time delay of 2 ns by introducing the mirror, one finds on the order of 20 μm spatial displacement due to plasma evolution yet this is for 1:1 imaging smaller than the 26 μm linear dimension of two vertically combined pixels.

The electron number densities for the hydrogen alpha and beta line was determined from the measured the full-width half maximum (FWHM) of the atomic emission spectra at a given time delay, using Equation (1) below [8]:

$$\mathrm{H}_\alpha : \ N_{e,\alpha}[m^{-3}] = (w_\alpha[nm]/1.098)^{1.47135} \times 10^{23}; \qquad \mathrm{H}_\beta : \ N_{e,\beta}[m^{-3}] = (w_\beta[nm]/4.8)^{1.46808} \times 10^{23}. \tag{1}$$

The FWHM measurement is facilitated by fitting of measured spectra. For hydrogen beta fitting, contributions from Lorentzian and Gaussian components of spectral broadening are included. The shape of the Stark broadened line varies with N_e which also corresponds with changes of the asymmetry. Measurement of the Stark profile FWHM is reasonable approach to determine N_e. For our purposes, we use the FWHM of the fitted Voigt profile, and use empirical formulae [8], tabulated profiles [9] or results from convergent theory [10, 11].

In addition, we evaluated use of the asymmetric line shapes for the purpose of determining electron density using Equ (2) [12]:

$$\mathrm{H}_\beta : \ N_{e,\beta}[10^{23} \ m^{-3}] = 0.14(\rho[\%])^3; \qquad \rho = 2(I_B - I_R)/(I_B + I_R), \tag{2}$$

where the blue I_B, and red, I_R, intensity values denote the lower and higher wavelength peaks of the Stark-broadened hydrogen beta profile. Details of this study are elaborated in Ref. [12] and show that for hydrogen beta lines, the accuracy of N_e determination is much improved when using the FWHM approach.

In the analysis of recorded H_α spectra, ionized nitrogen lines at 648.2 nm and 661.1 nm can be used as well [7] to determine electron density. These two lines can be seen in Figures 2 (a) and (b) at the low and high wavelength side of the 656.28 nm hydrogen Balmer Series line. In turn, for the H_β line, the neutral nitrogen line at 493.51 nm can be used, see Figs. 2 (c) and (d). Inferred electron densities for a delay of $\Delta\tau = 0.3$ μs from optical breakdown (from only H_α profiles) amount to 14 to 20 $\times$ 10^{23} m^{-3}. Yet N^+ diagnostics [7] indicate electron densities of $\simeq$ 12 $\times$ 10^{23} m^{-3}. Consequently, self-absorption is evident for the 0.3 μs data set. For a time delay $\Delta\tau$ of 0.4 μs, hardly any self-absorption occurs; for time delays of 3.0 μs self-absorption effects appear to be insignificant as well. For the 3.0 μs data sets, the inferred electron densities from hydrogen alpha and hydrogen beta agree within experimental error, and amount to 1.1 $\times$ 10^{23} m^{-3} and 0.96 $\times$ 10^{23} m^{-3}, respectively.

3. Conclusions

Determination of the electron density from the FWHM of the Balmer H_α and H_β line proves to be more accurate than from asymmetry of the line shapes. Difficulty of determining the asymmetry parameters for especially incomplete profiles increases the uncertainty. For larger electron densities, or for shorter time delays from optical breakdown, the uncertainties become too large for determining N_e from the asymmetry parameter. The electron density FWHM diagnostic shows the effects of hydrogen alpha self-absorption for the very early time delays of 0.3 μs, or when hydrogen alpha emerges from the free electron background. Hydrogen alpha and hydrogen beta measurement results of electron density agree for time delays of 3.0 μs within experimental error margins, and appear to only show insignificant effects due to self-absorption. The studies reported here focus on optically generated plasma in laboratory air; however, it is expected that self-absorption effects are more readily observable during laser ablation processes due to significantly higher electron density for laser-radiation interacting with solids.

Acknowledgments

The authors thank for support of this work by the Center for Laser Applications at the University of Tennessee Space Institute, and we gratefully acknowledge discussions with Dr. Eugene Oks on Stark-broadened line shapes.

References

[1] Parigger C G 2010 *Int. Rev. At. Mol. Phys.* **1** 129
[2] Parigger C G 2013 *Spectrochim. Acta Part B* **79-80** 4
[3] Parigger C G, Woods A C, Witte M J, Swafford L D, Surmick D M 2014 *JoVE* **84** E51250
[4] Moon H, Herrera K K, Omenetto N, Smith B W, Winefordner J D 2009 *Spectrochim. Acta Part B* **64**, 702
[5] Konjević N 1999 *Phys. Rep.* **316** 339
[6] El Sherbini A M, El Sherbini Th M, Hegazy H, Cristoforetti G, Legnaoili S, Palleschi V, Pardini L, Salvetti A, Tognoni E. 2005 *Spectrochim. Acta Part B* **60** 1573
[7] Konjević N, Lesage A, Fuhr J R, Wiese W L 2002 *J. Phys. Chem. Ref. Data* **31** 819
[8] Konjević N, Ivković M, Sakan N 2012 *Spectrochim. Acta Part B* **76** 593
[9] Griem HR 1964 *Plasma Spectroscopy*, McGraw-Hill, New York, 1964
[10] Oks E 2006 18th ICSLS: AIP Conf. Proceedings **874** 19
[11] Parigger C G, Plemmons D H, Oks E 2003 *Appl. Opt.* **42** 5992
[12] Parigger C G, Swafford L D, Woods A C , Surmick D M, Witte M J 2014 *Spectrochim. Acta Part B* **99** 28

Journal of Physics: Conference Series **548** (2014) 012050 doi:10.1088/1742-6596/548/1/012050

Effect of Plasma Compositions on Growth and Field Emission Properties of a Spherical Carbon Nanotube (CNT) tip

A Tewari and S C Sharma

Department of Applied Physics, Delhi Technological University, Shahbad Daulatpur, Bawana Road, Delhi-110 042, India

aartitewari1386@rediffmail.com

Abstract. The effect of plasma compositions (i.e., different concentration of participating ions) on growth, structure and field emission properties of spherical Carbon Nanotube (CNT) tip has been theoretically investigated. In plasma, two different kinds of positively charged ions with heavy to light ion mass ratio of 11.5 are considered and the effect of the different fractional concentrations of participating ions on the growth and field emission properties of CNT is studied. Numerical calculations of the radius of spherical CNT tip for different fractional light ion concentrations have been carried out for the typical glow discharge plasma parameters. It is found that on increasing fractional light ion concentration, the radius of spherical CNT tip decreases and consequently the field emission factor for spherical CNT tip increases.

1. Introduction

Levchenko*et al.* [1] have studied the growth of carbon nanocone arrays on metal catalyst and the effect of carbon flux ionization. They have found that a high degree of ionization of neutral gas in plasma leads to a better uniformity of metal catalyst saturation and nanocone growth and thereby contributes to the formation of more height – uniform arrays of carbon nanostructures. Lee *et al.*[2] have investigated the effect of plasma treatment on surface morphology and field emission characteristics of CNT and revealed that the plasma treatment can modify the surface morphology and enhanced the field emission characteristics of the CNTs. Srivastava *et al.*[3] have synthesized carbon films by microwave plasma enhanced chemical vapour deposition (MPECVD) using mixture of methane and argon gases on Ni coated and Si substrates and found that the increase in microwave power causes more ionization of the gas, which increases the density of plasma species of relatively higher energy. Moreover, increased nucleation of graphitic clusters is expected to occur and this leads to formation of carbon petals of relatively smaller size and higher density at increased microwave power.

2. Model

We consider plasma containing electrons, positively charged ions of type A and B, neutral atoms of type A and B and CNT is grown in the presence of plasma. The two ions have considerable differences

Journal of Physics: Conference Series **548** (2014) 012050 doi:10.1088/1742-6596/548/1/012050

in their masses such that we consider ion A to be a light and ion B to be a heavy ion. The ratio of heavy to light ion masses is 11.5.

We have developed a model whereby changing the concentration of light positively charged ion, the corresponding change in heavy positive ion concentration is automatically obtained.

The initial radius of spherical CNT tipa_0 can be estimated by equating the accretion of electrons and light and heavy positively charged ions on the CNT, i.e.,

$$n_e\left(\frac{T_e}{m_e}\right)^{\frac{1}{2}}\exp\left(-\frac{e^2}{a_0 k_B T_e}\right)=\left(1+\frac{e^2}{a_0 k_B T_i}\right)\left[n_{ilA}\left(\frac{T_i}{m_{ilA}}\right)^{\frac{1}{2}}+n_{ihB}\left(\frac{T_i}{m_{ihB}}\right)^{\frac{1}{2}}\right], \tag{1}$$

where n_e and T_e are the electron density and temperature, respectively. The ion temperature is denoted by T_i, the Boltzmann constant by k_B, and the electron charge by e.The densities of light and heavy positive ions are indicated by n_{ilA} and n_{ihB} with masses $m_{ilA}$and m_{ihB}, respectively. The plasma density, n_{p0}, equals the electron density, n_e; consequently, with the fractional densities α_l and α_h of light and heavy positive ions, and α_h=1-α_l, one finds $n_{ilA}=\alpha_l n_{p0}$ and $n_{ihB}=\alpha_h n_{p0}$.

2.1 Growth rate equation of the mass of CNT

The growth rate can be computed using

$$\frac{dm_{ct}}{d\tau}=\left(m_A\gamma_A n_{Act}+m_{ilA}\gamma_{ilA}n_{ilAct}\right), \tag{2}$$

where m_{ct} is the mass of the CNT of radius a. With the density of the CNT, ρ_{ct}, the mass of the CNT can be expressed in terms of the density$m_{ct}=4/3\ \pi\ a^3\rho_{ct}$. The sticking coefficients for the ionic species and the light positive ions are γ_A and γ_{ilA}, respectively.The two terms in Equation (2) correspond to gain in mass due to collection of atomic and ionic species A.

3. Results and discussion

We have carried out calculations to study the dependence of the radius of spherical CNT tip on the various compositions of ions in plasma by varying n_{ilA0} and n_{ihB0}.It is the accretion of neutral atoms and positively charged ions on the CNT that leads to its growth. Different plasma compositions signify the different concentrations of participating ions. In the present analysis we have varied the concentrations of participating positively charged ions and studied its effect on radius of spherical CNT tip.The other $\alpha_l$parameters used in the calculations are the same as in Tewari*et al.*[4].

Figure 1 illustrates the variation of normalized radius a/a_0 of spherical CNT tip with time for different fractional concentrations of light positively charged ions, i.e.,α_l= 0.1,0.3,0.6 & 0.9. From Fig. 1 it can be seen that the normalized radius of CNT first increases with time and then attains a saturation value. It also shows the decrease of normalized radius a/a_0 with fractional concentrations of light positively charged ions, α_l, because for larger values more and more neutral atoms ionize to produce positively charged ions and electrons. Since it is the accretion of neutral atoms on the CNT which leads to its growth, therefore by increasing the fractional concentrations of light positively charged ions the radius of the CNT decreases.

Figure 2 shows the variation of normalized ionic density of type A in plasma with time for different values of sticking coefficients γ_A= 0.1, 0.3, 0.5 & 1.0. It can be seen in Fig. 2 that with increasing fractional concentrations of light positively charged ions, the positively charged ion of type Adecays faster. With plasma treatment and increase in plasma power, ions of type$A$ will decay faster leading to decrease in CNT radius which is consistent with the experimental observations of Lee*et al.*[2]and Srivastava *et al.*[3].

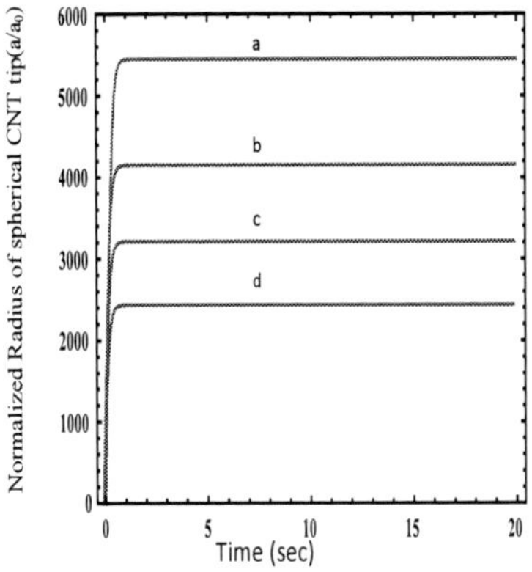

Figure 1. Variation of the normalized radius a/a_0 for spherical CNT tips for different fractional concentrations of light positive charged ions. The indicated a, b, c, d, responses correspond to α_l = 0.1,0.3,0.6 & 0.9, respectively.

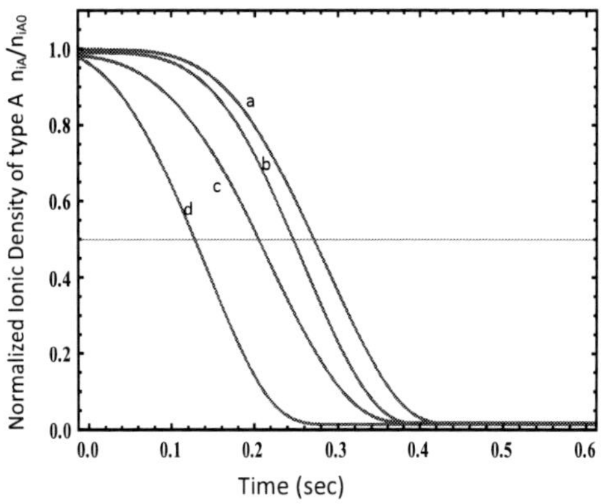

Figure 2. Variation of the normalized ionic density with time for different sticking coefficients of atomic species a, b, c, d, corresponding to γ_A = 0.1,0.3,0.5 & 1.0, respectively.

References

[1] Levchenko I, Ostrikov K, Khachan J, Validimrov S V 2008 *Phys. Plasmas* **15**, 103501
[2] Srivastava S K, Vankar V D, Sridhar Rao D V, Kumar V 2006 *Thin Solid Films* **515** 1851
[3] Lee S F, Chang Y-P, Lee L-Y 2009 *J. Mater. Sci. Mater. Electron* **20** 851
[4] Tewari A,Walia R, Sharma S C 2012 *Phys. Plasmas* **19** 013502

Journal of Physics: Conference Series **548** (2014) 012051 doi:10.1088/1742-6596/548/1/012051

Advanced Plasmadynamic Modelling for Fluid Plasmas

R J Thompson and T M Moeller
University of Tennessee Space Institute, Department of Mechanical, Aerospace & Biomedical Engineering, 411 B. H. Goethert Pkwy, Tullahoma, TN 37388, USA

E-mail: rthompso@utsi.edu

Abstract. High-fidelity plasma simulations can be accomplished using microscopic models. However, this is often infeasible for dense plasma problems, in which case continuum models are more economical. Our work has investigated construction of high-fidelity fluid plasma models that retain effects such as displacement current, charge separation effects and plasma wave behaviour.

1. Introduction

Plasmadynamic modelling and simulation plays a significant role in furthering the understanding of the basic underlying physics, and in predicting the behaviour of the plasma under certain conditions. Broadly, two classes of models may be employed to simulate the behaviour of plasma: microscopic models, which include kinetic, Boltzmann and particle methods, and continuum approaches, characterized by statistical averaging. The former more naturally captures anisotropic and tensorial behaviours as well as micro-physics; however, for dense plasmas characterized by a very low Knudsen number, these models may be intractable on contemporary hardware, and the latter approaches are then far more economical for simulation of the overall plasma evolution.

Our work has focused on the exploration of higher-fidelity physical models and the development of numerical methods to construct approximate solutions to such models with only relatively modest computational effort. This paper briefly describes some of these approaches and outlines essential results. First, we present the physical model used in these investigations. We outline the numerical methods developed, and then present some typical results that show excellent agreement with theory.

2. Physical model

The union of the Navier-Stokes and Maxwell equations can be compactly written as a coupled, nonlinear system of first-order hyperbolic partial differential equations,

$$\frac{\partial}{\partial t}\begin{bmatrix} \varrho \\ \varrho\mathbf{u}+\mathbf{R} \\ \varrho e+U \\ \mathbf{B} \\ \mathbf{E} \end{bmatrix} + \partial_k \begin{bmatrix} \varrho u \\ \varrho\mathbf{uu}\ +p\mathbf{1}-\mathbf{T} \\ \varrho H\mathbf{u}+\mathbf{S} \\ \mathbf{1}\times\mathbf{B} \\ -c^2(\mathbf{1}\times\mathbf{E}) \end{bmatrix} = \begin{bmatrix} 0 \\ 0 \\ 0 \\ 0 \\ -\frac{\mathbf{j}}{\epsilon_0} \end{bmatrix}, \qquad (1)$$

where ϱ is the mass density, $\mathbf{u}$ is the fluid velocity, e is the total specific energy, H is the total enthalpy, p is the fluid pressure, $\mathbf{E}$ and $\mathbf{B}$ are the electromagnetic fields, and $\mathbf{j}$ is the current density. Here, $\mathbf{R} = \mathbf{S}/c^2$, where $\mathbf{S}$ is the Poynting vector, U is the electromagnetic energy density, and $\mathbf{T}$ is the Maxwell stress tensor. $\mathbf{1}$ is the identity matrix. The current density is calculated using Ohm's law, and the ideal gas relation is used for relating the pressure and specific energy.

3. Numerical methods

The numerical solver is a parallelized, high-performance research code that supports fully unstructured meshing in one, two or three dimensions. It is capable of discretizing and solving any number of coupled, nonlinear, first-order hyperbolic partial differential equations using an advanced treatment of the finite volume method.

An important issue is the treatment of the divergence constraints (scalar Maxwell's equations). These equations are constraints placed on the initial conditions; analytically, if the initial conditions satisfy these constraints, then the solution will for all time. However, numerically, this is not the case; divergence error frequently accumulates unchecked, and pollutes the solution into an unphysical one. To avoid this, the divergence constraints were cast into a hyperbolic form by adding a first-order time derivative in a numerical potential. These hyperbolic equations were driven to a pseudo-steady-state each time step, eliminating the divergence error.

The generic system of equations can be written in a generic compact vector form as

$$\frac{\partial Q}{\partial t} + \nabla \cdot F = H, \tag{2}$$

and can be solved implicitly using preconditioned pseudo-time schemes to accelerate the convergence.

4. Results

Typical results for basic science simulations have included investigations of plasma turbulence and instability effects [1-4]. Several investigations have characterized Kelvin-Helmholtz plasma instability under the influence of a magnetic field. The magnetic field can be shown to suppress the smaller-wavelength instabilities (see Figure 1).

Simulations have also investigated compressible plasma turbulence behaviour. A common benchmark problem for this case is the Orszag-Tang problem, which has been tested under the new solution approach, and the solution is in excellent agreement with the typical magnetohydrodynamic solution (see Figure 2).

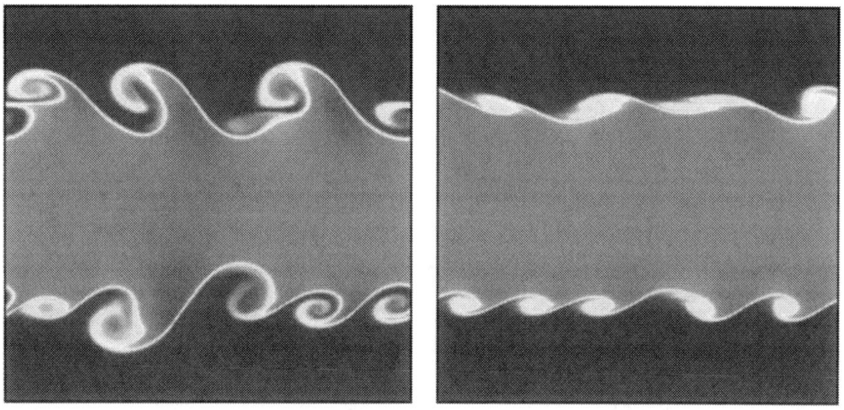

Figure 1. Results of a Kelvin-Helmholtz instability simulation with identical initial conditions and at an identical simulation time [1]. (Left) Without an applied magnetic field. (Right) With a weak applied magnetic field.

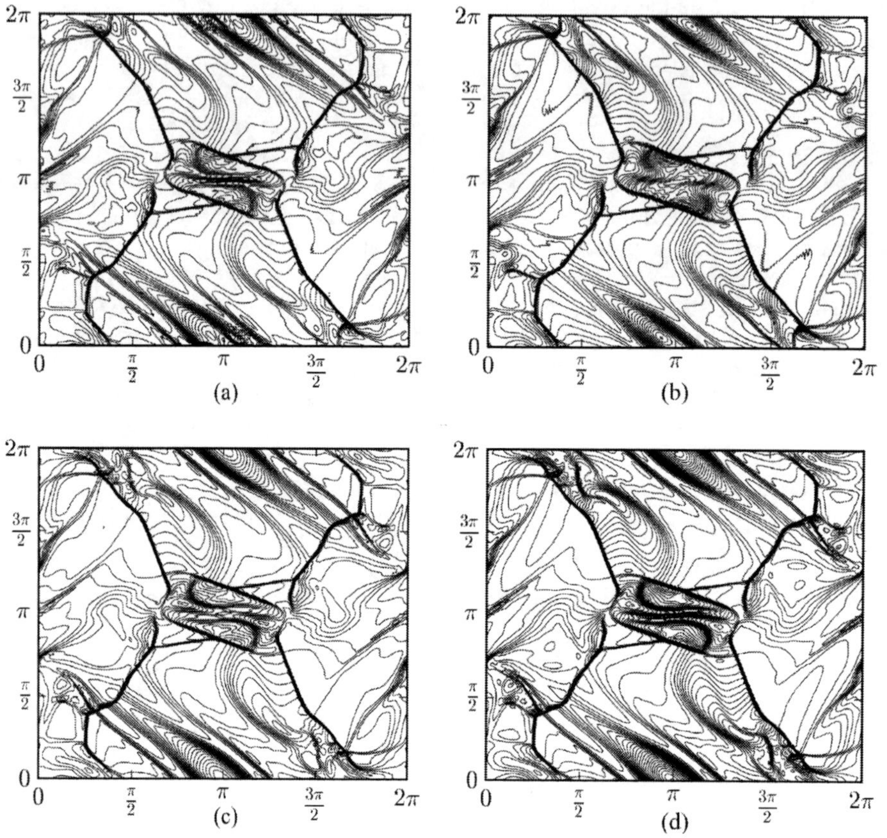

Figure 2. Solution of the density for the Orszag-Tang problem on different grid resolutions [4]. (a) 64×64 (b) 128×128 (c) 192×192 (d) 256×256.

5. Conclusions and future work

The physical model has already been extended now to fully relativistic, multiple-species plasma dynamics. Future work will continue this trend of extending this approach for high-fidelity fluid modelling of dense plasmas, and applying it to explore cases of both basic plasma science and engineering device applications.

References

[1] Thompson R, Wilson A, Moeller T, Merkle C 2014 Advected Upstream Splitting Method for the Coupled Maxwell and Navier–Stokes Equations *AIAA Journal* (Accepted)
[2] Wilson A, Thompson R, Moeller T 2014 Plasma Modeling with Mixed Wave/Diffusion Behavior *IEEE Trans Plas Sci* (Accepted)
[3] Thompson R, Wilson A, Moeller T, Merkle C 2014 *Journ Comp Phys* **258** 431
[4] Thompson R, Wilson A, Moeller T, Merkle C 2014 52nd AIAA Aerospace Sciences Meeting Paper AIAA-2014-1191

Laser-Induced Spectroscopy of Graphene Ablation in Air

M J Witte and C G Parigger

University of Tennessee Space Institute, 411 B.H. Goethert Parkway, Tullahoma, TN 37388, USA

E-mail: mwitte@utsi.edu, cparigge@utsi.edu

Abstract. Carbon Swan spectra are observed following laser ablation of graphene in laboratory air. Previous experiments showed temperatures that ranged from 4500 to 7500 K for the $\Delta v = 0$ transition and 4200 to 4500 K for the $\Delta v = -1$ transition for time delays on the order of 1.6 μs to 70 μs. This experiment explored in greater detail time delays > 10 μs for both molecular bands. Temperatures were found to be similar, ranging from 4500 to 6700 K for the $\Delta v = 1$ transition and 3200 to 5500 K for the $\Delta v = -1$ transition. Investigation is also made into spatially resolving the plasma emissions along the slit height. In addition, efforts are made to investigate the applicability of the local thermodynamic equilibrium (LTE) assumption. Comparisons are discussed in view of previous work that utilized Stark broadening of the H_β line, confirming LTE for delays < 10 μs, yet further research needed for later delays.

1. Introduction

Laser-induced breakdown spectroscopy (LIBS) is a technique utilized for analyzing a substance's composition [1]. It is a non-invasive and non-contact technique that only requires a small amount of the substance to be useful while still allowing for accurate determination of unknown material [2]. This becomes especially useful when the substance itself or its environment is hazardous in nature. However, in order to be of use as a diagnostic tool, known models are required. Previous investigations had been made [4, 3, 5] into C_2, though limited. To that end, investigations are made into carbon Swan spectra emissions from laser-induced plasma.

2. Experimental approach

The experiment utilizes a Nd:YAG pulsed laser operating at 1064-nm with 13 nanosecond, 190 mJ/pulse to generate laser-induced breakdown. The emitted spectra were dispersed with an 1800 grooves/mm grating in a HR640 Jobin-Yvon 0.6 m spectrometer and then recorded with an Andor iSstar ICCD. Measurements were time-resolved with varying time delays (10 μs to 100 μs) and corresponding gate widths (1 μs to 20 μs). Spectra were sensitivity and wavelength calibrated with a standard tungsten and mercury lamp, respectively.

3. Computational approach

For diatomic molecules, spectral predictions require a temperature along with line-strengths [6], for the allowed spectral transitions. In order to fit the spectra, the Nelder-Mead algorithm is utilized [7] through the so-called Nelder-Mead-Temperature (NMT) program.

The NMT program generates a single temperature spectral fit for the recorded data. This fit, within the accuracy and precision bounds, is also used to refine the calibration the data. This is repeated several times, with each run increasing from a linear, quadratic, to finally a cubic calibration. After that, the program is run three times in succession to ensure accuracy of the computed temperature.

4. Results

Figures 1 and 2 are representative of the results from the NMT program for carbon Swan spectra of the $\Delta v = 0$ and $\Delta v = -1$ bands. Seen in both are the clear vibrational peaks and underlying rotational structure. No atomic lines are present. Note, there is good matching between the experimental and theory spectra as the count difference between both is generally within the computed background line.

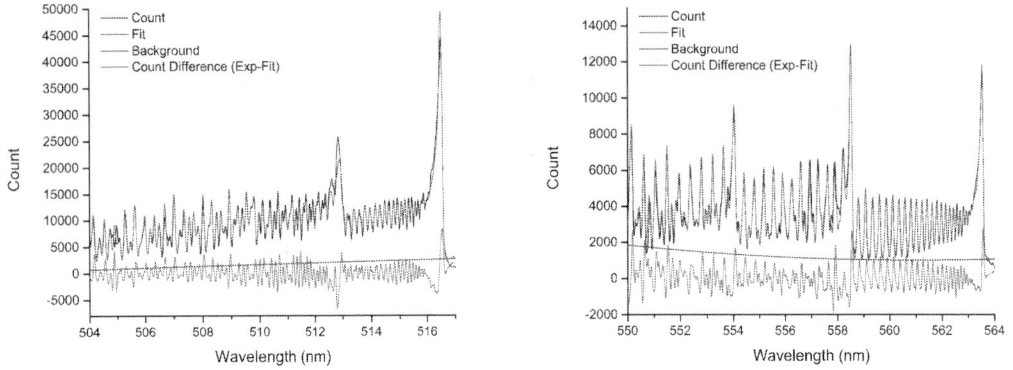

Figure 1. 40 μs delay, 5 μs gate width. Temperatures are 5500 K and 4400 K, left and right.

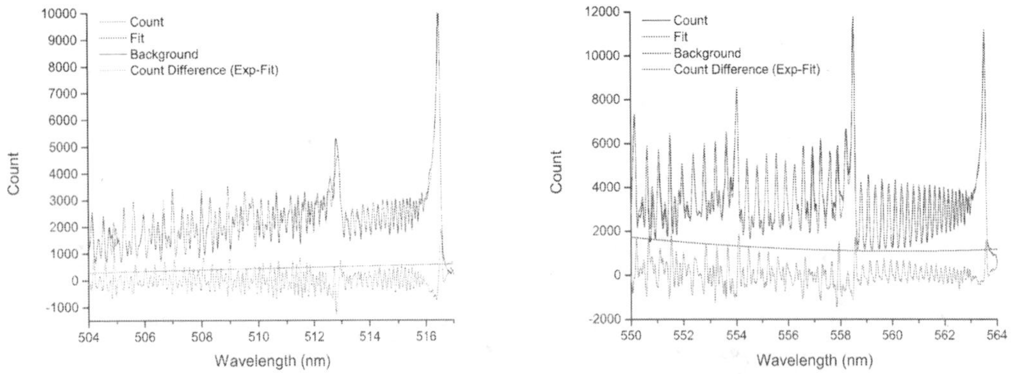

Figure 2. 70 μs delay, 10 μs gate width. Temperatures are 5800 K and 3200 K, left and right.

Figure 3 shows a 2-D image of the C_2 spectra for the $\Delta v = 0$ band. Calculated temperatures range from 6600 K to 7700 K, increasing towards 7 mm, with a dip just under it at 6 mm. Also of note is the slight vertical symmetry.

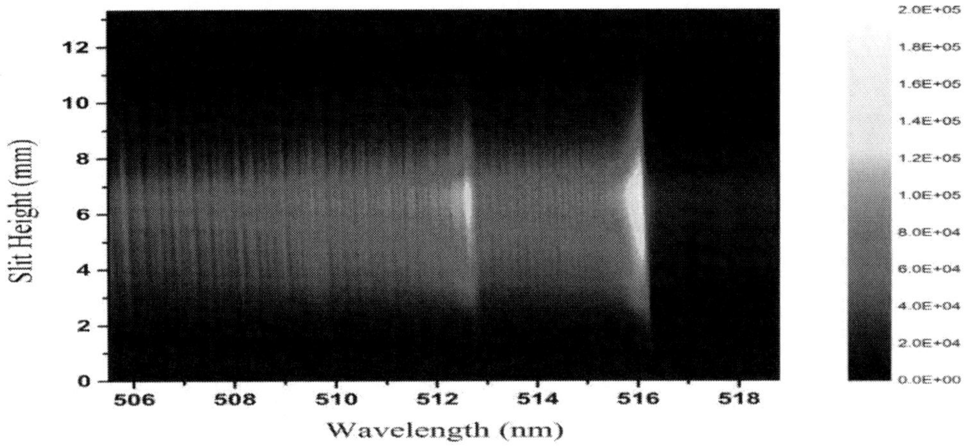

Figure 3. Raw image of two dimensional C_2 spectra at 15 μs delay and 5 μs gate width. The FWHM averaged to 0.13 nm for each of the individual spectral lines.

5. Conclusions

For our theory models, we have used a single temperature for fitting with assumption of LTE. For early time delays on the order of 1 μs with large laser energy per pulse, time-delayed spectra indicate LTE due to the presence of H_β [3].

Past that, there is some indication for LTE at later delays: The experimental spectra having good agreement with computed theory spectra within background variations. However, there is no definite indication of LTE at later delays. As the electron density diminishes, H_β emissions disappear in the measured spectra past 10 μs. There is some indication that LTE is absent since there appears to be different temperature ranges for the different wavelength bands, as seen in comparing the $\Delta v = 0$ and $\Delta v = -1$ bands.

From the recorded 2-D data, we conclude that the plasma shows a higher temperature towards the top of the recorded emission spectra near 7 mm but is cooler away from it. Just below that region, there appear to be significant variations in inferred temperature. These temperature differences could indicate non-LTE conditions. Consequently, exploration and utilization of Abel and Radon techniques [8] in better analyzing the plasma is called for.

Acknowledgments

This work is in part supported by the Center for Laser Applications at UTSI.

References

[1] Parigger C G 2012 *Spectrochim. Acta Part B* **78-79** 4
[2] De Lucia Jr F C, Harmon R S, McNesby K L, Winkel Jr R J and Miziolek A W 2003 *Appl. Opt.* **42** 6148
[3] Witte M J and Parigger C G 2013 *Int. Rev. At. Mol. Phys.* **4** 63
[4] Witte M J, Parigger C G, Bullock N A, Merten J A and Allen S D 2014 *Appl. Spectrosc.* **68** 367
[5] Parigger C G, Woods A A, Witte M J, Swafford L D and Surmick D M 2014 *J. Vis. Exp.* **84** E51250
[6] Hornkohl J O, Nemes L and Parigger C G 2011 *Spectroscopy of Carbon Containing Diatomic Molecules* in *Spectroscopy, Dynamics and Molecular Theory of Carbon Plasmas and Vapors* eds L Nemes, S Irle (Singapore: World Scientific Publishing) chapter 4 pp 113–165
[7] Woods A C, Parigger C G and Hornkohl J O 2012 *Opt. Lett.* **37** 5139
[8] Gornushkin I B, Merk S, Demidov A, Panne U, Shabanov S V, Smith B W, Omenetto N 2012 *Spectrochim. Acta Part B* **76** 203

Characterization of Titanium Laser Ablation

A C Woods and C G Parigger

University of Tennessee Space Institute, 411 B. H. Goethert Pkwy, Tullahoma, TN, 37388, USA

E-mail: cparigge@utsi.edu

Abstract. The atomic and ionic emission lines of titanium are often studied in laser-induced breakdown spectroscopy (LIBS) investigations, partly due to the abundance and luminosity of the lines and titanium's prominence in industry. In the current study, a 13 ns pulsed Nd:YAG laser with 160mJ per pulse ablates a titanium sample in laboratory air at 10 Hz. Ti III emission lines between 232 nm and 244 nm are observed at 200 ns after laser-surface interaction, utilizing a 6 ns window. Two-dimensional images are obtained, providing spectra emanating along the height of the ablation plume. A Boltzmann plot method is implemented in order to infer electron temperature as a function of height along the plume. The hottest region of the plasma tends to be further away from the sample surface and is on the order of 16000 K.

1. Introduction

Recent studies involving Ti III and Ti II ionic emission lines address electron density inferences in the modified semi-empirical (MSE) approach, as well as temperature inferences by means of a Saha-Boltzmann plot method [1, 2]. The current discussion serves to enhance the understanding of electron temperature in an ablation plume. To this effect, temperature of Ti III ionic spectral lines is obtained, utilizing a Boltzmann plot method [3], as a function of height in an ablation plume 200 ns after laser-induced plasma (LIP) formation.

2. Experiment

A Nd:YAG laser is utilized as the excitation source, operating with a 13 ns pulse width at 160 mJ per pulse. The beam is focused vertically downward to approximately a 1 mm spot size onto a flat Ti 6-4 sample at rest in laboratory air. The plume emission are 1:1 imaged onto the aperture of a Jobin-Yvon spectrometer such that the slit entrance is just overfilled. The slit of the spectrometer is oriented such that the height of the slit runs vertically parallel to the beam path. Attached to the exit aperture of the spectrometer is an Andor iStar ICCD. The ICCD provides a means of gated detection, in order to achieve time-resolved spectra imaged along the height of the plume. At a time delay of 200 ns, the Ti III lines from 232 nm - 244 nm are observed utilizing a 6 ns gate width. The Ti III spectral lines are then fit with Lorentzian line profiles. The area under the Lorentzian line shapes are used as the integrated intensities for each Ti III lines observed. The integrated intensities for the Ti III lines are then implemented in the Boltzmann plot technique in order to infer temperature as function of height along the plume.

The detector allows for two-dimensional imaging, consisting of 1024×1024 pixels. In the current arrangement, the horizontal axis represents wavelength, and the vertical axis represents the height along the detector. The detector can be utilized such that the 1024 horizontal tracks are consolidated into bins. This feature is implemented such that 8 tracks are integrated into a single bin. The resulting image provided by the detector then consists of 128 spectral measurements along the height of the detector.

3. Results

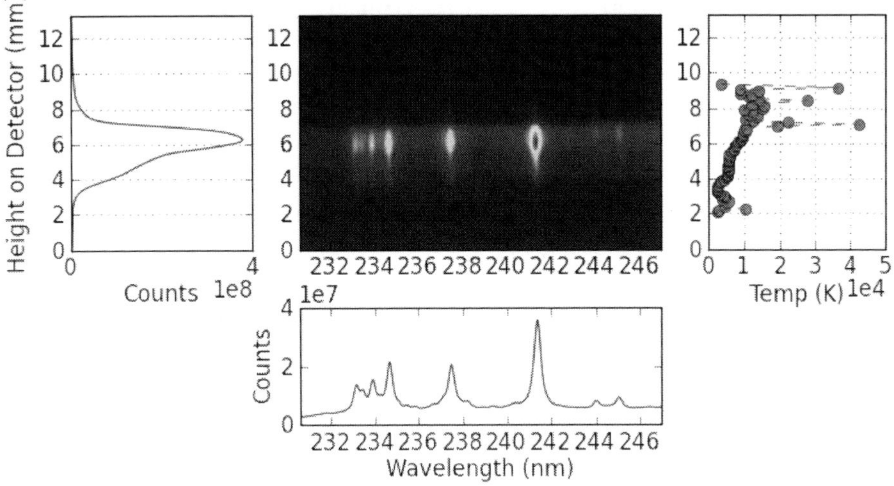

Figure 1. The above image represents data accumulated from 200 laser events at 200 ns after laser surface interaction.

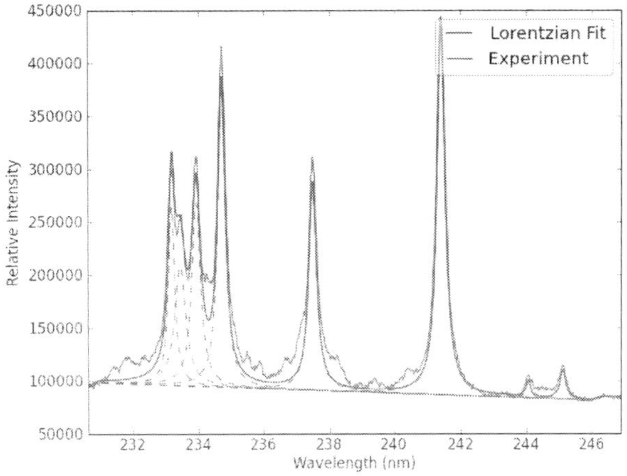

Figure 2. The above plot represents one horizontal slice of the image in Figure 1.

Figure 1 illustrates the sort of measurements obtainable by implementing a two-dimensional ICCD as a means for gated detection. To put the image in perspective, if the laser beam were to be present, the beam would propagate vertically downward from the top of the image to the titanium sample resting just below the image. The plot to the left of the image represents the intensity of the plume measured by the detector as a function of height on the detector. The plot immediately below the image represents the sum of the vertical pixels across the spectral region. It effectively provides the information obtained if the two-dimensional ICCD were to be used as a linear diode array. The plot to the immediate right of the image represents the inferred temperatures by performing the Boltzmann plot method, utilizing integrated intensity of the observed Ti III lines for each horizontal spectrum.

Figure 2 provides an arbitrarily selected horizontal slice of the image contained in Figure 1. Lorentzian profiles are fit to each Ti III and Ti II spectral line discernible. Table 1 provides information for each of the Ti III lines observed in the two-dimensional analysis. The integrated intensity of these Lorentzian line shapes for only the Ti III lines are then used in the Boltzmann plot method to provide the temperatures presented in Fig. 1. If this procedure is performed with the plot just below the image in Fig. 1, the Boltzmann plot method for Ti III lines yields a temperature of 16,000 K. The two Ti II lines observed at 244.01 nm and 244.79 nm have nearly identical energy levels and are ignored because they would not significantly aid in a Saha-Boltzmann plot technique [2].

Table 1. Atomic data for Ti III lines that were observed [4].

λ (nm)	A (10^7 s^{-1})	E_l (cm^{-1})	E_u (cm^{-1})	Configurations	Terms
233.166	12	38,064.35	80,939.19	$3p^63d4s$ - $3p^63d4p$	^3D - $^3P^o$
233.434	7.7	38,198.95	81,024.47	$3p^63d4s$ - $3p^63d4p$	^3D - $^3P^o$
233.9	30	38,198.95	80,939.19	$3p^63d4s$ - $3p^63d4p$	^3D - $^3P^o$
234.678	33	38,425.99	81,024.47	$3p^63d4s$ - $3p^63d4p$	^3D - $^3P^o$
237.4986	40	41,704.27	83,796.86	$3p^63d4s$ - $3p^63d4p$	^1D - $^1P^o$
241.3989	38	41,704.27	83,116.93	$3p^63d4s$ - $3p^63d4p$	^1D - $^1F^o$
244.0165	5.1	12,628.8455	53,597.2664	$3d^3$ - $3d(^2D)4s4p(^3P^o)$	b^2D2 - w^2D^o
244.7897	0.37	12,758.2597	53,597.2664	$3d^3$ - $3d(^2D)4s4p(^3P^o)$	b^2D2 - w^2D^o

4. Discussion

The resulting temperature inferences show increase as a function of height from the ablation surface. By comparing the inferred temperature of 16,000 K obtained by analyzing the spectrum produced by summing the vertical pixels of the ICCD with the temperature inferences along the height of the detector, we notice that inferred temperature from the sum corresponds to the temperature of the hottest region of the plasma. This would correspond to a measurement made using a linear diode array detector. The temperature inferences above 20,000 K appear to be outliers and can be the result of poorly fitting Lorentzian curves. In applying the Boltzmann plot technique, the observed portions of the plasma are assumed to be in local thermodynamic equilibrium (LTE) and optically thin. While LTE should be satisfied by McWhirter's criterion in this regime, a test for optical thickness can be readily implemented with the addition of a mirror to the experimental apparatus [5].

References

[1] Chen CS, Man BY, Liu D, Song X, and Chen XJ 2013 *J. Phys. B: At. Mol. Opt. Phys.* **46** 015701
[2] Parigger CG, Woods AC, Surmick DM, Swafford LD, and Witte MJ 2014 *Spectrochim. Acta Part B* **99** 15
[3] Gornushkin IB, Shabanov SV, Merk S, Tognoni E, Penne U 2010 *J. Anal. At. Spectrom.* **25** 1643
[4] Kramida A, Ralchenko Y, Reader J, and NIST ASD team 2013 *NIST Atomic Spectra Database* `http://physics.nist.gov/asd`
[5] Moon H-Y, Herrera KK, Omenetto N, Smith BW, and Winefordner JD 2009 *Spectrochim. Acta Part B* **64** 702

Journal of Physics: Conference Series **548** (2014) 012054 doi:10.1088/1742-6596/548/1/012054

Collision induced phenomena in the autoionizing spectrum of Ba and Sr

Y Makdisi, J Kokaj, K Afrousheh, J Mathew, N Thomas and G Pichler

Physics Department, Kuwait University, P.O.Box 5969, 13060 – Safat, Kuwait

Email: pichler@ifs.hr

Abstract. Broadening and shift of spectral lines in dense Ba and Sr vapor have been studied below and above their first ionization limits. The latter are well-known autoionizing spectral lines that exhibit many different spectral features. However, at larger pressures of noble gases some previously unobserved or very weak spectral phenomena become visible and their studies might be also linked to energy conversion purposes. Originally, we used nanosecond laser which was scanned in a certain spectral region where the two-photon excitation of autoionization lines occurred. This pulsed laser was overlapped by a mode-locked femtosecond laser and a very sensitive thermionic detection was used to observe autoionizing spectral lines. We also presented our preliminary results with hot strontium vapor excited to the resonance level at 460.7 nm.

1. Autoionizing lines perturbed by noble gases

In the past few years we have been working on a two photon excitation of the levels in barium atoms below and above the first ionization limit [1,2,3]. We measured the shift and broadening parameters of many barium spectral lines. However, above the first ionization limit where the autoionization spectral lines prevail, the shifts could be measured, whereas the broadening have been difficult to measure accurately because of diminishing intensities of spectral lines with increasing noble gas pressure. In addition, some spectral features increased in intensity with increasing noble gas pressure (collision induced bands), disturbing the observation of the autoionization spectral lines [4,5]. Furthermore, we studied the influence of the noble gas pressure on the ionization continuum by simultaneously applying the femtosecond laser overlapping the nanosecond laser beam [6].

1.1. Thermionic detection

Thermionic detection of the barium ions produced by nanosecond and femtosecond laser excitation in a special heated oven was carried out in a number of different experiments. This was achieved when the femtosecond laser beam was spatially overlapped by excimer pumped dye laser beam (Fig.1). The latter nanosecond laser was scanned in a certain spectral region where the two-photon excitation of autoionization lines occurred. The broadband ultrashort laser light (mode-locked femtosecond laser) was cantered at 427 nm, which is also exciting barium atoms above the first ionization limit after the two photon excitation. The bias voltage between the cell body and the tungsten rod (set at either 9 or 0 Volts) was used to collect electrons, after the barium ions have been created by multiphoton ionization. The overall background of the thermionic signal was appreciably elevated due to the two-photon ionization by the broadband femtosecond laser. Thermionic signal with and without femtosecond laser overlap was also measured. A thermionic diode ion detector is the most sensitive tool for the detection of highly excited Rydberg states [7]. It is a positive ion detector with high sensitivity which helps to detect even a single ion. When the thermionic diode performs in the space charge limited mode, the existence of N ions in the potential well of the thermionic diode lowers the space charge and causes an increase of the diode current by Δj. This will have a consequence of an amplification factor of $\Delta j/N$. Thus with a very high thermal emission of electrons from the cathode surface, the amplification factor could reach the values in the range of 10^4 - 10^6.

Figure 1. Sketch of the thermionic diode and the laser beams overlap.

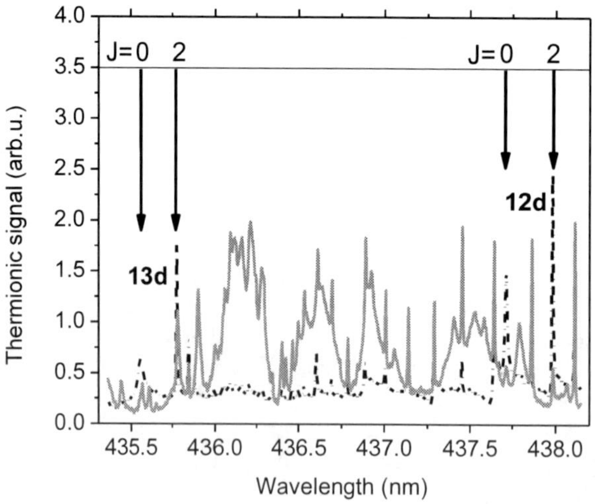

Figure 2. Comparison of the two spectra at 5 mbar He (black dotted trace) and 50 mbar Xe (red trace).

The effects of the elevated noble gas pressure are not only seen in the broadening and shift of the Ba autoionizing spectral lines, but also in the appearance of the bands that are faint at lowest noble gas pressures used in the present experiments (5 mbar). The overall intensity of the autoionizing lines diminish with the pressure. The ratio between collision induced new bands and the autoionizing lines is increasing with the noble gas pressure (Fig. 2).

2. Effect of the femtosecond laser

The effect of the femtosecond laser was appreciable enhancement of the background ionization continuum. This was especially visible in the presence of noble gases at a pressure of 50 mbar. Argon (helium) produced the largest (smallest) enhancement in the background continuum. In addition, we observed a few broad spectral features of presumably collision induced nature. In our experiment we used femtosecond laser oscillator with a repetition frequency of 80 MHz and 100 femtosecond pulse duration. It means that the multiphoton ionization was present whenever the femtosecond laser was used in the experiment. Unfortunately, the bandwidth of the femtosecond laser was not sufficiently broad in order to overlap the nanosecond excitation laser in the same scanned spectral range. Such spectral overlap will be hopefully performed in the near future.

3. Experiment with strontium

We intend to perform similar experiments with strontium. The experimental conditions should be the same, but the temperature of the oven should be slightly higher, and the analogous spectral range will be shifted towards the lower wavelengths. However, our preliminary measurements were done by resonantly excited strontium vapor into the first excited resonance level at 460.7 nm. A partial energy term diagram of strontium atom is shown in Fig. 3. The experimental setup is shown in Fig. 4. This setup allows the measurement of thermionic signal, transmission signal and a laser induced fluorescence. Thermionic signal is caused by the barium ions produced by resonantly enhanced multi photon ionization. The measurement of transmission of the laser beam during the scan of the laser wavelength enables the determination of the absorption coefficient in the line wings of the Sr 460.7 nm resonance line. Laser induced fluorescence could be measured in the backward and the forward direction, while the laser wavelength is scanned.

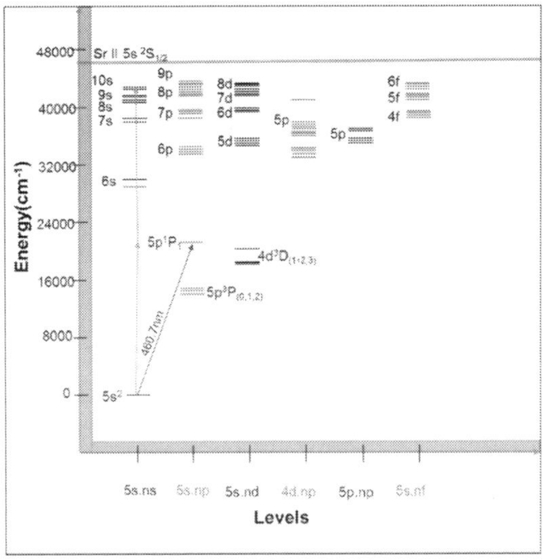

Figure 3. Partial energy term diagram with two photon excitation of Sr autoionizing levels and single photon excitation of the resonance level by 460.7 nm laser.

We could fix the laser wavelength at some line wing position and then measure forward and backward fluorescence by means of the digital spectrometer. In order to collect the fluorescence light in front and behind the oven we put the mirrors with holes to allow the laser beam to pass through. An example of our results may be seen in Fig. 5.

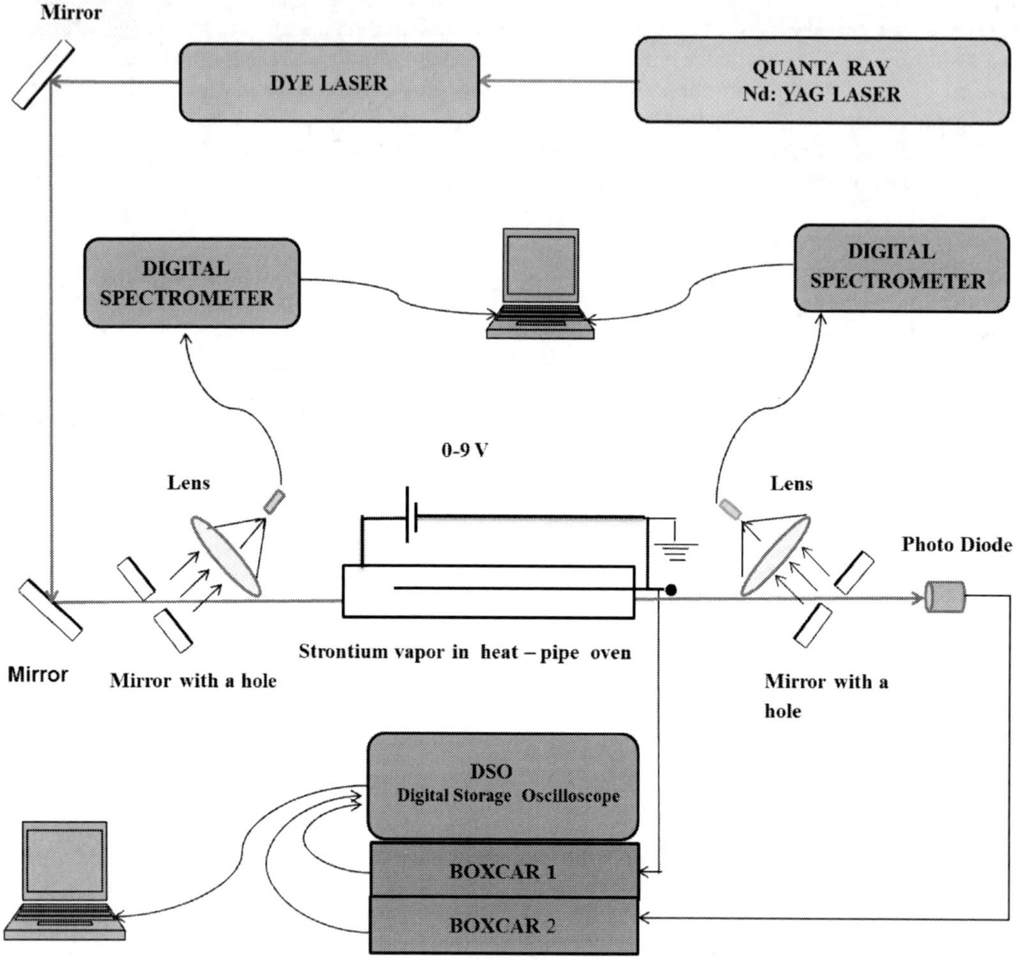

Figure 4. Experimental setup with strontium vapor.

We may compare the different signals in Fig. 5. In 5(a) we present thermionic signal, when the laser was scanned over the resonance line at 460.7 nm. In 5(b) transmission signal when the laser beam was scanned over the resonance line. In 5(c) and 5(d) we show the forward and the backward laser induced fluorescence which were spectrally resolved by the digital spectrometer. The laser wavelength was fixed and located close to the line center.

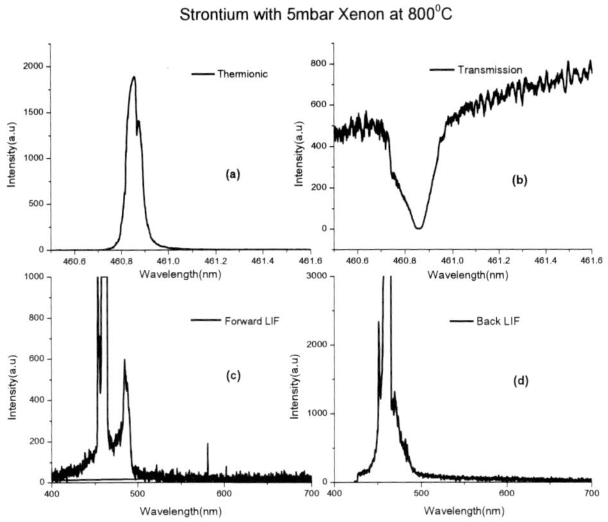

Figure 5. Experiment with strontium vapor at 800 °C and 5 mbar of Xe, (a) thermionic signal, (b) transmission, (c) forward laser induced fluorescence, (d) backward laser induced fluorescence.

In Fig. 6 we present spectrally resolved laser induced fluorescence at argon pressures of 5, 100 and 200 mbar. The molecular band, presumably from the ArXe excimer is diminishing toward higher argon pressures. In addition, the peak of the excimer emission is also shifting toward shorter wavelengths.

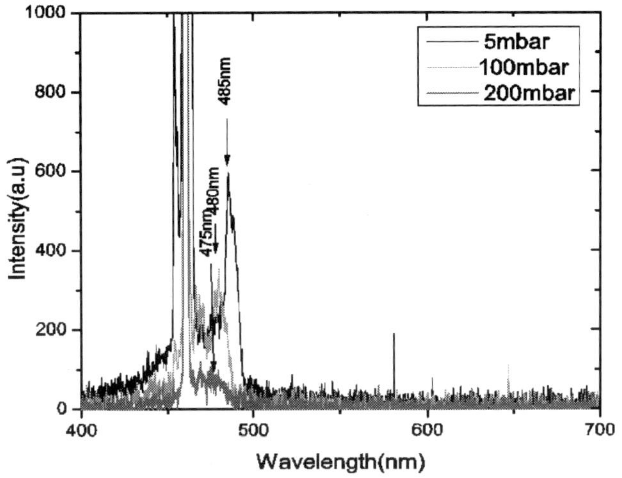

Figure 6. LIF spectra observed with laser beam at 460.9 nm with argon at pressures of 5, 100 and 200 mbar. The oven temperature was 800 °C and the laser power before the oven was 550 nW.

In Fig. 7 we present the back LIF spectra made with He and Ar as buffer gases at 5 mbar pressure. Different laser excitation wavelengths are denoted in each panel. The comparison shows that SrHe and SrAr excimers have very similar structure of the excited potential energy curves.

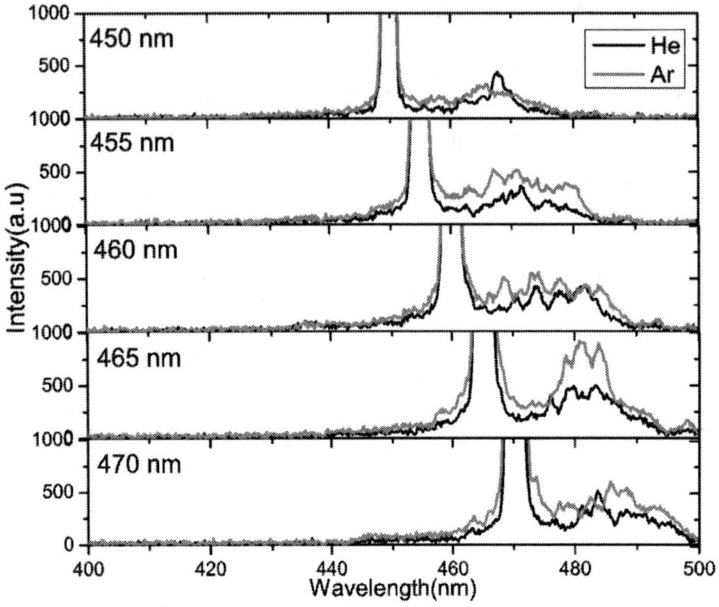

Figure 7. Back-LIF spectra with He and Ar as buffer gases at 5 mbar pressure.

Similar experiments with barium vapor with resonance excitation at 553 nm will be performed in the near future. We expect to observe similar excimer bands in the red wing of the Ba resonance line. In addition we intend to carry out similar experiments with mixtures of Ba and Sr, and some other specific gases. For all these new experiments we already prepared specially designed heat-pipe ovens with inserted electrodes. Our hope is to bring about new results which will contribute our understanding of the behavior of the autoionizing spectra under the influence of different noble gas pressures. These new results might also illuminate new possibilities in the field of thermionic energy conversion.

Acknowledgements
We thankfully acknowledge Kuwait Foundation for the Advancement of Sciences (KFAS) for funding this project (Code:2012151301). General Facility projects GS03/01 and GS01/08 of Kuwait University are also gratefully acknowledged.

References

[1] Marafi M, Suji Z, Bhatia K, Makdisi Y, Mathew J 2007 *J. Phys. B: At. Mol. Opt. Phys.* **40** 4211
[2] Marafi M, Suji Z, Bhatia, K, Mathew J 2008 *J. Phys. B: At. Mol. Opt. Phys.* **41** 205203
[3] Afrousheh K, Marafi M, Kokaj J, Makdisi Y, Mathew J 2012 *Phys. Rev. A* **85** 052517
[4] Makdisi Y, Kokaj J, Afrousheh K, Mathew J, Nair R, Pichler G 2013 *Opt. Commun.* **290** 95
[5] Afrousheh K, Makdisi Y, Kokaj J, Marafi M, Mathew J, Pichler G 2013 *Eur. Phys. J. D* **67** 262
[6] Makdisi Y, Kokaj J, Afrousheh K, Mathew J, Nair R, Pichler G 2014 *J. Quant. Spectrosc. Radiat. Transf.* **148** 221
[7] Niemax K 1985 *Appl Phys B* **38** 147

Combustion diagnosis for analysis of solid propellant rocket abort hazards: Role of spectroscopy

W Gill[1], A A Cruz-Cabrera[2,6], A B Donaldson[3], J Lim[4], Y Sivathanu[4], E Bystrom[2], A Haug[3], L Sharp[3] and D M Surmick[5]

[1]Sandia National Laboratories, PO box 5800, MS1135, Albuquerque, NM 87185-1135, USA
[2]Sandia National Laboratories, PO box 5800, MS1139, Albuquerque, NM 87185-1139, USA
[3] New Mexico State University, 1780 E University Ave, Las Cruces, NM 88003, USA
[4]En'Urga Inc, Business and Technology Center, 1291-A, Cumberland Avenue, West Lafayette, IN 47906, USA
Center for Laser Applications
[5]University of Tennessee Space Institute, 411 B.H. Goethert Parkway, Tullahoma, TN 37388, USA

E-mail: aacruzc@sandia.gov

Abstract. Solid rocket propellant plume temperatures have been measured using spectroscopic methods as part of an ongoing effort to specify the thermal-chemical-physical environment in and around a burning fragment of an exploded solid rocket at atmospheric pressures. Such specification is needed for launch safety studies where hazardous payloads become involved with large fragments of burning propellant. The propellant burns in an off-design condition producing a hot gas flame loaded with burning metal droplets. Each component of the flame (soot, droplets and gas) has a characteristic temperature, and it is only through the use of spectroscopy that their temperature can be independently identified.

1. Introduction

For most of the deep space probes and some of the landing craft on mars, NASA has used radioisotope thermal generators (RTGs) to power and to provide heat for the proper operations of their instruments. Most launches of these probes have used launchers with solid rocket boosters (SRBs) as the stage before insertion to earth orbit. In the event of an accident, liquid stages can be shutoff and the autodestruct mechanism allowed to burn and/or disperses the liquid propellant. The fragments of the solid propellant will continue to burn and drop to the ground with a risk of falling near or even on top of the RTG. The main concern is the maximum temperature from the plume impinging on and potentially breaching the RTG.

[6] Author to whom any correspondence should be addressed: Sandia National Laboratories, PO box 5800, MS1139, Albuquerque, NM 87185-1139, USA, aacruzc@sandia.gov

2. Experimental Setup

Sandia was tasked with measuring the temperatures of a solid propellant plume in an off-design situation. The objective was to define the thermal-chemical-physical environment in and around a plume formed from a burning fragment of solid rocket fuel at atmospheric pressure. Figure 1(a) shows an off-design test burn at Sandia. Figure 1(b) shows three analysis volumes where different chemical dynamics are expected: Region I where the aluminum and the ammonium perchlorate from the propellant burn. Region II is populated by hot emitting gases: hydrogen chloride (HCl), carbon dioxide (CO_2), and water (H_2O). In this region some burning particles like aluminum monoxide (AlO), alumina (Al_2O_3), and other particles exist. Region III is where the flame from the plume will most likely interact with the hazardous material in the payload. This situation could lead to more complex interactions in Region II if there is a breach and subsequent release of plutonium-238.

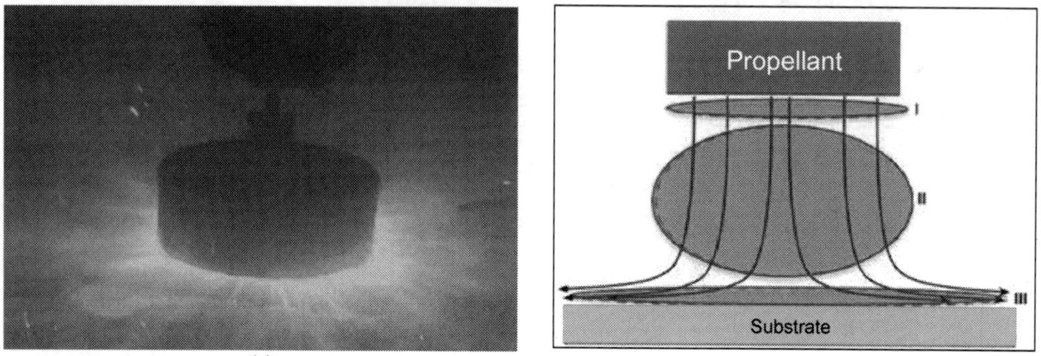

(a) (b)

Figure 1. (a) Downward setup for off-design burn of solid propellant (b) the plume in the analysis is divided into three regions: I - where aluminum and ammonium perchlorate burn; II - populated by hot emitting gases (HCl, CO_2, H_2O) and particles(AlO/ Al_2O_3, etc.); III - where flame will most likely interact with the hazardous material in the payload.

To obtain a better picture of the temperatures seen in the propellant plume, several spectrometers were fielded to measure emissions from the gases (CO_2, H_2O, and HCl) and alumina (AlO). Two angle-scanning spectrometers were placed orthogonally to each other, to measure CO_2 and H_2O; one Fourier transform infrared (FTIR) spectrometer measured for HCl; and a grating snapshot fiber connected spectrometer measured AlO emissions.

3. Experimental complexity

The solid propellant burn will generate multiple gases from the reaction of the ammonium perchlorate (NH_4ClO_4), including CO_2, H_2O and HCl. The gas temperatures should be lower than the temperatures associated with the burning of aluminum particles as seen in Figure 2. It is expected that the temperatures from the soot will be lower than the temperatures of the gasses and the AlO.

Journal of Physics: Conference Series **548** (2014) 012055 doi:10.1088/1742-6596/548/1/012055

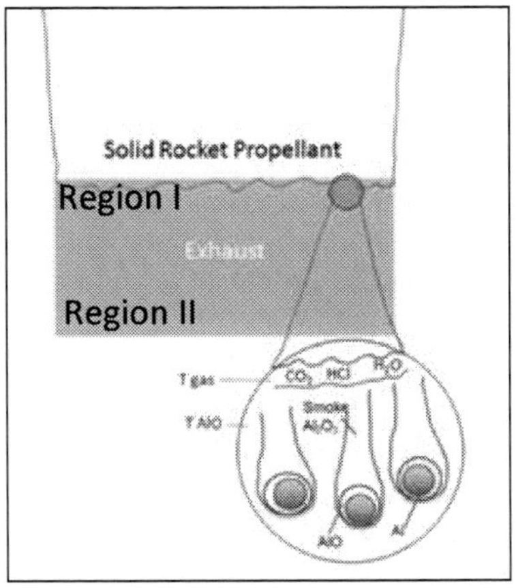

Figure 2. Region I shows the burning aluminum and ammonium perchlorate. Multiple temperatures are present in Region II, complicating measurement and models. AlO is believed to have some of the highest temperatures in this region. Other oxides may be present but they are not being measured. The smoke and soot (non-aluminum related) have the lowest temperature in Region II, probably conforming to a grey body profile. The gases (CO_2, H_2O, and HCl) fall in the middle of temperature ranges.

4. Aluminum monoxide particles

To observe AlO optical emission spectra from the plume, a probe was inserted into the flame so that specific regions of the plume could be observed. The experiments were performed in a 6.6 by 6.6 meter chamber at ambient conditions, i.e. atmospheric pressure and temperature, analogous to those of an aborted launch. The probe consisted of a steel pipe attached to a stepper motor that moved in regular intervals. One end of the steel pipe was inserted into the plume while a collimating lens coupled to a bifurcated fiber optic cable was attached to the other. Optical emission spectra were recorded with two Ocean Optics fiber spectrometers. One of the spectrometers (HR2000) had narrowband sensitivity in the UV wavelength region of 420-560nm while the other (HR4000 CG) had broadband sensitivity in the wavelength region of 200-1100nm. The average resolution was 0.5nm for the narrowband spectrometer and 0.75nm for the broadband spectrometer. The purpose of the narrowband instrument was to collect aluminum monoxide emission spectra to infer the temperature of the diffusion flame surrounding aluminum particles in the propellant plume. The purpose of the broadband instrument was to collect thermal continuum radiation from the propellant plume to infer the flame temperature. The amount of distance traversed by the steel pipe varied from test to test. Distances from the edge of the propellant ranged from 7.6cm to 38.1cm, with an average spacing between measurements of approximately 0.25cm. To infer the temperature of the propellant plume from optical emissions, aluminum monoxide spectra were fit to theoretical calculations of the AlO spectra and the continuum thermal radiation was used to infer the temperature of the flame.

4.1. AlO results

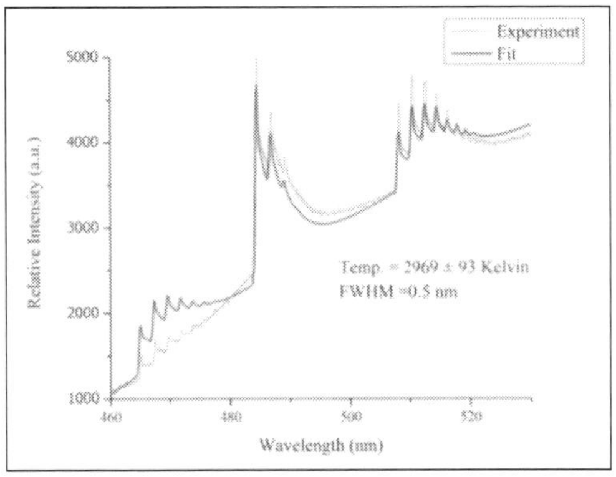

Figure 3. Fit to experimental AlO propellant spectra.

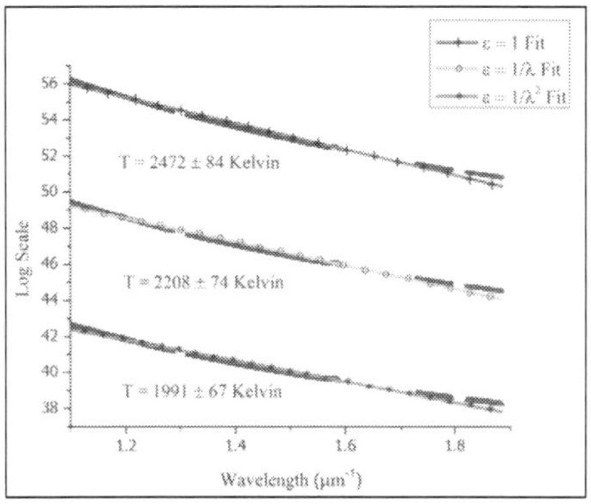

Figure 4. Planck fitting results for a downward burning propellant.

To infer the AlO temperature, the experimental spectra are fit to theoretical calculations of the AlO $B^2\Sigma^+ \rightarrow X^2\Sigma^+$ transition. The theoretical AlO spectra are calculated from accurately compiled line strength tables [1, 2]. The line strength is a popular method for calculating diatomic spectra given the symmetric nature of the quantity. The line strength is also easily calculated in a factorized form from the electronic transition strength, Hönl-London factors, and Franck-Condon factors [3]. The comparisons between experimental and theoretical spectra are made with the use of a Nelder-Mead algorithm. The Nelder Mead algorithm is an optimization technique in which a geometric simplex is created and reduced in size until a specified tolerance level is reached [4, 5]. In order to fit AlO spectra from downward burning propellants, the parameters to be minimized were the temperature, the spectral resolution, and a variable baseline offset. Quadratic baseline offsets were considered for all fits and were found to be linear in nature. This is expected given the AlO emissions are superimposed on thermal continuum emissions and the wavelength region of the AlO emissions. Though the spectral resolution is a largely determined quantity, it was varied to determine the error of the AlO temperature. Though both spectrometers recorded AlO emissions, the narrowband instrument was the primary source for AlO temperature inferences. Values of 0.35 and 0.65nm were used to over- and underestimate the resolution of this instrument and were corroborated by allowing the algorithm to determine the best possible resolution. A sample fit of AlO spectra is shown in Figure 3 and corresponds to downward burning propellant on a graphite substrate. The measurement was made 7.6cm from the propellant edge. The inferred temperature for this position was 2969 ± 93K. Typical AlO temperatures were found to be in the 2700-3000K range with typical errors of 95-115K.

4.2. Planck fitting

The temperature of the flame is determined from spectroscopic measurements of the propellant plume and is inferred from broadband emissions in the 500-950 nm wavelength region. The region of analysis is determined from the sensitive range of the broadband spectrometer which was primarily used to infer the flame temperature. The thermal emissions are expected to follow Planck's radiation law and are considered with various emissivity models. The emissivity describes the ratio of actual power emitted to the power emitted by a black-body radiator. As such, an emitting source following a black-body emission model will have a constant emissivity of one, $\varepsilon = 1$. So called grey emitters in the plume, such as large aluminum agglomerates and alumina soot particles, will have constant emissivities that are less than one as black-body radiation represents an ideal emission model. Three emissivity models are considered in this work: 1) black-body emission, 2) $1/\lambda$ wavelength dependent emissions, and 3) $1/\lambda^2$ wavelength dependent emissions. The difference between grey and black-body emissions is expected to be relatively small and, as such, a black-body emission model is used. When the size of an emitting body is on the order of the wavelength of light being emitted it is possible for the emissivity to become wavelength dependent, $\varepsilon = \varepsilon(\lambda)$, when the classical Lorentz oscillator model is considered (6). Specifically $\varepsilon(\lambda) = 1/\lambda$ and $\varepsilon(\lambda) = 1/\lambda^2$ emissivity models are used where $1/\lambda$ dependency may be caused by molten aluminum agglomerates and $1/\lambda^2$ may be caused by the presence of micron sized or smaller aluminum and alumina particles [7] [8]. Wien's displacement law may also be developed using wavelength dependent emission models through the use of poly-logarithm functions [9] as given below

$$\lambda_{max} T = \frac{hc}{k_B} \frac{1}{5 + W_0(-5e^{-5})} = 2.898 \times 10^6\, nmK \tag{1}$$

Where h is Planck's constant, k_B is Boltzmann's constant, c is the speed of light, and W_0 is the Lambert function. For sufficiently small arguments the Lambert function may be approximated by its argument, $W_0(x) = x$, and leads to the Wien approximation. When using $1/\lambda$ and $1/\lambda^2$ emission models, the resulting temperatures show a 5/6 and 5/7 reduction as shown below.

$$\bar{\lambda}_{\max}\bar{T} = \frac{hc}{k_B}\frac{1}{6+W_0(-6e^{-6})} = 2.415\times10^6\,nmK \tag{2}$$

for 1/l dependence and

$$\bar{\bar{\lambda}}_{\max}\bar{\bar{T}} = \frac{hc}{k_B}\frac{1}{7+W_0(-7e^{-7})} = 2.070\times10^6\,nmK \tag{3}$$

for $1/\lambda^2$ dependence. To further illustrate this effect, consider a black-body emitter with a maximum wavelength emission of 1000nm which has a corresponding emission temperature of 2898K. Wavelength dependent emissivity models will have temperatures of 2415K and 2070K for $1/\lambda$ and $1/\lambda^2$ emissions, respectively.

The maximum wavelength emission for the spectra observed from the propellant is expected to be in the 1000-1300nm range. Given the decreased sensitivity in the nominal 1000nm wavelength region, the limited range of the instrument and the expected temperatures from propellant plume direct peak fitting with Wien's Displacement law is not preferred for determining the flame temperature. Rather fitting with a semi-log approximation of the Planck equation is performed [7, 8] where the approximation is given by

$$\ln\left(C*\frac{I(\lambda,T)*\lambda^5}{\varepsilon(\lambda)}\right) = -\ln\left(\exp\left[\frac{hc}{k_B}\frac{1}{\lambda T}\right]-1\right) \cong -\frac{hc}{k_B}\frac{1}{\lambda T} \tag{4}$$

where C is a constant, $I(\lambda,T)$ is the recorded spectrometer intensity, and $\varepsilon(\lambda)$ is the emissivity model. Using this fitting method the natural logarithm of the collected data multiplied by λ^5, λ^6, and λ^7 for constant, $1/\lambda$, and $1/\lambda^2$ emissivity models are plotted against the inverse wavelength and are fit using linear least squares methods. The slope of the fitted line is used to determine the temperature by way of the approximation in equation 4. A typical fit of this kind is shown in Figure 4 which corresponds to the same propellant burn shown in Figure 3 for AlO emissions. Contributions from atomic and molecular spectral emissions have been removed to reduce the error of the fit. Such emissions are AlO, sodium, potassium, aluminum, and iron. The inferred temperatures in Figure 4 are 2472 ± 84K, 2208 ± 74K, and 1991 ± 67K for constant, $1/\lambda$, and $1/\lambda^2$ emissions, respectively. The errors are inferred from the systematic error in the slope associated with the linear least squares fitting and also include a three percent error that accounts for the approximation. Typical temperatures from a downward burning propellant are found to be in the nominal 2200-2500K range with typical errors falling into the 65-80K range.

5. Carbon dioxide and water gases

Gas (CO_2 and H_2O) and broadband radiation were measured using two ES100 Spectraline spectrometers from En'Urga Inc. They were placed orthogonal to each other with the propellant burn at the center of view of both spectrometers; see Figure 5. This setup allows emission tomography measurements in multiple angles of the propellant exhaust. The two spectrometers measured in a band of 1.3 to 4.8μm, sampling 256 wavelengths with 0.018 μm resolution. Each spectrometer scans 128 angles at a rate of 1.32KHz for full 128 angles measurement at 10.3Hz. The use and placement of two spectrometers is intended to determine the transmissivity distribution around the propellant.

Temperatures at each angle are calculated by looking at local spectral radiation intensity measurement, where the spectral specific emission coefficient, $\alpha(\lambda)_{nm}$, for that path is calculated for each homogeneous region using a radiative transfer equation [10,11].

$$I(\lambda)_n = \sum I(\lambda)_{BBm}\cdot\left(1-e^{-k(\lambda)_{nm}\Delta_{nm}}\right) = \sum I(\lambda)_{BBm}\alpha(\lambda)_{nm} \tag{5}$$

Where n is the angle being analysed, and m is the region where the blackbody temperature and plume transmissivity are quasi-homogeneous. $k(\lambda)_{nm}$ is the spectral specific emission coefficient for the angle n, and exhaust region m. Δ_{nm} is the path length in that region. All 128 equations are then linearized:

$$\log(I(\lambda)_n) \approx \log(I(\lambda)_{BBm}) + \log(1 - \tau(\lambda)_{nm}), \quad \text{Where } \alpha(\lambda)_{nm} = 1 - \tau(\lambda)_{nm} \qquad (6)$$

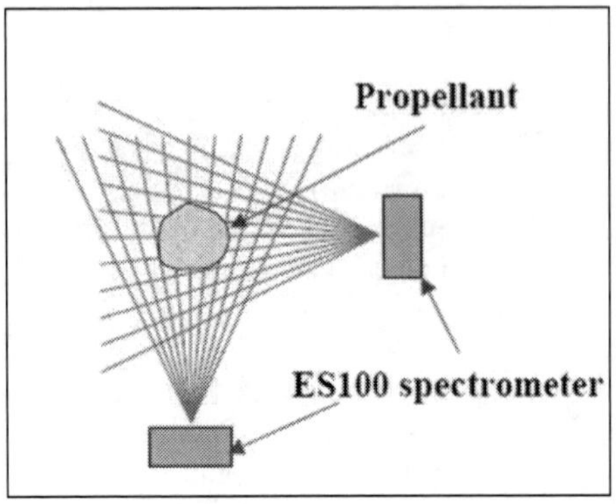

Figure 5. Setup for measuring CO_2 and H_2O gas molecules. Two orthogonal spectrometers with 128 scanned angles each. Each spectrometer measures from 1.3 to 4.8μm at a rate of 1.32KHz with a full spectra set (all angles) at a rate of 10.3Hz.

The equation is set up to solve for two linear equations:

$$\log(I(\lambda)_{BBm}) \approx C0_m + C1_m T_m \qquad (7)$$

$$\log(1 - \tau(\lambda)_{nm}) \approx C2_m + C3_m X(\lambda)_{CO_2m} + C4_m X(\lambda)_{H_2Om} + C5_m T_m + C6_m f_m \qquad (8)$$

Equation (5) can be written:

$$\log(I(\lambda)_n) \approx C0_m + C2_m + C3_m X(\lambda)_{CO_2m} + C4_m X(\lambda)_{H_2Om} + (C1_m + C5_m)T_m + C6_m f_m \quad (9)$$

Where C3, C4 and C6 are larger than zero, C1+C5 > 0, $\log(I(\lambda)_{BBm})$-C0-C2 > 0 and all the constants are non-negative. $X(\lambda)_{H20m}$, $X(\lambda)_{CO2m}$ are the volume fractions for H_2O and CO_2 respectively, and f_m is the volume fraction for soot [12,13]. C2, C3, C4, C5 and C6 are found in the RADCAL database [14]. The linear equations require at least six wavelength measurements, and it can be solved by the maximum likelihood estimate (MLE) method [15]. The algorithms guess the transmittance of the regions in the propellant, in this case rings, and use the MLE to calculate the local intensities, which recalculates the local transmittances until the algorithm converges.

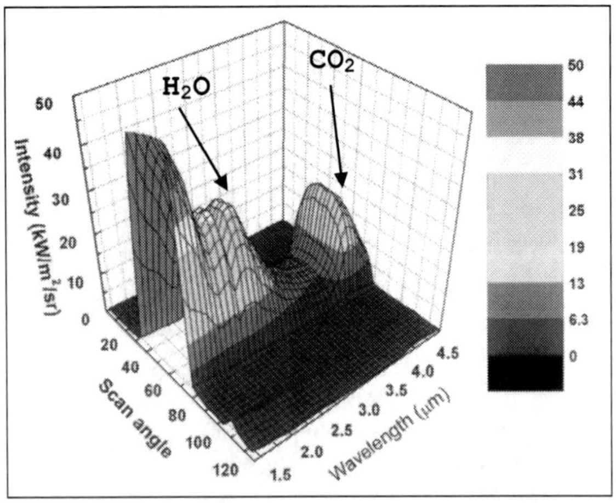

Figure 6. Typical intensity gas spectrum from the one Spectraline system.

Figure 6 shows an intensity distribution obtained from one ES100 Spectraline units. The surface istribution is a function of angle and wavelength, showing the spectral locations of CO_2 and H_2O from an upward burn test. This measurement is a snapshot during the steady burning. Figure 7(a) shows the setup for a downward setup and the location of the spectrometers with respect to the propellant, a substrate (in this case a graphite plate) and the plume. The diagram is not to scale and the propellant to substrate gap starts at 50.8mm, and increases as the propellant is consumed. The line of sight of the spectrometers was set at 50.8mm from the substrate. Figure 7(b) is the instant measured temperatures for the two gases and the soot. The propellant diameter in this test was 508mm in diameter. Gase temperatures were greater than 2500K, and soot temperatures were approximate 1800K.

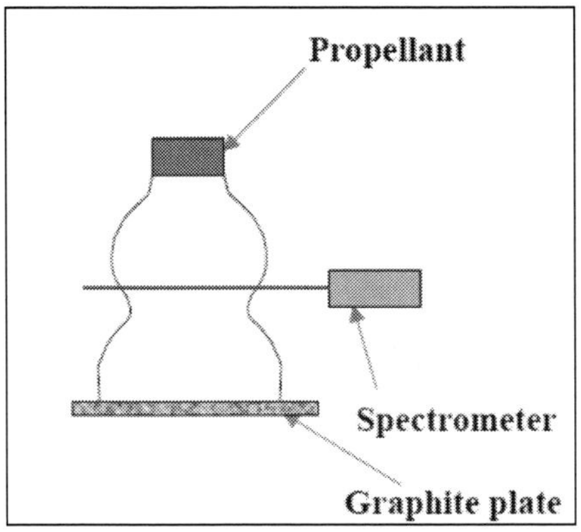

(a)

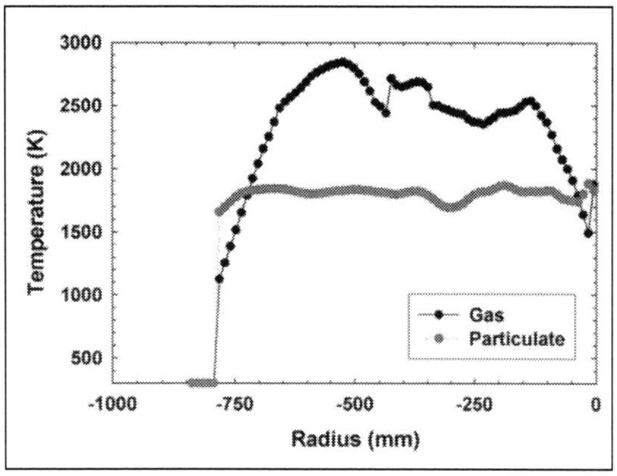

(b)

Figure 7. (a) Downward burn basic schematic with propellant 50.8mm above graphite plate (b) Temperature measurement for gases and particles for a downward burn; the abscissa is the radius from the center of the propellant.

6. Hydrogen chloride gas

Hydrogen chloride (HCl) gas is one of the byproducts of burning aluminum and ammonium perchlorate, the main ingredients in some solid propellants. HCl is a diatomic molecule with emission spectra in the mid-wave infrared (2µm - 5µm) showing a rotational-vibrational spectrum centered at λ_0= 3.4598µm (2890.34cm^{-1}). This wavelength is the transition from ground state to the first excited vibrational state. There is not emission at this wavelength, see Figure 8, but rotational emission lines are seen at both sides of that wavelength following a Boltzmann distribution.

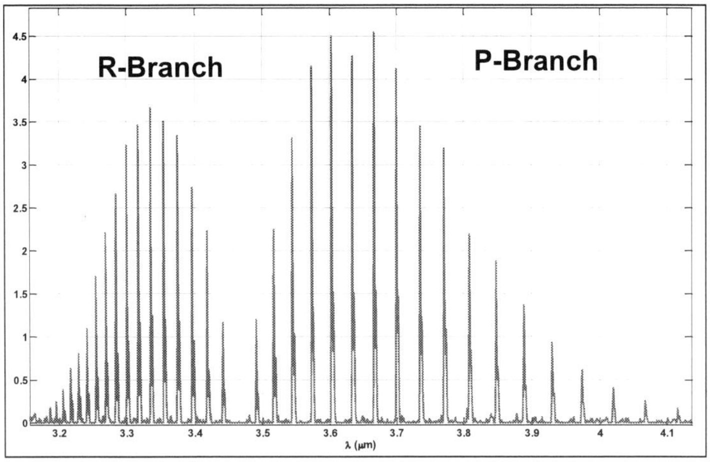

Figure 8. Measured HCl spectrum at the mid-wave infrared. Notice the P and R Branches with the rotation-vibrational emissions with an envelope following a Boltzmann equation. The shape and location of the peak of the envelope is determined by the temperature of the gas.

The measurements of HCl were performed with a Nexus 670 Fourier transform infrared (FTIR) spectrometer. An FTIR is a modified Michelson interferometer; see Figure 9, which has a moving mirror in one of its arms. Every scan of the mirror generates an interferogram that will be Fourier transformed to generate a spectrum. The length of the scan is inversely proportional to the resolution of the spectrum measured. This means that for higher resolution measurements the mirror will take longer to conduct a scan. A 1cm^{-1} measurement typically takes 3 seconds, and a 0.25 cm^{-1} measurement will take approximately 12 seconds, if there are no large changes in the light flux. Otherwise, the auto-gain step can add 4 more seconds. The 1cm^{-1} resolution was selected because it balances the requirement for high speed measurement and the ability to resolve the HCl lines.

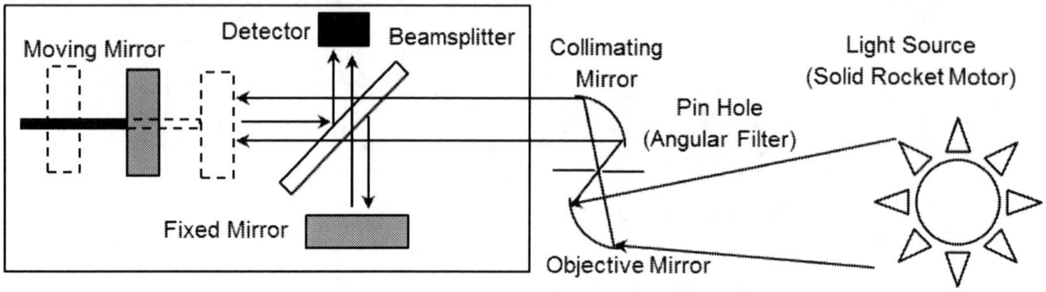

Figure 9. FTIR basic spectrometer schematic with light collecting optics.

The instrument was used in single beam mode without a reference blackbody source. Light from the propellant was gathered though an external port. The spectral response of the system is delimited by the liquid nitrogen cooled mercury cadmium telluride (MCT) detector (2-5μm). Light from the propellant was collected using two 76.2 mm off-axis parabolic mirrors. The one in front (objective mirror) was the main gathering optical components. The light was focused into an iris where it was spatially filtered and then a second off-axis parabolic mirror collimated the light into the FTIR. There were cases were the pin hole was not used, but the gathered light was considered collimated when it was coming through a 1016mm long alumina tube. The collecting optics and the FTIR were always aligned at the center of the gap between propellant and substrate. For experiments where the gap was 12.7mm or less, the data were extremely noisy and no spectrum fit could reach $r2 > 0.9$, making the data unusable. This was later overcome with experiments using the 1016mm long alumina tube. The alumina tube was used to avoid the soot and dust from the propellant exhaust.

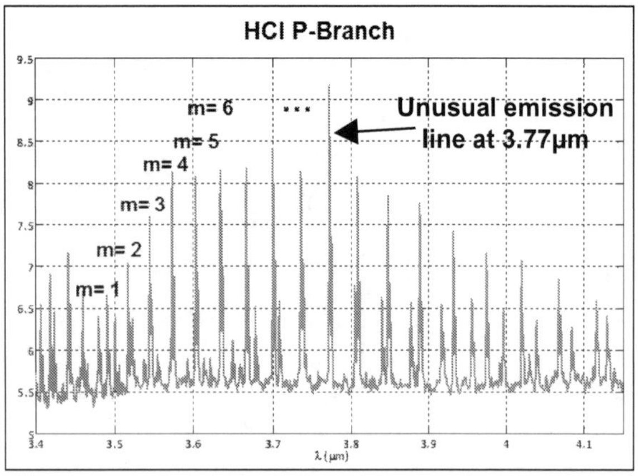

Figure 10. HCl P-branch spectrum showing the fine structure of the vibrational rotational emission. The envelope of the peaks should fit a Boltzmann distribution from which a temperature can be calculated. Notice the peak at 3.77μm that does not fit the Boltzmann distribution. It is believed to be another HCl vibrational-rotation emission peak that cannot be discerned due to the resolution of the FTIR for these measurements.

Most of the preliminary analysis on the HCl was done on the P-branch. From a careful look at the spectrum of the P-branch, it is possible to see the emission peaks from Cl_{35} and Cl_{37}. The R-Branch seems to overlap with the water emission spectra. There are some water emission lines in the P-branch that reach 3.5μm, but they do not overlay with the HCl lines, see Figure 10. However, the P-branch shows a higher than normal emission line around 3.771μm. Some modeling is underway to verify that this line is a rotational-vibrational emission from nearby vibrational excitation and the line is close but does not overlap the line in 3.771μm. The FTIR is, at this instance, unable to resolve that extra line on high resolution measurements (0.25 cm^{-1}).

Temperatures were calculated by a modified version of the method used by Zhang et al. [16], and Herzberg [17]. An equality relates the spectral running number, m, of each line in the P and R-branches to a linearized Boltzmann distribution:

$$\ln\left(|m|\frac{\vartheta^4}{I_r}\right) = \frac{B_\vartheta hc}{kT} m(m+1) + b \tag{10}$$

for integer m= 1, 2, 3,.. From measured spectra, the frequency of each peak is elevated to the power of four, v^4, multiplied by the magnitude of the running number, |m|, and divided by the line corresponding peak. The right side of the equality is the linearized form of the Boltzmann distribution multiplied by the rotational constant[7], B_v= 10.5934 cm^{-1}, times the running number m(m+1), where c is the speed of light, k is the Boltzmann constant, T is the temperature, and h is the Planck constant. Plotting the right side of equation 10, versus the m(m+1) relationship, provides a series of points that can be fit to a straight line, see Figure 11.

The slope, A, of the line is equated to the linearized Boltzmann distribution, from which the temperature of the gas is calculated:

$$T = \frac{B_\vartheta hc}{kA} \tag{11}$$

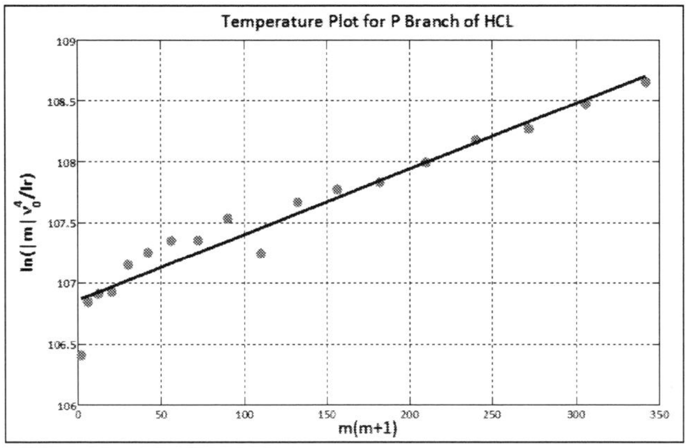

Figure 11. The linearizing of peaks for the fine emission spectra of the P-Branch can be fit to a line and its slope used to calculate the temperature of HCl. A more sophisticated code uses the P & R branches.

Preliminary work focused on the P-branch because it is the less noisy of the two branches. The calculated temperatures were high, usually in the order of 3000K. These numbers are hard to dismiss as previous measurements showed similar temperatures. The numbers from the R-branch were extremely low, on the order of 1200K to 1500K, and are suspect because the R-branch straight line fits tended to be lower than the fits from the P-branch. Further analysis was done using temperature measurements from both branches and correlated to laboratory measurements of HCl emissions inside an oven. The set temperatures of these tests ranged between 963K and 1863K, for a total of 10 temperatures, with equal 100K steps. The measurements showed that the calculated temperatures could be 600K higher than the oven temperature settings when using only the P-Branch. The error reduces to 100K when using the average temperature from both P and R-branch. If we use both

[7] NIST: http://cccbdb.nist.gov/exp2.asp?casno=7647010

branches in the calculation for all propellant tests, then the average temperature is 2247K with an error of +360K/-270K. These numbers are restricted to measurements with line fits, r^2, greater than 0.9.

7. Conclusions

This paper presents the complex conditions for off-design solid propellant tests, where multiple gases, aluminium monoxide and soot provide different temperatures and conditions that can mask the effects of each other. Three spectral measurement methods were presented, each looking at a different species (AlO, HCl, H_2O, and CO_2). Each method is in a different state of development. Where the spectral tomography measurement of H_2O, and CO_2 has had years of refinement, the work on FTIR measurement of HCl is still in the initial stages, and will require refinements to eliminate a 100K difference between the test oven reported temperatures and the spectrally calculated ones. Understanding the different temperatures with respect to the different species should provide a better picture of what an RTG will be exposed to in an accident scenario.

Acknowledgement

The authors acknowledge the contributions of Jill Suo-Anttila. Funding for this research was provided by the National Aeronautics and Space Administration under contract FI015110718. Sandia is a multiprogram laboratory operated by Sandia Corporation, a Lockheed Martin Company, for the United States Department of Energy's National Nuclear Security Administration under contract DE-AC04-94AL85000.

References

[1] Parigger C G and Hornkohl J O 2011 *Spectrochim. Act.* A **81** 404
[2] Dors I G, Parigger C G and Lewis J 1998 *Opt. Lett.* **23** 1778
[3] Hornkohl J O, Parigger C G and Nemes J A 2005 *Appl. Opt.* **44** 3686
[4] Nelder J A and Mead R 1965 *The Comp. J.* **7** 308
[5] Lagarias J C, Reeds J A, Wright M H and Wright P E 1998 *SIAM J.* **9** 112
[6] Boren C F and Huffman D R 1983 *Absorption and Scattering of Light by Small Particles* (New York: Wiley)
[7] Parigger C G, Woods A C, Surmick D M, Donaldson A B and Height J L 2014 *J. Appl. Spectrosc.* **68** 362
[8] Surmick D M, Parigger C G, Woods A C, Donaldson A B, Height J L and Gill W 2012 *Int. Rev. Atom. Mol. Phys.* **3** 137
[9] Stewart S M 2012 *J. Quant. Spectrosc. Radiat. Transf.* **113** 232
[10] Siegel R, Howell J R and Howell Z 2001 *Thermal Radiation Heat Transfer* (London: Taylor and Francis)
[11] Lim J, Sivathanu Y, Ji J and Gore J 2004 *Combustion and Flame* **137** 222
[12] Konopka W L, Reed R A, Calia V S 1983 *AIAA* **83** 1568
[13] Duval R, Soufiani A and Taine J 2004 *J. Quant. Spectrosc. Radiat. Transf.* **8** 513
[14] Grosshandler W L 1994 RADCAL: A Narrow Band Model for Radiation *NIST Technical Note 1402* (Gaithersburg, MD: NIST)
[15] Vardi Y and Lee D 1993 *J. R. Statist. Soc.* B **55** 569
[16] Zhang L, Zhang L, Li Y, Liu B and Wang J 2006 *Propellants, Explosives, Pyrotechnics* **31** 410
[17] Herzberg G 1950 *Molecular Spectra and Molecular Structure I. Spectra of Diatomic Molecules* (New York: Van Nostrand Reinhold) p 94

Journal of Physics: Conference Series **548** (2014) 012056 doi:10.1088/1742-6596/548/1/012056

Combustion Diagnostics with Femtosecond Laser Radiation

S Couris[1,2], M Kotzagianni[1,2], A Baskevicius[3], T Bartulevicius[3], V Sirutkaitis[3]

[1]Department of Physics, University of Patras, Greece
[2]Institute of Chemical Engineering Sciences (ICE-HT), Foundation for Research and Technology–Hellas (FORTH), Patras, Greece
[3]Department of Quantum Electronics and Laser Research Center (VU LRC), University of Vilnius, Lithuania

Email: couris@physics.upatras.gr & couris@iceht.forth.gr

Abstract. In the present work, the potential of Laser Induced Breakdown Spectroscopy employing femtosecond laser pulses (fs-LIBS) for fuel-air equivalence ratio measurements in premixed methane-air and propane-air flames is presented. A Ti-Sapphire laser system (100 fs, 10 Hz, 800 nm) was used as an excitation source for the plasma creation, while a spectrometer was employed to record the plasma emission spectra. The concentration of the investigated methane-air and propane-air mixtures were expressed by the fuel mole fraction X_{fuel} and varied from only air – $X_{fuel}=0$ ($\varphi=0$) to only fuel – $X_{fuel}=1$ ($\varphi=\infty$). The spectral characteristics of the fs-LIBS spectra are discussed, while the time and energy dependence of the main spectral features are presented. Moreover, from the analysis of fs-LIBS spectra collected at different fuel mole fractions X_{fuel}, it was found that the fuel variations could be very well correlated with the variation of the intensity of some spectral lines and/or their ratios. The prepared calibration curves of the fuel mole fraction X_{fuel} versus the atomic line total intensity ratios (H_α 656.3 nm and O (I) 777 nm) and molecular lines total intensity ratios (C_2 516.5 nm and CN 388.3 nm) suggest the high potential of using fs-LIBS for the determination of the local fuel concentration and its temporal variations.

1. Introduction

Over the years, Laser Induced Breakdown Spectroscopy (LIBS) has been established as a versatile laser based diagnostic technique in the field of combustion diagnostics [1-5]. So far, LIBS has been used to perform measurements of the equivalence ratio [6], determine the employed hydrocarbon among various hydrocarbons of a fuel [7], measure the flame temperature [8] and describe the fuel distribution within the flame volume [9]. However, all these studies have been focused on fuel lean combustible mixtures since energy saving and reduction of soot emission were targeted. Nevertheless, an overall fuel lean mixture can be also accompanied by severe changes of fuel concentration locally in the flame. Seeing this, Kotzagianni et al [10] and Yuan et al [11] have developed a new calibration scheme able to determine all potential fuel concentrations in methane-air and heptane droplets-air spray flames.

Simultaneously, in the majority of LIBS experiments related to combustion, nanosecond lasers have been used, while only few publications using femtosecond lasers have been reported up to now [12, 13]. Despite the limited published works so far, the use of fs-laser pulses in the field of combustion could hold great advantages especially for the mapping of the fuel distribution within the flame volume. In more detail, the high energetic content of the fs laser pulses leads to the breakdown of the gases even in the case of the hot products allowing the determination of the concentration of fuel everywhere in the flame, while the reduced volume of the laser plasma in comparison to the plasma induced by ns laser source offers the opportunity to perform more accurate local measurements of the amount of fuel with higher spatial resolution. Thus, deeper understanding of fundamental aspects of plasma emission induced by fs pulses in flames is required, while the prospects and limitations of fs-LIBS for combustion diagnostics need further study and evaluation. In order to achieve these goals, fs-LIBS has been applied in combustible mixtures of air with hydrocarbons, namely methane (CH_4) and propane (C_3H_8), with variable fuel concentrations ranging from only air to only fuel. Thus from the acquired emission spectra, the most significant spectral features are assigned and the influence of the laser energy and of the detection time on their shape (FWHM) and duration (lifetime, τ) are examined. Finally, the potential of the quantitative correlation of the intensity or the ratio of intensities of atomic and/or molecular origin spectral lines to the fuel mole fraction is investigated for combustion diagnostics.

2. Experimental arrangement

For the needs of the present study, the laser employed was a 100 fs Ti-sapphire laser system operating at 800 nm and at repetition rate of 10 Hz. The laser beam passed through a $\lambda/2$ polarizer to adjust its energetic content and was then expanded ×2 by means of a telescope. The expanded laser beam was reflected by a dichroic mirror onto a 100 mm focal length parabolic mirror in order to create plasma in the centreline of the flame whilst the plasma emission was transmitted through the mirror which was then coupled into an optical fibre. The optical fibre was attached to a 75 mm focal length Czerny-Turner type spectrometer (AvaSpec-2048-USB2), equipped with a 300 lines/mm grating and a 2048 pixel CCD detector array. The spectrometer was able to acquire emission spectra extended from 175 to 1100 nm, while the minimum integration time of its CCD detector was 1.1 ms.

The flames used were stabilized on the rim of a stainless steel Bunsen burner having 9 and 11.5 mm inner and outer diameters respectively, and 350 mm length in order to ensure uniform mixtures. The fuels methane and propane, were supplied from bottles of 99.90 % purity while synthetic air was employed (20.5 % oxygen, 79.5 % nitrogen, 5 ppmv humidity). The flow of the gases was regulated by calibrated flow meters. Although often in LIBS experiments the amount of fuel in a flame is expressed by the equivalence ratio (φ), in the present work the fuel mole fraction (X_{fuel}) is used since it is considered to be more suitable for the interpretation of the results. The values of X_{fuel} were varied from 0, which corresponds to only air, up to 1, which corresponds to only fuel. For stoichiometric methane- and propane-air flames, the corresponding mole fractions X_{CH4} and X_{C3H8} are 0.1 and 0.04, respectively. Mixtures with lower fuel mole fraction are characterized as fuel-lean, while in the opposite case are fuel-rich ones.

3. Results and discussion

3.1. fs-LIBS Emission Spectra in different hydrocarbons

Typical emission spectra arising from the fs laser induced spark in air and in various fuel-air mixtures with different mole fractions are shown in Figure 1. Each spectrum is the average of 100 laser shots detected with a time delay of 170 ns after the plasma creation, while the laser energy was 2 mJ. In the case of fs-LIBS in air, the spectrum is dominated by atomic lines arising from fragments of the constituent molecules of air, namely N_2 and O_2, and are peaked at N (I) 744 nm, N (I) 821.6 nm, N (I) 871.2 nm, and O (I) 777 nm. These atomic lines were also observable in the spectra of stoichiometric fuel-air mixtures (i.e. at X_{fuel}=0.1). However, in this case, they were weaker in intensity, while some new spectral features assigned to atomic hydrogen (H_α 656.3 nm and H_β 486.1 nm) and to cyanogen molecular band CN B^2X^+-Σ^2X^+ ($\Delta\upsilon$=1, 0, -1 at 359.0 nm, 388.3 nm and 421.6 nm respectively) were also presented. The formation and appearance of these spectral features in the spark emission spectra are directly linked to the presence of fuel (CH_4 or C_3H_8) which is atomized and excited/ionized under the intense laser radiation at the focus.

Shown next in Figure 1 are the fs-LIBS spectra in the case of laser spark formed in fuel-air mixtures with increasing fuel concentration. From the analysis of the spectra, it was found that at X_{CH4}=0.25 for methane and at X_{C3H8}=0.15 for propane, the C_2 $d^3\Pi_g$-$a^3\Pi_u$ ($\Delta\upsilon$=1, 0, -1 at 473.7 nm, 516.5 nm and 563.5 nm, respectively) molecular band started to appear in the fs-LIBS spectra. The presence of the C_2 band at lower mole fraction in the case of propane is most probably related to the larger number of carbon atoms of this hydrocarbon in comparison to methane. In addition, the atomic lines of nitrogen and oxygen were found to decrease significantly and finally disappear completely at richer mixtures, i.e. for fuel mole fractions higher than 0.5 for methane and 0.2 for propane.

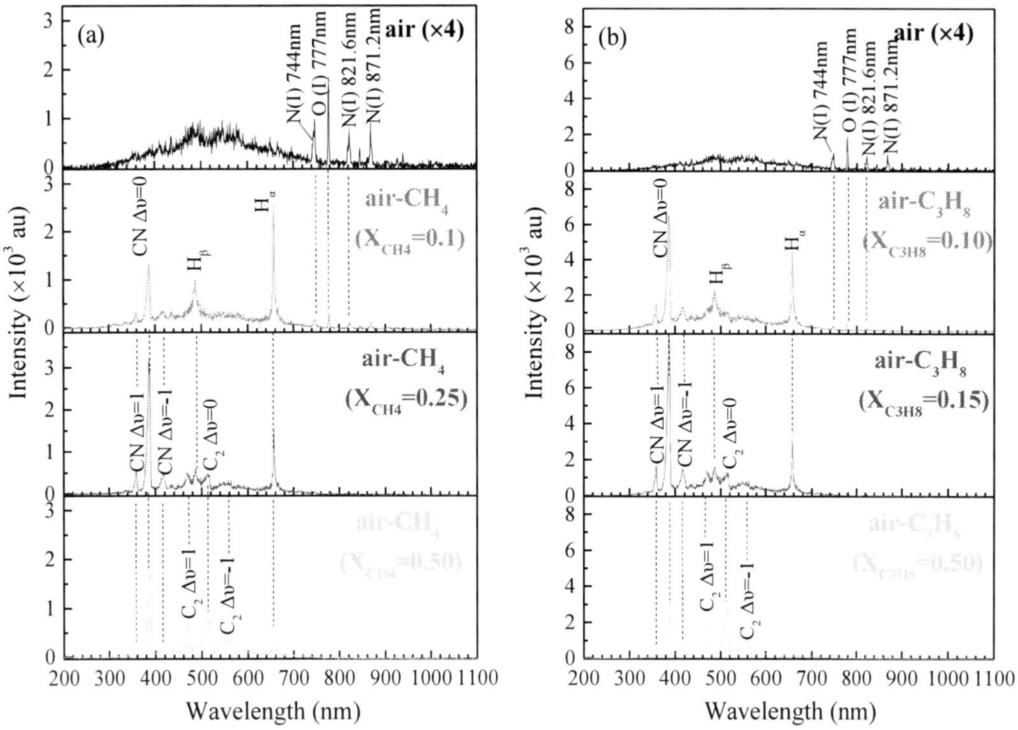

Figure 1. fs-LIBS spectra in various a) CH_4-air and b) C_3H_8-air fuel mole fraction flames.

3.2. Temporal evolution and energy dependence of fs-LIBS spectra

In this section, the temporal evolution of the plasma emission is discussed. The emission spectra shown in Figure 2 correspond to the formation of plasma in a methane-air mixture with a fuel mole fraction X_{CH4} of 0.3 while the acquisition parameters were: t_w=1.1 ms and t_d= 40, 100, 170, 230 and 300 ns. As can be seen from this Figure, at the initial stages of the plasma formation (t_d=40 ns), a strong continuum corresponding to the Bremsstrahlung radiation is extended through the visible region of the spectrum, whilst some spectral lines which appear are superimposed on the continuum and severely broadened.

It is obvious that by increasing the delay of the detection time, the intensity of the continuum drops significantly, and at about 170 ns after the laser spark it is extinguished completely. At the same time, the broadening has been reduced significantly while the intensity of the spectral lines follows this also decrease, as expected. The atomic lines of hydrogen remain significant up to 230 ns while the molecular bands of CN and C_2 remain clearly observable up to 300 ns after the plasma formation. The full width at half maximum (FWHM) and the lifetime of the most important spectral features observed in the fs-LIBS spectra were obtained from the quantitative analysis of the spectra shown in Figure 3. Presented in more detail in Figure 3 are the variation of the total intensity of the 383.3 nm CN band, and of the lines H_α (656.3 nm) and H_β (486.1 nm) of hydrogen and O (777.3 nm) of oxygen as a function of the delay time. For all excitation energies used, ranging from 1 to 3 mJ, all emission lines were found to exhibit an exponential decay, their intensity becoming negligible after 200 ns, whereas the intensity of the molecular band remains observable up to 400 ns. The FWHM of the lines together with the lifetimes are presented in Table 1.

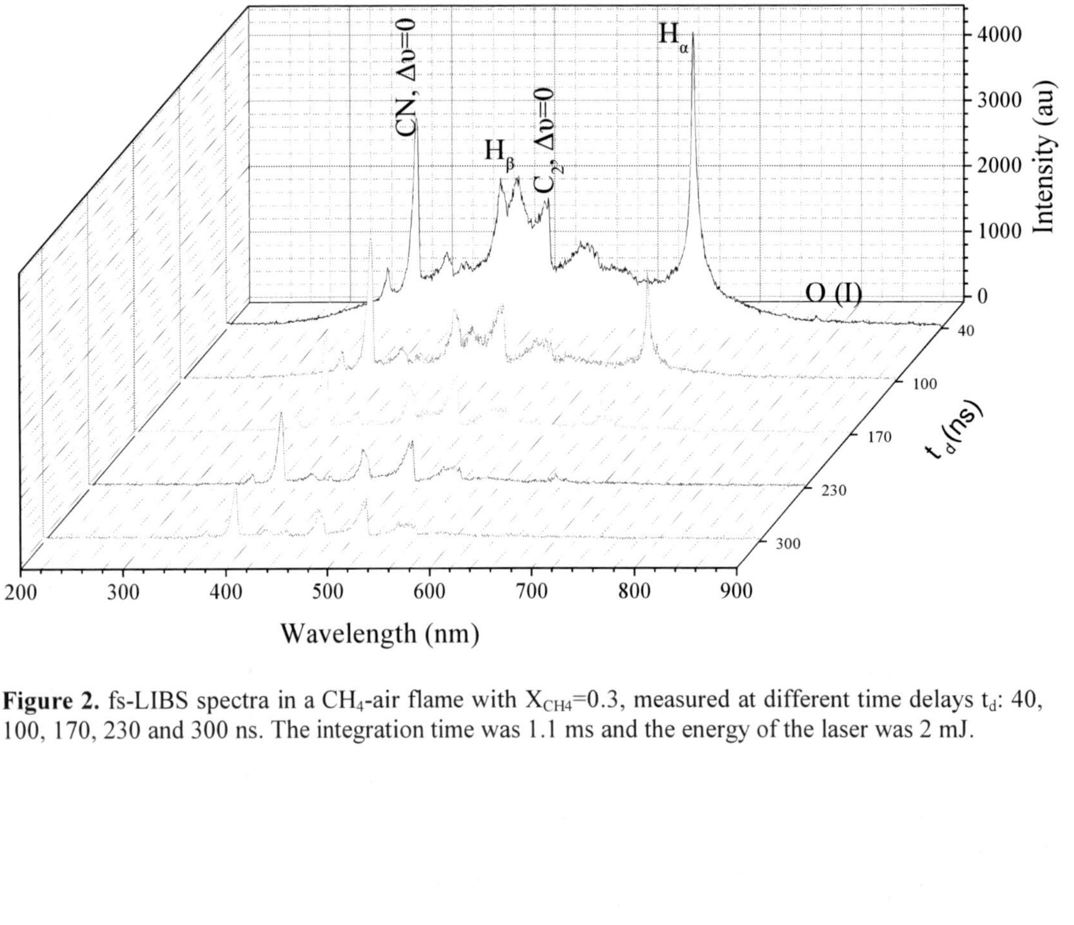

Figure 2. fs-LIBS spectra in a CH_4-air flame with $X_{CH4}=0.3$, measured at different time delays t_d: 40, 100, 170, 230 and 300 ns. The integration time was 1.1 ms and the energy of the laser was 2 mJ.

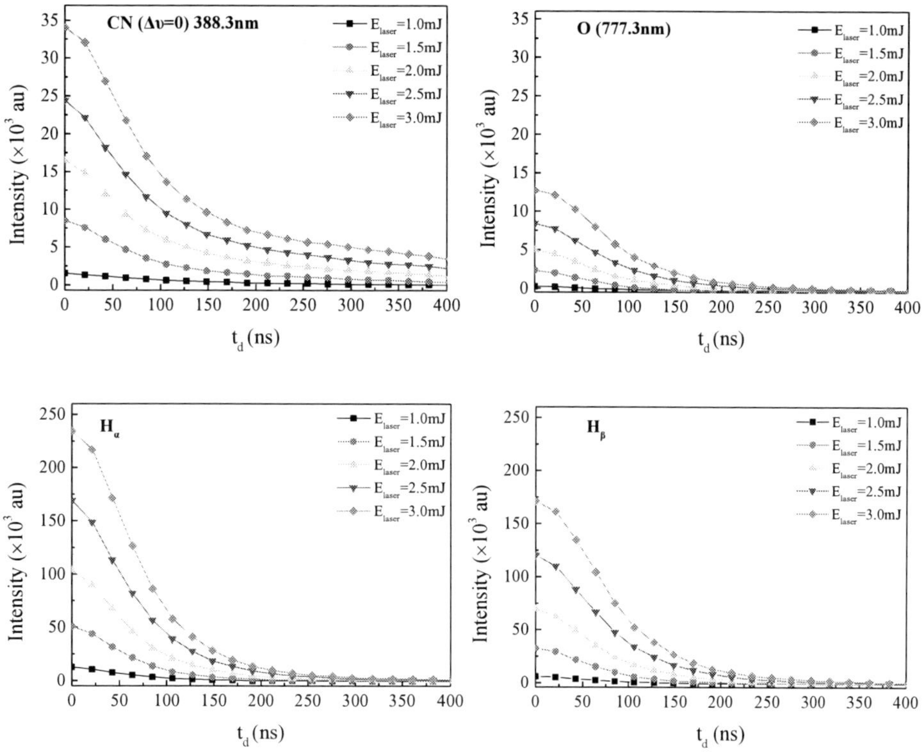

Figure 3. Temporal evolution of the intensity of the: CN (388.3 nm) band, the H_α (656.3 nm) and H_β (486.1 nm) and the O (777.3 nm) atomic lines obtained from fs-LIBS spectra occurring in CH_4-air flame (X_{CH4}=0.3) at different laser excitation energies.

Table 1. The lifetime and the FWHM of CN, H_α, H_β and O lines as a function of E_{laser}

E_{laser} (mJ)	τ (ns)				FWHM (nm)			
	CN	H_β	H_α	O	CN	H_β	H_α	O
1	70	55	54	34	6.8	17.7	4.5	2.9
1.5	72	60	55	52	6.3	19.9	4.7	2.6
2	74	65	58	63	7.2	19.6	4.6	2.6
2.5	81	71	62	73	6.5	19.3	4.6	2.8
3	85	77	69	77	6.5	19.7	4.8	2.6

3.3. Dependence of the CN intensity and of the H_α/O and C_2/CN ratios upon the fuel mole fraction in CH_4- and C_3H_8-air flames

Finally, the dependence of the intensity of CN (388.3 nm) and the ratios H_α/O (777 nm) and C_2 (516.5 nm)/CN (388.3 nm) are presented in Figure 4. These choices have been recently suggested [10-13] and in the present work it is intended to investigate if they can be used in fs-LIBS spectra in very rich fuel mixtures. As can be easily observed, the intensity of CN is increasing up to X_{CH4}=0.25 and of X_{C3H8}=0.17 for methane and propane-air flames respectively, while the same linear increase is also observed in the case of H_{α}/O (777 nm) ratio. Further increase of fuel leads to formation of C_2 and decrease of the CN band intensity, which seems to be related to the decrease of N_2 molecules in the mixture. Therefore, the C_2/CN ratio exhibiting a monotonic increase with the fuel mole fraction can be a useful tool for diagnostics in fs-LIBS.

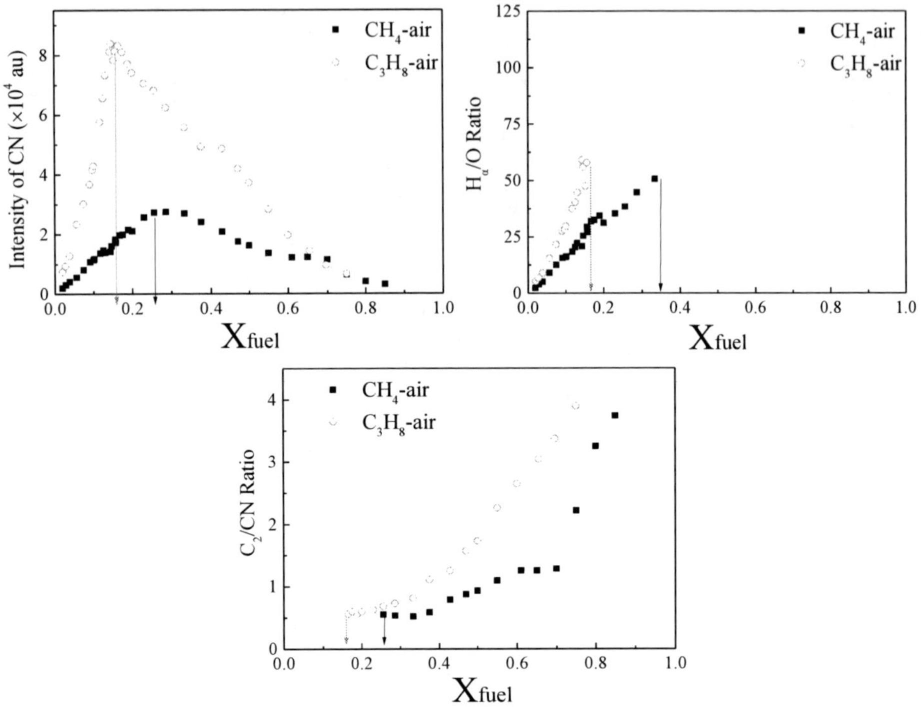

Figure 4. Dependence of the: a) CN intensity, b) H_{α}/O ratio and c) C_2/CN ratio versus the fuel mole fraction X_{fuel} in some CH_4-air and C_3H_8-air flames.

4. Conclusions

Femtosecond Laser Induced Breakdown Spectroscopy (fs-LIBS) has been successfully employed in hydrocarbon-air flames of different compositions in order to use the light emitted from the plasma to obtain useful information about the fuel content and its distribution within the flame and demonstrate the potential of fs-LIBS for combustion diagnostics as well. Moreover, the temporal evolution of the plasma emission was used to derive valuable information about the conditions of formation of species like C_2, CN, etc., while the influence of the laser energy on the spectral lines intensity was also studied. Finally, the H_{α}/O ratio was found to vary linearly with the fuel mole fraction for values up to 0.17 for C_3H_8-air mixtures and up to 0.35 for CH_4-air ones while the ratio of C_2/CN was found to increase monotonically with X_{fuel} for values greater than 0.17 and 0.35 for C_3H_8 and CH_4 mixtures respectively. The present results can be applied to prepare appropriate calibration curves for CH_4- and C_3H_8-air mixture allowing for the fast determination of fuel concentration in such flammable hydrocarbon-air mixtures and providing a spectroscopic tool for *in situ* diagnostics.

Acknowledgements

This research has been co-financed by the European Union (European Social Fund – ESF) and Greek national funds through the Operational Program "Education and Lifelong Learning" of the National

Strategic Reference Framework (NSRF) - Research Funding Program: THALIS (ISEPUMA). Investing in knowledge society through the European Social Fund. The research leading to these results has received also funding from LASERLAB-EUROPE (grant agreement no. 284464, EC's Seventh Framework Programme). (Prop N: VULRC001879). S.C. and M.K. acknowledge the Department of Quantum Electronics and Laser Research Center (VU LRC), University of Vilnius, Lithuania where the femtosecond laser experiments have been carried out and also the "Optics characterization" group for the excellent collaboration, the substantial technical help and the fruitful discussions.

References

[1] Panne U, Hahn D 2006 Analysis of aerosols by LIBS eds 2006 *Laser Induced Breakdown Spectroscopy (LIBS): Fundamentals and Applications* eds W Miziolek, V Palleschi, I Schechter (New York: Cambridge University Press) chapter 5 pp 194 - 253
[2] Yueh FY, Singh JP 2007 LIBS Application to Off-Gas Measurement *Laser Induced Breakdown Spectroscopy* eds J P Singh, S N Thakur (Amsterdam: Elsevier) chapter 9 pp 199 - 222
[3] Ferioli F, Buckley S G 2006 *Comb Flame* **144** 435
[4] Stavropoulos P, Michalakou A, Skevis G, Couris S 2005 *Spectrochim. Acta Part B* **60** 1092
[5] Kiefer J, Tröger J W, Li Z S, Alden M 2010 *Appl. Phys. B* **103** 229
[6] Stavropoulos P, Michalakou A, Skevis G, Couris S 2005 *Chem. Phys. Lett.* **404** 309
[7] Michalakou A, P. Stavropoulos P, Couris S 2008 *Appl. Phys. Lett.* **92** 081501
[8] Lee T-W, Hegde N 2005 *Comb. Flame* **142** 314
[9] Phuoc T X, White F P 2002 *Fuel* **81** 1761
[10] Kotzagianni M, Yuan R, Mastorakos E, Couris S 2014 Laser-induced breakdown spectroscopy measurements in turbulent methane flames *52nd Aerospace Sciences Meeting (AIAA Sci. Tech., doi: 10.2514/6.2014-1351)*
[11] Yuan R, Kotzagianni, Couris S, Mastorakos E 2014 Laser-induced breakdown spectroscopy measurements in spray flames *17th International Symposium on Applications of Laser Techniques to Fluid Mechanics (Lisbon, Portugal, 07-10 July, 2014)*
[12] Kotzagianni M, Couris S 2013 *Chem. Phys. Lett.* **561–562** 36
[13] Kotzagianni M, Couris S 2012 *Appl. Phys. Lett.* **100** 26410

Journal of Physics: Conference Series **548** (2014) 012057 doi:10.1088/1742-6596/548/1/012057

Frequency comparison of lattice clocks toward the redefinition of the second

T Ido[1]
Research manager, *National institute of information and communications technology*, 4-2-1 Nukui-kitamachi, Koganei, Tokyo, 184-8795, Japan

E-mail: ido@nict.go.jp

Abstract. Strontium is the most popular species for optical lattice clocks. Recent reports of the accuracies from Boulder, U.S. and Tokyo reach 10^{-18} level, which is better than state-of-the-art caesium clocks more than one order of magnitude. While this achievement accelerates the discussion to redefine the second, the agreement of frequencies in separate laboratories is of critical importance. For this context, intercontinental comparison of Sr lattice clocks were demonstrated between Japan and Germany using a satellite-based technique. The frequency difference was consistent with zero with an uncertainty of 1.6×10^{-15}.

1. Introduction

Optical frequency standards are one of major applications of laser spectroscopy. When lasers are invented, it was quite natural to consider making highly stable oscillator in optical region instead of microwave region. There has been tremendous effort for this goal in this half century. It is generally said that laser is coherent radiation. But the coherence time is normally much shorter than 1 second, whereas the stable microwave oscillator seldom cause phase jump. Therefore, laser is much more incoherent when it is compared to microwave. In addition, precision spectroscopy in optical region is affected by photon recoils since momentum of photons is much larger than that of microwave. However, the coherence time of laser radiation is lately extended to more than one second using a highly stable optical cavity [1]. The development of the technique to laser-cool and pin-down ions or neutral atoms has allowed us to eliminate Doppler and recoil shifts. These efforts have finally achieved optical clocks that clearly surpass the state-of-the-art microwave-based clocks in accuracy as well as stability [2-4]. This achievement inspires the community of time and frequency standards to seriously discuss the redefinition of the SI second because current realization of the second does not have the least uncertainty to scale time and frequency. The accuracy of 10^{-18} level, on the other hand, has not been fully trusted at this point in the community since the evaluation is based on the total uncertainties of known systematic shifts. There could be unknown systematic effect which may shift the atomic resonance from laboratory to laboratory. Thus, it is required for the redefinition of the second that the optical reference frequencies in various laboratories were confirmed to be identical.

2. ^{87}Sr lattice clock: the most major optical clock

The necessity to confirm the identical frequencies in separate laboratories is receiving attention since optical lattice clocks based on strontium-87 atoms are now operated in seven laboratories [2-8] with

[1] To whom any correspondence should be addressed.

systematic uncertainties of 10^{-16} level or below. Furthermore, PTB has accurately evaluated sensitivity of the resonance frequency to black body radiation (BBR), which allowed the total accuracy of clocks to be below 10^{-16} [9]. Furthermore, innovations to manage ambient temperature of strontium atoms have recently occurred. One realized at JILA is to equip highly accurate thermometer inside vacuum chamber which accurately characterizes the BBR of room temperature environment [2]. Another one lately achieved in Katori's group at RIKEN is cryogenic lattice clock, where the atoms were transferred into a small ceyo-cooled cavity of 95 Kelvin for the interrogation, drastically reducing the BBR [3]. These claim the accuracies in 10^{-18} level, which are currently the state-of-the-art of optical clocks.

Figure 1 summarizes the results of absolute frequency measurements of ^{87}Sr clock transition in various laboratories. You may find there is still discrepancy of frequencies in 10^{-15} level, which clearly shows that the SI second is not a good scale any more, and only direct frequency comparison can evaluate the frequency reproducibility of Sr lattice clocks among laboratories. Note that recent frequencies measured at SYRTE (Systèmes de Référence Temps-Espace, Observatoire de Paris) [5] and PTB (Physikalisch-Technische Bundesanstalt, Germany) [6] are in good agreement in 10^{-16} level. This is because both laboratories operate caesium fountain clocks which realized the SI second at the 10^{-16} level. Such accurate SI second is only available in limited number of laboratories. Rather large uncertainties of Japanese clocks are due to the lack of the good realization of the SI second.

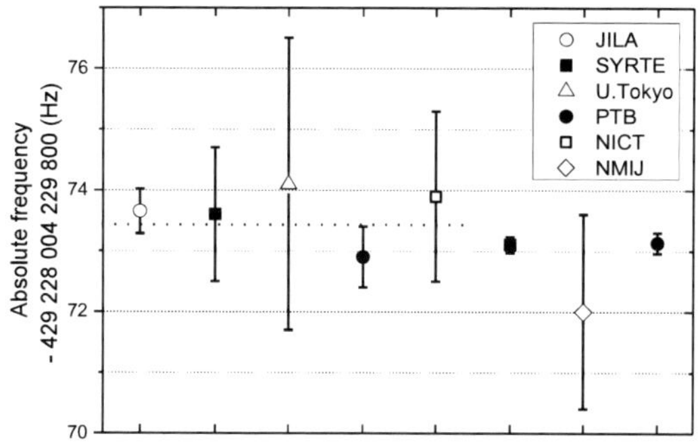

Figure 1. Results of absolute frequency measurements in various laboratories.

3. Two methods of direct comparison: fiber link and satellite link

Direct comparison of optical clocks is straightforward solution to confirm the identical frequency, where we don't express the frequency as numerical number based on the SI second. Thus, fractional frequency difference of two locally available clocks were evaluated in Ref. [1-3], resulting in 10^{-17} level of frequency agreement. For the comparison between physically separated clocks, on the other hand, we need a method to transfer the highly stable frequency to other places. For this purpose, optical fiber is probably the ultimate as the transfer media. For global comparisons, however, satellite-based techniques are the only way at this point since fiber transfer normally extends up to 200km.

Journal of Physics: Conference Series **548** (2014) 012057 doi:10.1088/1742-6596/548/1/012057

3.1 Fiber links

In the case that two clocks are on the same continent and not far separated, an optical fiber link is a strong tool. So far two fiber-based direct comparisons of optical clocks were demonstrated, one is between a Sr clock at JILA and neutral Ca clock at National Institute of Standards and Technology (NIST) with a 4-km-long optical fiber link [11], and the other is between two ^{87}Sr lattice clocks at National Institute of Information and Communications Technology (NICT) and the University of Tokyo (UT) using a 60- km long optical fiber [12]. Optical fiber links are promising at large distances, as demonstrated up to 1840 km [10]. These demonstrations also claim the capability to disseminate standard frequency. However, it is unrealistic that optical fiber network is dedicated only for the delivery of standard frequency. The conventional method of transfer adopted in these work requires dark fiber. In addition, note that 1840km transmission [13] was demonstrated using fiber cables that are calmly buried in the ground. On the other hand, some countries, for instance Japan, hang the optical fiber cables in the air. Such cable suffers considerable vibration noise, which prevents the accurate transfer of reference frequency. Indeed, the NICT-UT comparison [12] often did not work on windy days.

3.2 Satellite –based link

The global positioning system (GPS) is a strong tool of clock synchronization. While modulation code of ~Mbps is utilized for the synchronization of commercial GPS clocks, the use of carrier-phase (GPSCP) improves the accuracy of time transfer for orders of magnitude. However, an averaging time of more than a day is still required to reach the 10^{-16} instability [10]. Another satellite-based technique is two way satellite time and frequency transfer (TWSTFT), which also has the code transfer as well as carrier phase transfer. Using the carrier-phase in TWSTFT might be a quicker route to 10^{-16} instability. Two-way carrier-phase technique (TWCP) was first demonstrated by USNO [14] and characterized for various ranges of the baselines up to 1 000 km [15]. The characterization has been extended to 9 000 km baseline between NICT and PTB [16].

4. NICT-PTB direct comparison using a TWCP technique

NICT and PTB have an ^{87}Sr lattice clock in operation for each. Thus, the two-way carrier phase technique (TWCP) was applied to perform real-time frequency comparison of lattice clocks in intercontinental scale. The experimental scheme is depicted in Fig. 2. The geostationary satellite (AM2) locates at the longitude of 80 degrees east. The elevation angles of the satellite from the earth station at NICT and PTB are 16.0 degree and 3.7 degree, respectively. The transponder is available for approximately 13 hours (from 10:05 to 22:59 in UTC) per a day. The correction due to the dispersion of up-down link was included referring a VTEC map provided by NICT [17] and Royal Observatory of Belgium [18]. Figure 2(b) shows schematic diagram of the comparison. During the measurement campaign, the clock frequencies were measured referring to respective local hydrogen masers (H-masers). The frequency difference of the two H-masers was simultaneously evaluated by the TWCP technique. An Yb^+ clock is also operated in PTB to bridge offline-times of the Sr clock at PTB, and thus reduced the statistical uncertainty of the measurement. We performed the comparison for four days in late June 2013. The direct Sr lattice clock comparison worked for ~70 000 s, extended to 84 000 s by the Yb^+ ion clock. The total measurement time leads to a statistical uncertainty of 1.2×10^{-15}. Systematic uncertainty of the fractional frequency difference is summarized in Table 1. Frequency link is currently the source of the dominant uncertainty. This uncertainty is the experimental result of TWCP-GPSCP comparison reported in [16]. The uncertainty of ionospheric delay could be a dominant source of this uncertainty. Progress of the evaluation of ionospheric effect may reduce systematic as well as statistical uncertainty. Considering the systematic uncertainties, the total uncertainty was determined to be 1.6×10^{-15}, whereas the overall average of the fractional frequency difference was 1.1×10^{-15}, indicating the agreement of clock frequencies.

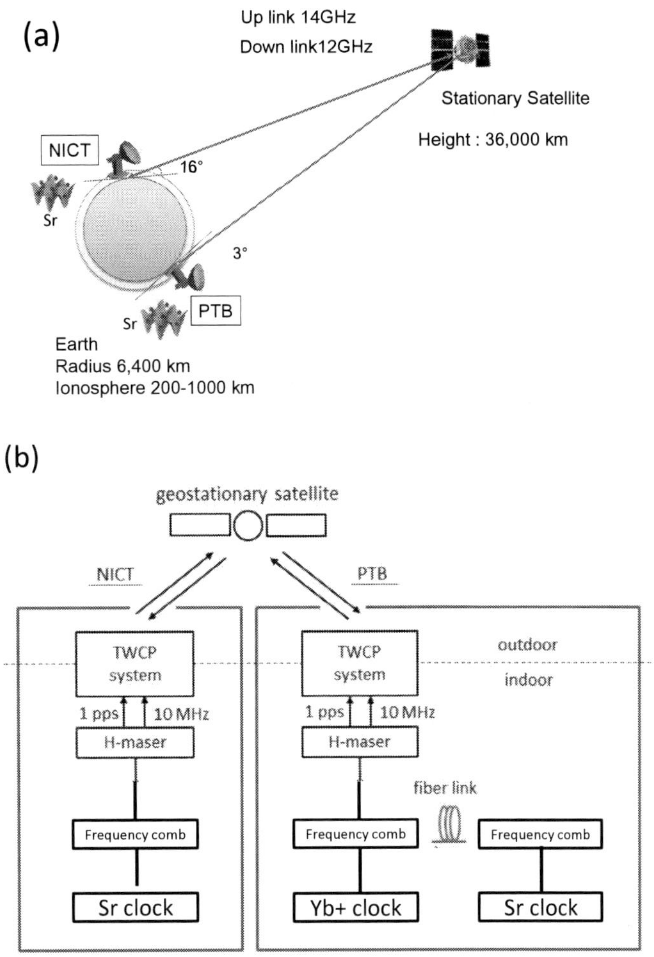

Figure 2. (a) Elevation angle and distance from two earth stations to the satellite (b) Schematic diagram of the experiment

Table 1. Uncertainty budget of the frequency comparison

Contributor	uncertainty ($\times 10^{-16}$)
Statistical	12
Systematics	
TWCP link[a]	10
Sr clock @ PTB[b]	0.4
Sr clock @ NICT	2
Gravitational red shift	1
Total	16

[a] Ref. [16]
[b] Ref. [6]

5. Conclusion

Following the fiber link experiment of two Sr lattice clocks in Tokyo area, the frequency agreement was also confirmed on intercontinental scale between Japan and Germany by using the satellite-based TWCP technique. The frequencies agreed with uncertainty of 1.6×10^{-15}, which is dominated by the uncertainty contribution of the link.

6. Acknowledements

The author thanks NICT and PTB members for the intercontinental comparison, including M. Fujieda, H. Hachisu, S. Nagano, and T. Gotoh from NICT, and St. Falke, N. Huntemenn, C. Grebing, B. Lipphardt, Ch. Lisdat, and D. Piester from PTB.

References

[1] Kessler T, Hagemann C, Grebing C, Legero T, Sterr U, Riehle F, Martin M, Chen L, Ye J 2012 *Nat. Photon.* **6** 687
[2] Chou C, Hume D, Koelemeij J, Wineland D, Rosenband T 2010 *Phys. Rev. Lett.* **104** 070802
[3] Bloom B, Nicholson T, Williams J, Campbell S, Bishof M, Zhang X, Zhang W, Bromley S, Ye J 2014 *Nature* **506** 71
[4] Ushijima I, Takamoto M, Das M, Ohkubo T, and Katori H 2014 *Preprint* physics/1405.4071
[5] Le Targat R *et al* 2013 *Nat. Commun.* **4** 2109
[6] Falke St *et al* 2014 *New J. Phys.* **16** 073023
[7] Yamaguchi A *et al*, *Appl. Phys. Express* **5** 022701
[8] Akamatsu D, *et al Opt. Express* **22** 7898
[9] Middleman T, Falke St, Lisdat Ch, Sterr U 2012 *Phys. Rev. Lett.* **109** 263004
[10] Bauch A *et al* 2013 *Phys. Rev. Lett.* **111** 110801
[11] Ludlow A *et al* 2008 *Science* **319** 1805
[12] Yamaguchi A *et al* 2011 *Appl. Phys. Express* **4** 082203
[14] Fonville B *et al* 2004 *Proceedings of 36th Annual PTTI Meeting* (U.S. Naval Observatory) 149
[15] Fujieda M *et al* 2012 *IEEE Trans. Ultrason. Ferroelectr. Freq. Control* **59** 2625
[16] Fujieda M, Piester D, Gotoh T, Becker J, Aida M, Bauch A 2014 *Metrologia* **51** 253
[17] http://wdc.nict.go.jp/IONO/gps-tec/tecv/
[18] http://www.gnss.be/Atmospheric_Maps/ionospheric_maps.php

Journal of Physics: Conference Series **548** (2014) 012058 doi:10.1088/1742-6596/548/1/012058

Line Shape Effects in the Operation of Thulium and Erbium Doped Fiber Amplifiers

O Mahran and G D Roston

Department of Physics, Faculty of Science, Alexandria University, PO Box 21511, Alexandria, Egypt

dr.gamal_daniel@yahoo.com

Abstract. This work describes the effect of Gaussian line shape on the comparison of the amplification characteristics (gain) and the Noise figure (NF) of the Thulium and Erbium in the host materials which the Yttria Alumina-Silica glass. The gain using this host material covers the range 1.45-1.65 μm. Thulium-doped fiber amplifiers operated in the region of wavelength (1480-1510 nm) which is called S-band. The main pump source is 1.04 and 1.55 μm which creates population inversion between 3F4 (upper laser level) and 3H4 (lower laser level), and Erbium-doped fiber amplifiers operated in the region (1510-1650 nm) which is called L –band with the pump wavelength 980 nm. It is found that the erbium doped yttria-alumina silicate fiber amplifier exhibits a maximum gain of 40.3 dB at the central wavelength 1540 nm and minimum noise figure 14 dB, but the broadening in the gain curve is 20 nm. Also it is found that the thulium doped yttria-alumina silicate fiber amplifier exhibits a maximum gain of 27.5 dB at the central wavelength 1467 nm and minimum noise figure 2.5 dB, with the broadening in the gain curve is 41 nm. Gain flatness was investigated and the results strongly confirm the feasibility of using this hosts' glass doped with Thulium in practical ultra large capacity wavelength division multiplexing networks.

1. Introduction

In wavelength division multiplexing (WDM) systems, overall transmission capacity depends significantly on the spectral characteristics of the optical amplifiers, such as flatness, bandwidth, and the magnitude of the gain. Erbium-doped fiber amplifiers (EDFAs) have provided an efficient optical gain in the 1.5μm communication windows in conventional single-mode fibers (SMF) [1]. When the population is highly inverted, the stimulated emission cross section of erbium ions in silica provides ample gain over the 1520-1560 nm range, called the conventional band or C band. Increasing demands on the capacity of WDM transmission system now require newly developed transmission windows beyond the amplification band width supported by erbium doped fiber amplifiers (EDFAs). So Thulium doped fiber amplifiers (TDFAs) for S-band have been studied extensively in recent years, as the candidates for the next generation amplifiers competing / com - promising L-band EDFA's, Raman amplifiers [2], [3].Thulium doped fiber amplifiers are promising amplifiers for the 1400 nm band, they are reported to have high gain and low noise characteristics in the 1450-1480 nm region [4], Thulium-doped fiber amplifiers (TDFAs) are a promising candidate for the S-band amplification because the amplification bandwidth of the TDFA is centered at 1470 nm [5], which falls within the S-band. A 1050 nm pumping scheme can be used to obtain the population inversion in a Thulium-doped fiber (TDF) [6]. The TDF length and pump power are the important parameters that determine the attainable gain and noise figure in TDFA, which provides gain in the S-and S+ band for such systems. Many different pumping schemes have been proposed to exploit the complex energy level transitions of Thulium in the -band, either by single-wavelength up conversion pumping or dual-wavelength pumping [8]. The most efficient scheme has been a combination of 800 and 1050 nm, which takes advantage of the strong ground state absorption (GSA) at 800 nm [8]. However, it has been reported in [8] that GSA in thulium at 690 nm is even stronger than that at 800 nm.

2. Model

The mathematical model for erbium and thulium was given in Refs. [9, 10].

3. Results and discussion

The emission and absorption cross section data fitted to Gaussian shape for both erbium and thulium in yttria glass are given in Tables 1 and 2, respectively.

Table I: The emission and absorption for erbium doped yttria fiber

	Emission			Absorption		
(i)	a_i	λ_i(nm)	$\Delta\lambda_i$(nm)	a_i	λ_i(nm)	$\Delta\lambda_i$(nm)
1	0.03	1462	50	0.05	1462	30
2	0.10	1499	42	0.20	1482	35
3	0.26	1525	32	0.24	1499	36
4	0.57	1533	11	0.43	1522	34
5	0.555	1553	31	0.42	1532	13
6	0.02	1561	10	0.40	1547	30
7	0.17	1586	60	0.15	1562	30

Table II: The emission and absorption for Thulium doped yttria fiber

	Emission			Absorption		
(i)	a_i	λ_i(nm)	$\Delta\lambda_i$(nm)	a_i	λ_i(nm)	$\Delta\lambda_i$(nm)
1	2.674	1732	25.1	0.6557	1451	65.67
2	6.303	1761	65.08	2.47	1601	58.12
3	6.299	1701	24.54	0.7948	1784	168.2
4	6.001	1788	103.7	3.004	1662	89.62

Figures 1 and 2 show the absorption and emission Gaussian fitted values for the erbium and thulium doped yttria alumina silicate glass fiber amplifier, respectively.

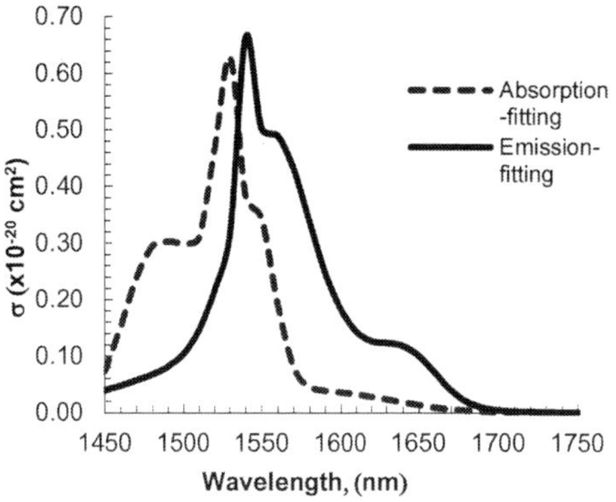

Figure 1. Absorption and emission cross section with Gaussian fitted values for erbium doped yttria alumina silicate glass fiber amplifier.

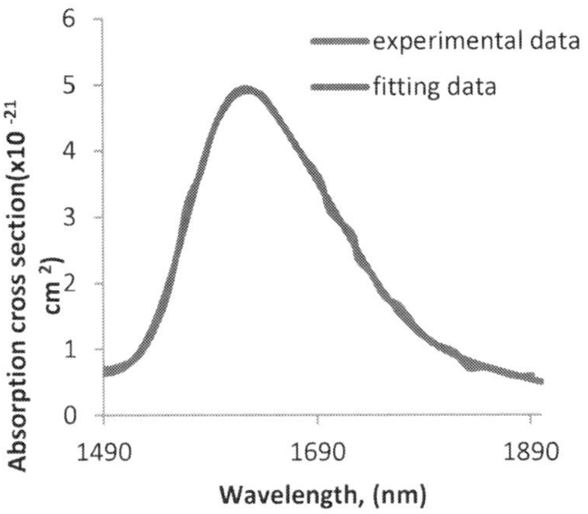

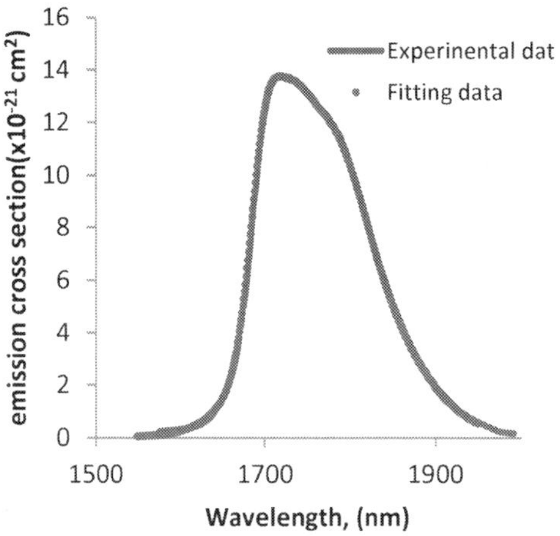

Figure 2. Absorption and emission cross section with Gaussian fitted values for thulium doped yttria alumina silicate glass fiber amplifier.

Figure 3 shows the gain in dB as function of the signal wavelength in nm at input pump power 100mW (for Tm the second pump [dual pump]1550 pump power also at 100mW) and fiber length 34 m for both Er and Tm doped tellurite glass fiber amplifier. From the figure the central wavelength for Er is 1530 nm at which the gain maximum 20 dB, with bandwidth 20 nm, while for Tm the central wavelength is 1474 nm at which the maximum gain is 9.29 dB with bandwidth 42 nm, this mean that erbium doped gives higher gain than thulium doped but the gain flatness in thulium is higher than erbium for the same glass fiber doped.

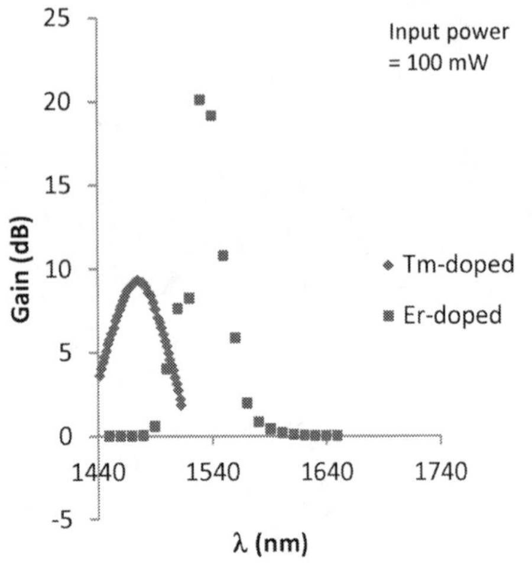

Figure 3. The gain spectrum for Er and Tm doped tellurite glass fiber amplifier

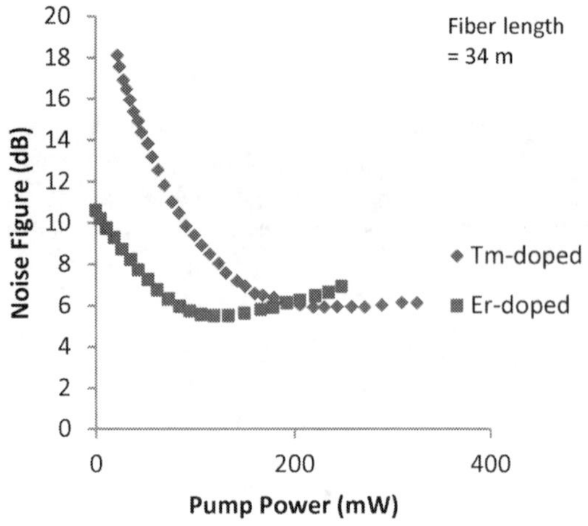

Figure 4. Noise figure against the pump power for Er and Tm doped tellurite glass fiber amplifier

From the figure, the noise figure is 2.38 dB at the central wavelength 1474 nm for thulium doped while 8 dB at the central wavelength 1530 nm for the erbium doped tellurite glass fiber amplifier. Finally the noise figure as a function of the pump power at 34 m fiber length for the two signal wavelength 1530 and 1474 nm for both Er and Tm doped tellurite glass fiber amplifier respectively is plotted in Fig.4.As in figure, the noise figure decreases with pump power increases for both erbium and thulium but the decrease in noise for erbium is higher than thulium doped with the given glass.

4. Conclusion

In this paper, we studied the amplification characteristics (gain) and the Noise figure (NF) of the Thulium and Erbium in yttria host materials through the cross section calculation using Gaussian fitting . The gain using these host material covers the range 1.45-1.65 μm. It is found that the erbium doped yttria-alumina silicate fiber amplifier exhibits a maximum gain of 40.3 dB at the central wavelength 1540 nm and minimum noise figure 14 dB, but the broadening in the gain curve is 20 nm. Also it is found that the thulium doped yttria-alumina silicate fiber amplifier exhibits a maximum gain of 27.5 dB at the central wavelength 1467 nm and minimum noise figure 2.5 dB, with the broadening in the gain curve is 41 nm.

References

[1] Mears R J, Reekie L, Jauncey I M, Payne D N 1987 *Electron. Lett.* **23** 1026
[2] Min B, Yoon H, Lee W J, Park N 2000 *IEEE Photo. Techno. Lett.* **12** 480
[3] Kani J, Jinno M 1999 *Electron. Lett.* **35** 1004
[4] Komukai T, Yamamoto T, Sugawa T, Miyajima Y 1995 *IEEE J.Quantum Electron.* **31** 1880
[5] Kozak M M, Caspary R, Kowalsky W 2004 Thulium doped fiber amplifier for the S-band *Proc. 6th International Conference on Transparent Optical Networks* **6** pp 51–54
[6] Yam S S H, Akasaka Y, Kubota S Y, Inoue H, Parameswaran K 2005 *IEEE Photon. Technol. Lett.* **17** 1001
[7] Sakamoto T, Shimizu M, Kanamori T, Terunuma Y, Ohishi Y, Yamada M, Sudo S 1995 *IEEE Photon. Technol. Lett.* **7** 983
[8] Aozasa S, Masuda H, Ono H, Sakamoto T, Kanamori T, Ohishi Y, Shimizu M 2001 1480–1510 nm band Tm doped fiber amplifier (TDFA) with high power conversion efficiency of 42% *Optical Fiber Communication Conference* (Anaheim, CA, 2001) paper PD1
[9] Mahran O, Shahat M EL, Samahy A EL, Helmi M S 2012 *Intern. J. Resear. Rev. Appl. Sciences* **10** 314
[10] Mahran O, Shahat M EL, Hagar W 2013 *Intern. J. Scient. Engin. Resear.* **4** 1815

Spectral signature of Egyptian crude oils

Z F Ghatass[1], A W Nashed[2], I H Saleh[1] and M M Mohmed[3]

[1] *Department of Environmental Studies, Institute of Graduate Studies and Research, 163 Horrya Avenue, P.O. 832 Shatby, Alexandria, Egypt*
[2] *Department of Material Science, Institute of Graduate Studies and Research, Alexandria University, Egypt*
[3] *Department of Biophysics, Medical Research Institute, Alexandria University, Egypt*

Email: z_ghatass@yahoo.com

Abstract. Crude petroleum oils are complex mixtures of diverse hydrocarbons, in widely varying compositions, that originate from a variety of geological sources. Fluorescence emission spectra have been measured for two types of Egyptian crude petroleum oil, its light and heavy products over a broad range of excitation and emission wavelengths. Both types of crude oil products are characterized by spectral signatures with a differing topography: the number of fluorescent peaks, their coordinates (λ_{ex}, λ_{em}) on the plane of the three dimensions spectrum, and the shape of the bands formed by the contour line density, changeable in either direction. The refined light oil shows emission spectra at λ_{max} between 350 and 500 nm according to the excitation wavelength. The refined heavy oil shows very broad unstructured emission spectra with λ_{max} >400 nm. As a group, they could certainly be distinguished from the light oil samples and most of the crude oil.

1. Introduction

Physical methods are now used widely in environmental science to analyze environmental data, estimate environmental impact of production and consumption, investigate the consequences of policy measures and help prepare government regulations [1]. Accurate quantitative analysis of the composition of the surface water, the soil or the atmosphere is crucial in our assessment of the quality of the environment and our judgment of the relative success of measures taken to reduce pollution. Fluorescence techniques represent one of the most powerful and promising tools for the monitoring of aquatic media [2-6]. Crude petroleum oils (CPO_s) are complex mixtures of diverse hydrocarbons, in widely varying compositions, that originate from a variety of geological sources [7-11]. The ultimate fate of the majority of this oil to a large extent remains an unresolved mystery. Accordingly, the final impact of this oil on the environment and mankind is an unsettling and anxious question [12]. The extensive pollution of the seas by oil has resulted in an increasing demand for an accurate teledetection system. The information that such a system should provide can be summarized as: is the water polluted with oil, and if it is, with what kind of oil? [13-19]. In this work, we demonstrate the total luminescent spectroscopy (using different wavelength for excitation) for representative fingerprint of different types of oil and its products.

2. Material and method

Two types of crude petroleum oil Samples (Balaaiem and West Desert oil), its light (Triptane, Jet A_1 and Kerosene) and heavy (Balaaiem and West Desert Solar) products were provided by Alexandria Petroleum Company. Fluorescence measurements of samples under study were carried out by using Spectrofluorometer LS 55 (Perkin Elmer). Excitation and emission slits were set to 10 nm width. All measurements were made using the real, non–degassed samples in quartz cell. Each sample was excited at different wavelengths and fluorescence emissions were recorded for each excitation.

Journal of Physics: Conference Series **548** (2014) 012059 doi:10.1088/1742-6596/548/1/012059

3. Results

3.1. Fluorescence spectra of crude oil samples

The fluorescence spectra of crude oil samples were recorded. For example, Figure 1 shows the West Desert crude oil spectra, its shown that the broad unstructured emission spectra with maximum intensity at wavelength (λ_{max}) between 400 and 530 nm according to the excitation wavelength. Generally, emission intensity of crude oil increases with increasing of excitation wavelength up to 440 nm and its emission spectra shift to right with increasing excitation wavelength. Spectra of the two types of crude oil show a relatively small difference through the spectral range considered here. Thus under field conditions, the two crude oil samples would not be distinguished by means of their fluorescence emission spectra. As an example, Figure 2; show the corresponding three dimensions fingerprints and contour lines of West Desert crude oil. The contour lines of West Desert crude oil has two maximum emission peaks around an excitation - emission pair of (393, 450 nm) and (440, 500 nm), while the contour lines of Balaaiem crude oil has one maximum emission peak around excitation-emission pair of (410, 470 nm).

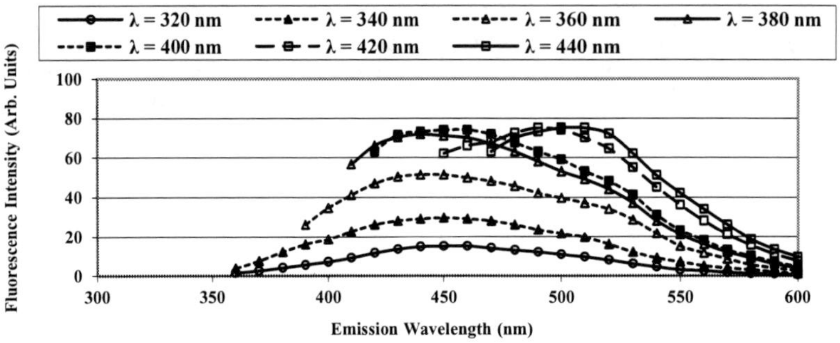

Figure 1. Fluorescence spectra of West Desert crude oil at different excitation wavelengths.

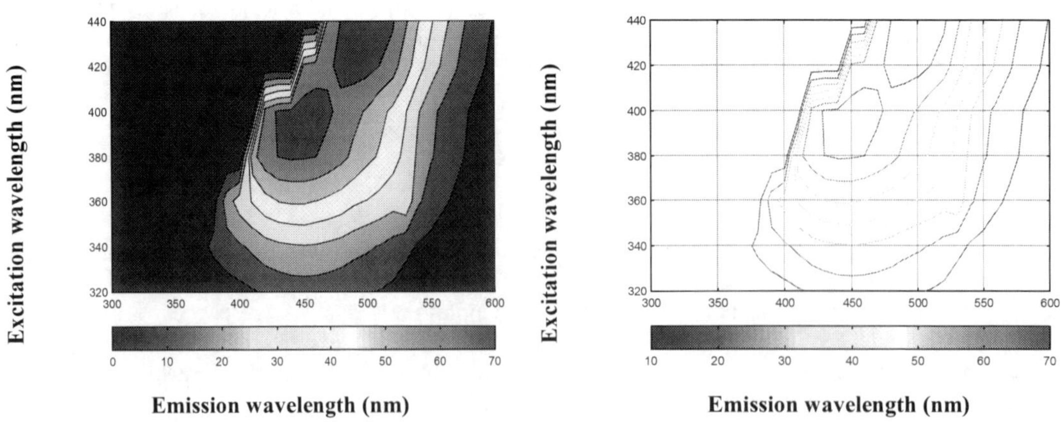

Figure 2. Three dimensions fingerprint and contour lines of West Desert crude oil.

3.2. Fluorescence spectra of oil products

3.2.1. Light products. The three dimensions fingerprints and contour lines of triptane, Jet A1, and kerosene were constructed. As an example Figure (3) shows the three dimensions fingerprints and contour lines of triptane. The contour lines of triptane has one maximum emission peak around (320, 390 nm), Jet A1 also has one maximum peak around (340, 370 nm), and kerosene has two maximum emission peaks around (320, 355 nm) and (347, 407 nm).

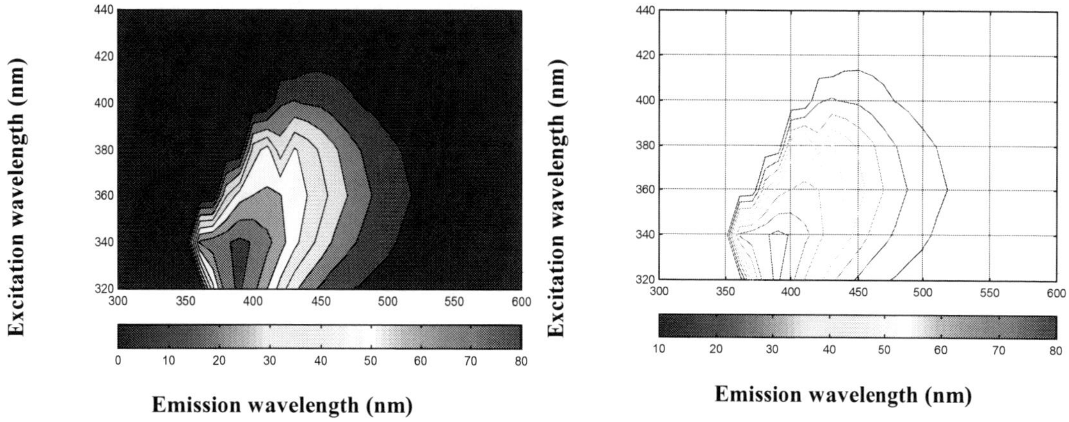

Figure 3. Three dimensions fingerprint and contour lines of Triptane.

3.2.2. Heavy products. The three dimensions fingerprints and contour lines for West Desert and Balaaim solar were constructed. Figure (4), shows the corresponding three dimensions fingerprints and contour lines of West Desert solar. The contour lines of West Desert solar has two maximum emission peaks around an excitation – emission pair of (400, 450 nm) and (440, 520 nm). The contour lines of Balaaiem solar, has two maximum emission peaks around (400, 457 nm) and (440, 505 nm).

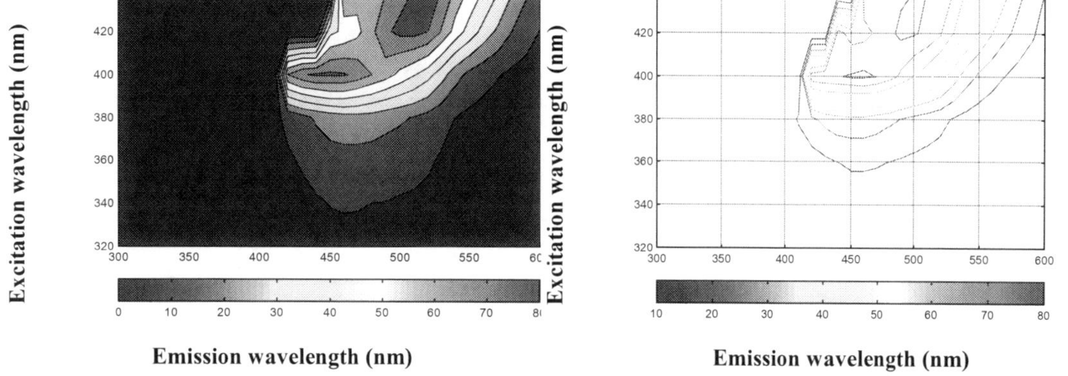

Figure 4. Three dimensions fingerprint and contour lines of West Desert solar.

4. Discussion

In the present work, the various crude oil, light and heavy products are characterized by spectral signatures with a differing topography: the number of fluorescent peaks, their coordinates (λ_{ex} , λ_{em}) on the plane of the three dimensions spectrum, and the shape of the bands formed by the contour line density, changeable in either direction. By counting the coinciding coordinates of peaks, it is possible to differentiate between the various samples as shown in Table (1).

Table 1. Catalog list of maximum coordinates (λ_{ex}, λ_{em}) on three dimensions diagrams

No.	Matter	(λ_{ex}, λ_{em}); (nm)
1	West Desert crude oil	(393; 450) and (440; 500)
2	Balaaiem crude oil	(410; 470 nm)
3	Triptane	(320; 390)
4	Jet A_1	(340; 370)
5	Kerosene	(320; 355) and (347; 407)
6	West Desert solar	(400; 450) and (440; 520)
7	Balaaiem solar	(400; 457) and (440; 505)

References

[1] Boeker E and Grondelle RV 1995 *Environmental Physics* (New York: John Wiley)
[2] Camagni P, Colombo G, Koechler C, Pedrini A, Omenetto N and Rossi G 1988 *IEEE Transactions on Geoscience and Remote Sensing* GE **26** 22
[3] Dudelzak AE, Babichenko SM, Poryvkina LV and Saar KJ 1991 *Appl. Opt.* **30** 453
[4] John P and Soutar I 1976 *J. Anal. Chem.* **48** 520
[5] Visser H 1979 *Appl. Opt.* **18** 1746
[6] Lloyd J F 1974 *Analyst* **99** 729
[7] Ryder A G, Glynn T J and Feely M 2003 *Proc. SPIE – Int. Soc. Opt. Eng* **76** 1188
[8] Hegazi E, Hamdan A and Mastromarino J 2005 *The Arab. J. for Sci. and Eng.* **30 B** 3
[9] Patra D, Sireesha K L and Mishra A K 2001 *Ind. J. Chem.* A. **40A** 374
[10] Patra D and Mishra A K 2001 *Appl. Spectrosc.* **55** 338
[11] Patra D and Mishra A K 2002 *Anal. Chim. Acta.* **454** 209
[12] Kung R V and Itzkan I 1976 *Appl. Opt.* **15** 409
[13] Visser H 1979 *Appl. Opt.* **18** 1746
[14] Vodacek A and Philpot WD 1987 *Rem. Sens. Envir.* **21** 83
[15] Measures R M 1984 Laser Remote Sensing (New York: John Wiley)
[16] Hoge F E and Swift R N 1988 *Appl. Opt.* **27** 3969
[17] Fadeev V 1992 *SPIE, Conf. Laser study of macroscopic bio-system* 410
[18] Rayner D M and Szabo A G 1978 *Appl. Opt.* **17** 1624
[19] Camagni P, Colombo A, Koechler C, Omenetto N and Rossi G 1991 *Appl. Opt* **30** 26

Journal of Physics: Conference Series **548** (2014) 012060 doi:10.1088/1742-6596/548/1/012060

Particle Image Velocimetry in Confined Vortex Flows

Brian A Maicke[1] **and Joseph Majdalani**[2]

[1] The Pennsylvania State University, 777 W. Harrisburg Pike, Middletown, PA 17057, USA
[2] Auburn University, 211 Aerospace Engineering Building, Auburn, AL 36849, USA

E-mail: `bam49@psu.edu, joseph.majdalani@auburn.edu`

Abstract. In this paper, the challenges of using Particle Image Velocimetry as a diagnostic tool for confined swirling flows are discussed. To highlight these challenges, results from a PIV experiment are compared to a pair of analytical models. While there is generally good qualitative agreement, the velocity measurements degrade in the core of the vortex. Techniques to mitigate the reduced fidelity in this region will be explored and alternative visualization methods will be presented.

1. Introduction

There has been a renewed interest in confined cyclonic motions in the context of aerospace engineering development. Applications for this technology range from hybrid rockets [1, 2] to purely liquid-fueled engines [3, 4]. The benefits of swirl driven flows are also realized in other ways: swirl injectors are often used to increase the combustion stability of flows [5] and swirl-based combustors often have reduced thermal loads on the walls of the chamber [6].

Characterizing the velocity in such confined vortex devices is often a challenging endeavor. Early investigations relied on intrusive measurement devices such as Pitot probes [7] to make measurements but these devices impact the ensuing flow field. The effect is especially significant in confined vortex flows where the fluid motion has components in all three orthogonal directions. Laser Doppler Velocimetry (LDV) and Particle Image Velocimetry (PIV) have reduced the intrusiveness of these measurements but challenges remain, especially in swirling flows where the seeding quality in the core is often inadequate [8]. In this paper, we discuss some of the challenges of laser-based velocity measurement by way of a comparison with a pair of analytical models.

2. Experimental and Analytical Methods

For the sake of geometric flexibility, a modular cylindrical chamber is fabricated from quartz and provides a range of aspect ratios from 2.8 to 8.8. Gaseous nitrogen enters tangentially at the base of the chamber through configurable injectors to provide pressure drops ranging from 10% to 30% of the chamber pressure. A smoke generator seeds the flow with liquid particles tuned to a 0.2 μm diameter.

The imaging system consists of a 250 mJ/pulse Nd:YAG laser with a typical pulse separation of 1 μs to adequately resolve the velocities in the r–θ plane. A LaVision 1280 x 1024 Flowmaster 3 camera captures the particle images at 1 μs increments spaced out over a thirty second run time. The images are then cross-correlated to deduce the swirl velocity in the chamber. The optics used to generate the laser sheet are movable, allowing investigation of the $r - \theta$ plane at three separate axial locations. For additional information on the PIV experiment, the reader is referred to the study by Rom [9].

In order to assess the results of the PIV experiment, two analytical models are compared with the data. The first is a laminar core model derived from the tangential momentum equation with second-order viscous terms retained. Solving the perturbed boundary-layer equation [10] yields an approximate solution of the form

$$u_\theta \simeq \frac{1}{r}\left[1 - \exp\left(-\tfrac{1}{4}Vr^2\right)\right] \tag{1}$$

The second comparison is with the constant shear stress model [11] which assumes a balance between the shear and pressure forces in the core and a free vortex of the $u_\theta = 1/r$ type in the outer region. The solution is a piecewise function, reminiscent of Rankine's [12], but with the benefit of matching both the value of the velocity and the slope where the inner and outer solutions meet. The resulting swirl velocity expression is

$$u_\theta = \begin{cases} \dfrac{r}{X^2}\left[1 - \ln\left(\dfrac{r^2}{X^2}\right)\right]; & r \le X \\ \dfrac{1}{r}; & r > X \end{cases} \tag{2}$$

A modified least-squares regression provides the location of the matching radius, X.

3. Results

The vortex Reynolds numbers observed in these experiments ($V \approx 35000$) are larger than a typical laminar solution. However, as the confined vortex still behaves as a laminar structure, it is possible to correlate the models to the experiment by way of an eddy viscosity ratio, $\ell_t = \mu/\mu_t$, with mu_t denoting a turbulent viscosity. For the constant shear stress model, we write an expression for the matching radius of the constant shear model in terms of the vortex Reynolds number [11]. The resulting vortex Reynolds number definition and matching radius relation may be written as

$$V = \frac{\rho U A_i}{\mu L} = \frac{\rho U A_i}{\mu_t \ell_t L} = \frac{V_t}{\ell_t}; \quad X = \frac{X_0}{\sqrt{V_t}} \tag{3}$$

A least-squares algorithm optimizes ℓ_T and X_0 to produce a matching radius that incorporates the experimental vortex Reynolds number. Average values of $X_0 \approx 50$ and $\ell_t \approx 150$ provide good agreement with the experimental data.

Fig. 1 depicts the experimental data with the analytical models. In the outer region, the data follows the free-vortex shape predicted by both models but the quality of agreement deteriorates in the core. To accommodate the scatter, the fits for both theoretical models are weighted such that they provide an envelope near the centerline rather than a strict regression fit through the data. This weighting is chosen specifically because of the increased scatter that artificially lowers the predicted velocities.

In most cases, the PIV technique constitutes a powerful non-intrusive method for measuring velocities, but the viscous core of swirling flows is difficult to accurately assess even with properly sized particles. There are two significant considerations in the core: increased drag owing to high azimuthal velocities and a tendency for the particles to separate via centrifugal entrainment. This is visible in Fig. 1, as the majority of the data points appear in the inviscid, outer vortex with markedly fewer points in the inner region. Adding to the uncertainty is the fact that the axial velocity is highest at the centerline, perpendicular to the measurement plane; the $r - \theta$ plane measurements are more difficult to correlate since particles that were illuminated in the

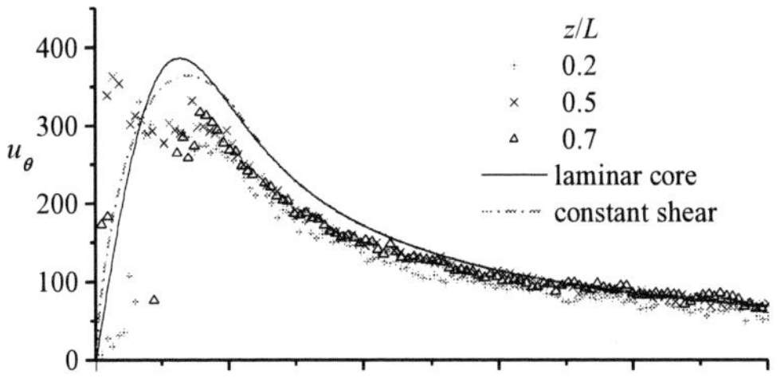

(a) $V = 47150$

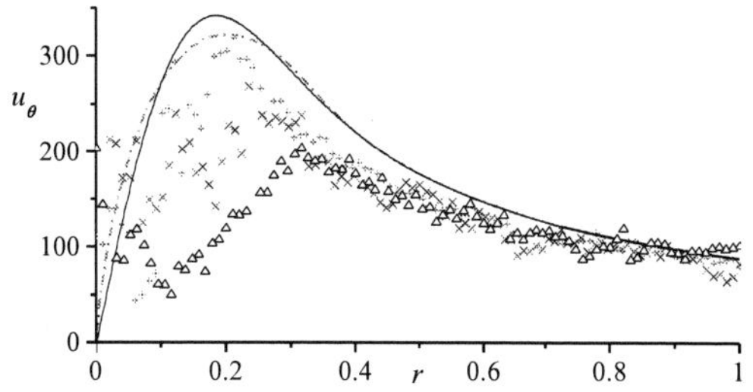

(b) $V = 22370$

Figure 1. Comparison of laminar and constant shear stress models with experimental data.

first exposure may have passed below the light sheet by the second. The combined effect of these factors is decreased confidence in the correlated data near the center of the vortex [13].

Aside from the measurement challenges near the centerline, fully characterizing the structure of this flow requires knowledge of all three velocities. Traditional PIV techniques are unable to simultaneously obtain all three velocity measurements since a traditional PIV setup cannot measure components perpendicular to the light sheet. This deficiency can be mitigated through the incorporation of a second PIV camera and additional optics, resulting in a stereoscopic PIV system. The second camera provides the system with depth of field information, but at the expense of increased equipment cost and longer processing times as images from both cameras require correlation. Additionally, the perpendicular velocity is usually captured at a lower resolution than its in-plane counterparts. Though not as common, it is possible for a single camera to behave as a stereoscopic PIV with the expense of additional optics, though research to improve the technique is ongoing [14].

Recent research has seen an increase in the use of plenoptic, or light-field, cameras in PIV experiments. A plenoptic camera adds an array of micro-lenses before the image sensor which allows the camera to capture both the position and angle of propagation of the incoming light [15]. While the concept is sound, improving the accuracy of the reconstructions from the light field data is still an active avenue of research [16]. Plenoptic PIV is a developing technology but as specialized cameras become more available and data reconstruction improves, it will begin to see more widespread use.

4. Conclusion

Despite its drawbacks, Particle Image Velocimetry offers a vital resource in the quantification of confined vortex flows. Since these flows require fully three-dimensional velocity measurement, improving existing stereoscopic or plenoptic techniques will provide researchers with valuable tools to characterize these structures. Though increased particle drag coupled with centrifugal separation make accurately assessing velocities in the center of these structures a challenge, it is possible to refine existing models to better predict measured swirl velocities.

Acknowledgments

This material is based on work that was supported partly by the National Science Foundation and partly by the Auburn Alumni Engineering Council Endowed Professorship, Auburn University.

References

[1] Gloyer P W, Knuth W H and Goodman J 1993 *CSTAR Fifth Annual Symposium* n96-16953

[2] Knuth W H, Bemowski P A, Gramer D J, Majdalani J and Rothbauer W J 1996 Gas-fed, vortex injection hybrid rocket engine SBIR Phase I Final Technical Report NAS8-40679 NASA Marshall Space Flight Center
[3] Chiaverini M J, Malecki M J, Sauer J A and Knuth W H 2002 *38th AIAA Joint Propulsion Conference* AIAA 2002–4149
[4] Matveev I, Matveeva S and Serbin S 2007 *43rd AIAA Joint Propulsion Conference* AIAA 2007–5628
[5] Huang Y and Yang V 2009 *Prog. Energ. Combust.* **35** 293–364
[6] Majdalani J and Rienstra S W 2007 *Z. angew. Math. Phys.* **58** 289–308
[7] Vatistas G H, Lin S and Kwok C K 1986 *AIAA J.* **24** 635–642
[8] Yan L, Vatistas G H and Lin S 2000 *J. Therm. Sci.* **9** 10–22
[9] Rom C 2006 *Flow Field and Near Nozzle Fuel Spray Characterizations for a Cold Flowing Vortex Engine* Master's thesis University of Wisconsin-Madison Madison, WI
[10] Majdalani J and Chiaverini M J 2009 *Phys. Fluids* **21** 103603–15
[11] Maicke B A and Majdalani J 2009 *P. Roy. Soc. Lond. A Mat.* **465** 915–935
[12] Rankine W J M 1858 *A Manual of Applied Mechanics* 9th ed (London, UK: C. Griffin and Co.)
[13] Hu L Y, Zhou L X, Zhang J and Shi M X 2005 *AIChE J.* **51** 740–749
[14] Bardet P M, Peterson P F and Savas O 2010 *Exp. Fluids* **49** 513–524
[15] Lynch K, Fahringer T and Thurow B 2012 *50th AIAA Aerospace Sciences Meeting* AIAA 2012-1056
[16] Fahringer T W and Thurow B S 2014 *AIAA SciTech* AIAA 2014-0398

X-Ray Diffraction Studies on Material Corrosions in Renewable Energy Storage Electrolyzers

J Mo, S M Steen III and F-Y Zhang

Nanodynamics and High-Efficiency Lab for Propulsion and Power
Department of Mechanical, Aerospace & Biomedical Engineering
UT Space Institute, University of Tennessee, Knoxville, USA

Email: fzhang9@utk.edu

Abstract. As a core component of the proton exchange water electrolyzer system, membrane electrode assemblies degrade due to the corrosion of the material. This creates a loss of interfacial contact necessary for the electron transports and electrochemical reactions, thus decreasing the performance. X-ray diffraction has been demonstrated to be an effective method that readily provides quantitative information about the phase-composition of solid materials. In this study, a group of materials have been selected and tested in the standard conditions for investigating the corrosion mechanisms with X-ray diffraction. The material lattice parameter and the crystal size were examined by X-ray diffraction spectrum.

1. Introduction

With the development of global energy shortage and climate change associated with the use of fossil fuels, it is extremely urgent to replace fossil fuels by a renewable energy source. Hydrogen, as a high energy and environment friendly fuel, is expected to be one of the most potential energy carriers in the near future. Water electrolysis, the most fundamental way, can split water into hydrogen and oxygen by using electric power. The integration of sustainable energy and water electrolysis is very attractive because of its high efficiency and pure production [1, 2].

In the nature of high efficiency, renewable, environment friendly and zero-close emissions, renewable energy storage will play an important role in our future energy economy. Compared to traditional technologies, polymer electrolyte membrane water electrolyzers have several main advantages, including higher energy efficiency/density, faster charging/discharging, and a more compact design. Water electrolysis can be conducted by the proton exchange membrane water electrolyzer (PEMWE) with Ru/Ir and Pt employed as anode and cathode catalysts respectively [3] PEMWEs use proton exchange membranes as the electrolyte that permits the protons to transfer from anode to cathode. Currently, the performance of PEMWE is not sufficient and highly depends on properties of membrane electrode assembly (MEA). To improve the performance and durability of PEMWEs, the catalyst-coated membrane (CCM) of the PEMWE has been studied and developed [4, 5]. For the time being, there is no research that has been done towards the corrosions in the gas diffusion layer (GDL) and MEA which affects the performance of PEMWE.

Carbon paper is widely used as the GDL in fuel cells, which can't be run in the anode side of PEMWEs because it is easily corroded at high positive potentials during water electrolysis operation [6]. In this study, stainless steel mesh will be used as the anode side GDL.

Journal of Physics: Conference Series **548** (2014) 012061 doi:10.1088/1742-6596/548/1/012061

The object of this study is to validate X-ray diffraction (XRD), which can be a tool to quantify the corrosions, choose alternative materials suitable for using in PEMWE and develop a high performance PEMWE.

2. Experimental detail

In the PEMWE as shown in figure 1, both end plates are made of aluminum. The anode current distributor with a parallel flow field is fabricated from a titanium plate, while the cathode current distributor is fabricated from copper and coated with nickel. The cathode flow field is also a parallel flow field that is fabricated from graphite. Both anode and cathode gaskets are made from PVC sheets. The cathode GDL is Toray 090 carbon paper treated with 5% PTFE. The anode GDL is stainless steel 316L mesh. The CCM is Nafion 115 film with Ru/Ir and Pt employed as anode and cathode catalysts loading of 3 mg/cm^2 respectively. Eight evenly distributed bolts assembled the cell to torque of 40 lb/in. The Teflon pipes and fittings were used throughout the system. While the cathode piping was merely intended to safely exhaust hydrogen gas, a diaphragm liquid pump from KNF Neuberger was used to circulate water at a constant volumetric flow rate of 40 ml/min through the anode. The electrolyzer with an active area of 5 cm^2 was operated for 15 hours at room temperature. Before and after the water electrolyzer testing, the leak and crossover tests were performed with air. The airtightness quality of the used CCM was still the same as the fresh one. No pinholes were found on the used CCM.

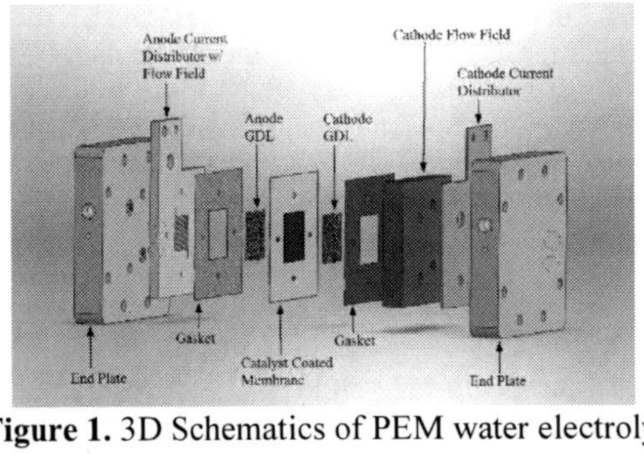

Figure 1. 3D Schematics of PEM water electrolyzer.

XRD is a tool used for identifying the atomic and molecular structure of a crystal, in which the crystalline atoms cause a beam of incident X-rays to diffract into many specific directions. The characterization of structure and identification of the phase were carried out by XRD with a Philips X'Pert materials research diffractometer (45kV, 40mA), controlled by PANalytical's XRD software, in Bragg Brentano reflection geometry with Cu Kα radiation (λ=1.5418 Å) and a 2θ scan from 20° to 100° (at 0.01° per 5 second). The diffraction patterns are analyzed by software MDI Jade9. The test system mainly consists of Panalytical Philips X'Pert MRD system as show in Figure 2.

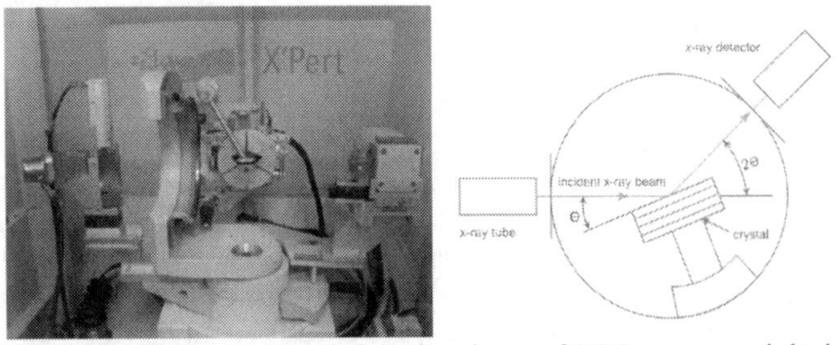

Figure 2. Philips X'Pert MRD System (left is the photo of XRD system, right is the schematic).

The applications of XRD, which related to this study, are as follows. The first one is to identify the composition of material. The sample for anode GDL is stainless steel mesh, which is made of stainless steel 316. XRD pattern can identify the major element in this sample which are the same as described in the reference [7]. The second one is to determine crystal structural properties. The grain size can be calculated with the Scherrer Equation.

$$\tau = \frac{K\lambda}{\beta cos\theta} \quad . \tag{1}$$

Here, τ is the mean size of the ordered domains, which equal to the grain size, K is a dimensionless shape factor, has a typical value of about 0.94, λ is the X-ray wavelength, β is the line broadening at half the maximum intensity and θ is the Bragg angle.

3. Results and discussion

As mentioned above, the gran size of major crystal in each layer can be calculated by Scherrer equation. In the XRD scanning, all necessary data for the major crystals for the calculations of grain sizes have been collected. The data are shown below in Figure 3, Figure 4 and Figure 5, respectively.

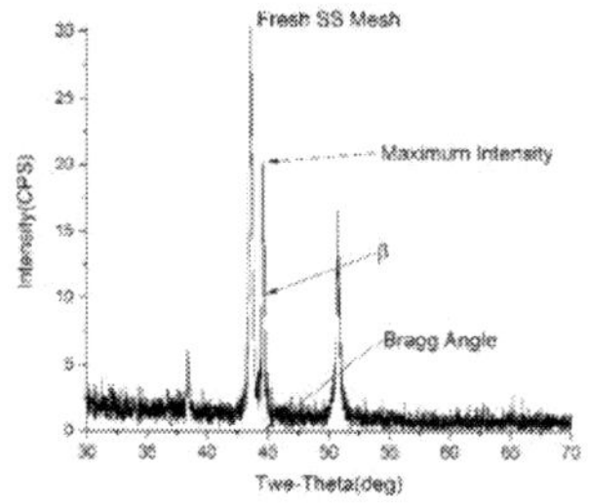

Figure 3. XRD pattern for anode side GDL stainless steel mesh.

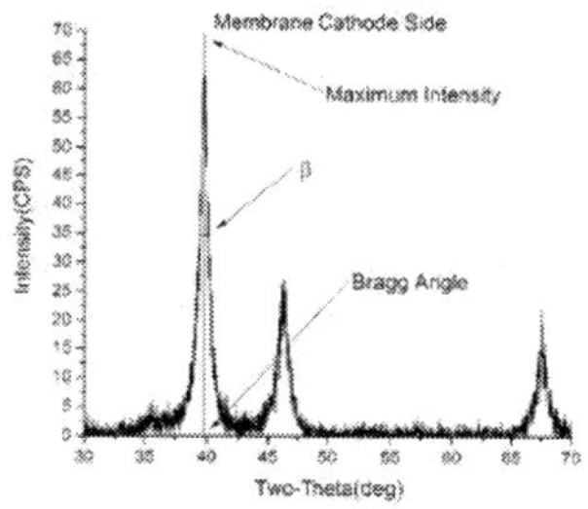

Figure 4. XRD pattern for anode side of CCM.

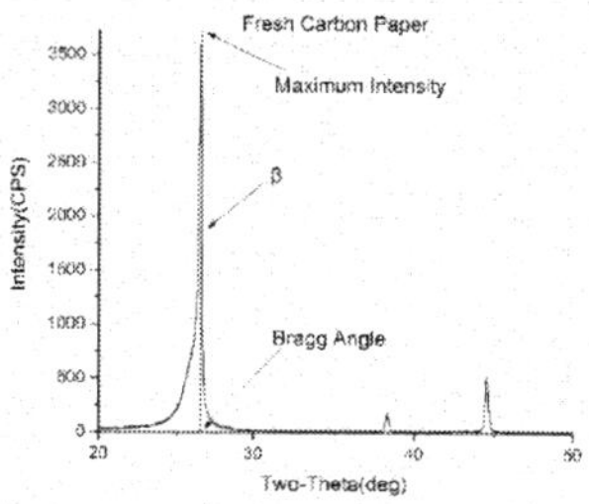

Figure 5. XRD pattern for cathode side GDL fresh carbon paper.

Based on the above data, the grain size of major crystal in fresh sample can be calculated as shown in Table 1.

Table 1. The grain size of major crystal in each layer.

Major Crystal	Grain Size
Graphite in Carbon Paper	34.12nm
Iron in Stainless Steel Mesh	29.81nm
Pt in Cathode side of Membrane	11.32nm

The major crystals at anode side GDL fresh stainless steel mesh is Fe, Cr and Ni. After testing in the PEMWE, the major crystal in carbon paper GDL on cathode side is graphite-2H.

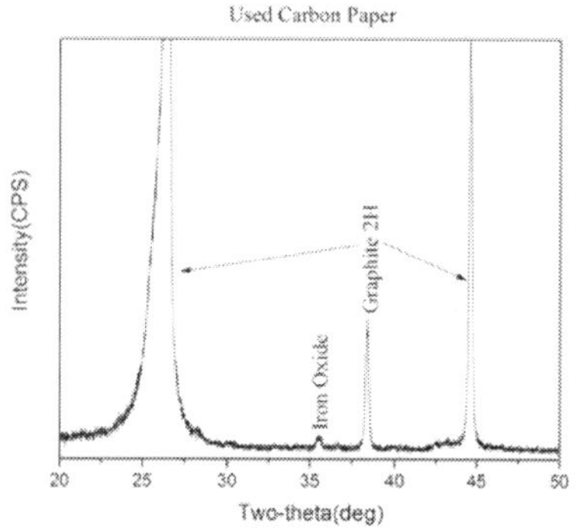

Figure 6. XRD pattern of cathode GDL use carbon paper.

The occurrence of corrosion at the anode side GDL is due to the iron exposed in a high oxidation and high potential environment; however, the mechanism of corrosion in the cathode GDL is unclear. Since the environment in cathode side is not the same as anode side. It can be assumed that the iron oxide diffused from the anode to the cathode and attached onto the surface of the carbon paper GDL. Based on the assumption, the XRD was used to further examine the material.

After testing all of the materials in the MEA, the XRD patterns have been compared respectively for anode GDL, catalyst coated membrane and cathode GDL. As shown in Figure 6, there is some iron oxide that exists in the used carbon paper compared to fresh one. The corrosion occurred in the anode GDL, this is due to in the extremely harsh environment, the iron changes to iron oxide. In this water electrolysis process, the anode side GDL stainless steel mesh is exposed in the flow to DI water. The iron oxide is easily detached from the substrate [7]. This phenomenon also verified the assumption that the iron oxide not only washed away by water flow but also diffused from the anode side to the cathode side through the membrane.

4. Summary

XRD technology has been demonstrated to quantify the major phases of materials before and after testing, to characterize the effect of oxidation on GDLs, to verify the diffusion of iron oxide through the membrane, has also successfully tracked corrosion pathways from the anode side to the cathode side through the membrane. To find more applications of XRD on PEMWE research, more works will be conducted in the future.

Acknowledgements

The authors acknowledge the support from U.S. Department of Energy under Award DE-FE0011585. The authors also wish to express their appreciations to Douglas Warnberg, Dr. Bo Han, Rong Chen and Aaron Liu for their help.

References

[1] Grigoriev S A, Porembsky V I, Fateev V N 2006 *International Journal of Hydrogen Energy* **31** 171
[2] Marshall A, Borresen B, Hagen G, Tsypkin M, Tunold R 2007 *Energy* **32** 431
[3] Xu W, Scott K, Basu S 2011 *Journal of Power Sources* **196** 8918
[4] Song S, Zhang H, Ma X, Shao Z, Baker R T, Yi B 2008 *International Journal of Hydrogen Energy* **33** 4955
[5] Ma L R, Sui S, Zhai Y C 2009 *International Journal of Hydrogen Energy* **34** 678
[6] Song S D, Zhang H M, Ma X P, Shao Z G, Zhang Y N, Yi B L 2006 *Electrochemistry Communications* **8** 399
[7] Luo X, Tang R, Long C S, Miao Z, Peng Q, Li C 2008 *Nuclear Engineering and Technology* **40** 147

Journal of Physics: Conference Series **548** (2014) 012062 doi:10.1088/1742-6596/548/1/012062

Energy Dispersive X-Ray and Electrochemical Impedance Spectroscopies for Performance and Corrosion Analysis of PEMWEs

S M Steen III and F-Y Zhang[1]

Nanodynamics and High Efficiency Lab for Propulsion & Power
Department of Mechanical, Aerospace and Biomedical Engineering
UT Space Institute, University of Tennessee, Knoxville

[1]Email: fzhang9@utk.edu

Abstract. Proton exchange membrane water electrolyzers (PEMWEs) are a promising energy storage technology due to their high efficiency, compact design, and ability to be used in a renewable energy system. Before they are able to make a large commercial impact, there are several hurdles facing the technology today. Two powerful techniques for both *in-situ* and *ex-situ* characterizations to improve upon their performance and better understand their corrosion are electrochemical impedance spectroscopy and energy dispersive x-ray spectroscopy, respectively. In this paper, the authors use both methods in order to characterize the anode gas diffusion layer (GDL) in a PEMWE cell and better understand the corrosion that occurs in the oxygen electrode during electrolysis.

1. Introduction

Proton exchange membrane water electrolyzers (PEMWEs) are a promising energy storage technology, highlighted by their high efficiency (theoretical maximum of 94%), compact design, rapid charging/discharging, and ability to be a part of a renewable and regenerative energy storage system [1]. With the onset of climate change, the need for clean energy is greater than ever before, and PEMWEs are a promising source of clean energy storage. A PEMWE is operated by electricity, and uses a catalyst coated membrane (CCM) in order to split water into hydrogen and oxygen [2-3]. These elements may then be stored easily using existing methods, and be used as the fuel and oxidizer for fuel cells, respectively.

Currently, there are several challenges the development of commercial scale PEMWEs are faced with. A vital component of the membrane electrode assembly (MEA), the gas diffusion layer (GDL) is responsible for facilitating the transport of reactants/products, electrons, and heat to and from the catalyst layer with minimal voltage, current, thermal, interfacial, and fluidic losses [4-6]. The GDL is also needed to provide structural integrity to the cell, and needs to be a porous material while still maintaining a good contact area with the catalyst in the membrane. For PEM fuel cells, carbon-based materials (graphite electrode, carbon cloth, and carbon paper) are desired and well suited as the material of choice for the GDL. However, the oxygen electrode (anode) in a PEMWE has a harsh, corrosive environment due to the high overpotential. In this environment, the carbon-based materials described above are ill-suited, as the material will corrode and degrade easily, leading to poor performance in the cell. As such, highly conductive and corrosion-resistant materials are desired for use as the anode GDL in a PEMWE. To decide which materials may be desirable, a better understanding of the mechanics of the corrosion at the oxygen electrode is needed.

For our research, there are two powerful methods of analysis that may be used to investigate the corrosion that occurs in a PEMWE. They are electrochemical impedance spectroscopy (EIS), and energy dispersive x-ray spectroscopy (EDX). For this study, both of these methods were employed in order to perform *in-situ* and *ex-situ* analysis on a PEMWE developed in lab, and from the results of the analysis were able to reach a greater understanding of the mechanisms of corrosion that occur in a PEMWE.

2. Experimental details

For the *in-situ* investigation performed, a test system that consisted of an electrolyzer test stand, EIS, a flow control and measurement system, and a hydrogen evacuation setup was used. The cell was attached to the control system with a current range up to 100A and a voltage range up to 5V. The hardware was connected and used in tandem with Bio-Logic's EC-Lab software. This was used to conduct performance and efficiency testing, as well as EIS. For the flow control, a system of plastic tubing was connected to the PEMWE. While the cathode piping was for the safe exhaustion of hydrogen gas that formed during electrolysis, the anode piping used a diaphragm liquid pump from KNF Neuberger to circulate deionized (DI) water at a constant volumetric of 40 ml/min through the anode.

The PEMWE cell was assembled using parts fabricated in lab. Two end plates were fashioned from commercial grade aluminum and used to provide an even compression pressure to the cell. The anode bipolar plate used a titanium plate with a parallel flow field which also served as the current distributor. The cathode bipolar plate used a copper plate as the current distributor, and a graphite plate with a parallel flow field. The cathode GDL was Toray 090 carbon paper treated with 5% PTFE, while the anode GDL was a mesh of 316L stainless steel. The membrane was Nafion 115 with RuIrOx and PtB catalysts with a 3 mg/cm^2 loading on the anode and cathode, respectively. The active area of the membrane was 5 cm^2. Gaskets for both GDLs were fashioned from PVC in order to improve the sealing of the cell. Once assembled, eight bolts were used to evenly tighten the cell to 40 lb-in. of torque. The cell was operated at room temperature.

3. Results and discussion

3.1 Electrochemical impedance spectroscopy

For the *in-situ* testing, galvanostatic EIS (GEIS) is desirable, because it is a macroscopic electrochemical technique that may be used while the cell is performing electrolysis. It is the most useful for studying the electrode-electrolyte interface as the well as analyzing corrosion that may occur in a PEMWE. From performing GEIS, the ohmic, activation, and transport resistances in a specific electrolyzer may be determined respectively. In a typical GEIS test a frequency range from 10 kHz – 0.1 Hz is performed, with the current applied to the cell remaining constant. The ohmic resistance typically dominates the losses in performance for the PEMWEs of interest. The activation impedance can be derived from the slope of the arc in the Nyquist Plot in the middle of frequency range, and the transport resistance may be found from the difference between the values of the x-axis intercepts at the low frequency range. Figure 1 below shows an example of data that may normally be obtained from GEIS analysis, plotted in the complex plane.

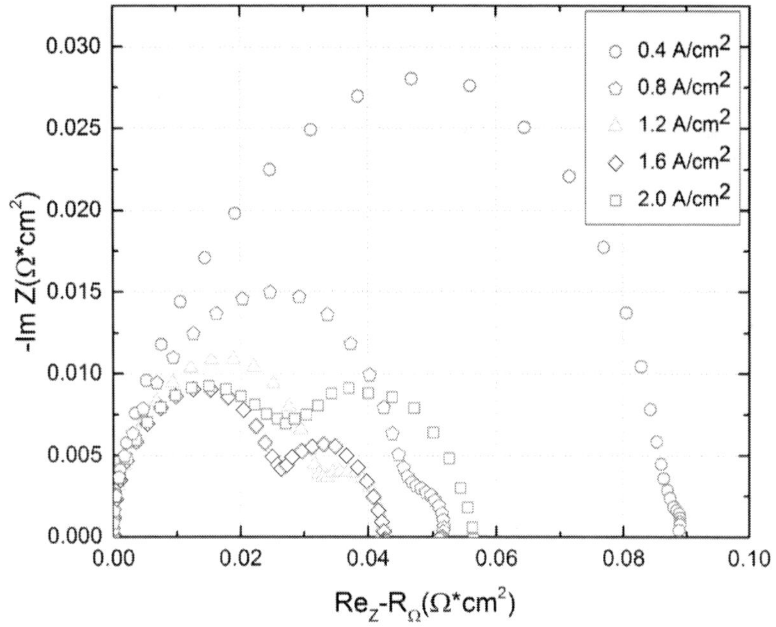

Figure 1. Nyquist plot showing the same GDL in a PEMWE measured under different operating conditions

In the figure, the real and imaginary axis are normalized by multiplying the values measured by the active area of the cell. There is one more step of normalization, where the real axis values are further normalized by subtracting each value by the ohmic resistance. In a Nyquist plot, the ohmic resistance is the value of the leftmost x-axis intercept at high frequency end.

GEIS allows for conclusions regarding the nature of the GDL in the anode of a PEMWE to be reached. As the interfacial contact increases and the thickness of the GDL decreases, the ohmic impedance greatly decreases. From other GEIS analysis performed during testing it is shown that even though some GDLs may have a low porosity (<40%), their favorable interfacial contact leads to a low ohmic resistance and thus a higher performance.

Figure 1 shows that the transport resistance in a PEMWE varies depending on the current density the cell is run at. The transport resistance is denoted as the rightmost x-axis intercept in low frequency range. The reason for the change in transport resistance at different current density is because there is a greater liquid water consumption and the difference in H_2/O_2 production between each current density.

3.2. Energy dispersive x-ray spectroscopy

After testing of the PEMWE has concluded, the cell may be disassembled and a further investigation of the GDL may occur. This can be done using SEM microscopy in tandem with EDX. The EDX detector used is from EDAX, and uses their TEAM software to perform both image processing and mapping of the sample of interest. EDX allows for viewing the physical characteristics of the materials before and after testing, and an analysis of the elements that make up the sample. Figure 2 shows images of the cathode GDL before and after testing. As seen in Figure 2B, there is a large amount of contamination on the used sample of carbon paper. From EDX analysis, it was found that this contamination was iron oxide. In the PEMWE used for testing, the only possible source of iron oxide was from the anode GDL.

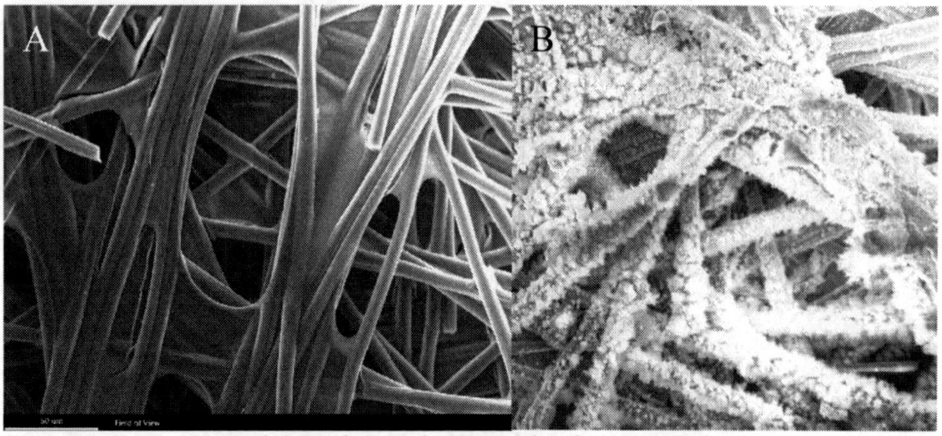

Figure 2. Toray 090 CP (A) Fresh Sample (B) Used and corroded sample

Table 1 below shows the atomic concentration of elements found in both the fresh and used GDL samples. An example of the spectrum found during EDX analysis is shown in Figure 3. As mentioned above and seen in Table 1 below, there is a large amount of iron and oxygen found on the used carbon paper. After testing concluded, testing was performed on the membrane to test for any leakage, but none were found. This leads us to believe that iron contaminants from the anode GDL migrated through the membrane and attached to the cathode GDL. While the mechanism of this reaction is unknown, further research is planned to better understand the phenomenon.

Table 1. Atomic concentration of elements in fresh & used GDLs

Test sample	Atomic concentration					
	Iron	Carbon	Fluorine	Oxygen	Chromium	Nickel
Fresh SS	60.27	0	0	12.67	15.63	11.43
Used SS	18.21	26.28	0	47.29	5.55	2.67
Fresh CP	0	95.8	4.2	0	0	0
Used CP	28.45	19.25	.89	51.41	0	0

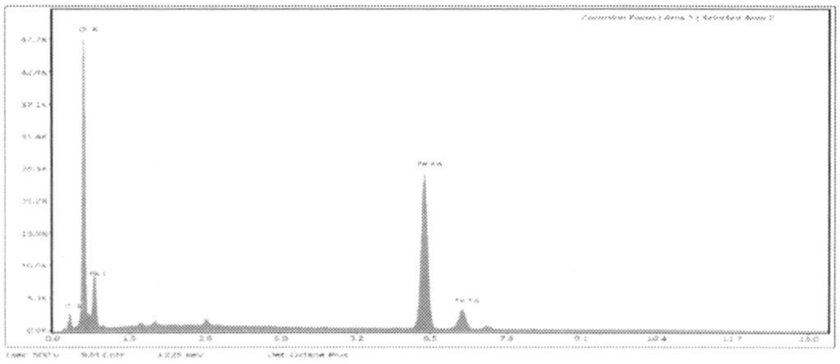

Figure 3. Example of spectrum found during EDX analysis

4. Conclusion

From the testing conducted, impedance spectroscopy allows for the *in-situ* investigation of the effects of porosity and thickness of the anode GDL on the performance and efficiency of a PEMWE cell. In addition, GEIS has been demonstrated to characterize the interfacial contact between the GDL and the CCM. SEM and EDX spectroscopy, when used together, characterize the effects of oxidation on

GDLs, and provide power tools to verify the diffusion of iron oxide through the membrane. The exact diffusion mechanism will be explored in future research.

Acknowledgements
The authors acknowledge support from the U.S. DOE under Award DE-FE0011585. The authors also wish to express gratitude to Douglas Warnberg, Jingke Mo, Dr. Bo Han, Rong Chen, Dr. Lino Costa, and Kate Lansford for their help.

References

[1] Ayers K E, Dalton L T, Anderson E B 2012 *Electrochemical Synthesis of Fuels* **41** 27
[2] Park S, Lee J-W, Popov B N 2006 *Journal of Power Sources* **163** 357
[3] Marshall A, et al. 2007 *Energy* **32** 431
[4] Wilde P M, et al. 2004 *Fuel Cells* **4** 180
[5] Eom K, et al. 2013 *International Journal of Hydrogen Energy* **38** 6249
[6] Prasanna M, et al. 2004 *Journal of Power Sources* **131** 147

Ultra-narrow spectroscopic cells in atomic spectroscopy: reflection, transmission, fluorescence, and nonadiabatic transitions at the walls

A Pazgalev[1], D Sarkisyan[2], S Cartaleva[3], S Przhibelskii[4], T Vartanyan[4]

[1]Ioffe Phys.-Technical Institute of the Russian Academy of Sciences, Politekhnicheskaya st. 26, St. Petersburg, 194021, Russian Federation
[2]Institute for Physical Research, NAS of Armenia, Ashtarak-0203, Armenia
[3]Institute of Electronics, BAS, boul. Tzarigradsko shosse 72, 1784 Sofia, Bulgaria
[4]ITMO University, Kronverkskiy pr. 49, St. Petersburg 197101, Russian Federation

E-mail: Tigran.Vartanyan@mail.ru

Abstract. Ultra-narrow cells with the thicknesses in the range from several wavelengths to the small fractions of the wavelength brought a number of new opportunities for atomic spectroscopy. Depending on the cell thickness, spectral lines recorded in ultra-narrow cells are either Doppler-free or Doppler-broadened. With careful selection of the cell thickness hyperfine structure may be easily resolved without resorting on the multibeam nonlinear optical techniques. Moreover, frequent collisions with the walls leads to the important modifications of velocity selective optical pumping resonances. Finally, ultra-narrow cells provide with the unique opportunity to study collisions of the excited atoms with the solid surfaces. In this contribution several examples of the use of the ultra-narrow spectroscopic cells filled with the alkali atomic vapour is presented. First, we discuss general aspects of the transient polarisation that defines all peculiarities of an ultra-narrow cell as a spectroscopic tool. Second, we demonstrate the resolution of the magnetic sublevels in the transition from Zeeman to Paschen-Back regime in the Cs hyperfine structure. Third, new aspects of velocity selective optical pumping resonances in reflection and transmission of resonant radiation by the 6 wavelengths thick cell filled with Cs are discussed. Forth, the experimental evidences of the nonadiabatic transitions between excited states of Rb atoms in the course of collisions with the sapphire surface are presented.

1. Transients in the atomic spectroscopy with ultra-narrow cells

Transient behavior of the atomic polarization under resonance electromagnetic excitation is a very well understood process that gave rise to a number of well-known time-resolved spectroscopic techniques. The role played by the transients in the stationary spectroscopy is much less appreciated although it was first clearly mentioned in 1954 by Cojan who observed sub-Doppler features in the reflection spectrum of resonant radiation from the mercury vapor. His interpretation [1] of this unexpected result was based on the statement that the atoms departing from the surface are in a transient state rather in the steady state. Hence, there are no reasons to describe their response by the steady state polarization corresponding to their Doppler-shifted resonance frequency. In the following year the first "ultra-narrow" pill-box shaped cell was built by Romer and Dicke [2] for ammonia interacting with microwave radiation. It was 6.2 mm thick, exactly one halve of the resonant wavelength. In this cell sub-Doppler line widths were obtained. Although the transient behavior of atomic polarization was not calculated in Ref. [2], it contains an important statement that would the cell be thicker than one halve of the wavelength the phase of the field acting on the moving atoms will have opposite phase in different parts of the cell.

The role of transients in reflection spectroscopy was further elucidated in the linear regime in Ref. [3]. It was stressed that due to their transient behavior after collision with the wall departing atoms mimic the contribution of arriving atoms that have same absolute value of their velocity, thus effectively reducing the vapor polarization to the doubled contribution of arriving atoms. Nonlinear theory of selective reflection with the full account of transient effect was given in Ref. [4].

Theoretical treatment of reflection and transmission of the narrow cell filled with the resonant atomic vapors with the full account for the transients was given in Ref. [5]. It was shown that the line shapes depend periodically on the cell thickness. The period of these changes is twice as large as compared to the period of Fabri-Perrot resonances.

First experimental realization of a really ultra-narrow cell suited for optical spectroscopy was reported in Ref. [6]. It has variable thickness of the Cs atomic vapor layer in the range of 150-300 nm and enables full resolution of hyperfine transitions that are usually covered by Doppler broadening. Thickness dependence of the narrowing effect was studied in more detail in Ref. [7].

2. Giant modification of atomic transitions probabilities induced by magnetic field

Ultra-narrow cell filled with Cs vapour was used to study the magnetic field-induced giant modification of probabilities for seven components of $6S_{1/2}$, $F_g=3 \rightarrow 6P_{3/2}$, $F_e=5$ transition of Cs D2 line forbidden by selection rules. Figure 1 plots the recorded absorption spectrum of Cs in an ultra – narrow cell. The fourteen transition components appear with ≅100 MHz linewidth, thus being completely frequency resolved except for transitions 5, 6 and 7′ resolved partially, and 6′ and 7, which are fully overlapped. As it is seen from the inset, the amplitudes of $F_g=3$, $m_F=-3 \rightarrow F_e=5$, $m_F=-2$ and $F_g=3$, $m_F=-2 \rightarrow F_e=5$, $m_F=-1$ transitions (labeled 7′ and 6′,respectively) are the largest ones among the group of fourteen lines.

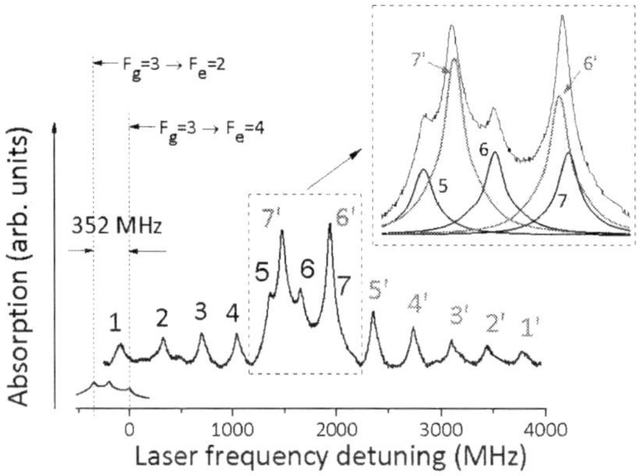

Figure 1. Absorption spectrum of Cs cell with $L=\lambda/2=426$ nm for magnetic field B = 920 G and σ+ laser excitation. Seven transitions $F_g=3 \rightarrow F_e=5$ are labeled **1'-7'**, while other seven transitions $F_g=3 \rightarrow F_e=4$ are labeled **1-7.**

The lower curve is absorption spectrum of the reference cell with $L=\lambda/2$, showing the positions of the Cs $F_g=3 \rightarrow F_e=2,3,4$ transitions for B = 0.

The inset shows the result of the fitting of the spectrum fragment enclosed in a dashed rectangle, where atomic transitions are partially overlapped.

3. Comparison of reflection and transmission spectra in Cs vapor

High resolution laser spectroscopy of Cs vapor confined in optical cell with longitudinal dimension L=6λ (λ=852 nm) provides good opportunity to study the sub-Doppler-width resonances like Velocity Selective Optical Pumping (VSOP) resonance by single beam spectroscopy. For such a cell, the mean time of atomic flight between the cell windows is close to the lifetime of the excited atomic levels. In addition, the slight enhancement of the cell thickness from L=0.5 λ to L=6 λ provides with the possibility for strong reduction of the used laser light intensity and allows working at lower atomic source temperature [8]. Both are of significant importance for the VSOP resonance study.

Journal of Physics: Conference Series **548** (2014) 012063 doi:10.1088/1742-6596/548/1/012063

Here, we present experimental results related to some peculiarities in VSOP resonance behavior observed for closed hyperfine transition on the D_2 line of ^{133}Cs. Most of the optical transitions involved in this line are open, suffering atomic population loss due to the fluorescence decay to hyperfine or Zeeman ground-state levels, which do not interact with the exciting laser light. Only the F_g=4→F_e=5 transition is a closed one, i.e. it exhibits fluorescence decay only to ground state levels that interact with the laser light. The two types of transitions demonstrate different saturation behaviors [9].

Figures 2 and 3 illustrate the transmission and reflection spectra, respectively. Surprisingly, it was observed that for the same (very low) light intensity and similar temperature interval, in the absorption spectrum observed by the reflected light the VSOP resonance sign at the closed transition is opposite, namely absorption dip, with centre at the $F_g = 4 \rightarrow F_e = 5$ transition (Fig.3). Moreover, the contrast of the VSOP dip is very good, even at the lowest used atomic density. Note that the shape of the resonance at the closed optical transition differs from those formed at both open transitions. More particularly in the first case, the resonance profile is asymmetric.

The difference in the absorption spectra of transmitted and reflected beams could be related to the fact that in the reflected light the contribution of atoms flying within the two layers with thickness L=λ, which are located right next to the cell window surfaces is larger than for the case of transmitted light beam. It could be assumed that such atoms undergo stronger depolarization of the excited F_e=5 state due to the influence of dielectric surface. This can result in registration of reduced absorption VSOP resonance even at lower temperature, for the absorption spectrum measured by the reflected beam.

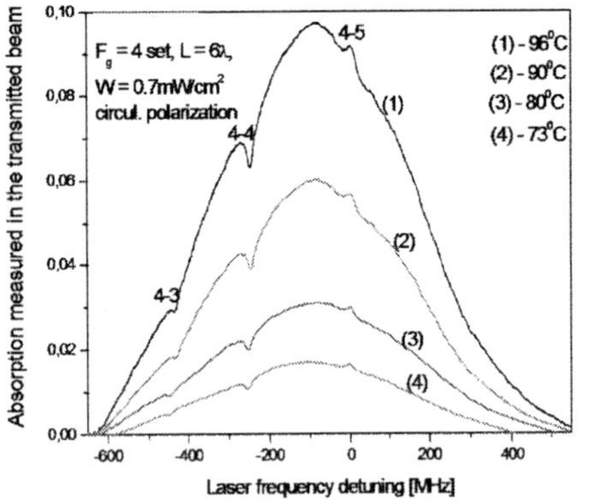

Figure 2. Absorption spectra measured by the laser beam transmitted through the L=6 λ cell, for different Cs source temperatures. VSOP resonances of reduced absorption occur for both open F_g=4 → F_e=3,4 transitions, while absorption peak is present at the closed F_g=4 → F_e=5 transition.

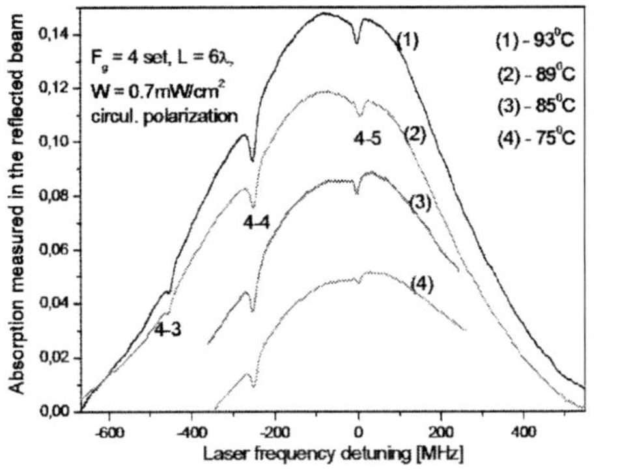

Figure 3. Absorption spectra observed by measuring the signal reflected from both inner surfaces of cell windows. Here all 3 resonances exhibit reduced absorption features with sub-Doppler spectral widths. Curves (1,2) present the full absorption profiles, while the (3,4) are zooms registered at related lower Cs source temperatures.

4. Nonadiabatic transitions between excited states of rubidium atoms

Ultra-narrow cell filled with Rb vapour was excited by two diode lasers at the wavelengths resonant to the transitions between the ground state $5^2S_{1/2}$, the first excited state $5^2P_{3/2}$ and the upper excited state $5^2D_{5/2}$ in sequence. An intense luminescence at the wavelength of 420 nm corresponding to the transition from $6^2P_{3/2}$ state to the ground state was observed. The luminescence intensity at the wavelength of 420 nm was found to be proportional to the product of the intensities of both laser sources. In macroscopically thick cells this luminescence follows after the radiative transition between $5^2D_{5/2}$ and $6^2P_{3/2}$ [10] at the wavelength of 5.2 μm. Taking into account that this radiative process takes 707 ns, that the time of flight of an atom between the walls in the ultra-narrow cell is only 1 ns, and that the probability of quenching of electronic excitation upon collision with the solid surface is very near to unity [11], one concludes that the nonadiabatic transition takes place in the course of the atom-wall collision. Fig. 4 displays luminescence intensity versus vapour thickness.

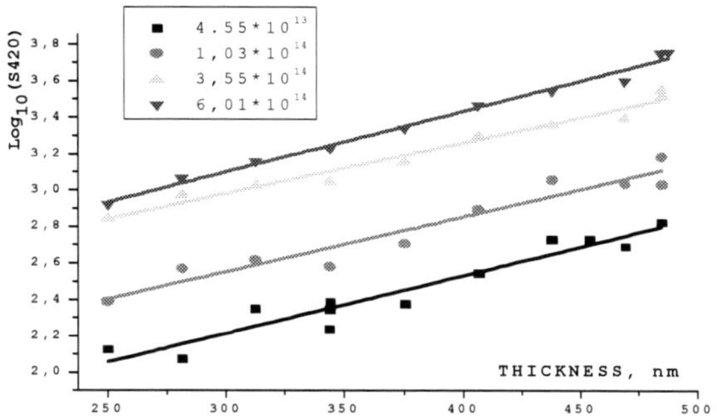

Figure 4. The thickness dependence of the intensity of luminescence at 420 nm measured for several densities of rubidium vapour. The data may be approximated by the power low dependence with the exponent of about 2.3. This highly nonlinear dependence on the layer thickness is in accord with the large probability of quenching at the surfaces and the transient character of the excitation process.

Acknowledgments

This work was supported jointly by the State Committee for Science, Ministry of Education and Science of the Republic of Armenia (grant no. 13 RF-066) and the Russian Foundation for Basic Research (project no. 13-02-90601 Arm_a), by Government of Russian Federation (Grant 074-U01) and Marie Curie Staff Exchange Scheme (PIRSES-GA-295264).

References

[1] Cojan J L 1954 *Ann. Phys.* (Paris) **9** 385
[2] Romer R H and Dicke R H 1955 *Phys. Rev.* **99** 532
[3] Schuurmans M F H 1976 *J. Phys.* (Paris) **37** 469
[4] Vartanyan T A 1985 *Sov. Phys. JETP* **61** 674
[5] Vartanyan T A and Lin D L 1995 *Rhys. Rev.* A **51** 1959
[6] Sarkisyan D, Bloch D, Papoyan A, and Ducloy M 2001 *Opt. Comm.* **200** 201
[7] Sarkisyan D., Varzhapetyan T., Sarkisyan A., Malakyan Yu., Papoyan A., Lezama A., Bloch D. and Ducloy M. 2004 *Rhys. Rev.* A **69** 065802
[8] Cartaleva S, Krasteva A, Moi L, Sargsyan A, Sarkisyan D, Slavov D, Todorov P and Vaseva K 2013 *Quantum Electronics* **43** 875
[9] Cartaleva S, Saltiel S, Sargsyan A, Sarkisyan D, Slavov D, Todorov P and Vaseva K 2009 *J. Opt. Soc. Am. B* **26** 1999
[10] Akulshin A, Budker D, McLean R, 2014 *Optics Letters* **39** 845
[11] Przhibelskii S G and Khromov V V 2000 *Optics and Spectrosc.* **88** 17

XXII International Conference on Spectral Line Shapes 2014 IOP Publishing
Journal of Physics: Conference Series **548** (2014) 012064 doi:10.1088/1742-6596/548/1/012064

Hydrodynamic Stability Analysis of Particle-Laden Solid Rocket Motors

T S Elliott[1,2] and J Majdalani[3]

[1] Department of Mechanical, Aerospace, and Biomedical Engineering, Ph.D. Candidate, University of Tennessee Space Institute, Tullahoma, TN 37388, USA
[2] Department of Mechanical Engineering, Adjunct Instructor, University of Tennessee at Chattanooga, Chattanooga, TN 37403, USA
[3] Department of Aerospace Engineering, Auburn Alumni Engineering Council Endowed Professor and Chair, Auburn University, Auburn, AL 36849, USA

E-mail: trevor-elliott@utc.edu ; joe.majdalani@auburn.edu

Abstract. Fluid-wall interactions within solid rocket motors can result in parietal vortex shedding giving rise to hydrodynamic instabilities, or unsteady waves, that translate into pressure oscillations. The oscillations can result in vibrations observed by the rocket, rocket subsystems, or payload, which can lead to changes in flight characteristics, design failure, or other undesirable effects. For many years particles have been embedded in solid rocket propellants with the understanding that their presence increases specific impulse and suppresses fluctuations in the flowfield. This study utilizes a two dimensional framework to understand and quantify the aforementioned two-phase flowfield inside a motor case with a cylindrical grain perforation. This is accomplished through the use of linearized Navier-Stokes equations with the Stokes drag equation and application of the biglobal ansatz. Obtaining the biglobal equations for analysis requires quantification of the mean flowfield within the solid rocket motor. To that end, the extended Taylor-Culick form will be utilized to represent the gaseous phase of the mean flowfield while the self-similar form will be employed for the particle phase. Advancing the mean flowfield by quantifying the particle mass concentration with a semi-analytical solution the finalized mean flowfield is combined with the biglobal equations resulting in a system of eight partial differential equations. This system is solved using an eigensolver within the framework yielding the entire spectrum of eigenvalues, frequency and growth rate components, at once. This work will detail the parametric analysis performed to demonstrate the stabilizing and destabilizing effects of particles within solid rocket combustion.

1. Introduction

It is generally accepted that the work of Varapaev and Yagodikin [1] has led the way in hydrodynamic stability investigations of rocket chambers. Over time several dedicated researchers have focused on hydrodynamic stability models and their ability to predict oscillations in solid rocket flowfields, such as Casalis, Avalon, and Pineau [2], Ugurtas *et al.* [3], Féraille and Casalis [4], Elliott, Batterson, and Majdalani [5], and Boyer *et al.* [6, 7]. Instabilities such as parietal, obstacle, and angle vortex shedding have been identified in these models where a link has been made between experimentally reported frequencies and stability eigenmodes. Cold-flow measurements obtained using two experimental facilities, VECLA (*Veine d'Etude de la Couche Limite Acoustique*) and VALDO (*Veine Axisymmétrique pour Limiter le Développement des Oscillations*), have been used to validate findings of the aforementioned hydrodynamic stability analyses [8]. The most noteworthy of these solid rocket motor (SRM) instability analyses can be attributed to Chedevergne and Casalis [9-11] with their biglobal framework.

The present work aims to advance previous biglobal stability analyses through the creation of a framework which can analyze particle-laden rocket flows. It has been long understood that the addition of particles increases specific impulse and suppresses high frequency instabilities. Pioneering work in this area can be attributed to Saffman [12], who extended the conventional single-phase stability theory. Following Saffman, Féraille and Casalis [4] have applied two-phase theory to the local stability approach using Stokes drag force to couple the gaseous and particle phases. The present work addresses the particle mass concentration in a different manner, thus resulting in a semi-analytical solution. In what follows, the biglobal equations are developed using the semi-analytical solution for particle mass concentration and mean flow using a self-similar form. The result is a framework to analyze particle-laden rocket internal motion. In short, the addition of particles will be shown to have both stabilizing and destabilizing effects as it will be described in the results and discussion section.

2. Physical geometry

The solid rocket core flow field is the primary geometry of interest for this study as illustrated in Figure 1. The motor is idealized as a cylindrical chamber of length L and radius a. Spherical and chemically-inert particles are injected with combustion gases from the sidewall. Furthermore, it is assumed that the density of the gaseous phase is sufficiently smaller than the particle phase, thereby permitting the use of the Stokes approximation as performed by Elliott and Majdalani [13].

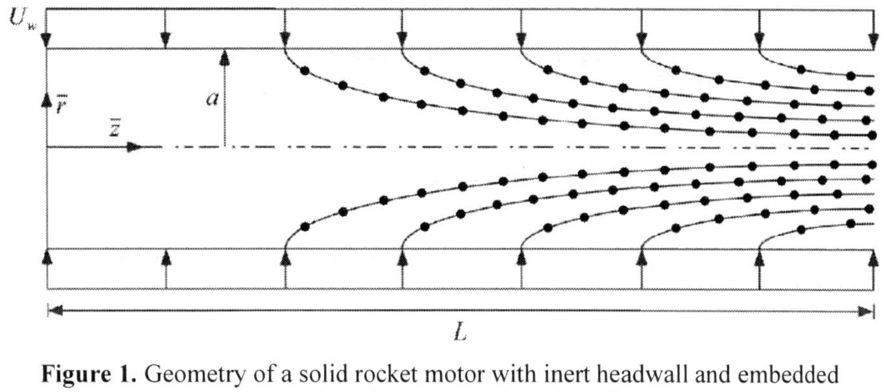

Figure 1. Geometry of a solid rocket motor with inert headwall and embedded particles released from the sidewall.

3. Biglobal equations

As with past biglobal analysis, the fundamental equations are normalized using standard terminologies, where coordinates are normalized by the chamber radius and velocities by a constant injection speed [5, 13]. The instantaneous parameters are then decomposed, leading to the first order perturbation, $\tilde{M} = M + \breve{m}$. Here M is the mean flowfield, or bulk fluid parameter, and $\breve{m}$ is the hydrodynamic wave. With the mean flowfield known, the $O(M)$ terms are omitted and to reduce the system the second order $O(\breve{m}^2)$ terms are neglected. The biglobal Ansatz is then applied, namely $\breve{m} = m(r,z)\exp\left[i(q\theta - \omega t)\right]$, and the resulting two-phase biglobal stability equations are obtained:

$$\left(\frac{\partial}{\partial r}+r^{-1}\right)u_r+\left(iqr^{-1}\right)u_\theta+\left(\frac{\partial}{\partial z}\right)u_z=0, \tag{1}$$

$$\begin{aligned}&\left\{U_r\frac{\partial}{\partial r}+\frac{\partial U_r}{\partial r}+iqU_\theta r^{-1}+U_z\frac{\partial}{\partial z}+\frac{X_P}{S}-\varepsilon\left[\frac{\partial^2}{\partial r^2}+r^{-1}\frac{\partial}{\partial r}-(1+q^2)r^{-2}+\frac{\partial^2}{\partial z^2}\right]\right\}u_r\\&+\left(2iq\varepsilon r^{-2}-2U_\theta r^{-1}+r^{-1}\frac{\partial U_r}{\partial \theta}\right)u_\theta+\left(\frac{\partial U_r}{\partial z}\right)u_z+\left(\frac{\partial}{\partial r}\right)p-\frac{X_P}{S}u_{rP}=(i\omega)u_r,\end{aligned} \tag{2}$$

$$\begin{aligned}&\left(\frac{\partial U_\theta}{\partial r}+U_\theta r^{-1}-2iq\varepsilon r^{-2}\right)u_r+\left\{U_r\frac{\partial}{\partial r}+U_r r^{-1}+iqU_\theta r^{-1}+r^{-1}\frac{\partial U_\theta}{\partial \theta}+U_z\frac{\partial}{\partial z}+\frac{X_P}{S},\right.\\&\left.-\varepsilon\left[\frac{\partial^2}{\partial r^2}+r^{-1}\frac{\partial}{\partial r}-\left(1+q^2\right)r^{-2}+\frac{\partial^2}{\partial z^2}\right]\right\}u_\theta+\left(\frac{\partial U_\theta}{\partial z}\right)u_z+\left(iqr^{-1}\right)p-\frac{X_P}{S}u_{\theta P}=(i\omega)u_\theta,\end{aligned} \tag{3}$$

$$\begin{aligned}&\left(\frac{\partial U_z}{\partial r}\right)u_r+\left(r^{-1}\frac{\partial U_z}{\partial \theta}\right)u_\theta+\left[U_r\frac{\partial}{\partial r}+iqU_\theta r^{-1}+\frac{\partial U_z}{\partial z}+U_z\frac{\partial}{\partial z}+\frac{X_P}{S},\right.\\&\left.-\varepsilon\left(\frac{\partial^2}{\partial r^2}+r^{-1}\frac{\partial}{\partial r}-q^2r^{-2}+\frac{\partial^2}{\partial z^2}\right)\right]u_z+\left(\frac{\partial}{\partial z}\right)p-\frac{X_P}{S}u_{zP}=(i\omega)u_z,\end{aligned} \tag{4}$$

$$\begin{aligned}&\left(\frac{1}{r}U_{rP}+\frac{\partial U_{rP}}{\partial r}+\frac{\partial U_{zP}}{\partial z}\right)x_P+U_{rP}\frac{\partial x_P}{\partial r}+X_P\frac{\partial u_{rP}}{\partial r}+\left(\frac{\partial X_P}{\partial r}+\frac{1}{r}X_P\right)u_{rP},\\&+\frac{1}{r}\left(\frac{\partial X_P}{\partial \theta}+iqX_P\right)u_{\theta P}+U_{zP}\frac{\partial x_P}{\partial z}+\frac{\partial X_P}{\partial z}u_{zP}+X_P\frac{\partial u_{zP}}{\partial z}=(i\omega)x_P,\end{aligned} \tag{5}$$

$$\left(\frac{\partial U_{rP}}{\partial r}+U_{rP}\frac{\partial}{\partial r}+U_{zP}\frac{\partial}{\partial z}+S^{-1}\right)u_{rP}+\left(\frac{\partial U_{rP}}{\partial z}\right)u_{zP}+\left(\frac{\partial U_{rP}}{\partial \theta}\right)u_{\theta P}r^{-1}-u_rS^{-1}=\left(i\omega\right)u_{rP}, \tag{6}$$

$$\left(U_{rP}\frac{\partial}{\partial r}+U_{zP}\frac{\partial}{\partial z}+S^{-1}\right)u_{\theta P}-u_\theta S^{-1}=i\omega u_{\theta P}, \tag{7}$$

$$\frac{\partial U_{zP}}{\partial r}u_{rP}+\left(U_{rP}\frac{\partial}{\partial r}+\frac{\partial U_{zP}}{\partial z}+U_{zP}\frac{\partial}{\partial z}+S^{-1}\right)u_{zP}+\frac{\partial U_{zP}}{\partial \theta}u_{\theta P}r^{-1}-u_zS^{-1}=i\omega u_{zP}. \tag{8}$$

4. Mean flow field

Following past work by Majdalani and Saad [14], the extended Taylor-Culick solution is used to represent the mean flowfield for the gaseous phase. The mean flowfield for the particle phase is chosen following Féraille and Casalis [4] with exception to the particle mass concentration. In their work they use a self-similar solution for particle mass concentration resulting in a PDE that accounts only for variation in the radial direction. Using what their generalized form we retain the axial variation and in so doing determine a semi-analytical solution to the particle mass concentration.

4.1. Particle mass concentration with axial variation

Assuming steady-state conditions and axisymmetric properties for the particle mass concentration, the normalized particle continuity equation becomes

$$U_{rP}\frac{\partial X_P}{\partial r}+U_{zP}\frac{\partial X_P}{\partial z}+\left(\frac{\partial U_{rP}}{\partial r}+\frac{U_{rP}}{r}+\frac{\partial U_{zP}}{\partial z}\right)X_P=0. \tag{9}$$

Equation (9) can be rewritten using the known characteristic similarity functions using

$$J(r)\frac{\partial X_P(r,z)}{\partial r}+zG(r)\frac{\partial X_P(r,z)}{\partial z}+\left[J'(r)+\frac{J(r)}{r}+G(r)\right]X_P(r,z)=0\ , \tag{10}$$

where at the sidewall, where $r=1$, the particle mass concentration is assumed to smoothly transition to a maximum concentration, X_0, at a chamber length, z_0, as noted in Eq. (11)

$$X_P(1,z)=X_{Pw}(z)=\begin{cases}\frac{1}{2}X_0\left[1-\cos(\pi z/z_0)\right]; & 0\le z<z_0\\ X_0; & z\ge z_0\end{cases}. \tag{11}$$

Dividing Eq. (10) through by $J(r)$ and letting $R(r)=G(r)/J(r)$, we find

$$\frac{\partial X_P(r,z)}{\partial r}+zR(r)\frac{\partial X_P(r,z)}{\partial z}+\left[\frac{J'(r)}{J(r)}+\frac{1}{r}+R(r)\right]X_P(r,z)=0. \tag{12}$$

In this form we find that Eq.(12) can be turned into an ODE using the method of characteristics. To this end, we allow $K(r)=X_P(r,z(r))$ and choose $z=z(r)$ to satisfy the method. Differentiating with respect to r we obtain

$$K'(r)=\frac{\partial X_P}{\partial r}+z'(r)\frac{\partial X_P}{\partial z}\ . \tag{13}$$

Upon inspection of Eq. (12) we find that the first two terms collapse into $K'(r)$, when $z'(r)=zR(r)$ is chosen. The result is a separable ODE which has an initial condition at the wall, $z(1)=z_0$, and renders

$$z(r)=z_0\exp\int_1^r\frac{G(x)}{J(x)}\mathrm{d}x\ . \tag{14}$$

The original equation now collapses into

$$K'(r)+\left[\frac{J'(r)}{J(r)}+\frac{1}{r}+R(r)\right]K(r)=0\ . \tag{15}$$

Given that $K(1)=X_P(1,z_0)=X_{Pw}(z_0)$, the general solution becomes

$$K(r)=K(1)\exp\left\{-\int_1^r\left[\frac{J'(x)}{J(x)}+\frac{1}{x}+R(x)\right]\mathrm{d}x\right\}=X_{Pw}(z_0)\frac{J(1)}{rJ(r)}\exp\left[-\int_1^r R(x)\mathrm{d}x\right]. \tag{16}$$

With Eq. (14) substituted into Eq. (16), we find the following semi-analytical solution

$$X_P(r,z)=X_{Pw}\left[zI(r)\right]\frac{J(1)}{rJ(r)}I(r)\quad\text{where}\quad I(r)\equiv\exp\left[-\int_1^r\frac{G(x)}{J(x)}\mathrm{d}x\right]. \tag{17}$$

5. Results and discussion

As performed in past studies [5, 15-18], Chebyshev collocation methods are utilized for discretization followed by the selection of judicious boundary conditions [13]. The result is a well-posed eigenvalue problem, which is solved to find the characteristic eigenmode frequencies and the eigensolutions corresponding to the fluctuating parameters. Figures 2–3 illustrate these results for the simulated SRM under investigation. The stabilizing effects of particles are shown in Figure 1 along with a possible evolution of eigenmodes, which compares favorably with the findings of Féraille and Casalis [4]. Conversely, Figure 2 illustrates the destabilizing effect of particles, where arrows are used to indicate the evolution of the eigenmodes.

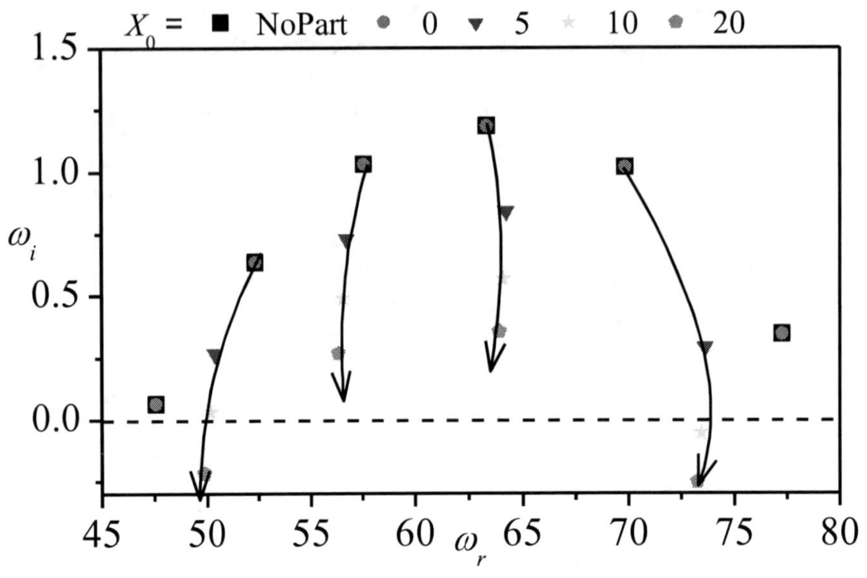

Figure 2. Possible evolution of two-phase eigenmodes. Illustration of *stabilizing* effects when $Re = 10^3$, $S = 10^{-1}$, $\phi = 0.1$, $q = 0$, and $u_h = 0.0$.

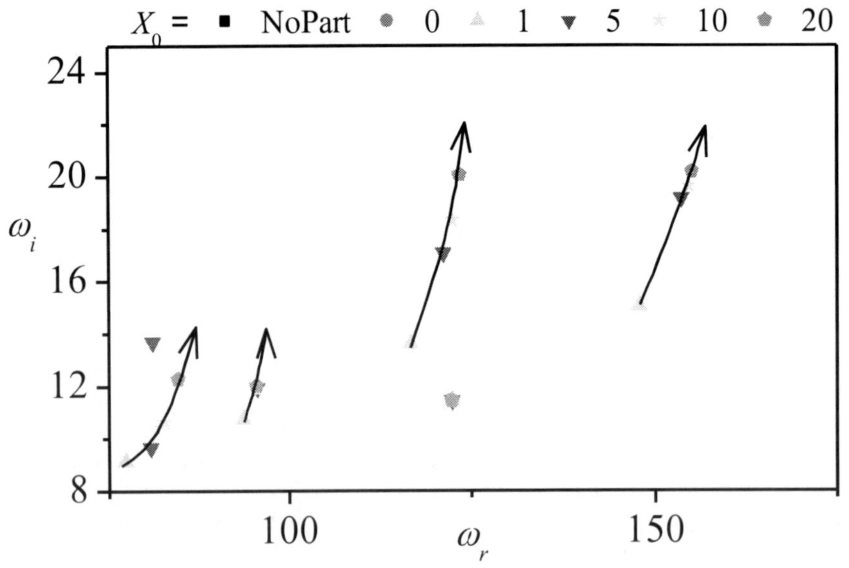

Figure 3. Possible evolution of two-phase eigenmodes. Illustration of *destabilizing* effects when $Re = 10^3$, $S = 10^{-3}$, $\phi = 1.0$, $q = 0$, and $u_h = 0.0$.

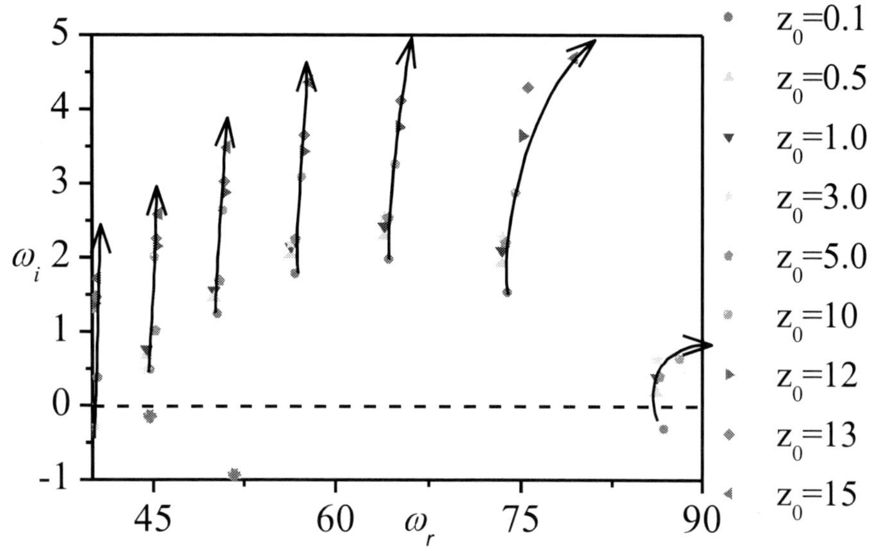

Figure 4. Possible evolution of two-phase eigenmodes. Illustration of stabilizing effects.

The effects of varying the chamber length from the headwall, z_0, where the particle mass concentration smoothly transitions to its maximum were also investigated. The findings, which are shown in Figure 4, agree with experimental data obtained from the VECLA facility by Avalon *et al.* [8]. We find that when the transition reaches its maximum at or after $z_0 = 5$ the range of unstable eigenmodes increases. This is the location where the experimental studies indicate flow breaks down. Avalon *et al.* [8] also find further flow disturbances after a distance equivalent to $z_0 = 10$ in the present study. We find that after this critical chamber length the eigenmodes are significantly amplified, more than three times that of values found for smaller values of z_0.

6. Conclusions

This work details the development of a semi-analytical solution for the particle mass concentration in two-phase rocket flows. The solution is added to a biglobal stability framework, which is used to understand the effects of particles in particle-laden solid rocket motors. The particles are found to have both stabilizing and destabilizing effects on the flowfield, mostly attributed to large particle-to-gas injection speeds. Results are also found that show amplification of instabilities when the chamber length, z_0, is increased to and beyond lengths where flow breakdown occurs according to Avalon *et al.* [8]. Ongoing analysis of eigenmodes involves the addition of headwall injection to simulate solid rocket motors with reactive headwall and hybrid rocket engines.

In future work, considerations will be made to vary particle properties such as reactivity, regression, size, and shape. In addition, compressibility effects will be investigated using a merger between this framework and the framework by Akiki *et al.* [17, 18]. Other areas of interest include sensitivity to noncircular cross-sections, swirling mean flow, and code parallelization to increase efficiency of simulations.

Acknowledgments
This material is based on work that was supported partly by the National Science Foundation and partly by the Auburn Alumni Engineering Council Endowed Professorship, Auburn University.

Journal of Physics: Conference Series **548** (2014) 012064 doi:10.1088/1742-6596/548/1/012064

References

[1] Varapaev V and Yagodkin V 1969 *Fluid Dyn.* (Izv. AN SSSR. MZhG) **4** 91
[2] Casalis G, Avalon G and Pineau J-P 1998 *Phys. Fluids* **10** 2558
[3] Ugurtas B, Avalon G, Lupoglazoff N and Vuillot F 1999 *Numerical computations of hydrodynamic instabilities inside channels with wall injection* AIAA (Los Angeles, CA)
[4] Féraille T and Casalis G 2003 *Phys. Fluids* **15** 348
[5] Elliott T S, Batterson J W and Majdalani J 2012 *Biglobal stability of cylindrically-shaped hybrid and solid rockets with injecting or reactive headwalls* AIAA (Atlanta, GA)
[6] Boyer G, Casalis G and Estivalèzes J L 2013 *Phys. Fluids* **25** 084109
[7] Boyer G, Casalis G and Estivalèzes J L 2013 *J. Fluid Mech.* **722** 618
[8] Avalon G, Casalis G and Griffond J 1998 Flow instabilities and acoustic resonance of channels with wall injection *AIAA Conference* (Cleveland, OH, 1998)
[9] Chedevergne F and Casalis G 2005 Thrust oscillations in reduced scale solid rocket motors, part II : A new theoretical approach *AIAA Conference* (Tucson, AZ, 2005)
[10] Chedevergne F and Casalis G 2006 Detailed analysis of the thrust oscillations in reduced scale solid rocket motors *AIAA Conference* (Sacramento, CA. 2006)
[11] Chedevergne F, Casalis G and Féraille T 2006 *Phys. Fluids* **18** 014103
[12] Saffman P 1962 *J. Fluid Mech.* **13** 120
[13] Elliott T S and Majdalani J 2014 Two-phase flow stability of cylindrically-shaped hybrid and solid rockets with particle entrainment *AIAA Conference* (Cleveland, OH, 2014) paper 2014-3611
[14] Majdalani J and Saad T 2007 *Phys. Fluids* **19** 093601
[15] Batterson J W and Majdalani J 2011 Biglobal instability of the bidirectional vortex. Part 1: Formulation *AIAA Conference* (San Diego, CA, 2011)
[16] Batterson J W and Majdalani J 2011 Biglobal instability of the bidirectional vortex. Part 2: Complex lamellar and Beltramian motions *AIAA Conference* (San Diego, CA, 2011)
[17] Akiki M, Batterson J W and Majdalani J 2013 Biglobal stability of compressible flowfields. Part 1: Planar formulation AIAA Conference (San Jose, CA, 2013)
[18] Akiki M, Batterson J W and Majdalani J 2013 Biglobal stability of compressible flowfields. Part 2: Application to solid rocket motors *AIAA Conference* (San Jose, CA, 2013)

Spacetime Discontinuous Galerkin FEM: Spectral Response

R Abedi[1], O Omidi[1], P L Clarke[1]

[1] Department of Mechanical, Aerospace & Biomedical Engineering, University of Tennessee Space Institute, 411 B.H Goethert, Tullahoma, TN, 37388, USA

E-mail: rabedi@utsi.edu,oomidi@utsi.edu,pclarke1@utsi.edu

Abstract. Materials in nature demonstrate certain spectral shapes in terms of their material properties. Since successful experimental demonstrations in 2000, metamaterials have provided a means to engineer materials with desired spectral shapes for their material properties. Computational tools are employed in two different aspects for metamaterial modeling: 1. Mircoscale unit cell analysis to derive and possibly optimize material's spectral response; 2. macroscale to analyze their interaction with conventional material. We compare two different approaches of Time-Domain (TD) and Frequency Domain (FD) methods for metamaterial applications. Finally, we discuss advantages of the TD method of *Spacetime Discontinuous Galerkin* finite element method (FEM) for spectral analysis of metamaterials.

1. Introduction

Since about a decade ago researchers have been able to design materials with particular repeating microstructure that demonstrate desirable spectral properties. Some notable advancements in electromagnets include materials with negative permittivity [1], permeability [2], or both [3]. Some of their applications are in electromagnetic and acoustic cloaking [4], perfect absorbers [5,6] sub diffraction imaging [7], and memory metamaterials [8]. In this paper, we discuss some approaches and challenges in numerical spectral analysis of metamaterials.

2. Numerical methods for spectral analysis and computation of metamaterials

Figure 1 shows the micro to macro transition of metamaterial properties and the use of computational tools. In Fig.1(a) we observe an electromagnetic metamaterial formed by an array of *Split Ring Resonators* (SRRs). The feature size and spacing of microstructures must be much smaller than wavelengths of interest. Computational tools are employed in two scales. At microscale, computational tools analyze a unit cell, *cf.* 1(b), with periodic boundary to obtain spectral properties of metamaterial. As shown in Fig.1(c) such forward analysis tools can be combined with an optimization scheme to enhance metamaterial properties. Once a metamaterial with desired spectral is designed, its interaction at macroscale with other material can still be model with computational tools as shown in Fig.1(d).

3. Computational challenges for metamaterials

In electromagnetics, acoustics, and electrodynamics most solvers are designed for conventional materials and are not suited for metamaterial applications. In Fig. 2 shows resonance phenomena and operation mechanism for metamaterials, for a *Split Hollow Sphere* (SHS) and

Journal of Physics: Conference Series **548** (2014) 012065 doi:10.1088/1742-6596/548/1/012065

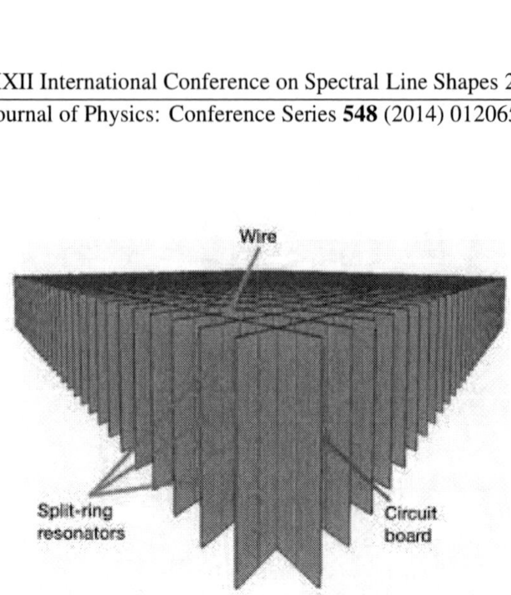

(a) A structured array of microstructures (Wikipedia).

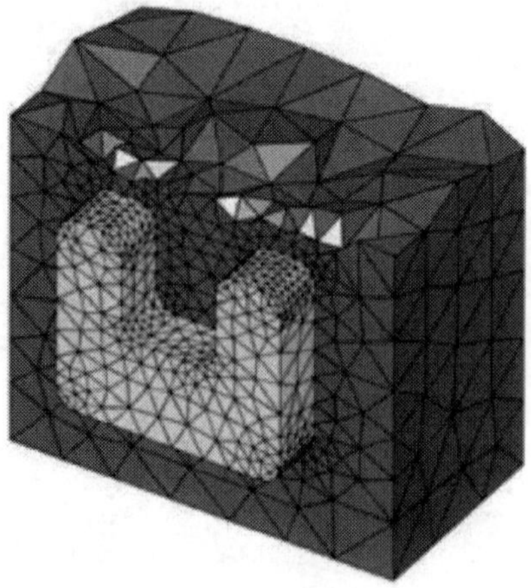

(b) Unit cell computational spectral analysis [9].

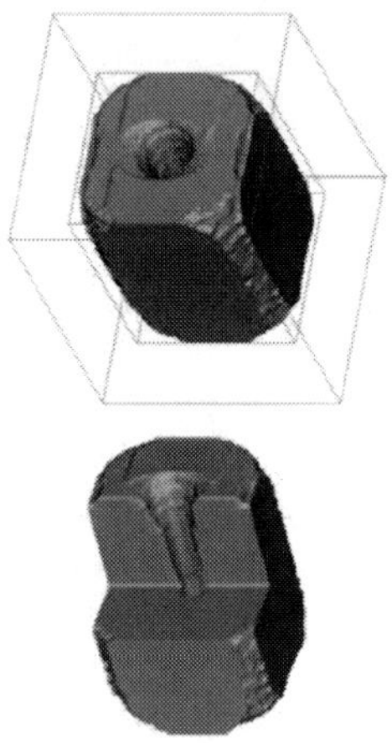

(c) Microstructure optimization [10].

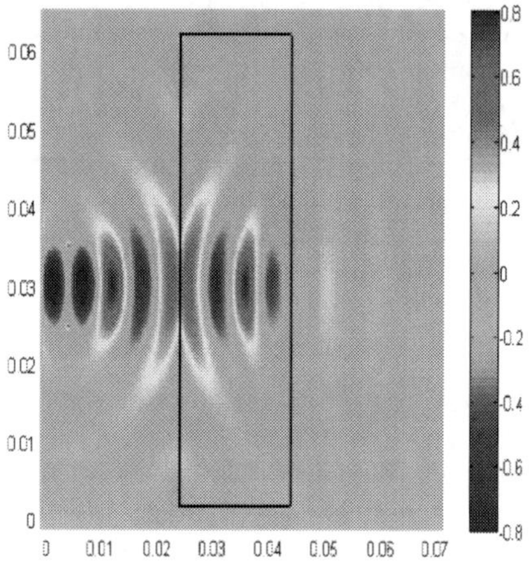

(d) Macro interaction of metamaterial with other materials [11].

Figure 1. Computational tools for analysis of metamaterials: (a) A sample array of microstructures, SRRs, in metamaterials ; (b) A unit cell computational domain to determine spectral properties of a SRR; (c) Topology optimization of microstructure to minimize effective dielecric permeability; (d) Backward wave propagation in metamaterial region (enclosed in a rectangle) using a time domain finite element method.

the corresponding acoustic transmission spectrum. Figure 3 shows intensity distribution for a silver V shape. Jumps in material properties and very strong fields both at microscale (figs. 2-3) and macroscale at material/metamaterial interface (Fig. 1(d)) call for numerical fields that can efficiently and robustly resolve these fields. In addition, metamaterial microstructure designs are inherently multiscale, *cf.* Fig. 4, a major computational challenge; *cf.* sec. 5.

4. Time Domain (TD) vs. Frequency Domain (FD)

Spectral lines and material response are obtained differently in Time Domain (TD) and Frequency Domain (FD) methods. FD analyze the problem in FD for various frequencies while in TD a Fourier transform of the response is employed. While FD may be more appropriate for small computational domains [9], TD has the following advantages: 1) Material nonlinearities, as

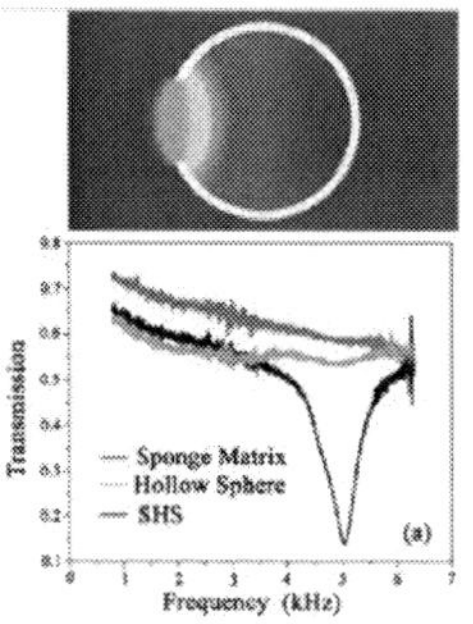

Figure 2. Resonance & transmission spectrum (SHS; [12]).

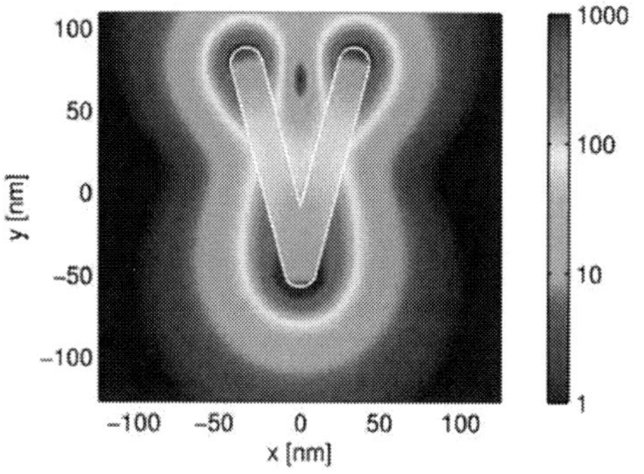

Figure 3. High gradient solutions in microscale [13].

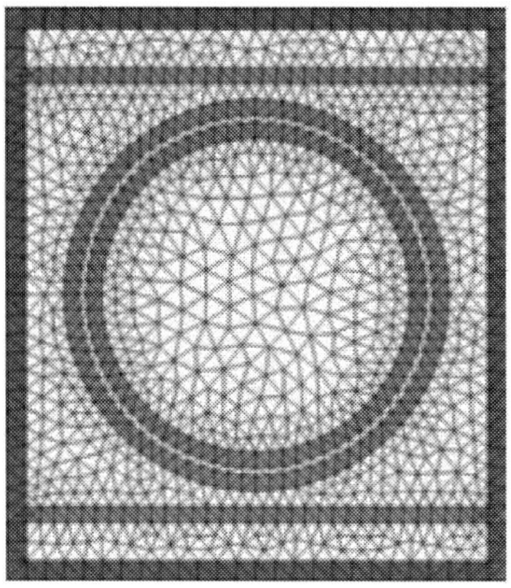

Figure 4. Multiscale geometries [9].

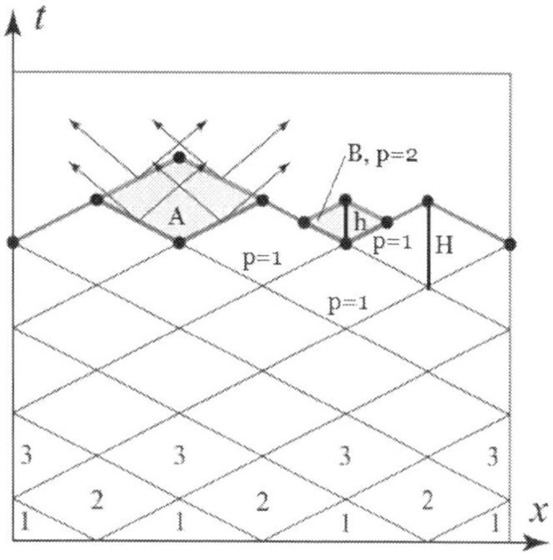

Figure 5. Schematic of SDG method (*cf.* sec. 5)

commonly encountered in metamaterials, are more naturally handled in TD; 2) Instead of several FD analyses, Fourier transformation to only one broad-band signal is sufficient in TD [13]; 3) Unlike the global spatial coupling in FD methods, in some TD methods the problem is local and solution scales linearly versus number of unknowns/elements; 4) Quasi-static FD analysis of stable state is not be sufficient when the early unsteady relaxation process is important [12].

5. Spacetime Discontinuous Galerkin (SDG) Finite Element Method

As mentioned, many TD approaches such as *Discontinuous Galerkin* (DG) methods [9] have a linear solution scaling versus number of elements, a major advantage over FD methods. However, there are still two concerns with TD methods. First, in linear scaling an explicit time integration method is needed. However, as shown in Fig. 6(a) for multiscale domains which are common in these applications, small elements severely limit the performance of these methods. While IMEX methods and subcycling in figs. 6(b-c) alleviate the problem, explicit time marching methods still perform poorly for multiscale domains. Second, frequency-dependent material properties induce a convolution term in TD. For example, $\int_{-\infty}^{\infty} \rho_0(x, t - t_0) \frac{\partial v}{\partial t'} \mathrm{d}t' + \nabla p = 0$ is the acoustic equation for frequency dependent density. The convolution term poses severe difficulties in TD.

Two approaches are used to eliminate the convolution term. First as in [14] by *Auxiliary Differential Equations* (ADE) additional fields are added to the system. Second, we maintain micro and macro fields in the formulation as opposed to homogenization to frequency-dependent metamaterials. For example in [15] convolution term is eliminated by preserving both microscale

Journal of Physics: Conference Series **548** (2014) 012065 doi:10.1088/1742-6596/548/1/012065

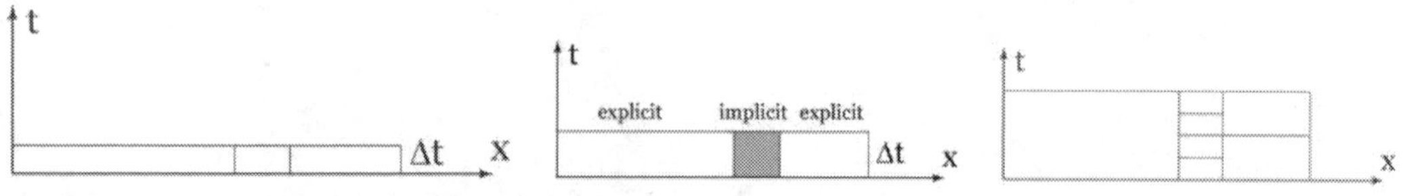

(a) Small time step for Multiscale domains. (b) Implicit-Explicit method. (c) Multistep time marching.

Figure 6. Multiscale domains, limitations (a) & remedies (b-c) for explicit Time Domain analyses: a) Smallest elements limit global time step; (b) Using Implicit time integration for small elements and explicit elsewhere (c) Time steps are adjusted based on element size.

(solid domain, undecorated) and microscale:

$$\bar{\rho}_m \frac{\partial^2 U_i}{\partial t^2} + \frac{\Sigma'_{\underline{ii}}}{R} - \frac{\partial \Sigma_{ij}}{\partial X_j} = 0 \tag{1a}$$

$$\frac{m_2}{A} \frac{\partial^2 U_{2i}}{\partial t^2} - \frac{\Sigma'_{\underline{ii}}}{R} = 0 \tag{1b}$$

Finally, the SDG method not only gracefully addresses the multiscale domain problem but also has several other advantages. In the SDG method, finiteness of the wave speed for hyperbolic problems, use of unstructured causal meshes, an use of discontinuous basis functions (a DG method) yields a local solution scheme [16, 17]. For example in figure 5, when the two inflows of element A are solved, it can be solved locally. The solution starts from any of the elements labeled 1 and continues to completion. Element B in Fig. 5 shows how element size and polynomial order can suddenly change. Some of distinct advantages of SDG over other TD methods are: 1) Linear cost vs. number of elements; 2) Excellent resolution of high gradient fields and discontinuities; 3) Arbitrary element size h and polynomial order p (element B); 4) Arbitrary high order in time as spacetime is directly interpolated instead of a separate finite difference integration in time (*cf.* element B); 5) Excellent for multiscale domains; local time step not affected by smallest size (*cf.* element B). While items 1-2 are common for DG methods, 3-5 are particular advantages of SDG over other DG methods.

References

[1] Schurig D *et al.* 2006 *Appl. Phys. Lett.* **314** 977
[2] Shelby RA *et al.* 2001 *Science* **292** 77
[3] Smith DR *et al.* 2000 *Phys. Rev. Lett.* **84** 4184
[4] Schuring D *et al.* 2006 *Appl. Phys. Lett.* **88** 1
[5] Landy NI *et al.* 2008 *Phys. Rev. Lett.* **100** 207402
[6] Tao H *et al.* 2008 *Phys. Rev. B, Condens. Matter Mater. Phys.* **78** 241103
[7] Fang N *et al.* 2005 *Science* **308** 534
[8] Driscoll T *et al.* 2009 *Science* **325** 1518
[9] Busch K *et al.* J 2011 *Laser Photonics Rev.* **5** 773
[10] Otomori M *et al.* 2012 *Comput Method Appl* M **237-240** 192
[11] Huang Y *et al.* 2013 *SIAM J Sci Comput* **35** B248
[12] Ding C *et al.* 2010 *J Appl Phys* **108** 074911
[13] Stannigel K *et al.* 2009 *Opt Express* **17** 14934
[14] Rodriguez-Esquerre VF *et al.* 2005 *Microw Opt Techn Let* **44** 13
[15] Huang HH and Sun CT 2012 *Mech Mater* **46** 1
[17] Abedi R *et al.* 2006 *Comput.Methods in Appl.Mech.Eng.* **195** 3247

Turbine engine rotor blade fault diagnostics through casing pressure and vibration sensors

J Cox and P Anusonti-Inthra

University of Tennessee Space Institute, 411 B H Goethert Pkwy, Tullahoma, TN 37388, USA

Email: ry_co@hotmail.com

Abstract. In this study, an exact solution is provided for a previously indeterminate equation used for rotor blade fault diagnostics. The method estimates rotor blade natural frequency through turbine engine casing pressure and vibration sensors. The equation requires accurate measurements of low-amplitude sideband signals in the frequency domain. With this in mind, statistical evaluation was also completed with the goal of determining the effect of sampling time and frequency on sideband resolution in the frequency domain.

1. Introduction

Rotor blade failures cause catastrophic damage to turbine engines, and therefore blade faults must be found before they are allowed to propagate. A common method of fault detection involves measuring shifts in the rotor blade natural frequencies, which can indicate blade damage. Current methods of rotor blade vibratory frequency identification are costly and overly intrusive, including drilling holes in the casing surface, or placing sensors directly on rotor blades, in the harsh environment of the engine internal flow path. A new, less intrusive method in development uses the engine casing vibratory response to determine rotor blade vibratory frequency.

Forbes and Randall determined that the turbine engine casing acts as a filter to the internal dynamic pressure, and so the frequencies contained in both signals are the same [1]. In a turbine engine fan, the casing internal pressure peak travels with the rotor blade tips. The resulting internal dynamic pressure signal contains the blade pass frequency (BPF), the number of blades multiplied by the rotor rotational frequency. Blade vibration causes modulation in the dynamic pressure signal, leading to sidebands in the frequency domain. The casing internal pressure fluctuation is not sinusoidal and therefore includes a number of harmonics. The possible harmonics include all integral multiples of the engine rotational frequency, not just the BPF.

Forbes and Randall showed that each harmonic of the rotor speed carrier frequency produces its own set of sidebands, as shown in Figure 1 [1]. To determine rotor blade vibratory frequency from the casing internal pressure response in the frequency domain, Forbes and Randall developed the following equation

$$\omega = \frac{\Delta f + q\Omega}{2}, \tag{1}$$

where ω is the blade vibratory frequency, Δf is the sideband spacing, Ω is the rotor rotational frequency, and q is an integer based on the possible sources of the sideband frequencies [1]. Equation

1 is linear in two variables (ω and q), and thus indeterminate. Additionally, the amplitude of the sidebands in the frequency response is on the order of the blade deflection amplitude, which may fall well below the noise floor.

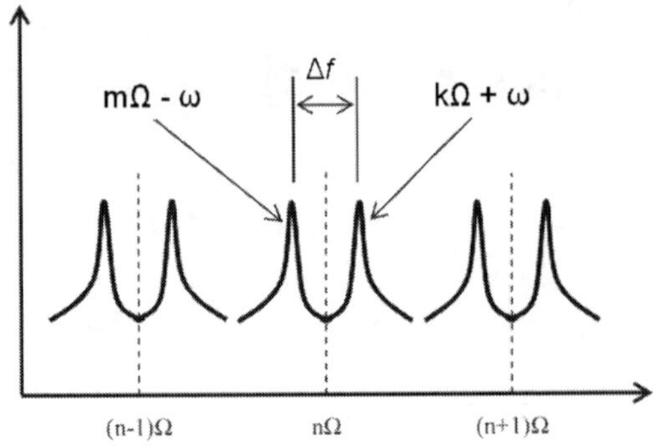

Figure 1. Blade vibration causes sidebands equally spaced around multiples of rotor speed, adapted from [1].

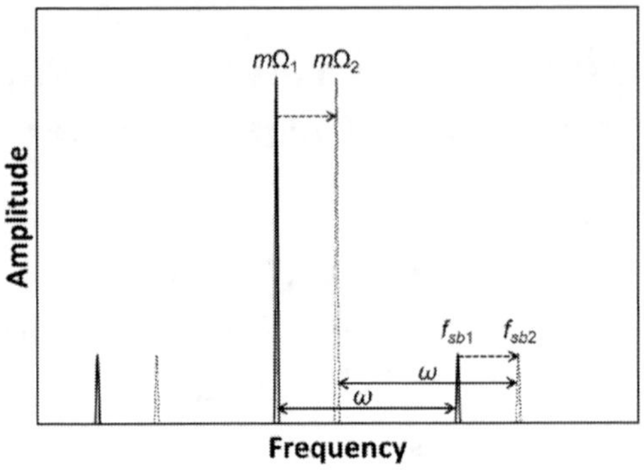

Figure 2. Assuming a constant modulation frequency, the sidebands shift with the carrier frequency.

2. Method

To overcome the indetermination problem of Equation 1, a rotor speed change was employed, such that a system of equations was formed. The system of equations was then solved to determine blade vibratory frequency.

An initial study of the amplitude problem was undertaken to determine the effect of a number of variables on the detectability of the sidebands in the frequency domain. Design of experiments (DOE) and analysis of variance (ANOVA) were used to determine influence of the blade vibratory amplitude, the signal-to-noise ratio, the sampling time, and the sampling frequency on the sideband and noise amplitudes in the frequency domain.

3. Results

The indetermination problem of Equation 1 is overcome by employing a rotor speed change (rotor frequency Ω_1 to Ω_2), assuming a constant rotor blade natural frequency, ω, as shown in Figure 2. The rotor speed change produces a set of linear equations,

$$m\Omega_1 \pm \omega = f_{sb1}, \tag{2}$$

$$m\Omega_2 \pm \omega = f_{sb2}, \tag{3}$$

where m is a multiple of rotor speed, and f_{sb1} and f_{sb2} are the sideband frequencies at rotor frequencies Ω_1 and Ω_2 respectively. Equations 2 and 3 are solved through substitution to obtain

$$\omega = \left|\frac{\Omega_1 f_{sb2} - \Omega_2 f_{sb1}}{\Omega_2 - \Omega_1}\right|. \tag{4}$$

The question of sideband amplitude was then studied using simulated data. An equation was derived to represent the pressure at a point on the internal surface of the fan casing, in the direct path of the fan blades. A simple modulated sine wave was used to represent the casing internal surface dynamic pressure, as shown

$$P_n = rand(-1,1) + \frac{(2SNR)^{1/2}}{6.6}\cos\left(2\pi b\Omega t + \epsilon cos(2\pi\omega t)\right), \tag{5}$$

where P_n is the noise-normalized pressure, *SNR* is the signal-to-noise ratio, b is the number of rotor blades, t is time, and ϵ is the rotor blade vibratory deflection amplitude. The objective of Equation 5 is to provide a means to determine the sampling time (*tf*) and frequency (*fs*) required to resolve the sideband amplitude with a given *SNR* and ϵ.

Principles from design of experiments (DOE) and analysis of variance (ANOVA) were used to determine the influence of each of the four main effects (*tf*, *fs*, *SNR*, and ϵ) and any interactions. The sideband amplitude was compared to the highest noise peak amplitude, as shown in Figure 3, and the percent difference was recorded as the output variable.

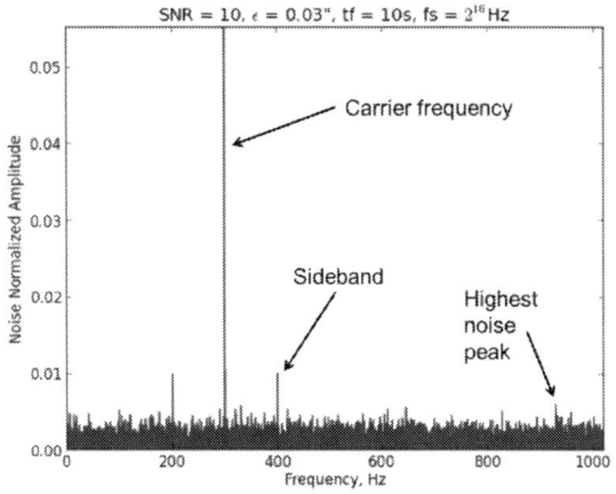

Figure 3. A typical output used to determine the percent difference in the sideband and highest noise peak amplitudes.

Through ANOVA, it was found that the greatest factors affecting sideband amplitude were, in order from highest to lowest: 1) ϵ, 2) *fs*, 3) *SNR*, and 4) *tf*, with two and three factor interactions contributing significantly less. The controllable factors, *fs* and *tf*, contributed approximately 75% and 61% (respectively) as much as the highest contributing factor, ϵ. As expected, noise in the FFT was reduced by using higher sampling frequencies and times.

4. Conclusions

In this study, an exact solution was determined for Forbes' and Randall's equation, such that the rotor blade frequency is determined by employing a shift in rotor speed, and then measuring the sideband frequency change as detected in the casing vibratory frequency response (or internal dynamic pressure frequency response). This equation works when the sideband amplitudes rise above the noise level.

Using simulated data, it was found that the rotor blade deflection amplitude has the largest influence on sideband detectability, followed by the sampling frequency, the signal to noise ratio, the sampling time, and various interactions.

Proposed future work includes 1) using a more realistic representation of the engine internal dynamic pressure, employing a saw tooth wave form and random rotor speed fluctuations for simulated data, and 2) collecting data from a fan in development at UTSI to analyze and verify the techniques proposed herein.

References

[1] Forbes G and Randall R 2013 Estimation of turbine blade natural frequencies from casing pressure and vibration measurements *Mech. Syst. and Signal Process*, http://dx.doi.org/10.1016/j.ymssp.2012.11.006.

CFD Modelling of a Quadrupole Vortex Inside a Cylindrical Channel for Research into Advanced Hybrid Rocket Designs

B Godfrey[1] and J Majdalani[2]

[1] Department of Mechanical, Aerospace, and Biomedical Engineering, Ph.D. Candidate, University of Tennessee Space Institute, Tullahoma, TN 37388, USA
[2] Department of Aerospace Engineering, Auburn Alumni Engineering Council Endowed Professor and Chair, Auburn University, Auburn, AL 36849, USA

E-mail: bgodfrey@utsi.edu; joe.majdalani@auburn.edu

Abstract. This study relies on computational fluid dynamics (CFD) tools to analyse a possible method for creating a stable quadrupole vortex within a simulated, circular-port, cylindrical rocket chamber. A model of the vortex generator is created in a SolidWorks CAD program and then the grid is generated using the Pointwise mesh generation software. The non-reactive flowfield is simulated using an open source computational program, Stanford University Unstructured (SU2). Subsequent analysis and visualization are performed using ParaView. The vortex generation approach that we employ consists of four tangentially injected monopole vortex generators that are arranged symmetrically with respect to the center of the chamber in such a way to produce a quadrupole vortex with a common downwash. The present investigation focuses on characterizing the flow dynamics so that future investigations can be undertaken with increasing levels of complexity. Our CFD simulations help to elucidate the onset of vortex filaments within the monopole tubes, and the evolution of quadrupole vortices downstream of the injection faceplate. Our results indicate that the quadrupole vortices produced using the present injection pattern can become quickly unstable to the extent of dissipating soon after being introduced into simulated rocket chamber. We conclude that a change in the geometrical configuration will be necessary to produce more stable quadrupoles.

1. Introduction

Hybrid rocket motors are motors where the fuel and oxidizer are stored separately in two different states. The typical configuration for a hybrid rocket motor (Figure 1) has the oxidizer stored as a liquid or gas and the fuel stored as a solid. A hybrid rocket motor has many advantages over liquid and solid fuel rocket systems. Hybrids tend to exhibit a simpler design than liquid fuel systems, as they require only one pumping system for the oxidizer, whereas liquid rocket motors require two pumping systems, one for each of the fuel and the oxidizer. Hybrids also have a higher specific density than liquid rocket engines, which enables them to produce more thrust energy for a given rocket volume. Unlike solid rocket motors, however, hybrid rockets can be throttled to some degree, and they can be shut down before the fuel is used up, thus increasing its mission capability and overall safety. Since most fuels for hybrid motors are more stable than solid fuel grains, a hybrid motor can be stored for longer periods of time without worrying about the stability of the fuel grain. With the oxidizer being stored separately, the fuel grain is not capable of igniting under non-operating conditions as is the case with solid rocket motors.

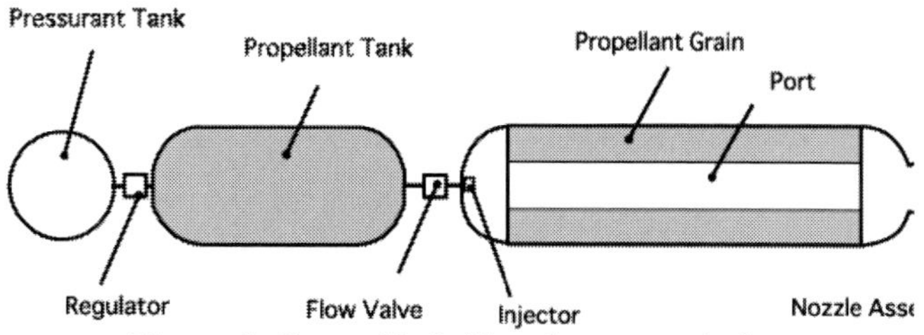

Figure 1. General hybrid rocket motor design.

1.1. Hybrid disadvantages

There are several disadvantages to hybrid rocket motors including the difficulty in ignition of the propellant and the low regression rate of the fuel grain. This paper is concerned with a flow configuration that seeks to enhance the regression rate of the fuel grain to the extent of improving the combustion efficiency and the thrust performance of the corresponding motor. To set the stage, several methods can be used to increase the fuel regression rate and thrust performance for hybrids. One such method consists of increasing the surface area of the fuel grain by incorporating multiple ports with a large exposed "wet" area at the beginning of the burn. However, this approach leads to complex fuel grains that are difficult to produce. Moreover, it leads to weaker grains as the grain can structurally fail during the firing, thus causing pieces of fuel to detach from the main grain, travel down the bore, and clog the nozzle entrance. Flow blockage can, in turn, result in motor detonation (e.g., the AMROC family of large scale motors). Increasing the fuel regression rate can also be accomplished by embedding energetic materials in the form of additives into the fuel grain. Most popular examples include ammonium perchlorate (oxidizer) and/or aluminum, which decrease the stability of the fuel grain and must be carefully formulated to mitigate the risks of self-ignition. Studies have shown that an addition of ammonium perchlorate and aluminum can lead to an increase of up to 170% in regression rate over unmodified fuel samples [1]. Another method for improving the regression rate consists of changing the internal ballistics of the gaseous motion within the motor by modulating the flow field. In this vein, a popular technique for enhancing the erosive capability of the mean flow while increasing the pressure at the fuel surface and prolonging the residence time of fuel particles within the chamber can be achieved by introducing a coaxially rotating, swirling oxidizer stream. At the outset, test results show that the introduction of unidirectional swirl can increase the regression rate by up to 200% as the swirl number is increased [2]. Nonetheless, the disadvantages of the unidirectional swirling flow technique are twofold. First, the angular momentum of the gas produces a torque on the rocket motor, which will have to be corrected using either active controls, or passive control surfaces on the airframe. Second, swirl may induce instabilities in the nozzle throat area where the angular momentum can cause the mass flow through the nozzle to decrease. This temporary disruption in effective mass flow can destabilize the vortex and lead to a pulsating motion throughout the motor. Third, by virtue of centrifugal action, the colder and, therefore, heavier gases are pushed outwardly against the grain walls whereas the lighter, hotter gases are pulled inwardly. This temperature stratification process reduces the flame temperature and, with it, the rate of heat transfer to the fuel grain. Naturally, suppressing the rate of heat transfer to the grain leads to a more sluggish burning rate.

1.2. Quadrupole vortex method

The method that is being investigated in this paper is a variation of the swirl/vortex method mentioned above (Figure 2). Instead of a single vortex, four vortices are set up within the flow in such a way that vortices are interlaced side-by-side to produce the same downwash at the interface between them (Figure 3). In this setup the angular momentum of each vortex is balanced by the presence of the other three, which eliminates the control issue and possibly the instability issues that are seen in single

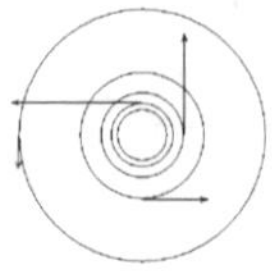

Figure 2. Monopole vortex.

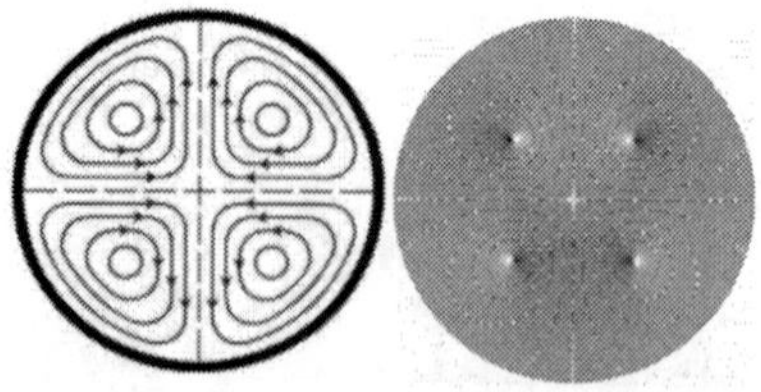

Figure 3. Quadrupole flow field.

vortex motors. It has also been shown that heat transfer rates can be improved by the introduction of circulatory motion within the fluid body, thus pulling gasses into the center of the chamber and allowing the hot gaseous products and, in some cases, the incoming oxidizer to mix vigorously before making contact with the walls of the chamber. As shown by Dake and Majdalani [3], heat transfer rates inside electronics enclosures can be visibly enhanced when the airflow across heat sinks is circulated using a quadruple vortex arrangement.

1.2.1 Motor design
A cold flow test article is currently being constructed to investigate the effectiveness of the quadrupole injection technique (Figure 4), especially that further analysis of the stability of the flow structure within the test article is needed. The experimental motor that is being evaluated is designed with 4 tangentially injected monopole vortex generators that are then inserted into the chamber in such a way to set up the quadruple vortex motion that we call "quadrupole." Through this arrangement, the flow first enters a reservoir in the injector assembly (Figure 5) so that a constant pressure can be maintained through the injector holes. The injector holes, located on the injector plate (Figure 6), are used to inject the oxidizer tangentially into the monopole vortex tubes. The latter consist of cylinders where the monopole vortex is allowed to develop. The vortex generator outlets into the main combustion

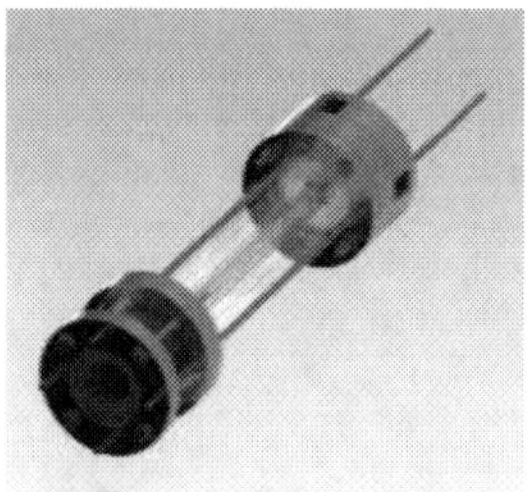

Figure 4. Motor design.

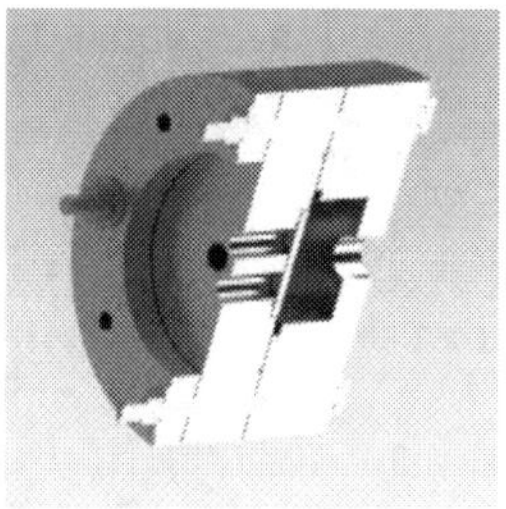

Figure 5. Injector assembly.

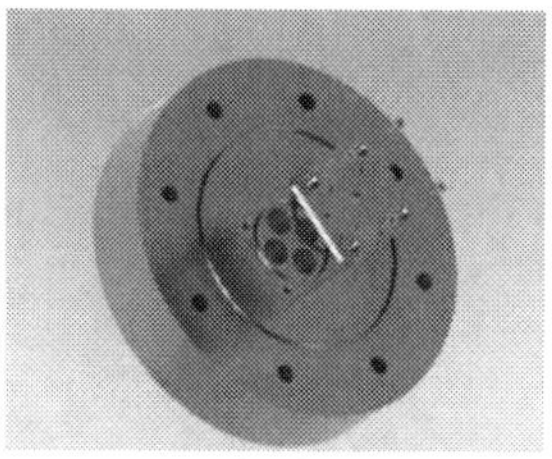

Figure 6. Injection plate.

chamber so that vortices, which are produced in a quad formation, rotate in opposite directions. This configuration will ensure that the interface between each vortex flows in the same direction, thus giving rise to a case of common downwash. The gas stream then flows across the combustion chamber where it will spiral around, burn with the fuel grain, and finally exit through the nozzle.

2. CFD methodology

2.1. Geometry
The geometry is generated by importing a simplified representation of the flow passages into the full motor design above into Pointwise (Figure 7). Surfaces are then created at the inlets and outlets of the flow path in order to close the path and generate a volume for the space that fills the flow passage. The rest of the structure is then removed from the geometric file since it is no longer needed in the remainder of the analysis. The resulting volume (Figure 8) is subsequently used to generate the mesh.

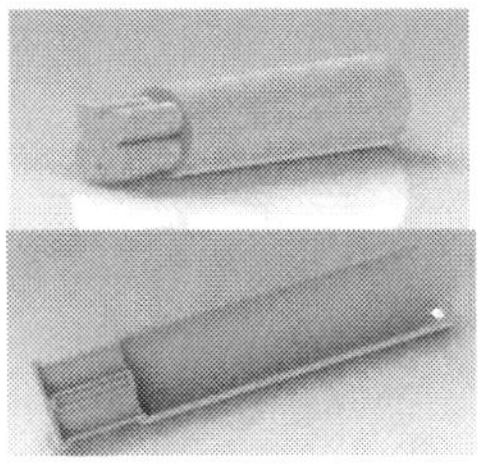

Figure 7. Simplified geometry.

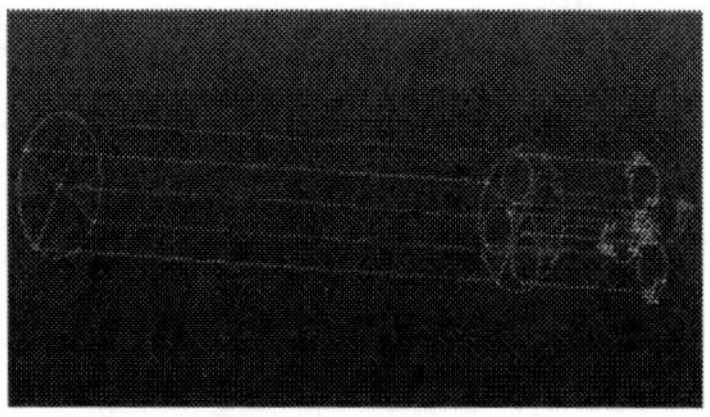

Figure 8. Geometry volume.

2.2. Grid Generation

The surfaces in the geometry are filled with unstructured grids (Figure 9). Baffles are used within the main chamber in order to refine the grid where the vortices interact (Figure 10). The walls in the analysis are assumed to be frictionless; therefore, an inviscid grid is generated within the volume. The resulting grid corresponds to an inviscid flow with frictionless walls (Figure 11).

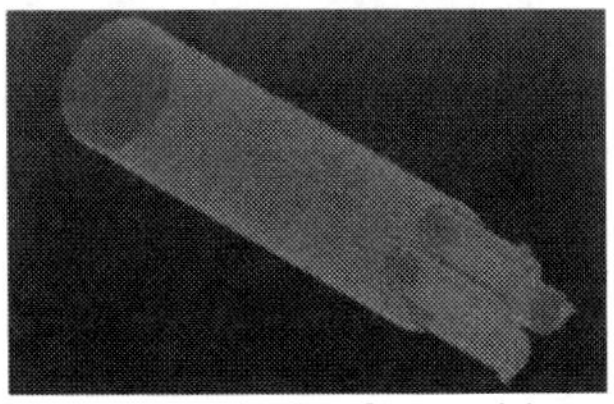

Figure 9. Surface grid.

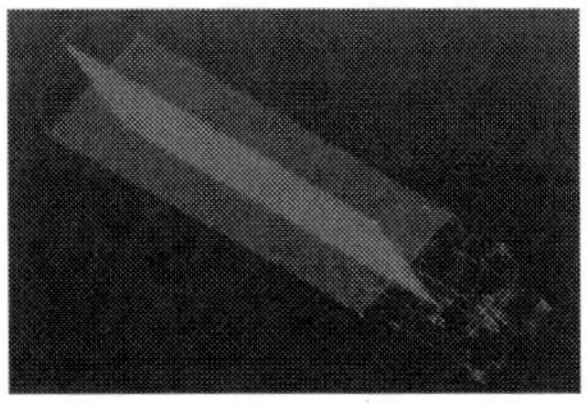

Figure 10. Baffle grid.

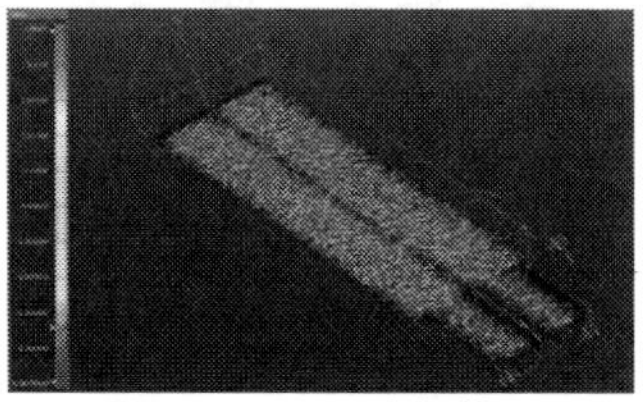

Figure 11. Grid slice.

2.3. Boundary conditions

The inlet to the problem is comprised of eight surfaces (Figure 12). The inlet condition consists of an inlet boundary, with a total temperature of 288K, a total pressure of 3.7MPa, and a velocity direction normal to the inlet surface. The outlet corresponds to the surface depicted at the end of the chamber tube (Figure 13). The outlet condition is set to be a pressure outlet with a prescribed static pressure of 3.5 MPa. The remaining surfaces are inviscid Euler walls.

Figure 12. Inlet surfaces.

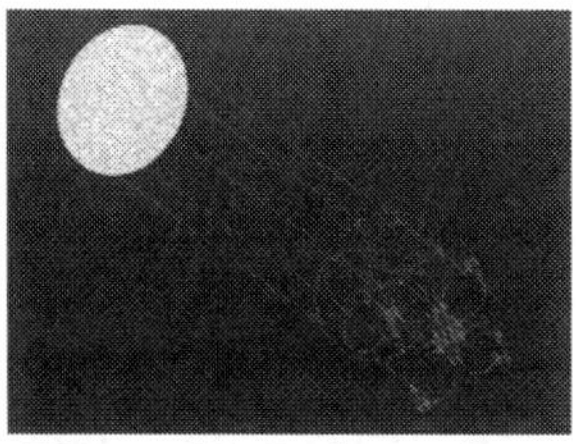

Figure 13. Outlet surface.

2.4. Numerical analysis

The second-order simulations were found to be too unstable to be solved directly from the initial conditions given above. Low pressure points would develop near the outlet, thus causing reverse flow through the outlet. For this reason, a first-order analysis was performed on the geometry using the ROE first-order method, and a CFL number of 4. The solution in this case was found to converge, with the residual decreasing by six orders of magnitude from its initial value, namely, with a final value of $10^{-11.2}$ for the Rho residual and $10^{-10.15}$ for the RhoE residual. The solution for the first-order case was then used as the initial condition for the JST Method with a CFL number of 4. This solution only converged to $10^{-7.83}$ for the Rho residual and $10^{-2.62}$ for the RhoE residual, due to the unsteady nature of the flow.

2.5. Data analysis

The data is analyzed using ParaView and the flow field is carefully examined to determine whether the desired flow configuration is achieved within the chamber. The difference in the *y*-component of the velocity on a cut plane perpendicular to the *y*-direction can be investigated to determine the intensity of the vortex at that point. Also vector plots of the velocity can be created on cut sections in the *z*-plain at different distances away from the inlet to the main chamber. These plots enable us to visualize the intensity and shape of the vortices within the chamber.

3. Results

Strong vortices are generated in the monopole vortex tubes as shown by the difference in the axial *x*-velocity on a section cut in the *x*-plane (Figure 14). However, there is dissipation of the vortices before they empty into the main chamber. We thus find that shortening of the monopole tubes will allow for a stronger vortex to enter the main chamber. However, the vortex generated in the monopole

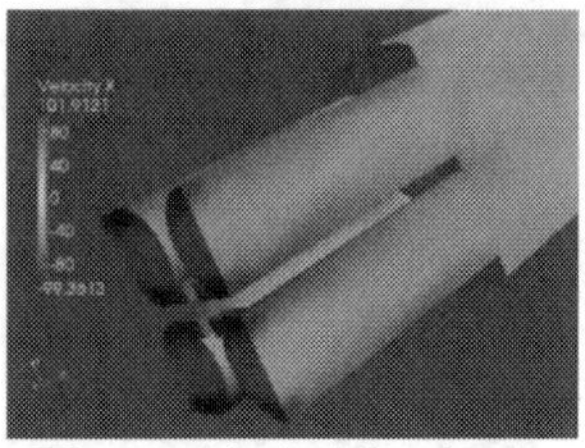

Figure 14. Axial velocity on section cut.

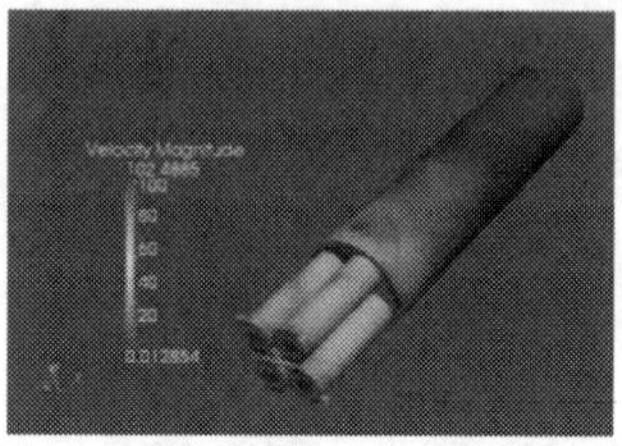

Figure 15. Velocity magnitude on surface.

tubes does not seem to be consistent due to the limited number of injectors creating a vortex at the injection tube outlet where the flow goes through a sudden expansion (Figure 15). This arrangement leads to instabilities in the flow after it has emptied into the main chamber, thus distorting the flow from its intended configuration. The dissipation of the quadrupole vortices soon after entering the main chamber can be seen in the destabilization of the individual vortices in cross-streamwise slices of the main chamber (Figure 16).

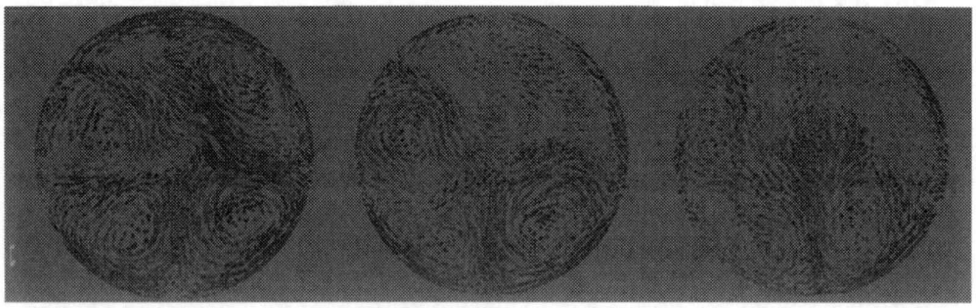

Figure 16. Flow pattern at 5mm, 10mm, and 15mm from chamber inlet.

4. Conclusions

In this paper, a basic simulation approach was considered to model the flow field associated with an innovative, quadrupole motor design. More work needs to address the geometry and boundary conditions that produce more stable quadrupole vortices that will remain coherent throughout the chamber. A significant amount of work must be performed to simulate this hybrid rocket motor thoroughly.

Future work will include modifying the geometry in such a way to produce smoother monopoles, while shortening the monopole tubes to create more intense monopoles. Then viscous walls will be added to the simulation to determine the effect of wall losses on the flow. Subsequently, the effects of wall pumping due to the sublimation of the fuel grain will be added by making the chamber walls porous, thus capable of injecting gas across the simulated surface boundary. A reacting flow simulation can also be carried out to study the heat transfer rate to the surface of the fuel grain. Then the porous boundary layer can be modified to capture the heat transfer rate that determines the regression rate and hence the mass flow rate at which the fuel may be injected into the chamber across the grain surface. Finally, the boundary layer analysis can used to model the physical regression of the fuel grain.

Acknowledgments
This material is based on work that was supported partly by the National Science Foundation and partly by the Auburn Alumni Engineering Council Endowed Professorship, Auburn University.

References
[1] Chiaverini M J, Serin N, Johnson D K, Lu Y-C, Kuo K K and Risha G A 2000 *Journal of Propulsion and Power* **16** 125
[2] Lee C, Na Y, Lee J-W and Byun Y-H 2007 *Aerospace Science and Technology* **11** 68
[3] Dake T J and Majdalani J 2009 Improving flow circulation in heat sinks using quadrupole vortices *Proceedings ASME 2009 InterPACK Conference collocated with the ASME 2009 Summer Heat Transfer Conference and the ASME 2009, 3rd International Conference on Energy Sustainability: American Society of Mechanical Engineers* pp 463-70

Cavity Flow Control Experiments and Simulations

A D Vakili[1], W L Fowler[1], R D Loewen[1], C L Thiemann[1], T P Fetterhoff[1], and A J Meganathan[2]
[1]The University of Tennessee Space Institute, Tullahoma, TN 37388, USA
[2]ESI-USA, Inc. Huntsville, AL, USA

Email: avakili@utsi.edu

Abstract. In this work we describe a summary of flow control research studies on active, passive and adaptive methodologies designed to attenuate large scale flow unsteadiness and the resulting pressure fluctuations in cavity flows. Spectral analysis of high frequency dynamic pressure measurements is used to determine the control effectiveness. Various control techniques, depending on their geometry and or distribution, can be advantageous in attenuating both the peaks and the broad spectral bands generated by flow unsteadiness. Increased effectiveness is associated with redistribution of the shear-layer vorticity. Combination of experimental and numerical results assists in understanding the underlying flow physics and interaction processes involved.

1. Introduction

Acoustic resonance, resulting from subsonic or supersonic flow over aircraft cavities, is a common occurrence in many engineering applications. The flows are known to create substantial pressure fluctuations within the cavity [1-4]. These fluctuations may cause damage to sensitive equipment inside the cavity or the cavity's structure. Our aim is the development of an adaptive flow-resonance suppression system that can be effective at different flow speeds and can be readily adapted to account for changes in the flow.

Many cavity flow studies have been carried out to consider the effects of upstream mass injection. Vakili and Gauthier [5] applied steady mass injection through a porous plate upstream of the cavity and obtained almost complete suppression of the cavity oscillations. Unfortunately, implementation of this solution for modern aircraft is challenging. To obtain the desired effect, it would require installation of a compressed air system capable of ejecting the required mass flow through a porous plate into the free stream. For example, use of supersonic micro-jets for flow control has been elaborated by Zhuang [6]. Note that most affected aircraft have little room to spare, and for this application a cumbersome ducting system to use engine bypass would not be a viable option as an add-on. It would have had to be considered in the aircraft design from the beginning.

The current research is motivated by the desire to find a solution with comparable results to the upstream mass injection but with a much simpler implementation potential. The basic concept is to use cylindrical rods placed vertically in the flow upstream of the cavity. Essentially, the rods would take the place of the mass injection jets providing a simpler solution to the implementation of cavity flow control that result in significant attenuation. The authors believe that this is a novel concept, and the authors are not aware of any such use in a previous study.

The goal of the present research was to investigate the effects of a set of configurations of cylindrical rods, or "pins", on the acoustic resonance, broadband noise level of a typical cavity and the shear layer above it. Moreover, objectives of this work include to determine not only whether specific configuration would lead to significant results, but also to study the flow mechanisms that cause attenuation.

2. Modelling background

Aeroacoustic interactions generated due to high speed flow over cavities have been extensively studied both using physical and numerical models. In shallow cavities, such as those used for weapons bays, longitudinal resonance oscillations are frequently encountered. These oscillations have been associated with interactions of the shear layers with the cavity. As the shear layer interacts with the trailing edge of the cavity a self-exciting feedback mechanism is created. Perturbations travel upstream in the cavity causing an excitation of the shear layer at the leading edge. This interaction can lead to severe amplifications of oscillations and resonance frequencies [7-9]. Rossiter [8] developed a semi-empirical formula for the prediction of possible oscillation frequencies for open cavity flows based on the Strouhal number. Heller, Holmes and Covert [9] adapted Rossiter's equation for application in the study of supersonic data.

Previous studies in active, adaptive and passive cavity flow control [5, 10-12] have shown successful results. The active flow control process [5] was able to greatly suppress the resonant tones but their implementation for certain systems applications could prove cumbersome or impractical.

3. Experimental approach

The experimental study is conducted in the High-Speed blow-down Wind Tunnel (HSWT) at the University of Tennessee Space Institute (UTSI) Gas Dynamics Laboratory. The wind tunnel test section consists of a 20 cm × 20 cm square duct that is 1.2 m long, with a test section Mach number of $M = 1.84$. Further details of the HSWT and the experimental setup are given in Ref. [13, 14].

4. Results and analysis

The experimental work utilized several different cavity configurations, including a baseline cavity test with no flow control, as a reference for comparison. Previous studies including determination of the baseline of the wind tunnel, where there was no cavity present, were performed; see the work by Fowler [11]. The broadband spectrum of the tunnel baseline is significantly lower and contains no resonant peaks.

Figure 1 illustrates acoustic spectra of the tunnel with a cavity installed and no pin plate for flow control. The first five modes are clearly visible; however, the fifth mode is small and disappears in all other configuration tests, so it is not further considered.

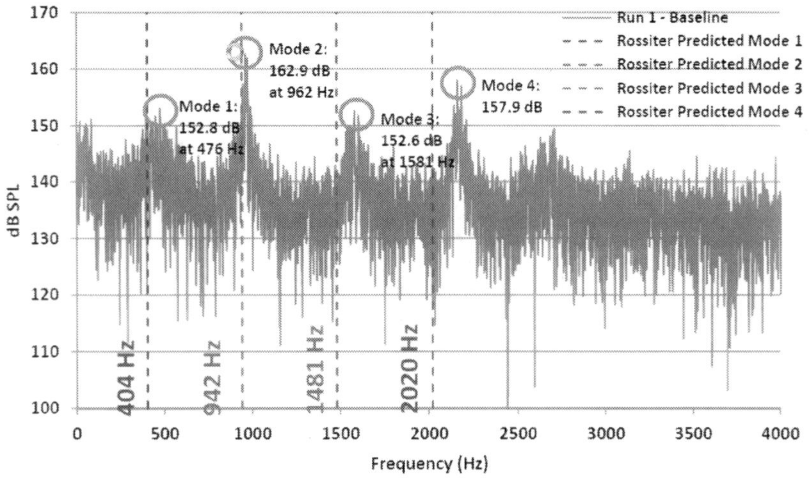

Figure 1. Config. 1: Baseline cavity acoustic spectra compared to the modes from the modified Rossiter Equation.

Distinct regions of fluid where density variations are present, such as boundary layers and shock waves, can be recorded using Schlieren flow imaging techniques. Video recordings of Schlieren images were taken in each experimental study, but here we illustrate only typical results.

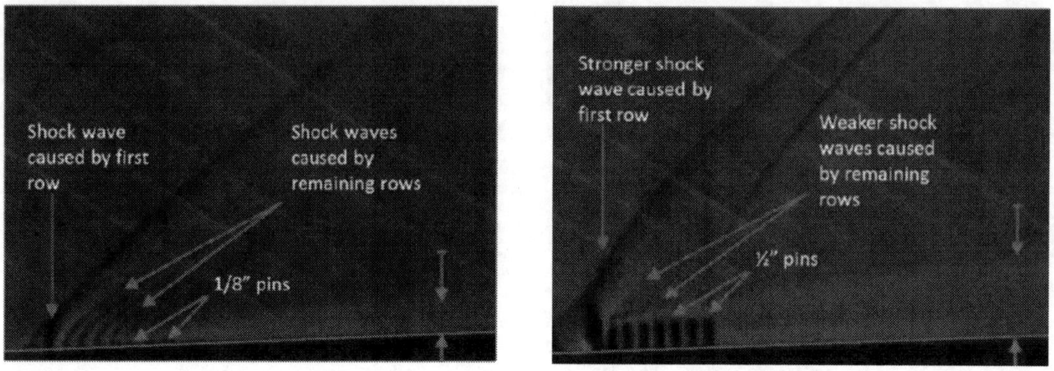

(a) **(b)**

Figure 2. Schlieren photographs of two pin heights filling the entire plate. (a) 3.2 mm (⅛") pins for configuration 3; (b) 12.7 mm (½") pins for configuration 13. The increased shear layer thickness over the cavity is indicated by the arrows at the bottom right of the images.

For the slightly rotated images, the tunnel floor is horizontal and is marked by a solid (red) line over the solid black floor. The incline is due to video camera's set up angle. In the baseline configuration with a solid plate (no holes) upstream of the cavity two important observations can be noted: First, as expected, there will be an oblique shock that forms at the leading edge. Second, a region of low speed flow (the boundary layer) is anticipated along the tunnel floor. When we replaced the solid plate with the empty pin plate (96 holes) the Schlieren image shows weak shocks forming for each row of holes, the strongest being at the front of the plate. Figure 2 shows the Schlieren images for two full plate configurations with different pin heights. A large number of pin configurations were studied. Details can be found in [13 &14]. Two selected pin distributions, with the best attenuation, were further studied.

Particle Image Velocimetry (PIV) data were obtained using 120 image pairs. Figure 3 shows typical results.

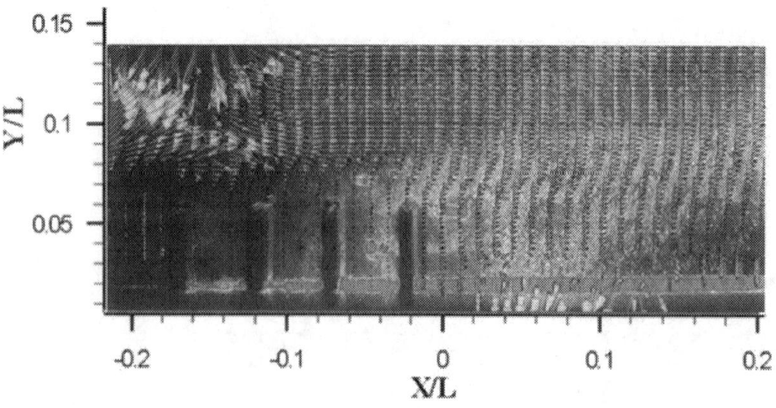

Figure 3. Typical PIV field of view with non-dimensional axes, pin config. 15.

The field of view for this image is roughly 5 cm (2 in.) of the leading edge, around 20% length of the cavity, as well as a similar distance of the upstream surface where the vertical pin configurations are located. Due to optical limitations of the PIV system, only 2.5 cm (1 in.) of the surface immediately preceding the leading edge is available. In Figure 4 the plots are on non-dimensionalized axes where zero is the leading edge.

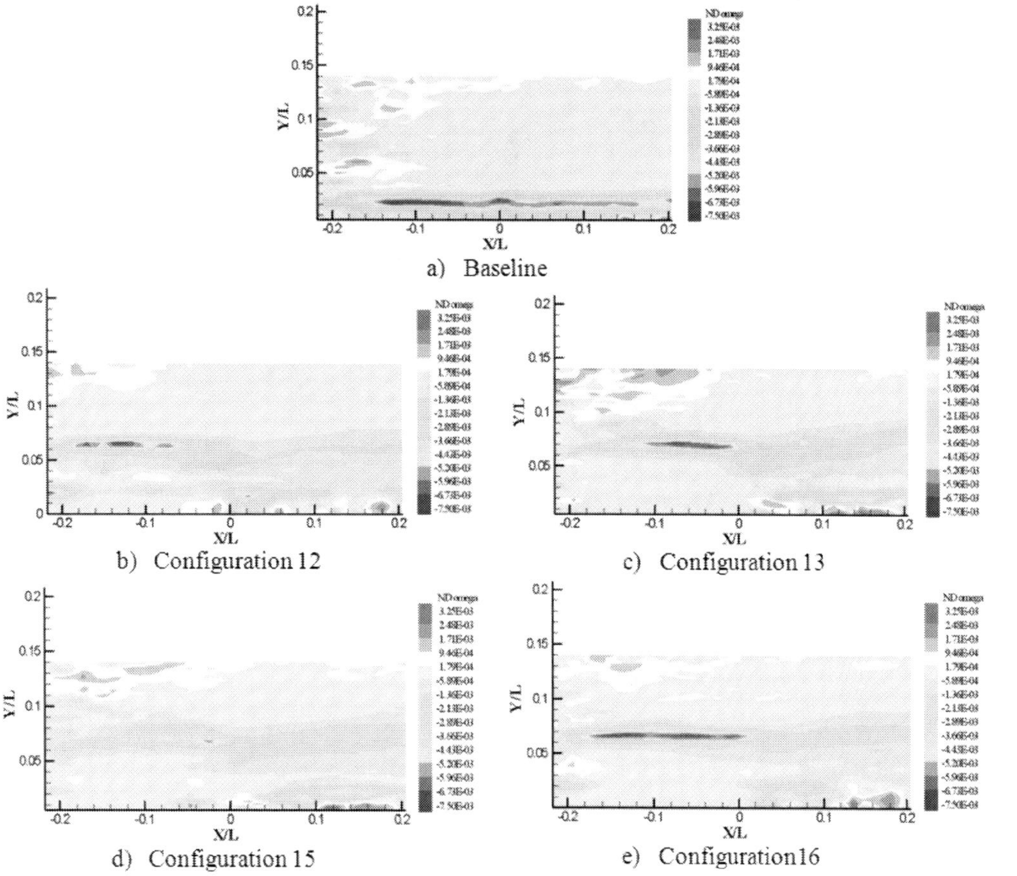

Figure 4. Averaged non-dimensional PIV vorticity magnitude contour plots; a) Baseline cavity, b) Configuration 12, d) Configuration 13, e) Configuration 15, f) Configuration 16.

Velocity and vorticity field plots revealed that the presence of vertical pins impacted the flow velocity distribution and hence the vorticity in the shear layer. Pins resulted in thickening of the shear layer over the cavity which changed kinetic energy exchange in flow's interactions with the cavity, resulting in varying amounts of suppression as measured by Sound Pressure Level (SPL). The largest suppression of the overall broadband spectra was determined to be connected to pin configurations which have staggered and widely spaced pins. Vorticity plots showed that these configurations seem resulted in more effective diffusion of the vorticity in the shear layer which results in improved peaks and broadband suppression. The various flow fields and attenuation effects resulting from the pin distributions are good indications for the effectiveness of this technique as a cavity flow control tool. Schlieren images, PIV velocity and vorticity data, including error analysis are for example discussed by Thiemann [13] and Meganathan [15].

Two selected pin distributions, with the best attenuation, were further studied. The effects of uncertainty in the experimental data and CFD predictions are considered collaboratively to help resolve potential ambiguities. Figure 5 displays an image obtained with CFD. Compared with the baseline, the pins have altered the flow distribution over the cavity, including vorticity and reduction of coherence in the shear layer vorticity. This resulted in thickening of the shear which then minimized coherent interactions with the cavity. Comparison between the 3-D numerical modelling and the experimental results help in understating of the uncertainties involved.

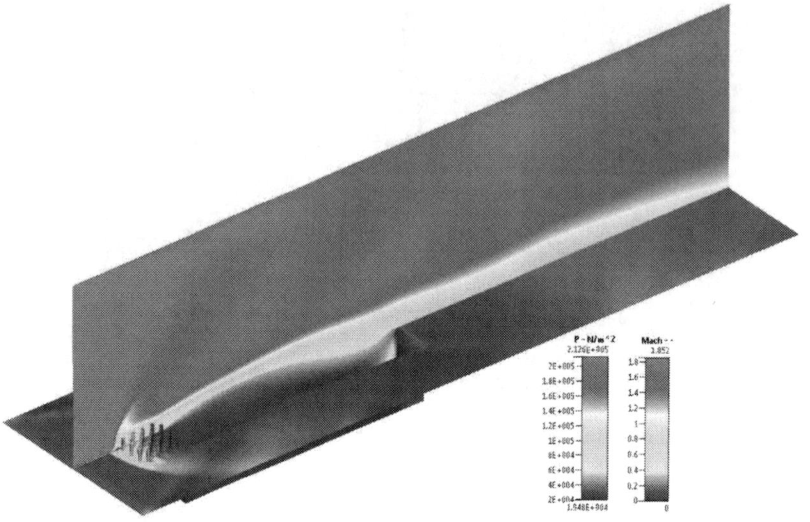

Figure 5. Typical CFD models, showing pressure and Mach number of the flow cavity field.

The effects of uncertainty in the experimental results and CFD predictions are considered collaboratively to help resolve potential ambiguities. The pins altered the flow characteristics over the cavity, including redistribution and dispersion of coherent vorticity in the shear layer. This resulted in thickening of the shear which reduced coherent interactions with the cavity.

5. Conclusions

The summary report reflects our ongoing efforts to develop cavity flow control techniques using experimental and complimentary CFD modeling studies. Selected pin configurations were tested and numerically modeled to help determine and understand the flow events within a cavity with $L/D = 4.89$ at $M = 1.84$ and $Re = 9.89 \times 10^6$ /ft. Cavity acoustics and sound pressure levels, and flow fields were measured and modeled. Pressure frequency spectra, were affected by the pins arrangements, in various distributions, upstream of the cavity. CFD results helped in analysis of flow results and determining the pins' interruption of the shear layer development as the mechanisms of effectiveness in cavity flow control. Further experiments with new pin spacing configurations are recommended, including Particle Imaging Velocimetry (PIV) measurements in the trailing edge of the cavity to record data in the cavity region near its rear bulkhead.

Acknowledgments
The authors thank Joel Davenport, Chris Armstrong, Andrew Davis and Jack LeGeune for their help in fabrication, installation and operation of the wind tunnel experiments. The authors would also like to thank Christian Parigger for his valuable editorial support.

References

[1] Roshko A 1955 *NACA TN 3488*
[2] Tam C K W and Block P J W 1978 *J. Fluid Mech.* **89** 373
[3] Plentovich EB, Stallings R L and Tracy M B 1993 *NASA Technical Paper* 3358
[4] Karamcheti K 1955 *NACA Digital Library TN 3487*
[5] Vakili A D and Gauthier C 1994 *AIAA Journal of Aircraft* **31** 169
[6] Zhuang Ning 2007 *Experimental Investigation of Supersonic Cavity Flow and Their Control* PhD Dissertation (Tallahassee: Florida State University)
[7] Heller H H and Bliss D B 1975 *AIAA Paper* 75–491
[8] Rossiter J E 1964 *Aeronautical Research Council R. & M. No. 3438*
[9] Heller H H, Holmes D G and Covert E E 1971 *Journal of Sound and Vibrations* **18** 545
[10] Givogue G A, Fowler W L and Vakili A D 2011 *AIAA 2011-3962*
[11] Fowler W L 2010 *Analysis and Comparison of Effects of an Airfoil or a Rod on Supersonic Cavity Flow* MSc Thesis (Knoxville: University of Tennessee)
[12] Loewen R D 2008 *Analysis of Cavity Flow and the Effects of a Rod in Crossflow* MSc Thesis (Knoxville: University of Tennessee)
[13] Thiemann C L 2013 *An Experimental Study of Supersonic Cavity Flow Control with Vertical Rods* MSc Thesis (Knoxville: University of Tennessee)
[14] Milne G J 2012 *An Experimental Investigation of Supersonic Cavity Flow Control Using Upstream Cylindrical Rods* MSc Thesis (Knoxville: University of Tennessee)
[15] Meganathan A J 2000 *An Experimental Study of Low Speed Open Cavity Flows* MSc Thesis (Knoxville: University of Tennessee)

AUTHOR INDEX

AUTHOR INDEX